AF323818

Conformational Concept for Synthetic Chemist's Use

Principles and in Lab Exploitation

Conformational Concept for Synthetic Chemist's Use

Principles and in Lab Exploitation

Anatoly M. Belostotskii

Bar-Ilan University, Israel

World Scientific

NEW JERSEY · LONDON · SINGAPORE · BEIJING · SHANGHAI · HONG KONG · TAIPEI · CHENNAI · TOKYO

Published by

World Scientific Publishing Co. Pte. Ltd.

5 Toh Tuck Link, Singapore 596224

USA office: 27 Warren Street, Suite 401-402, Hackensack, NJ 07601

UK office: 57 Shelton Street, Covent Garden, London WC2H 9HE

Library of Congress Cataloging-in-Publication Data
Belostotskii, Anatoly M.
    Conformational concept for synthetic chemist's use : principles and in lab exploitation /
Anatoly M Belostotskii, Bar-Ilan University, Israel.
        pages cm
    Includes bibliographical references and index.
    ISBN 978-981-281-409-8 (hardcover : alk. paper)
    1. Conformational analysis. 2. Chemical engineering. I. Title.
    QD481.B45 2014
    668.9--dc23

        2013023138

British Library Cataloguing-in-Publication Data
A catalogue record for this book is available from the British Library.

Typeset by Stallion Press
Email: enquiries@stallionpress.com

Printed in Singapore

To my parents and my family

Preface

Conformational theory is a concept that answers the apparently simple question: how do molecular backbones change without chemically changing? Organic and bioorganic researchers ought to be capable of answering this question. An idea, that organic compounds in solution are statistical ensembles of very rapidly interconverting stereoisomeric structures (conformers), prompts us to discover what organic molecules actually are at the moment before they start to react chemically. It is therefore not surprising that rational regio- or stereoselective organic synthesis launches from understanding conformational equilibria of reactants. Similarly, molecular biochemistry, i.e., the discipline which analyzes molecular mechanisms of biochemical reactions, looks for a partial answer in conformational theory when introducing the important meaning *biologically active conformation*.

At first glance, organic experimentalists have command of this fruitful concept of multifaced molecular shape. Graphical sketches of conformers are an attribute of many research articles on organic chemistry, and NMR, in the quantitative form of constants of nuclear spin-spin interactions of vicinal protons together with NOE enhancement cross-peaks, is the main support of conformational conclusions drawn there. At this point, the author needs to take a breath. I regret to inform the reader of these lines that this common practice is ill-suited to its goal, and the ability to conduct reliable conformational analysis by NMR is outside of competence of a majority of synthetic chemists. The related dolorous truth is that quite a number of academic and R&D organic researchers, not assuming this, know conformational concept at a primitive, "cartoon" level of textbooks. Not surprisingly, misinterpretations of NMR spectra, inadequate conformational analysis, and unsatisfactory explanations of reaction selectivity are not recognizable for many experts in organic synthesis and, of course, for their M.Sc and Ph.D. pupils. Should I not mention, for instance, the naive and, nonetheless, common belief in that one may impart to a conformationally mobile organic molecule a higher affinity to an enzyme/receptor by covalently locking the molecule in its biologically active conformation? Don't serious synthetic chemists, today's and future leading researchers in credible universities and R&D centers, feel that they should acquire some extra knowledge to become capable of judging conformational conclusions and performing valid conformational examinations as well?

Conformational Concept For Synthetic Chemist's Use provides exactly this knowledge. It is about what, in essence, conformers are and how to "reveal" them, both reliably and readily. Indeed, nowadays, synthetic chemists face a unique challenge. They probably know that computational chemistry is a powerful, in principle universal, tool of exploring molecular structure. What they probably do not know is that theoretical calculations have become fast, reliable, and user-friendly after appearing in successful tandem *theoretical chemistry – high performance computers*. The computer power is doubled approximately every 18–20 months, according to the rough empirical estimate, Moor's "law." Why not perform conformational analysis inside the walls of the synthetic organic laboratory by using a both "high resolution" and easily handled tool — computational chemistry of today? Thus, the focus of this book is *how non-specialists in computational chemistry, by themselves, can rapidly and reliably carry out conformational explorations by involving theoretical calculations (quantitative molecular modeling)*. It is intended to be certainly useful for synthetic organic and bioorganic specialists. Equal addressees are medicinal and material chemists, who non-computationally deal with structure-based design of small molecules: many of them will obtain the possibility to cardinally rationalize it and obtain much better results when utilizing the information given by the book. Also researchers, who specialize in instrumental methods of structural organic chemistry, in particular, NMR professionals, may find it useful. Natural product chemists will discover in this book a detailed explanation of the relatively new effective methodology of structure determination by means of *calculated chemical shifts*. A good graduate student level is sufficient to effectively read the book and extract operational information from it.

Conformational Concept is not a standard text on conformational theme for dummies. It is an unorthodox conformational account that delivers the material *de novo*, analyzes it, and advises the reader. The book does not follow the traditional manner of presenting conformational notions by showing typical conformational scenarios. Implying that the reader is familiar with the "cartoon" conformational theory shown in examples by stereochemistry books, the author starts the book from a critical overview of inadequate approaches to predicting molecular geometry and defining conformers. After straightening them out, the account proceeds to consider the main explanatory theme — what, in fact, is understood under *conformer*. This matter is presented in light of the fundamental, but easily digestible, concept of the potential energy surface that supplies relevant insight into molecular flexibility. Along the text, the author explains misunderstood conformational ideas, which have taken root among many organic chemists, but have not been explicitly treated in chemical literature.

The practical discussion, which includes elementary theory of molecular modeling, starts from Chapter 4. This explicitly mentoring part of the book is a compact consistent elucidation of how to explore conformational space with new-fashioned, click computational chemistry. Assuming no considerable experience of organic experimentalists in computations, the author has tried to combine a pedantic academic account and a user guide style by merely indicating those calculation methods which are suitable for researchers from organic laboratories. In order to ginger up theoretical considerations too irking to the reader, fragments of great masters' paintings are used in some illustrations. Closing this description, I find it appropriate to augment *the Preface* with Isaac Asimov's phrase "I do not fear computers. I fear the lack of them."

Thus, modeling first: skills in computational chemistry are necessary for *effectively* studying organic matter. I do not feel uncomfortable stating that illiteracy in molecular modeling will become unacceptable for frontier chemistry masters. I hope that this book helps non-computational organic chemists to take first steps in upgrading to the 2015–20 version of Chemical Scientist.

Acknowledgements. I count myself very fortunate to have met, during my research career, chemical scholars whose input enabled me to make my conformational knowledge active; it is my pleasure to acknowledge them. First of all, I mention Academician Professor Ivan L. Knunyanz (also spelled Knunyants), the pioneer of fluoroorganic chemistry in the USSR and the founder of the related chemical industry there; Head of the ~60-member fluoroorganic laboratory at Organoelement Institute, Science Academy, Moscow. Thanks to this liberally thinking great chemical scientist and man, I had accomplished my Ph.D. studies in this purely synthetic group as an independent researcher dealing with both synthesis and conformational analysis of model compounds which had seemed to me conformationally interesting, i.e. as exactly as I had desired realizing that reacting molecules are conformer ensembles.

Dr. Anatoly B. Shapiro from the Institute of Chemical Physics, Science Academy, one of the first explorers of the chemistry of stable nitroxyl radicals, had been my long-term collaborator in Moscow. Also today, I am cordially grateful to my elder comrade for solicitously supporting my first steps in worthwhile research.

Professor Alfred Hassner, a worldwide reputed expert in synthetic methods, was not perturbed when asked by me, his new, just arrived post-doctoral researcher from the chemistry school known to him only faintly, for independent research space. My deep appreciation and many thanks to Fred, who, after discussing with the newcomer the chemistry proposed by him, kindly agreed to turn on the green light to my conformational and biomolecular studies not associated with the research run in his group.

Grateful thanks to Dr. Hugo Gottlieb, NMR unit head, Bar-Ilan University, for affably deliberating with me on DNMR — his expertise in this technique is merely impressive.

Finally, special thanks to my generous, keen-minded colleagues from the computational chemistry field, in the past and in the present — Tatiana V. Timofeeva, now Professor at New Mexico Highlands University, and Dr. Pinchas Aped, Allinger's scientific school fellow landed at the Bar-Ilan University — who both had spent non-negligible time to friendly advise me, a bioorganic synthetic chemist, in mastering methods of computational chemistry.

Anatoly M. Belostotskii

Contents

1

Geometries of Organic Molecules: Is There a Place for a Simple Look?

1.1 Building Blocks of Molecular Geometry: Elementary Fragments

Most multiatomic organic molecules have three-dimensional (3D) structures. This feature of invisible subnanometer matter is explicitly fascinating to organic experimentalists due to the existence of astonishing species that have the same elemental composition as well as the same chemical connectivity, and yet, are different individual compounds. Discovered by purely chemical methods, the phenomenon of non-constitutional isomerism was the first unambiguous message from the 3D molecular world. Characterized as individual species many years ago, carbohydrate stereoisomers are an impressive example. For instance, there are 32 aldopyranoses: α-D-allo- α-D-altro-, α-D-gluco-, α-D-manno-, α-D-gulo-, α-D-ido-, α-D-galacto-, α-D-talopyranose, eight corresponding β-D-saccharides as well as 16 pyranoses of the L-series. Within the limits of the covalence meanings, it is impossible to explain this number without assuming a 3D arrangement of atoms in pyranose molecules.

Isolation of stereoisomers for many organic compounds of quite different chemical classes shows that non-planar arrangements of atoms in molecules are common. Physical methods, X-ray diffraction, and electron diffraction, by "displaying" molecular geometry, directly confirm this; theoretical chemistry provides solid milestones for trustworthily modeling any 3D molecular structures. Fortunately for organic chemists, the

3D labyrinth of polyatomic constructions does not appear to be infinitely complex. A very general, abstract, description of molecular geometry of organic compounds may be formulated in two statements: (*1*) molecular frameworks are formed by 3D building blocks that possess certain geometries and (*2*) these geometries are of only several types. The characteristics "certain geometries" and "only several types" are the point here because they transform a chaos of arbitrary 3D geometrical possibilities into a relatively small set of actual molecular geometries. Indeed, digonal, trigonal, and tetrahedral carbon; trigonal, and pyramidal nitrogen; pyramidal, tetrahedral and trigonal bipyramidal phosphorus, as well as similar structural assignments for other "organic" elements, do not leave too many options for molecular geometry. When accepting the geometry of the four-coordinated carbon tetrahedral, the above aldopyranose set does not look frighteningly complicated.

Easy understanding of molecular geometry starts from considering the rigid 3D geometry of elementary molecular fragments. Such fragment $A(L_i)_N$ may be defined as a unit consisting of a central atom A and N attached atoms (ligands) L_i ($i = 1,2,3\ldots N$), where *attached* means that the distance between atoms A and L_i lies in the range of chemical bond lengths. Practically always, organic experimentalists build these simplest structural elements $A(L_i)_N$ exploiting the customary meanings of covalence, i.e., an approximation of chemical bonds as electron pairs shared between bonded atoms. The coordinate (dative) bond is approached herein as a particular case of covalent bonding. Unpaired (free valence) electrons of an atom are considered as localized at this atom and not participating in chemical bonding, while lone electron pairs are considered as more mobile: still belonging to an atom, they can shift to a covalently bonded neighbor increasing the order of the bond between these two atoms. We remember that this old-fashioned mechanistic approximation is still very successful in explaining the chemical functionality of different organic compounds and, often, their reactivities. For instance, routine structural formulae (Lewis structures), a graphical language of organic chemists, are actually a totally ubiquitous application of this model of chemical bonding. Could this shared electron pair approximation also describe the 3D structure of elementary molecular fragments $A(L_i)_N$?

1.2 Geometries of Elementary Molecular Fragments

At first glance, this convenient model of *chemical bonding* provides no information about the *geometry* of elementary molecular units. In fact, very often, the shared electron pair model does predict their 3D structure! When it is complemented with a "childishly" plain principle of maximal repulsion of electron pairs [the valence shell electron pair repulsion (VSEPR) model[1,2]],[a] the spatial organization of such atomic fragments may be revealed even without pencil and paper. According to this principle, bonds (the bond electron pairs) of atom A as well as its lone electron pairs (formal bonds) are arranged in the space around this atom with minimizing their reciprocal repulsion. Since, for given localized charges, charge–charge interactions are determined only by the intercharge distances, the resulting arrangement of the electron pairs satisfies their maximal remoteness. If the number of bonds $A-L_i$ is not more than six ($N \leq 6$), for each such N, there is a *single* spatial arrangement of these bonds that provides maximal bond remoteness (Fig. 1).

Of course, interligand distances L_i-L_j, which are clear and measurable parameters of molecular geometry, are more representative than arbitrarily defined interbond distances. Fortunately, from the geometrical point of view, these interligand distances are an adequate indicator of the bond remoteness because ligands $L_1 \ldots L_n$ are "termini" of the bonds which converge to a common point (atom A). Therefore, due to the VSEPR-postulated unambiguous correspondence of the bond number and spatial arrangement of atoms L_i around central atom A, a rough 3D geometry of elementary molecular fragment $A(L_i)_n$ may be extracted directly from the covalent chemical connectivity of A.

A lone electron pair, a double bond, and a triple bond are each a dummy single bond in this mechanistic approach. In VSEPR terms, the sum of all "bonds" is the *steric number* of the central atom in the surrounding unit of its ligands. Transparently, each value of steric number corresponds to a certain 3D geometry of an atomic unit (Fig. 1).

What are concrete, steric number-related geometries? The simplest nontrivial case is the two "bond" fragment that consists of central atom A,

---

[a] Also called the electron domain concept.

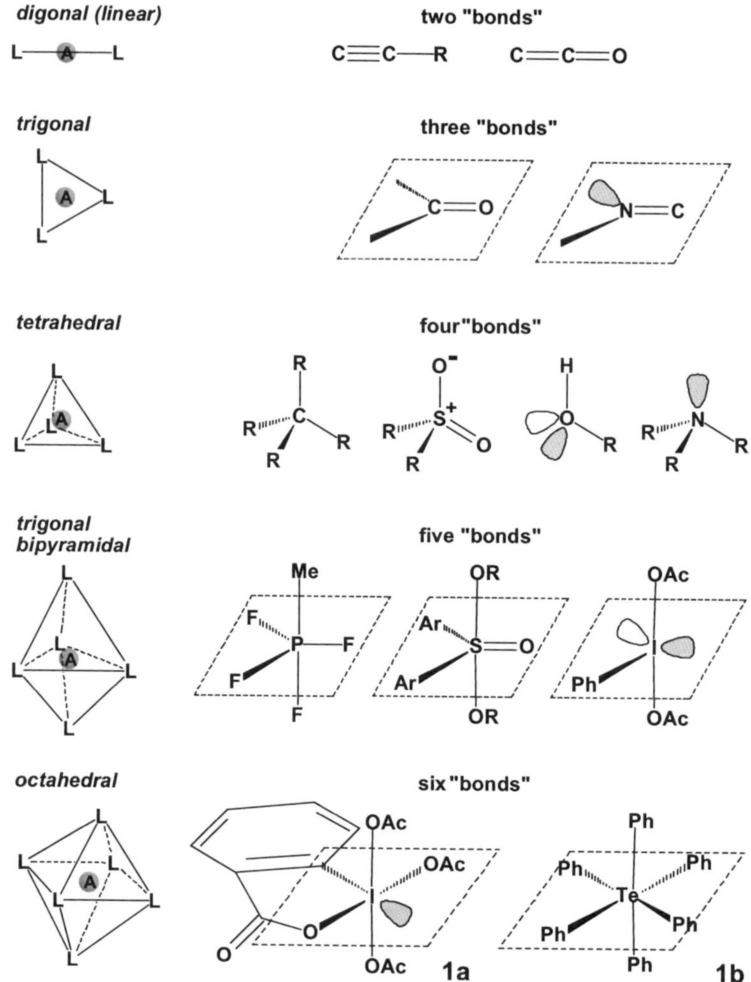

Figure 1. Molecular engineering by VSEPR. The geometry of elementary molecular fragments is determined by the number of the atom-centered "bonds" (**A** is the central atom and **L** represents its ligand).

which does not bear lone pairs, and two connected ligands L_i (as, e.g., for the R–C≡C or C=C=O fragment; Fig. 1). A linear geometry provides maximal remoteness of the "bonds," or, in other words, atoms L_i, one from the other.

Similarly, in the case of the three "bond" fragment (e.g., three ligands L_i and the central atom without lone pairs), a trigonal disposition of ligands $L_1,...L_3$, with atom A in the middle, fulfills the arrangement principle [as, e.g., for monomeric boranes BR_3, borates $B(OR)_3$, $R_2C=C$ and $R_2C=O$ fragments]. Sometimes, the parent triangle of ligands L is cryptic. For instance, the N-centered "angle" in azomethins $C-N=CR_2$ is actually the detectable rest of the hidden triangle. Each of its three apexes is located on one of three axes which is either of the single $C-N$ bond, or of the double $N=C$ bond, or of the dummy bond of the N lone pair.

A tetrahedral geometry corresponds to the four "bond" fragment. For instance, four ligands L occupy the apexes of a tetrahedron with the central atom A inside (e.g., as for an aliphatic carbon, the P or S unit of tetracoordinated phosphorus or sulfur compounds, respectively). Formally, tetrahedral geometry is also related to a molecule of H_2O or the oxygen-containing fragment of alcohols ROH as well as ethers R_2O. Four electron pairs, which surround the oxygen, connect it and the apexes of the virtual tetrahedron. The bent geometry of this O-centered unit is, in fact, the "visible" part of this tetrahedron. Similarly the pyramid of NH_3, PH_3, aliphatic amines NR_3 as well as phosphines PR_3 or triphosphites $P(RO)_3$ is a part (called trigonal pyramidal) of the virtual tetrahedron formed by three covalent bonds and the N or P lone pair.

Five and six "bond" units form bipyramids, one of which is trigonal, and the other is octahedral. If the steric number is five, ligands L are located at the apexes of the trigonal bipyramid, whereas, if this number is six, ligands L are located at the apexes of the octahedron. Two bonds $A-L_i$ and $A-L_j$ are perpendicular to the plane, in which all other ligands lie; these bonds, as well as the corresponding ligands L_i and L_j, are termed *apical*. Other ligands are *equatorial*. Numerous experimental data show that, for elementary molecular units of bipyramidal geometry, apical bonds are longer than equatorial bonds (this structural feature does not follow from VSEPR). Sulfurane oxides and phosphoranes (Fig. 1) are typical examples of compounds that contain a five "bond" unit. However, one should be also attentive in other cases. For instance, it is not difficult to recognize this geometry in bis(acetoxy)iodobenzene (BAIB), a well-known reagent used in oxidative addition to double bonds. Two lone electron pairs are equatorial ligands of the I-centered trigonal bipyramid, and the T-shaped planar backbone of this unit is actually the "visible" structural component of the "disappeared" bipyramid.

Octahedral geometry is rare in organic molecules; it is inherent to metal organic complexes and hypervalent organic compounds.[3a,b] This geometry could be exemplified by a "visible rest" of a phantomic bipyramide of Dess-Martin periodinane **1a** (a successful reagent for oxidation of alcohols that, in the last decade, has become the major synthetic tool among other reagents of this chemoselectivity), or a "real" octahedron of hexaphenyl tellurane **1b** (Fig. 1). In octahedrons of chemical elementary units with steric number six, apical and equatorial bonds are of approximately the same length.

Thus, VSEPR informs us that the set of possible geometries of elementary molecular fragments is small, and the geometry of any such chemical unit may be easily deduced from its Lewis structure. This structural concept appears as an astonishingly simple tool for predicting the geometry of elementary molecular fragments. However, could it be used with confidence? Molecular polyhedrons and polygons are usually not of rectilineal shape since the bond multiplicity and electronegativity of ligands **L** interfere with interatomic distances **A–L** as well as bond angles **L–A–L**.[1,2] Here, we meet the first, tough minor, obstacle in using VSEPR: its predictions of the distortion trend are often unreliable for organic species. For instance, it is postulated by this model that electron pair repulsion is decreased in the order: lone pair/lone pair, lone pair/bond pair, and bond pair/bond pair. According to this rule, the nitrogen lone pair of trimethylamine NMe_3 repulses the three C–N bonds more effectively than the fourth C–N bond of tetramethylammonium cation Me_4N^+ repels the other three C–N bonds in the cation. As a result, the Me groups of the amine should be closer to each other than the Me groups of the cation, or, in other words, the CNC angles in Me_4N^+ should be larger than those angles in NMe_3. However, for the molecule lacking a lone pair (i.e., Me_4N^+), these angles are even less ($\sim 109.5°$) than the CNC angles in the parent amine ($\sim 110.9°$).

Probably, organic researchers could tolerate such a rough idea of VSEPR in view of its prediction ability and inspiring simplicity which has proven to be true in many occassions. In fact, similarly to Lewis structures in describing chemical bonding, this tool appears far from being universal in predicting molecular geometry. Versatility of molecular features of *ligands* **L** (e.g., their bulkiness, chemical bonding of the ligands) are ignored by the VSEPR technique. However, its main flaw is its "theoretical" prerequisite that consists of mechanistically localizing electron pairs between certain atoms and arranging these pairs in the space. Such a consideration is actually not physical: although the density of valence electrons is essentially increased in the spatial domain between any bonded atoms and, for any chemically different pairs of such atoms, may be associated with electrons of certain quantum numbers,[a] there is no simple physical principle that would describe the entire 3D distribution of the valence electron density (the relative spatial disposition of the maxima of the valence electron density) as well as the total electron density.[b] The correlated motion of electrons can be described by solely using the formalism of the quantum mechanical wave function (Chapter 5), and this formalism does not consist of following a primitive principle of electron repulsion. That is, while indicating chemically bonded atoms, Lewis structures do not indicate the 3D distribution of the valence electron density; otherwise, quantum mechanics would be unnecessary at all. We can therefore anticipate that VSEPR misleads in predicting the geometry of some elementary molecular fragments, and our methodological question is what these fragments are. Since this prediction tool shows an undeniable success in modeling the geometry of covalently bound fragments of many "textbook organic compounds," one has to expect numerous failures of the VSEPR concept when it is applied to less trivial structures. Let us glance at examples of such molecules that may be divided into three groups (A), (B) and (C).

(A) Local geometry locked by the entire chemical framework. Sometimes elementary molecular fragments of trivial chemical structure are "embedded" into the molecular skeleton not quite properly for their typical 3D geometry. In other words, such a local 3D unit is distorted (and therefore strained) by a certain chemical connectivity of several other elementary fragments (the rest of the molecule) which are of "normal" geometry. Usually, distortions of elementary molecular fragments are not significant; however, this is not always the case. The first structures, which are recalled in this context, are three-membered cycles. This three-atom ring, of course,

---

[a]Within the limits of the abstract theoretical model. As we know, one cannot enumerate electrons (indicate concrete electrons) in a real polyelectron system.

[b]For the precise definition of the total electron density, see Section 5.2.5.

does not need geometry predictions. Nevertheless, any elementary molecular unit, which is centered at the ring atom, is a vivid demonstration of how inconsistent VSEPR estimates may be. Cyclopropane, with endocyclic angles of 60°, is an "on-duty example" of a very distorted carbon tetrahedron.

Deformation of a trigonal boron unit in boracyclopropane **2** (Fig. 2; this compound has not been synthesized yet) is even more evident. The CBC bond angle in cycle **2** is ~60°, i.e., one half of the value supplied by the VSEPR tool. Let us glance at another boron case. The boron fragment is pyramidal in 1-boraadamantane[3c] **3**, since nine severely fixed carbon tetrahedrons dictate the spatial structure for the boron unit. Ignoring the ligand structure, a stiff VSEPR technique suggests a planar trigonal arrangement for the three C–B bonds in **3**. This is a vivid example of how the canonical geometry of several individual elementary molecular fragments (nine tetrahedral carbons) causes an "incorrect" shape of the remaining (boron) fragment.

If, in the case of adamantane derivative **3** professional erudition of organic chemists is used to guess the correct boron geometry, the 3D structure of polycyclic aromatic hydrocarbons with one or more six-membered rings replaced with five-membered rings (so-called buckybowls) may seem to them to be planar. Moreover, VSEPR confirms this hasty conclusion. In fact, buckybowls are convex. The carbons of the middle rings are not planar; instead, they are slightly pyramidal as, e.g., in a "piece" of C_{60} fullerene, corannulene **4a** (Fig. 2). In this molecule, the central benzene ring is replaced with a five-membered ring that has larger endocyclic bond angles. This pyramidalization causes the surface curvature that is necessary for fusing the outer aromatic rings by means of five benzene cycles instead of six.

Geometry distortions due to the integrity of a strained molecular framework are not always guessable, and, of course, VSEPR only deorients. Olefinic as well as aromatic carbons are always planar in the VSEPR projection; of course, this "directive" is also related to *trans*-cycloheptene **4b**, [2.2.2](1,3,5)cyclophane-1,9,17-triene **4c** and [2.2](1,4)cyclophane **4d**. In fact, both the olefinic carbons are pyramidal in cycle **4b**. The experimentally and theoretically obtained evidences that the bridgehead aromatic carbons in each phenylene ring of metacyclophane **4c** lose the planarity while such carbons in paracyclophane **4d** keep it can puzzle VSEPR followers.

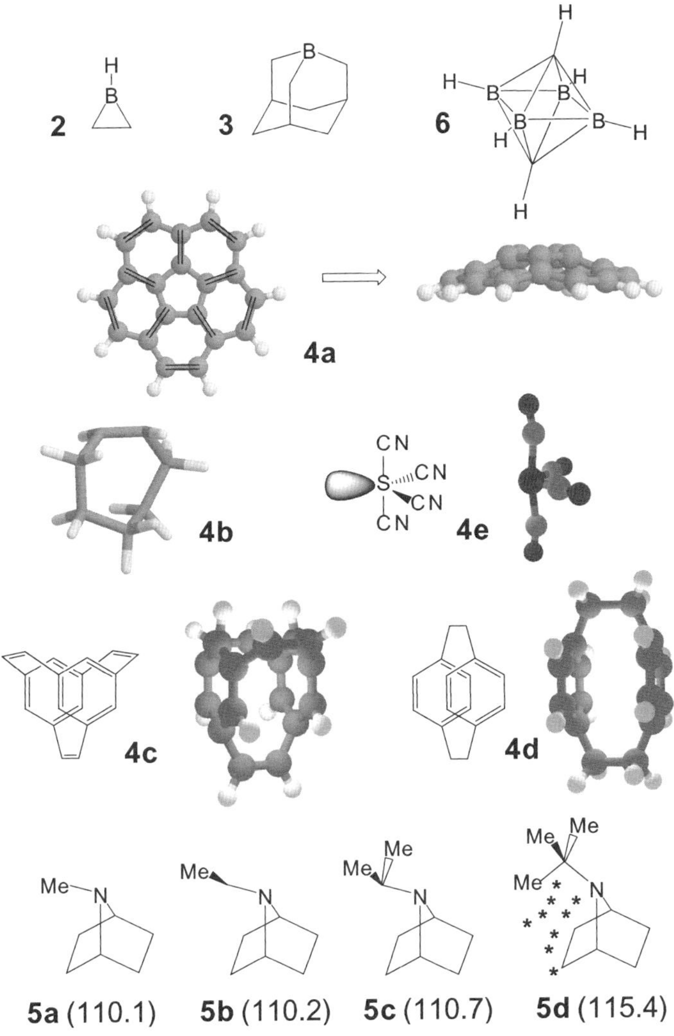

Figure 2. Examples of compounds with distorted geometries of elementary molecular fragments. The numbers in parentheses for azabicycles **5a–d** show the calculated values of the average CNC bond angle; stars indicate tight steric contacts.

(B) Ligands with tight steric contacts. Many organic ligands **L** have relatively large effective volumes. Therefore, owing to steric factors, the ideal fragment geometry is more or less disturbed for almost all organic compounds. As mentioned, these changes are invisible to

VSEPR. N-Bridged azabicycloheptanes **5a–d** (Fig. 2) illustrate this well. The VSEPR model suggests for all amines the same nitrogen geometry, a trigonal pyramid (a masked tetrahedron; Fig. 1) with close structural parameters. This is true for uncumbered alkylamines, as, e.g. triethylamine, adrenaline, morphine, etc. However, an increase of steric crowding of the N-fragment in alkylamines leads to its flattening, and N-substituents, which repel each other, diverge significantly in the space. This trend may be distinctly observed for azanorbornanes **5a–d**. Steric contacts between the exocyclic N substituent and the bicyclic backbone are approximately equal for the N-Me, N-Et and N-*iso*-Pr compounds, since the N-Et and N-*iso*-Pr molecules escape steric congestion by orienting the methyl tail of the N-substituent away from the bicycle. The average CNC bond angle is increased to an insignificant extent in this series. In sharp contrast, there is unavoidable proximity between the bulky N-*tert*-Bu group and the azabicycloheptane skeleton in **5d**, and the value of the corresponding average CNC bond angle "jumps up."[4] Of course, the VSEPR concept does not "see" this abrupt flattening. In extreme cases, the VSEPR-predicted geometry is fully degenerated, e.g., triisopropyl amine is planar in its N domain.[4a,5a] Similarly, regardless of the VSEPR rules, all four phosphorus atoms lie in one plane in sterically congested phosphane $P(t\text{-}Bu\text{-}P\text{-}t\text{-}Bu)_3$.[5b]

Obviously, steric factors, which are less important for monoatomic inorganic ligands **L**, play an essential role for the bulky organic substituents. This most exposed inaccuracy for organic compounds may be diminished if basic VSEPR geometries are corrected by means of correlations of the type *geometry–effective volume of ligand L*. However, this is scarcely reasonable. The convenience of quick, "hand-made" geometrical estimates by VSEPR is lost if computations are required. Moreover, it would be unjustified to utilize computers, which are very precise instruments, in indirect, approximate estimations. If computations are of concern, note that solidly based quantitative calculations, e.g., by means of molecular mechanics (Chapter 4), provide an accurate molecular geometry in seconds and do not "forget" to take into account steric interactions.

(C) Non-localized covalent or non-covalent A–L bonding. Insensibility of the VSEPR engineering to not quite ordinary chemical bonding **A–L** is clearly seen if one examines acylated amines. From the VSEPR point of view, the N fragment of amides that has four "generalized" bonds (four electron pairs in the valence electron shell of the nitrogen atom) is stereo-chemically similar to that of amines (a "phantomic" tetrahedron, i.e., actually a trigonal pyramid; Fig. 1). In fact, it never is.[a] The four–"VSEPR bond" nitrogen fragment in amides is trigonal owing to the involvement of the N lone pair in conjugation with the carbonyl unit. To satisfy VSEPR rules, the amide moiety should be represented as an imine salt $R(R)N^+=C(O^-)$. Of course, neither chemical reactivity nor any spectra of amides indicate such a betaine. The possible argument for VSEPR is that this technique may deal with molecular resonance structures and does not require exclusively real ones. It is, however, not valid: there are no indicators that could assist us in *a priori* selecting the "correct" resonance structure.

Moreover, resonance structures may not help at all. For instance, both resonance forms $R-N^--N^+\equiv N$ and $R-N=N^+=N^-$ of the three–nitrogen unit of the azido moiety have a linear geometry, according to the VSEPR suggestion. The fact that NNN bond angle is $\sim 170°$ in alkylazides leads us to recall that molecular geometry cannot be successfully built by mechanistically localizing electron pairs and then aligning them relative to each other. Genuine concepts of molecular physics should be involved in explaining this bent geometry of alkylazide molecules.

Perfluorinated trialkylamines $N(R_f)_3$ are an additional representative example. As indicated, VSEPR does not recognize the 3D distribution of the valence electron density. Let us suppose that it is possible to make some reasonable qualitative assumptions regarding this distribution along **A–L** bonds. For instance, it is clear that the electron density at the nitrogen is significantly decreased in perfluoro alkylamines relative to alkylamines. As we remember, the N unit of the latter compounds is pyramidal (Fig. 1). Could VSEPR establish whether the geometry of the N fragment of fluorinated amines is similar to that of parent alkylamines or is essentially different? The answer is obvious. This qualitative tool cannot "digest" the N-centered elementary molecular fragment with only three ligands and a dissipated lone pair. Interestingly, the nitrogen fragment of fluorinated amines was found to be almost planar.[6,7] For reproducing this geometry of, e.g., $N(CF_3)_3$, VSEPR would require an ionic resonance structure $(CF_3)_2N^+=CF_2\cdots F^-$. Such a speculative manipulation with resonance forms would be equivalent

---

[a] In the ground state.

to the use of the resonance structure $Cl_3C^+ \cdots Cl^-$ in modeling the molecular geometry of carbon tetrachloride. The molecule (of course, tetrahedral) appears planar from this formally uncorrupted perspective!

VSEPR meets even more serious problems when representation of the valence electron shell as shared or unshared (lone) electron *pairs* is not vigorous enough. This mechanistic geometrical concept is useless in dealing with open shell structures (molecules with a valence electron–deficient shell); simply put, the VSEPR rules are not applicable to radicals, radical-anions, radical-cations, hetero- and carbocations as well as carbenes and their heteroanalogs. For example, this oversimplified concept of molecular geometry is unable to treat both the singlet and triplet carbene CH_2. As kosher research tools, *ab initio* calculations (Chapter 5) show that these reactive intermediates have a bent geometry with the bond angle of $\sim 100°$ and $\sim 132°$, respectively; VSEPR prescripts the bond angle of $120°$ for the singlet carbene and cannot approach the triplet electron configuration of CH_2.

One can extend this understanding. We could expect an absolute feebleness of VSEPR in the case of the triplet methylene carbene if we pay attention that this prediction tool supplies valence electrons with a single, limiting characteristic: electron pairs. That is, this college–level concept does not "know" about different electronic configurations (distinct distributions of electrons in the same set of energy levels). Saying *electron pairs*, VSEPR actually implies only one case: a closed shell atom **A** (when bonded to ligands L_i) with the singlet electronic configuration and the ground state of the molecule. Alteration of any of these conditions expels any elementary molecular fragment $A(L_i)_n$ from the "VSEPR paradise."

Also, if chemical bond $A–L_i$ is not of the 2c–2e (two center–two electron) character,[a] the corresponding fragment $A(L_i)_n$ may be out of the VSEPR competence. While the electron-sufficient polycenter bonding ($c < e$; e.g., 3c–4e) does not necessarily mean violation of VSEPR rules [e.g., the I-centered fragment in the molecule of Dess-Martin periodinane has both the VSEPR-predicted geometry (Fig. 1) and a 3c–4e bonding in the O–I–O three–atom apical unit], the electron-deficient bonding ($c > e$) is always accompanied with overriding VSEPR prescriptions. Carboranes supply a good example of "anti-VSEPR" molecules that have both the closed shell

---

[a] I.e., it is not the classical localized covalent or dative bond.

and electron deficiency in chemical bonding. Let us glance at a typical carborane framework. The 3D structure of 1,6-*closo*-$C_2B_4H_2$ carborane **6** (Fig. 2) does not follow from any design based on classical 2c–2e covalent bonds. The description in terms of the three center–two electron (3c–2e) bonding leads to five-coordinated carbons and borons and, thus, to the octahedral bipyramidal cluster **6**. Note that elementary molecular units with unusual coordination of the central atom, five-coordinated carbon and boron, adopt neither VSEPR-dictated geometry. Such a tetragonal pyramid of C or B in **6** is absent among ordinary chemical structures with shared electron pair bonding (Fig. 1). That is, it could not be derived from the VSEPR stereochemistry.

Polycenter chemical bonding is not easily recognizable, and this is the next trap for "VSEPRgoers." For instance, all SCN elementary molecular units in sulfurane **4e** (Fig. 2) are undistinguishable for VSEPR and, according to its prescription, should have a linear geometry. This geometry should also be attributed to the $C_{apic}SC_{apic}$ fragment related to the apical substituents since, from the VSEPR perspective, these three atoms lie in the longest diagonal of the trigonal bipyramid stretched in the apical position direction (Fig. 1). As a result, the five–atom chain $N\equiv C_{apic}-S-C_{apic}\equiv N$ in **4e** should have a linear geometry. A "genuine" research tool, quantum mechanical calculations (Chapter 5), supports none of these predictions.[a] They show a bent geometry for each of these three–atom fragments (the CSN bond angle is 167.8° in the "axis" of the apical S-substituents, and the $C_{apic}SC_{apic}$ bond angle is 170.1°; Fig. 2). Hence, this chain $N\equiv C_{apic}-S-C_{apic}\equiv N$ adopts a stretched zigzag geometry instead of a straight line one. Only when analyzing this molecule within the limits of an orbital model,[b] we reveal that the apical S–C bonds are not covalent: while the bond order for a covalent single bond is 1, the calculated bond order for each apical S–C bond is 0.55. Similarly, an ideal $C\equiv N$ bonding has order 3, while the calculated bond order for the apical cyano groups in **4e** is 2.6. In simple words, an appreciable "portion" of each bond (understood as the involvement of an electron pair in bonding) between two atoms in the apically positioned substituents disappears. This "disappearance" means that the 2c-2e model

---

[a]Calculations at the MP2/aug-cc-pVTZ level.[9] What these symbols indicate is explained in Chapter 5.
[b]Concretely, the natural bond orbital (NBO) model.

of chemical bonding (covalent bonding) is inadequate when applied to the bonds in the considered five–atom chain of bonded atoms. Some electron pairs that include p electrons of the sulfur and two carbons of this chain bind more than two atoms. A universal Bersuker's rationalization[3d] (a concept of a very broad involvement of vibronic coupling in dictating molecular shape, i.e., a non-adiabatic, non-orbital, approach) probably leads us to conclude that the pseudo-Jahn–Teller effect actually is the primary reason for this geometrical "anomaly." VSEPR, as a mechanistic, theoretically oversimplified concept, can suggest neither delocalization of bonding, nor, of course, vibronic coupling.

A trivial case of limitation of covalent representation is bonds that are more ionic than covalent. Concerning molecular geometry, this means that the location of ligands **L** cannot be deduced from VSEPR rules, if the covalent character of the **A–L** bonding is essentially decreased (e.g., $\mathbf{L}_i$ is a cation, and **A** is an anion). In contrast to covalent bonds, ionic bonds have no favored orientation. In the extreme case of a 100% ionic bond, the corresponding anion and cation may be oriented arbitrarily, excluding, of course, the solid state.[a] For many organic anions, e.g., C-, Si-, N-, P-, or S-centered anions, the *anion–cation* bond is mainly ionic but the partial covalence of the $\mathbf{A}-\mathbf{L}_i$ bond remains. Therefore, even though a certain $\mathbf{A}-\mathbf{L}_j$ disposition (relative to other ligands $\mathbf{L}_i$) takes place, one cannot easily predict it by following a common rule.

As we can see, the weakness of the VSEPR technique is not manifested very often. Nonetheless, one has to be careful in order not to fall "victim" to this too enlightened approach. Let us consider some familiar synthetic reagents.

*Unordinary **A–L** bonding.* Diethylaminosulfur trifluoride **7** (DAST; Fig. 3) is a widely used mild agent for chemoselective fluorination of alcohols (OH is replaced with F). Its nitrogen is almost planar,[8] contrary to the VSEPR "prognosis" of a pyramidal (i.e., pseudotetrahedral) geometry. Most likely, a high electron acceptor ability of the SF_3 group results in substantial involvement of the N lone pair in the N–S bonding (a case of a highly delocalized lone electron pair).

---

[a]In the solid state, *intermolecular* forces provide a certain arrangement of anions and cations.

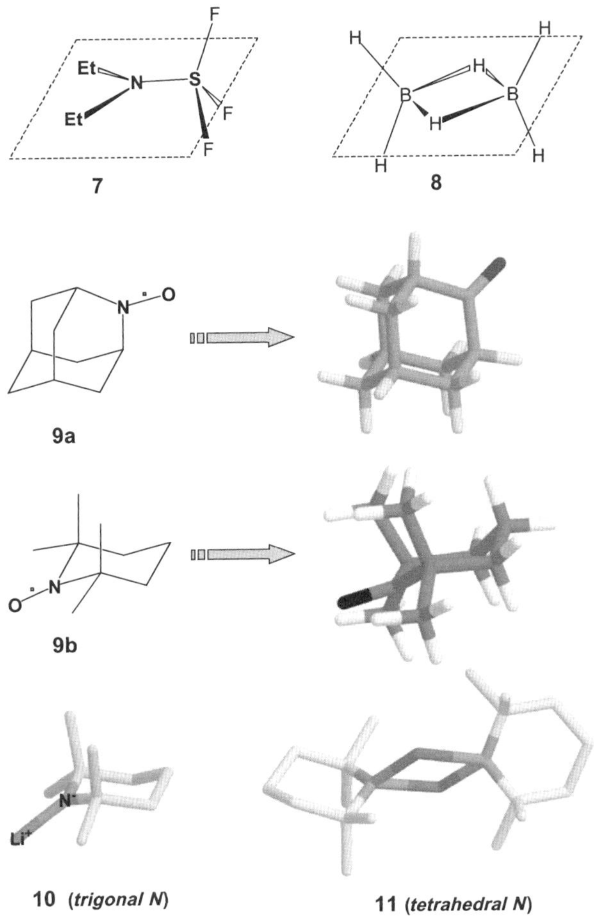

Figure 3. Spatial structure of some selected synthetic reagents (the *ab initio*-optimized geometry for structures **9a,b–11**). For clarity, protons are not shown for molecules **10** and **11**.

Hydroboration of olefins is mediated by boranes which are also effective reducing agents of many functional groups, sometimes more convenient than complex metal hydrides. Their preferred chemical form is dimeric, e.g., B_2H_6 (**8**) for the simplest borane. There are not enough electron pairs to trivially combine two borane monomers in a stable dimer, and the bonding is treated by means of the three center–two electron (3c-2e) formalism. As a result, the 3D structure of **8**, which is formed by electron-deficient B–H–B bonds (Fig. 3), is unavailable by VSEPR. The number of dummy bonds is insufficient to explain the tetrahedral geometry of each boron elementary unit as well as the angular geometry of the H bridges.

In synthetic chemistry, it is difficult to overestimate the significance of the Wittig reaction that smoothly converts carbonyl compounds to olefins. Ylides of phosphorus (alkylidenephosphorane ylides) are reactive intermediates in this reaction. In terms of Lewis resonance structures, these compounds may be represented as ylenes $R_3P=CR_2$ or, stressing their chemical reactivity, ylides $R_3P^+-C^-(R)_2$. Consequently, the VSEPR machinery supplies two alternative 3D geometries for the ylide carbon, tetrahedral and trigonal. In order to reveal the geometry of this elementary molecular fragment by still using these VSEPR predictions, we require the information on which of them is correct.

However, such an information could not be provided. The ylidic bond is not described by a primitive formalism of the shared and unshared electron pairs. In terms of orbital interactions, the ylidic bonding (in addition to the present σ-bond P–C) takes place because the donor orbital of the carbon atom yields π-type back-donation to the phosphorus atom while the lone pair orbital of the phosphorus provides σ-type bonding with the acceptor (the carbon atom).[10] Clearly, it is a peculiar bond. One should not be surprised that none of the two VSEPR-predicted structures reflects the ylide geometry. The true geometry of the ylidic carbon is between these two 3D alternatives: it is a significantly flattened pyramid.

Stable nitroxyl radicals N-oxyl azaadamantane **9a** and N-oxyl 2,2,6,6-tetramethylpiperidine **9b** (Fig. 3) are superior catalysts for the oxidation of secondary alcohols to ketones. For instance, excellent yields of carbonyl compounds are provided by using nitroxyl **9a**, even for sterically hindered alcohols. Is the VSEPR expertise suitable for predicting the nitrogen geometry in stable nitroxyls, compounds with many interesting chemical and magnetic properties?

At first sight, VSEPR should meet no problems in dealing with these chemical structures. According to the conventional Lewis structure of nitroxyl radicals, N–O·, four electron pairs form the valence shell of the N atom. By VSEPR, these four pairs of valence electrons yield a masked tetrahedron for the nitrogen, i.e., an N pyramid (Fig. 1). *Ab initio* calculations[9] show that the nitrogen pyramid in nitroxyl **9a** is very flattened (Fig. 3).

The calculated CNC angle is 114.4°, and each ONC angle is 118.6°; their sum is 351.6°.[a] The N unit is almost planar in monocycle **9b**. Calculations provide values 124.6°, and 115.4° for the same bond angles, respectively[9] [the corresponding experimental values (by X-ray diffraction) are 123.6° and 116.7°]; their sum is 355.6° (357.0° for the experimental values). What is the reason for this failure of VSEPR?

The problem is that the chemical structure of nitroxyl radicals cannot be described by this only Lewis structure. It is impossible to relate the unpaired electron either to the nitrogen or oxygen atom. In rough terms of Lewis structures, there is an unusual bond (N–O)·. If we assume that the resonance structure N–O· with the N-localized unpaired electron provides a major contribution to the chemical structure of the nitroxyl fragment, an "amine geometry" (Fig. 1) is assigned by VSEPR to the N unit. This molecular fragment, however, has another geometry (see the previous paragraph). In view of this failure, one can hypothesize that the alternative resonance structure $\cdot N^+ - O^-$ with the N-localized unpaired electron very significantly contributes to the chemical structure of the nitroxyl group. This assumption[b] appears useless: as we remember, VSEPR cannot operate with open shell structures, and, thus, VSEPR is incapable of predicting the geometry of the N-centered unit in these compounds even if it considers both resonance structures. Clearly, inability in recognizing unusual chemical bonding and dealing with it essentially restricts the use of VSEPR.

Increased ionicity of A–L bonds. As known, organic lithium amides are frequently used as C–H deprotonation agents in generating stabilized carbanions. From the VSEPR viewpoint, monomeric amides $LiNAlk_2$ and amines $NAlk_3$ have geometrically equivalent nitrogen fragments: a pyramidal (i.e., virtual tetrahedral) geometry for the nitrogen. There are exactly four VSEPR bonds for both N fragments, and, nevertheless, the VSEPR technique supplies irrelevant geometrical conclusions for such lithium amides. The real N units indeed are pyramidal in alkylamines, while they are planar in monomeric lithiated amides. For instance, *ab inito* calculations[c] show this planarity for monomeric Li tetramethylpiperidide **10**, one of the conventional non-nucleophilic deprotonation agents (Fig. 3).

---

[a]For a planar trigonal fragment, the sum of all three angles $L_i A L_j$ is 360°.

[b]It is incorrect, as a genuine *theoretical modeling* shows.

Interatomic distance Li–N in **10** is 0.178 nm, i.e., of the order of the length of an elongated covalent bond for elements of the second row. Therefore, one cannot claim that there is an ion pair $Li^+ \cdots {}^-NR_2$. Nevertheless, the increase of the length of the Li–N bond shows that this bond is essentially ionic, and the failure of VSEPR at these circumstances becomes understandable. Ionic bonds have no directionality; however, VSEPR is careless about this issue.

The real situation with Li dialkylamides is much more complicated. Depending on the solvent, concentration, and additives, they are different oligomers (often dimers) connected by N–Li–N bridges. In these dimers, the N fragment is tetrahedral (see, e.g., the structure of dimer **11** modeled by *ab initio* calculations[9]; Fig. 3). All four Li–N bonds in **11** (0.210 nm) are appreciably longer than the Li–N bond in **10**; this indicates that the bond type in **11** is dissimilar to a mixed covalent-ionic bond in monomeric amide **10**. Both lithium atoms are formally hypervalent in dimer **11**; in other words, the covalence/ionicity meanings cannot explain which bonding combines the two Li amide molecules in dimer **11**. These N–Li–N bridges should be considered as formed by electron-deficient bonds (3c-2e). Hence, the correspondence of the N geometry in diamide **11** to the VSEPR-predicted geometry is only formal. Indeed, the angular geometry of each Li-centered elementary unit in this compound nowise follows from VSEPR. In higher oligomers of organic Li amides, e.g., ladder-shaped structures, the nitrogen, forming similar N–Li–N bridges,[11,12] is a five-coordinated atom. The related N geometry is certainly not predictable by VSEPR.

Of course, it would be disrespectful to synthetic chemistry to disregard Li carbanions when assessing semi-ionic bonds and localization of negative charge. Probably, both such shelf reagents, and various *in situ* generated anionic syntons, are the most versatile modifiers of the carbon skeleton. As to their molecular geometry, this is more ambiguous. Let us discuss it briefly.

Although the "ionic/covalent ratio" for C–Li bonding is not constant for differently branched or differently functionalized molecular backbones, this

---

[c]At the B3LYP/6-31+G(d) level.[9] For explanation of *ab initio* methods, see Chapter 5.

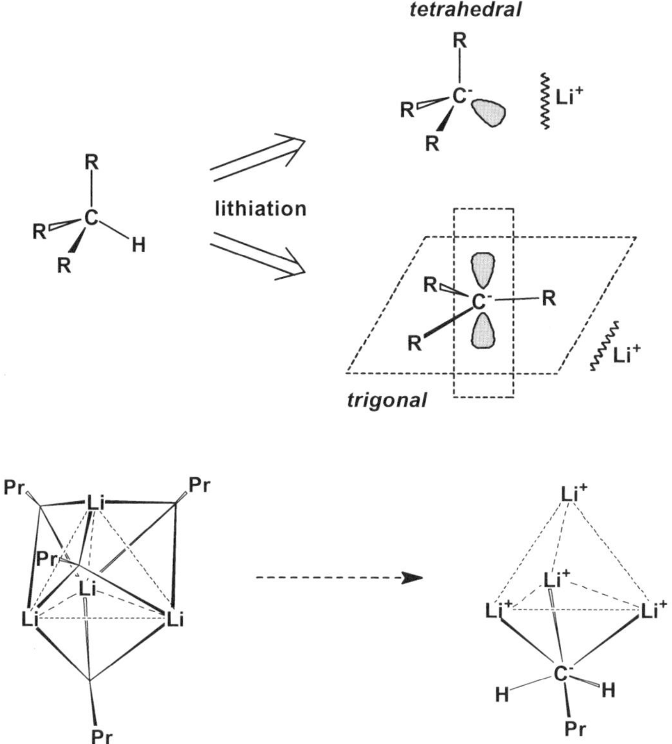

Figure 4. Ligand arrangement in alkyl carbanions (top): the C geometry is case-specific. Bottom: schematic representation of the BuLi tetramer. Dashed lines indicate the tetrahedron of four non-bonded Li cations. For clarity, one six-coordinated carbon is shown separately (bottom, right).

bond is always significantly more ionic than covalent.[11] The C-anionic center is more often pyramidal (Fig. 4); the solid angle of the C pyramid varies from compound to compound. VSEPR, of course, misses these variations in the carbon geometry. The VSEPR "four-bond" rule only constructs a tetrahedron with ligands Li, R, R^1, and R^2 positioned at its apexes.

Sometimes Li^+ is coordinated by neighboring functional groups and, therefore is essentially distanced from the anionic carbon.[11,12] Accordingly to VSEPR, this carbon is a masked $^-CR(R^1)R^2$ tetrahedron. However, Li carbanions do not follow this prescription that correctly predicts tetrahedral geometry for the parent C–H compounds. Often, even localized alkyl C-anions appear to be planar, unexpectedly for VSEPR users and structural chemists as well. VSEPR does not work for Li carbanions.

Similarly to N anions, C anions are, in fact, not monomeric. In solution, they exist as different oligomers at equilibrium, often with predominance of one of them.[11] The chemical structure of their dimers, tetramers, hexamers, and other oligomers is represented by multi-center electron-deficient bonds. For instance, the favored form of BuLi in hydrocarbon solutions is a tetramer (Fig. 4, bottom). Each anionic carbon is bound to three Li cations by a four center–two electron (4c-2e) bond. The four Li atoms form a rectilineal tetrahedron; these cations are, of course, not directly bonded, and the tetrahedron is held only by the four $Bu–Li_3$ bonds that "utilize" triads of Li atoms from the same set of four Li's. The entire 3D structure may be formally considered as a tetrahedron with four Li atoms in the apexes and n-Bu tails, each of which is leveled at the center of one of the tetrahedron edges. There is no question whether any primitive approach could supply the geometry of the anionic center. It is a six-coordinated carbon! Clearly, this carbon unit does not have the geometry which is prescribed by VSEPR for six-coordinated atoms (Fig. 1).

One should not forget that VSEPR provides correct predictions for a general 3D geometry of elementary molecular fragments (Fig. 1) with covalent (including coordination) bonding. This seems like a paradox, because this concept is ill-grounded theoretically (see below). What may be the reason for this success of VSEPR? Let us glance at the classical plot of van der Waals (VDW) potential that illustrates the dependence of the energy of two *non-bonded* atoms on their interatomic distance (Fig. 5).[a] This plot shows that, in terms of classical physics, repulsion between these atoms prevails over attraction at interatomic distances below r_0 (the sum of VDW radii of the atoms). What is the area of this plot which corresponds to *geminal* atoms? From experimental measurements and theoretical calculations for many molecules, we know that distances r_{gem} between geminal atoms in elementary molecular fragments $A(L_i)_n$ lie in the range $r_{VDW} < r_{gem}$

---

[a]Often, this dependence is approximated by the Lennard–Jones potential, an empirical function $E(r)$ that is an algebraic sum of two simple power functions. One of them represents the energy of attraction, while the other one deals with the repulsion energy (Section 4.2.2).

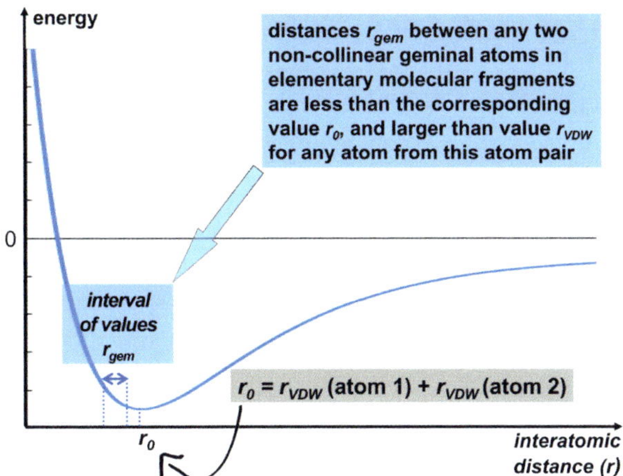

Figure 5. Energy of two chemically non-connected atoms as a function of interatomic distance r. The energy minimum corresponds to interatomic distance r_0 (the sum of VDW radii r_{VDW} of atoms 1 and 2; therefore, values r_0 vary for different chemical elements). The two-head arrow indicates the schematical range of interatomic distances r_{gem} between geminal atoms $\mathbf{L}_i$ and $\mathbf{L}_j$ in elementary molecular fragments $\mathbf{A}(\mathbf{L}_i)_n$.

$< r_0$, i.e., these interatomic distances are less than the sum of the VDW radii of these atoms while both the individual VDW radii are less than these distances r_{gem} (Fig. 5). Thus, the distance between any two geminal atoms corresponds to the distance range where the repulsion energy is the main component of the energy of interactions between these non-bonded atoms. Simply put, from experimental and theoretical studies, we know that geminal atoms in non-excited molecules repel each other; in other words, there is a tendency of geminal atoms $\mathbf{L}$ bonded to the central atom $\mathbf{A}$ to diverge in the space as far as possible, due to their mutual repulsion. At this point, we discern that the structural tendency deduced from analyzing VDW distances and the main tendency proposed by the other empirical generalization considered here, VSEPR, are actually the same.

It is difficult to "translate" the classical physics term *repulsion* applied here to geminal atoms into quantum mechanical interpretations of atom–atom interactions that cause this tendency of geminal ligands to distancing one from the other. Intuitively, we feel that there is no way to model the correlated motion of electrons by means of a primitive approach. However, it is clear that to summarize *empirical* data for molecular geometry as a valid rule may be possible. The described situation with interatomic distances and typical distances between non-collinear geminal atoms (Fig. 5) requires the "VSEPR geometries" of elementary molecular fragments (Fig. 1), excluding linear ones. Now, one can realize what VSEPR actually is: it is a satisfactory proven *empirical* tool.

It is worth noting that the traditional assignment of a theoretical basis to VSEPR is associated with exclusively considering electron–electron interactions. However, molecular energy and, hence, molecular geometry, are determined not only by electron energy; the energy of nuclei-including interactions is just as important. The energy of electrons includes the energy of electron–electron interactions as well as electron–nuclei interactions. Thus, there are three energy contributions to the total molecular energy. The first is of nuclear origin, the second belongs only to electrons, and the third is "mixed." It is clear that VSEPR implicitly includes only one component, the energy of electron–electron interactions, and ignores the two other components. Is this component of the total molecular energy crucial? Regarding this point, a useful but not readily available source[13] notices that electron energy of a non-tetrahedral methane is lower than the electron energy of the rectilinear methane tetrahedron. Illustrative *ab initio* calculations[a] confirm this. They show that electron-related energies (Coulombic, electron term and correlation energies) for the canonical tetrahedron of CH_4 (Fig. 6; each HCH angle is $109.28°$) have higher values than those for, e.g., a distorted tetrahedron with one of the HCH angles of the $99°$ value.

One should not think that the theoretical basis of VSEPR is Coulomb's law that is qualitatively applied to pairs of valence electrons considered as more or less localized electric charges of the same sign. The Pauli principle has been recruited for theoretically explaining this widespread stereochemical concept. The proposed quantum mechanical consideration places valence electrons of parallel spin as far away as possible. However, one cannot say that it is conceptually correct. The anti-symmetry of the polyelectron wave function (the generalized Pauli principle) does not specify *the most probable* location of electrons in molecular space. Model evaluations for a two electron system[14,15] illustrate this. At the first stage, this explanatory modeling deals with two hypothetical spinless electrons, i.e., a virtual fermion system that is described by the anti-symmetric wave function which contains only orbital (spatial) components. The consideration shows that the anti-symmetry requirement

---

[a]At the B3LYP/6-31+G(d,p) level.[9] Again, these acronyms are "decoded" later, in Chapter 5.

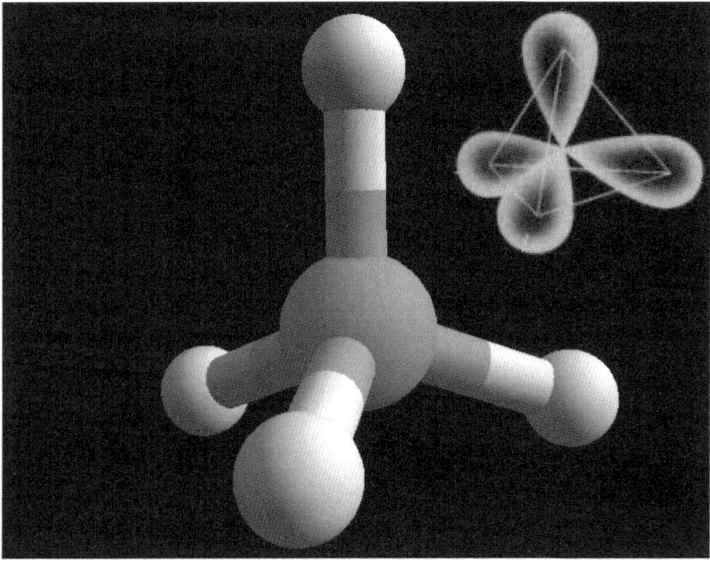

Figure 6.   The molecule of methane in a ball-and-stick representation and four $2sp^3$ hybrid orbitals of the carbon atom. Geometrically, this molecule is a tetrahedron with H atoms in its apexes, and these four orbitals reproduce this shape.

forces the electrons to tend to be further apart. Importantly, the separation between them (in the space) is not trivially determined by their spatial wave functions, in contrast to the VSEPR-dictated maximal remoteness. The next step approximates the system to a more real one in that spins are added to the electrons (i.e., the wave function is now composed of spin-orbitals, i.e., from the spatial and spin functions). Then this former "spinless" separation remains (it is not increased) if the total spin state is symmetric (the Pauli situation: the spins are parallel). If the spin state is anti-symmetric (the spins are antiparallel), the "spinless" separation is decreased. Also this decrease is determined by the spatial part of their wave function in a non-trivial way.[a] From this short digression we can extract a very general understanding. It is not easy to estimate the 3D domain of dissipation of any electron pair. Parallel spin electron pairs do not differ from other electrons in this aspect.

---

[a]Again, electron separation in space is exclusively determined by the spatial part of the antisymmetric wave function. The symmetry of the spin state only makes them to be "of Pauli" (i.e., they have the same spin number). That is, there are no specific or additional (here: repelling) interactions between "Pauli electrons." Therefore, it is incorrect to say, for instance, "electrons with parallel spins repel each another." The VSEPR terminology suffers from such a misusage of the term "repulsion," and many lierature and chemistry-oriented Internet sources are polluted with its derivative "Pauli repulsion."

As a final critical remark, one should add that although results of *ab initio* calculations for simple molecules may confirm the VSEPR-prescribed distribution of valence electron density,[16,17] they do not yield any support to the theoretical explanation (based on the Pauli principle) of the concept. We know that the valence electron density increases in the spatial area between bonded atoms, but this fact says nothing about the reason for a certain mutual disposition of these areas in intramolecular space. On the contrary, VSEPR-contradicting results for the electron density distribution would indicate the unsteadiness of this technique as a concept of the electron pair repulsion. There are such results obtained for an ordinary system, methanol (MeOH).[18] The VSEPR theoretical prerequisites presuppose localized molecular orbitals. For the oxygen in a molecule of MeOH, theoretical calculations show that the models with delocalized frontier molecular orbitals satisfy experimental data, in contrast to models with localized orbitals. Simply stated, the distribution of the valence electron density in molecules of MeOH is not reflected by the "VSEPR-localized" electron domains. The correct prediction of a general molecular geometry of MeOH fragments by VSEPR only demonstrates a solid empirical basis for this concept.

Obviously, there is a gap between VSEPR as a theoretical concept (that refers us to the Pauli principle)[a] and VSERP as a prediction tool. The former is incapable of predicting the geometry of any elementary molecular fragment while the latter supplies correct rough 3D structures of many chemically different elementary molecular fragments. What does this gap indicate? Obviously, it is an indication of a weak theoretical explanation *vs.* a good empirical basis of VSEPR.

Description of molecular shape is not reduced to discussing 3D geometry of organic molecules in terms of polygons and polyhedrons. Ligands can be formally distributed in polyhedron apexes in different manner, affording stereoisomeric structures, e.g. diastereoisomers (diastereomers). Clearly, diastereomeric structures have different thermodynamic stability. Which of them is more stable? Solving this problem (identification of the preferable stereoisomer) is outside of the VSEPR competence and, thus, this important stereochemical problem escapes a "simple look" solution.

Regrettably for organic experimentalists, there are no empirical rules or other simple tools at all for estimating the relative stability of diastereomeric elementary fragments. One could mention only an empirical generalization about apicophility. Apicophility (more often termed apicophilicity) is the tendency of ligand **L** to occupy an apical position in structures of a trigonal

---

[a]In addition, this oversimplified concept does not "see" both the Jahn-Teller and pseudo-Jahn-Teller effect that affect molecular geometry (see Ref. 3d for the recent relevant review).

bipyramidal geometry. The measure of apicophility for substituent **L** is the energy difference between diastereomers with apical and equatorial orientation $\mathbf{L}_{ap}$ and $\mathbf{L}_{eq}$, respectively. The apicophility "rule" holds that electron-withdrawing ligands tend to be apical. However, it is difficult to estimate how broad this empirical generalization is.

The ready reader may ask at this point: if VSEPR is sometimes problematic, can we treat "anti-VSEPR" systems by means of another simple tool? Unfortunately, we cannot. Excluding modeling by VSEPR, 3D structures of elementary molecular units cannot be predicted in any simple fashion. Then, how can we know, without resorting to experiments, whether the 3D geometry of an elementary unit obeys VSEPR rules; if this unit is an "anti-VSEPR" structure, what is its geometry? The above examples for the geometries of some reagents actually give an answer which is crucial for our choice of the general methodology. One could pay attention that the reference geometry is supplied by methods of computational chemistry for the most examples given above. Thus, the answer is as following: through the use of *theoretical calculations*. In stereochemical research, one has to leave the "hand-made" modeling which is so unreliable and, at the same time, so favored by synthetic chemists.

The example of nitroxyl derivatives **9a,b** (Fig. 3) is indicative of this. Primitive prediction tools in stereochemistry are actually a questionable selection of a structure from a very limited finished set of assumed molecular geometries. If this set does not contain the true structure, the researcher actually has no solution as, e.g., in the situation with structures **9a,b** and VSEPR. Moreover, if any structure is nevertheless selected from the pool, the nascent mistake cannot be discerned. For instance, the pyramidal N geometry, which is supplied by VSEPR for azacycles **9a,b**, might be accepted if solid modeling (here: *ab initio* calculations) had not rejected it. In contrast, theoretical calculations are capable of generating exactly the correct geometry, whether it was assumed or not, and ignoring an abstract infinite set of 3D geometries impossible for the given chemical structure. One can liken this elegant, successful tool to the conjurer who, in a tricky way, brings the rabbit out of the void. A near planarity geometry of the N fragment of nitroxyls **9a,b** is such a "rabbit" — it is appeared as a "mystic" result of calculations based on a strict physical theory, and that is all.

For the user of theoretical computational methods, the problem of obtaining the correct molecular geometry is solely methodological. It may be defined as following: what model and calculation method should be selected for reproducing the adequate molecular geometry of the compound of interest? In general, this problem is easily solvable for researchers even with minimal experience in theoretical calculations, and Chapters 4 and 5 explain what the methods of computational chemistry are and how organic experimentalists can operate with them.

Note that the use of theoretical modeling is a minimal risk strategy in structure determination. Synthetic chemists often come across a failure of apparently reliable synthetic methodologies when they are applied to new systems. The phrase "Our attempt to obtain compound X using reagent Y was unsuccessful" in a synthetic report would surprise nobody. A similar negative statement regarding optimization of the geometry of, e.g., a 50–60 atom molecule, scarcely can be found in today's scientific literature. With a proper calculation method in hand, the 3D geometry of *any* organic molecule of regular size can be modeled with a high accuracy. Unfortunately, the "absolute" success of theoretical calculations in the determination of molecular geometry is rather unknown in organic laboratories.

This short remark about theoretical modeling is noticeable in our account. For the first time herein, the role of strict theoretical calculations in determining structure is mentioned. In contrast to VSEPR, these calculations utilize different but well-grounded approximations of quantum mechanics (QM; see Chapter 5) or classical physics, sometimes in combination with QM-derived parameters (molecular mechanics, MM; Chapter 4). The fundamental element of QM, the wave function, does not supply electrons with any artificial "tags" as paired or unpaired, core or valence, or lone pairs, etc. As we remember from the basics of quantum chemistry, the wave function distinguishes between electrons supplying them with quantum numbers and treats electrons and nuclei only according to the fundamental properties of this function.[a] Therefore, proper QM calculations (say, a proper "imitation" of the wave function) provide a true 3D geometry *without any preliminary assumptions*.

In this connection, calculation tools, which could be used by synthetic organic chemists, are eventually the central point of our discussion, and we will describe them in considerable detail in the following chapters. In Chapters 1, 2, and 3 the accounts of the conformational concept are

---

[a]Although these mathematical properties are not considered here, we could recapitulate that the wave function, on its domain, is single-valued (schlicht), finite, continuous, and differentiable; it is also orthonormal and antisymmetric for electrons.

only accompanied by the results of theoretical calculations; beginning with Chapter 4, our focus will be the calculation methods themselves, with emphasis on their use in the organic laboratory.

The sign ▦ appears in the material below to indicate which particular problems concerned do not follow an empirical rule or a primitive "theoretical" approach (e.g., mechanical molecular models) and, at the same time, are not intricate for computational methods. This sign indicates: trustworthy theoretical calculations must be used in order to solve this stereochemical problem! The sign also is a reminder that the reader should refer to Chapters 4 and 5 for general looking through calculation methodologies.

1.3 Hybridization of Atomic Orbitals Is an Illustrative Concept

Are VSEPR and theoretical calculations the only alternatives (although unequal) for determining the 3D structure of elementary molecular fragments, or are other methods also suitable? Synthetic organic chemists often suppose that the concept of atomic orbital hybridization provides geometry of these fragments, similarly to VSEPR. The tradition of their education includes this simplest orbital approach; more specifically, its qualitative geometrical aspect. Indeed, considering the covalent bond as the spatial overlap of the *hybridized* outer shell orbitals of bonded atoms, this theory reflects the same point with which VSEPR deals and describes the 3D structure of the first surrounding of an atom in a molecular fragment. Let us consider this orbital geometry in more detail.

Any orbital approach operates with orbitals (some mathematical functions constructed for describing electron energy); the latter are differently defined in various orbital concepts. Let us recall what atomic orbitals (AOs) are. The s, p, d, f, and g AOs are certain mathematical functions that relate the energy of an electron in an isolated (non-bonded and non-interacting) atom to quantum numbers n, l, and m (n is the principal quantum number, l is the angular momentum quantum number, and m is the magnetic quantum number). That is, each AO corresponds to a certain set of quantum numbers n, l, and m. Different AOs have different symmetry relative to the x, y, and z axes in coordinate space. In practice, non-theoreticians do not know these functions in their exact mathematical form and, of course, do not analyze their symmetry. The s, p, and d AOs are recognized by organic chemists as graphical images: spheres of s orbitals, orthogonal dumbbells of p orbitals, rosettes of four d orbitals and the ring-dumbbell of the fifth d orbital (the latter is shown in Fig. 7). Those are the probability isosurfaces for AOs in Cartesian space which cover 3D domains of a certain probability for the location of an electron with the energy that corresponds to a certain AO.[a] What are *hybrid* AOs? They are *combinations* of AOs of different symmetry. These combinations (hybrid orbitals) are presented in introductory textbooks of organic chemistry in the same pictorial form of isosurfaces and, assuming that the reader well remembers these graphical images, are omitted here. Nevertheless, in order to remind us of these hybrid geometries, at least a tetrahedron of four C-centered sp^3 hybrid orbitals for the classic molecular unit, the molecule of methane, is presented (Fig. 6, top). This symbol of organic chemistry (a tetrahedral methane molecule or how four $2sp^3$ hybrid orbitals of the carbon atom are spatially arranged in this molecule) opens the hybridization series.

If all hybridization geometries are reviewed, one can see that the 3D organization of s, p, and d hybrid orbitals is of the same shape that the VSEPR rules provide. The above symbolic tetrahedron, a result of $2sp^3$ hybridization, is formally equivalent to the VSEPR "four-bond" tetrahedron of the carbon. Consistently, the sp and sp^2 hybridizations orient localized AOs in the same spatial directions which VSEPR prescribes for two and three dummy "bonds," respectively.

For bipyramids, hybridization involves d orbitals. Mixing five outer AOs of, e.g., a phosphorous atom, three distinct orbital geometries are obtained. These consist of two different trigonal bipyramids as well as one tetragonal bipyramid, which are $3dsp^3$, $3d^3sp$ or $3d^4s$ hybrids, respectively (the dsp^3 hybridization is shown in Fig. 7). The "five-bond" rule affords a trigonal bipyramid geometry of five-coordinated organophosphorous compounds (Fig. 1). The latter corresponds to the dsp^3 bipyramid of the orbital valence theory (Fig. 7), and this hybrid geometry has the lowest energy among the three hybridization states. Further, mixing of six valent AOs furnishes an octahedral bipyramid of d^2sp^3 hybridization, an orbital origin double of the pyramid supplied by the VSEPR "six-bond" rule. It appears that all classical

---

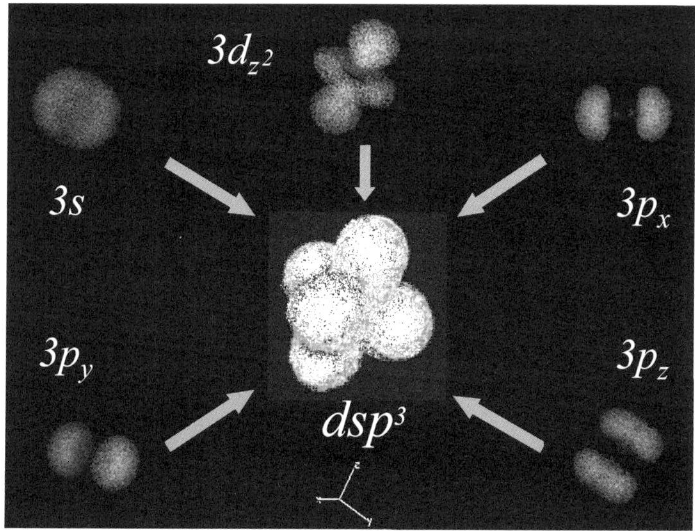

Figure 7. The $3dsp^3$ hybrid and the parent AOs: the hybrid orbital has a trigonal bipyramidal geometry. This familiar orbital representation indicates covering surfaces and not volumes i.e., it does not provide information about the distribution of electron density inside these closed surfaces.

geometries of molecular fragments are known to organic chemists, but their source of this information is the hybridization concept.

However, there is a remarkable discrepancy between the two simple descriptions of the geometry of elementary molecular units (VSEPR and hybridization of AOs). In contrast to VSEPR, the hybridization formalism cannot serve as a tool for predicting geometry. To the same extent, it is not a theoretical support of VSEPR. Organic chemistry textbooks do not clarify the simple point that molecular geometry is *not* a consequence of orbital hybridization (the opposite statement for molecular geometry expresses a widespread misunderstanding). Certain hybridization means certain geometry and vice versa. Hybridization is not a physical phenomenon; it is only a formal transformation of one mathematical form of AOs into another form upon some high symmetry constraints, thus, the final hybrid geometry is intentionally "pre-hidden" in the transformation type. Answering the question, "Why do we need this formal description?" one may be laconic: "For convenience." Though not determining the geometry, hybrid orbitals *reflect* the relationship between different geometries and distribution of valence electron density there. In addition, these orbital 3D fragments together with the *s*, *p*, and *d* AO descriptors, have become successful models in explanations of organic reactivity as well as reaction stereochemistry since these explanations operate with the correct geometry and reasonable distribution of the valence electron density in elementary molecular fragments.[a] As to the fragments themselves, their geometry, we repeat, determines the familiar geometry of hybrid orbitals and not vice versa.

The certainty in knowing what the geometry of elementary molecular fragments is does not mean that we know the 3D structure of organic molecules. When connecting these rigid fragments, one can find out that there are many distinct 3D molecular structures that may be assembled for the same chemical structure. Indeed, the number of relative orientations of connected elementary molecular fragments is infinite, and mechanistic assembly leads to organic molecules that have no certain shape. However, real molecules are characterized by certain individual geometries, i.e., there are some relative orientations of rigid elementary molecular fragments that provide thermodynamic stability to the entire molecular structure. How to determine these preferable orientations? That is, could the correct molecular geometry be deduced from the chemical structure in a simple fashion, without theoretical calculations or any support by experimental data?

1.4 Rotational Mobility of Elementary Molecular Fragments

Let us suppose for simplicity that the 3D structures of all elementary molecular fragments of a molecule are canonical chemical polyhedrons (Fig. 1). Then, the molecule is a chain of directly connected polyhedrons (including degenerated polyhedrons, i.e., polygons, if any fragments are linear or planar). If only the mass distribution is considered neglecting the electron mass, these polyhedrons are not solid. They consist of compact apexes (nuclei) and a compact center of mass (the central nuclei), i.e., they are "empty"

---

[a]Due to the seemed *geometry-hybridization* equivalence, the sp^3, sp^2 or sp terms are abused when these hybridization states are mentioned as synonyms of the corresponding molecular geometries. As indicated above, these meanings are not equivalent in the casuality aspect.

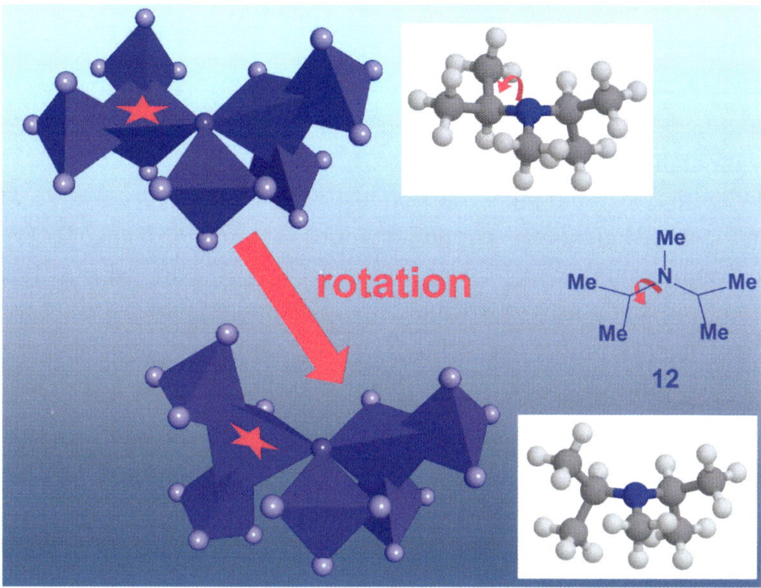

Figure 8. Left: schematic representation of two rotamers of amine **12** as sets of interpenetrating polyhedrons (here: tetrahedrons); the rotamers result from turning the certain tetrahedron (the star-labeled one) by ~100°. Right: the familiar ball-and-stick representation of these rotamers helps to see that they interconvert *via* rotation (indicated by arc arrows) of one certain *iso*-propyl group.

and even have no surface. For each connection of two polyhedrons, one of the apexes of each polyhedron is inside of the neighbor, being its center of mass, i.e., each of two directly linked polyhedrons interpenetrate each other.

For instance, a molecule of diisopropylmethyl amine **12** may be represented as a branched chain of interpenetrating tetrahedrons that contains also a VSEPR virtual tetrahedron of the nitrogen atom (Fig. 8). This poly-tetrahedron construction is obviously not rigid; separate fragments *can be rotated* around the axes which connect the central atoms of the neighboring fragments. The rotation keeps their own geometry unchanged since the solid apexes (the ligands) of the neighboring tetrahedrons do not come into tight contact at any rotation angle. From the geometrical viewpoint, this movement is the rotation of either ligand $\mathbf{L}_i$ or vicinal ligand $\mathbf{L}_j$ around the axis that connects two bonded atoms A_1 and A_2 of neighboring elementary fragments A_1L_i and A_2L_j. Because such rotation axes correspond to bonds, rotational freedom of connected polyhedrons (in strict terms, symmetry operation C_n) adopts a chemical connotation, i.e., it is understood as a rotation around the bonds.[a]

Thus, this "ball-and-stick" molecular model is rotationally flexible. However, the use of chemical terminology does not transform this mechanical model into a valid model of organic molecules. The model of a flexible chain of polyhedrons only advances a *hypothesis* that the same molecular skeleton may be shaped differently due to rotation of its units. Fortunately, this hypothesis is true. Microwave and IR spectroscopy have detected several rapidly interconverting stereoisomers for many substituted ethanes, and, remarkably, the spectra of these compounds are interpretable, representing molecular fragments as rotors of the symmetry related to these fragments.[19a] These numerous studies have shown that the number of stable forms is finite for each compound and, in addition, these forms may be of different stability. For instance, a methyl group corresponds to a C_3 rotor. The corresponding rotational forms (rotamers) are of identical geometry and, therefore, of equal energy. An *iso*-propyl group is an asymmetric rotor, and the corresponding rotamers have different stabilities. An additional evidence of the existence of rotamers has been supplied by NMR. As organic experimentalists well know, this physical method permits determination of parameters of molecular geometry; among them, torsion angles are usually determined most easily and, remarkably, essentially accurately. For some sterically hindered ethanes or their heteroanalogs, individual forms have been detected by NMR. Values of torsion angles obtained for the detected species show that these observable stereoisomers are related to each other as *rotamers* (and not as stereoisomers of any other geometry). Thus, experimental studies have led to a clear understanding that speculative rotation of polyhedrons does model real stereoisomerization processes in organic molecules.

Does this mean that real nuclei-in-molecules change their spatial disposition in such a way that the trajectory of their motion is a matter of discussion? The answer is probably yes, because these particles are much

---

more massive than electrons, and the quantum description of their motion (no accurate evaluation of the trajectory for states of accurate energy) may be reduced to classical physics (accurate coordinates and accurate energy for any particle at any moment). More exactly, among an infinite number of possible trajectories for the relative motion of vicinally positioned nuclei, the rotational trajectory has a probability near unity.

At this point, let us summarize our tentative conclusions about molecular shape. Taking into account the rotational freedom of organic molecules, we can conclude that the geometry of elementary molecular fragments does not determine the molecular shape unambiguously, i.e., a molecule may adopt different 3D geometries; fortunately, this 3D diversity is limited. *Static description of molecular shape is quite insufficient. Some rotational flexibility is inherent in a vast majority of organic molecules.* Therefore, our next question is, "Is information regarding the chemical structure sufficient for enabling to estimate how rotationally flexible a molecule or molecular fragment is in the ground state?"

One may start answering this question by considering rigid molecules. Rotation of elementary molecular units cannot destroy chemical connectivity in the molecule, because the cleavage of covalent bonds requires much more energy than internal rotation "contains." That is, the chemical connectivity in some molecules or molecular fragments may not allow rotational freedom at any physical conditions. One can characterize such molecules or fragments as absolutely rigid rotationally (in contrast to temperature-dependent hindered rotation addressed below). A plain example of a molecule of absolute rotational rigidity is adamantane. Even for an untrained eye, it is clear that formal rotation around any bond significantly twists the molecular framework and, if continued, disrupts its chemical connectivity.

However, this simple thought rotational experiment should not mislead. Identification of rotationally absolutely rigid systems by "simply looking" at the integrity of molecular structures when rotating their fragments (i.e., actually reducing molecular backbones to abstract ball-and-stick frameworks with rotational freedom for the neighboring ball-and-stick fragments as well as some longitudinal elasticity of sticks and mechanistically estimating whether internal rotation essentially disturbs the framework) is very far from being a universal tool. For instance, one can conclude that rotation of the phenylene rings of paracyclophane **4d** (synchronous rotation around two $C_{aliphatic}-C_{aromatic}$ bonds; Fig. 2) is impossible: it is visually obvious that there is no room for one of the aromatic rings to pass the other one while rotating. However, this "simple look" does not assist in answering the question whether the phenylene rings of a close homolog of [2,2]paracyclophane **4d**, [3,3]paracyclophane, have rotational freedom (of course, theoretical calculations have answered this question[19b]). One cannot estimate the "destructive" strain, which is developed in real molecules of [3,4]paracyclophane upon phenylene rotation, by pushing an abstract aromatic ring into the interphenylene cavity of an imaginary molecular model of unknown elasticity. ■

This primitive modeling of internal rotation often adopts the form of using different mechanical models of molecules that permit to rotate their elementary molecular fragment pieces one relative to the other. Sometimes virtual molecules generated by a molecular graphics program replace mechanical models. Rotational manipulations with mechanical molecular skeletons or on-the-screen molecules can seem to some organic researchers to be an adequate tool for drawing conclusions about the rotational mobility or rigidity of a molecular framework. These most primitive molecular models only work in obvious cases. For instance, using them, one cannot conclude whether rotation around the central C–C bond is possible in hexaphenylethane. The bond is extremely weak in this sterically overcrowded molecule, and rotation would dissociate it because of intolerably tight steric contacts of all vicinal substituents in the transition state of rotation. ■ Of course, we do not expect from molecule-modeling mechanical "toys" and molecular graphics programs to calculate bond energies. Therefore, with missing the rotation-induced breaking of chemical connectivity, they only indicate the necessity of using the genuine theoretical modeling.

Regrettably, arguments based on mechanical molecular models can still be encountered in organic literature. One should ignore them since these illustrative models are not research tools. Interactions of vicinally positioned substituents may be approximately described as a sum of interactions of valence electrons, which form the related vicinal bonds, and VDW interactions of these non-bonded groups (or atoms). It is reasonable to suppose that changes in interactions of the vicinal bond electrons have too low energy to cause the rupture of the central bond (the bond that connects the central atoms of two neighboring elementary molecular units) or another bond, when rotating these units one relative to the other.

The repulsive energy component of VDW interactions may drastically increase because of rotation-associated geometrical changes (changes that bring the distances between non-bonded atoms into the area of short atom–atom distances; Fig. 5, Section 1.2). Then, the chemical connectivity is broken if the energy of repulsion of non-bonded atoms is larger than the energy of a bond dissociation. It is absolutely clear that it is meaningless to compare these energies by rotating fragments in a mechanical molecular model or molecule image on the screen.[a]

Also, by mechanistically probing the integrity of molecular structure, one cannot conclude whether cyclic organic molecules are flexible. E.g., let us forget for a moment that cyclohexane undergoes ring inversion and try to conclude whether rotation around a ring bond is possible in the cyclohexane molecule. Similarly to the adamantine case, rotation around such a bond in the cyclohexane ring stretches the ring bonds and, even for relatively small rotation angles, separates the neighboring ring carbons to non-bonding distances. This reasoning is equally true for each bond of the ring, and one may conclude that the six-membered ring is rigid in real cyclohexane molecules. However, we remember that an essential component of ring inversion is rotation around the ring bonds.[b] Why does our thought experiment lead to the incorrect conclusion? "Simple look" misses cooperation of elementary molecular fragments in rotating them simultaneously, but with different magnitude, and distorting endocyclic bond angles accordingly. Therefore, from this mechanistic perspective, the cyclohexane ring seems rigid. Without unbiased judgment of theoretical calculations, it is difficult to indicate those concerted structural distortions which do not lead to a non-realistic increase of molecular energy of a whole system in the course of its 3D reorganization.

---

[a] In simple words, one must not use different mechanical molecular models as a probe for flexibility of molecular structures. This naive probing is similar to the "methodology" for easily measuring strain of molecules, which can be stated as: "Molecular strain is measured by the minimal height that is sufficient to break the mechanical model of a molecule by dropping it." This old joke of computational chemists cautions us against direct analogies between molecular and mechanical systems.

[b] In terms of intramolecular rotation, ring inversion (also termed ring interconversion or ring reversal) is a concerted, though restricted in magnitude, internal rotation around all endocyclic bonds accompanied with changes of endocyclic bond angles. In converting the initial ring invertomer into the resulting invertomer, both the ring torsion angles and endocyclic bond angles are altered, and distortions of these angles are maximal in the transition state of ring inversion. The magnitude of these alterations is larger for the torsion angles. Therefore, with some caution, one can approximate ring inversion in saturated and partially saturated cycles as a concerted restricted rotation around the ring-forming bonds.

If we figure out that such concerted geometrical transformations are in principle possible and include this point into considering possibility of rotation in cyclohexane, we will be incapable of making any relevant conclusion. There is no transferability of positive results (here: possibility of internal rotation) from a mechanistic molecular model to the real molecular system. Otherwise, one could assign intramolecular rotational flexibility to many chemical structures that, in fact, do not possess it. Phenathrene ([3]helicene) provides a simple explanatory example. Formally, one may twist an abstract molecule of phenanthrene and not disrupt the chemical connectivity, by changing endocyclic torsion angles of the central aromatic ring by reasonable values (i.e., destroying the ring planarity via restricted rotation around each ring-forming bond). This formal concerted rotation does not mean that phenanthrene is flexible: planar molecules of this compound are absolutely rigid; there is no internal rotation that would generate screwed stereoisomeric forms.

In practice, we know what the situation with the shape mobility of basic monocycles as well as many other cycles and polycycles is. However, structural analogies are not a reliable tool in probing molecular skeletons for absolute rigidity. We still cannot indicate for a cyclic system, which is not covered by our structural set, whether it has some rotational freedom in its ring(s). For instance, possessing information about the absence of rotational mobility in [3]helicene, one could assume that higher helicenes, e.g., [4]-, [5]-, and [6]helicene, are also rotational "freedom-deprived." This analogy-based prediction is absolutely incorrect: as ubiquitously known, interconversion of enantiomers of helicenes, or, in terms of rotational flexibility, restricted rotation around several aromatic bonds, is a fast intramolecular process.

The next example shows how unreliable even very close structural analogies may be when concerning absolute rigidity. Cyclohexane stereodynamics has been studied in detail.[a] With this knowledge in hand, let us try to decide whether the piperidine cycle of tropane (**13a**; Fig. 9) undergoes inversion and pseudorotation[b] or is absolutely rigid. It is clear that the

---

[a]Probably, the reader is well-versed in the general understanding of the stereodynamics of this conformationally basic system considered in any course on stereochemistry. Nonetheless, summarizing the basic knowledge on fast stereoreorganizations of the six-membered ring, Fig. 53 (Section 2.8) may refresh this understanding.

[b]One should not mix two different meanings of the same trivial term, pseudorotation. It is used for specifying both pseudorotaion in cyclic systems and pseudorotation in trigonal bipyramids (Section 1.7). Pseudorotation in six-membered rings is a change of the ring shape so that twist forms interconvert through boat-shaped forms (i.e., an elementary pseudorotation is a *twist–boat–twist* transformation; Figs. 37 and 53; for systematic nomenclature of structural transformations in six-membered rings, see

1,3-positioned dimethylene bridge blocks any rotation around the endocyclic bonds which would stretch this bridge.

Pseudorotation of the piperidine ring as well as ring inversion, which includes participation of a half-chair-shaped form (as a transition state; Fig. 53), involve bridge-stretching rotations, while ring inversion, which occurs through a sofa-shaped form (as another transition state), does not require such rotation. Also, the cyclohexane-related conversion of the *sofa* transition state into the *boat* transition state occurs without disturbing the dimethylene bridge in tropane (for conversion *transition state–transition state*, see Section 3.1). It follows from this short consideration that stereoreorganization *chair–sofa–boat* is possible in tropane. This prediction appears true: a fast diastereomerization *chair–boat* is a single ring-reshaping equilibrium in piperidine **13a** (Fig. 9; we omit discussing why the boat geometry that corresponds to a transition state in cyclohexane is stable in tropane). Thus, bicycle **13a** has a restricted rotational freedom for endocyclic bonds. The rotation is concerted and associated with altering endocyclic bond angles.

Now, let us look at tropinone (**13b**, Fig. 9). There is no doubt for our "simple look" that this molecule possesses the same rotational freedom as the parent piperidine **13a**. As *ab initio* calculations surprise,[9] the bicyclic backbone of amine **13b** is locked in the geometry of a chair-shaped six-membered ring. In contrast to the close structural analog **13a**, no minimum of molecular energy is located for a boat-shaped piperidine ring of **13b**, i.e., this geometry does not correspond to a thermodynamically stable structure. The lowest energy structure appears as a single stable 3D form of the piperidine ring. That is, the six-membered, chair-shaped ring is absolutely rigid in tropinone.

One can guess that "through space" interactions between the N lone electron pair and the π-electrons of the carbonyl moiety destabilize the boat geometry for tropinone (Fig. 9). We have come to this conjecture with computational results in hand, hence, it is only a *post factum* explanation of thermodynamically instability of the boat form of **13b**. One might say that this destabilizing electronic effect could be assumed by "simply looking," i.e., not

---

Ref. 38). Pseudorotation in five-membered rings is a change of the ring shape so that alternate twist and envelope forms are transformed one into another through intermediate geometrical forms (Fig. 61). Of course, pseudorotation in cyclic molecules is a restricted concerted rotation around all ring-forming bonds coupled with some alterations of all bond angles.

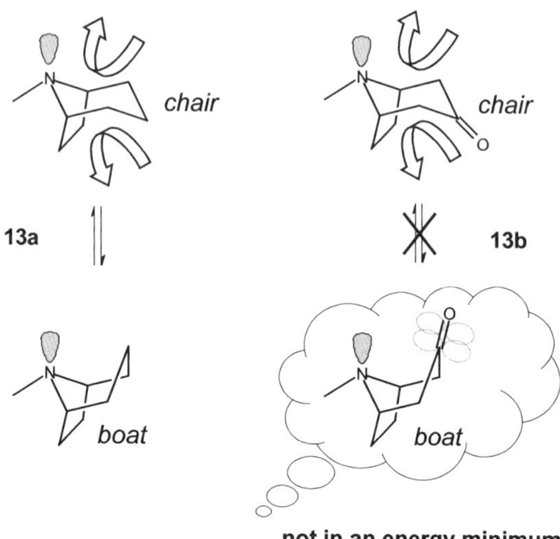

Figure 9. Two stable 3D structures (two different energy minima) of bicycle **13a** with distinct geometry of the piperidine ring (chair- and boat-shaped). These diastereomers are interconverted via disrotatory concerted rotation around four C–C bonds of this ring accompanied by some change of endocyclic bond angles (left). The same transformation in a close analog **13b** does not occur since the related stereoisomer of boat geometry of the piperidine ring does not correspond to an energy minimum (right). For this structure, the n orbital of the N atom and π orbital of the C=O group (both doubly occupied) are shown for indicating their unfavorable disposition.

performing theoretical calculations, at the boat geometry of the piperidone ring in **13b**. In fact, the pictorial argument does not put instability of the boat form, and, therefore, it does not put any restriction for flexibility in cyclic ketone **13b**. "Simply looking," destabilizing electron interactions may only be associated with an increase of molecular energy. As known, stability of a molecular structure is equivalent to its correspondence to an energy minimum. Clearly, this energy increase for the boat structure of **13b** does not indicate the absence of a *minimum* in the potential energy surface for this geometry (for a description of potential energy surface, see Chapter 3).

It is obvious that a solid conclusion about the rigidity of the hydrocarbon skeleton of tropinone could not be available without using theoretical calculations. They locate the energy minima (Chapters 4 and 5). Thus, a sole chemical structure does not always "report" whether the molecule is or is not absolutely rigid rotationally. One can only note that, by chemically bonding to remote fragments (e.g., by closing the molecular backbone in a ring), we decrease rotational freedom in the molecule. For instance, the octamethylene chain is thermodynamically less flexible, if closed in a cyclooctane ring.[a] The ring continues to lose a significant part of its rotational freedom (i.e., from pseudorotation and ring inversion) if any ring atoms are connected by, e.g., a dimethylene bridge. At some extent of bridging, any system becomes absolutely rigid rotationally.[b] Since, as we have seen, this structural "critical point" of the loss of rotational mobility is not quite obvious for a non-trivial system (e.g., when assessing new polycyclic compounds), "discovery" of rigid molecules is of great interest. Analysis of short-range intermolecular interactions, which develop in the course of a chemical reaction, as well as long-range intermolecular interactions in, e.g., absorption, solvation, crystal packing processes, and biorecognition, is incomparably easier for "frozen" molecular shapes.

1.5 Rotation Around One Bond: A Simple Post-Textbook Discussion

After discussing rotational rigidity, we can come back to the initial consideration of rotation of chemically bonded elementary molecular fragments. Keeping in mind that rotationally rigid systems exist, we, nevertheless, may accept that this simple motion is an attribute of intramolecular flexibility for most organic backbones independently on their chemical functionality. Indeed, absolute rotational rigidity requires a particular not quite ordinary chemical structure. Hence, other molecules in principle do possess rotational freedom. Due to this ubiquity, rotational flexibility is discussed here in detail. In the first turn, we are interested in answering the primary stereochemical questions, i.e., "What is the geometry of *stable* rotational forms?" and "Which among them are more stable?"

---

[a] I.e., the number of stable structures generated by rotation is less for cyclic structures.

[b] Absolute rotational rigidity does not mean an absolute molecular rigidity since rotation is not a single possible intramolecular motion (Section 1.7). For instance, nitrogen inversion is inherent to amines **13a** and **13b**.

One should note that organic specialists usually relate these questions to compounds that are not in the solid state. This is not only because organic reactions are almost always studied for occurring in the liquid as well as gas phases. In the solid phase, molecules are packed in a highly ordered supramolecular structure (crystal lattice). In such tight packing, intermolecular forces secure a certain 3D molecular structure, preventing any atom motion excluding their oscillations. Briefly stated, there is no internal rotation in molecules-in-crystals. Besides, intermolecular interactions are strong at short distances, as they are in the crystal lattice. This factor can essentially affect static rotational orientation of neighboring elementary molecular fragments of a molecule. As a result, rotational geometry as well as the preference of either rotational form in the molecule that is a part of the crystal lattice may differ from those for a free molecule. Indeed, "intrinsic" rotational stereochemistry, i.e., the molecular geometry of rotamers (concerned mainly with torsion angles between vicinal substituents), relative stability of these stereoisomers, and the rate of their interconversion, are associated with molecules in vacuum or in the ideal gas phase (conditions of no intermolecular interactions). In the liquid phase, intermolecular forces are much weaker than in the solid state, but nevertheless, they cannot be neglected because distances between solute and solvent molecules are still short in solvates. One may expect that 3D geometries and rates for rotamer interconversion in solution differ from those in vacuum if strong solvent–solute interactions, i.e., high energy H-bonding, take place.

Obviously, analysis of rotational flexibility is easier when using the model of free (non-interacting) molecules. In order not to overload this introduction by discussing intermolecular interactions, we can simplify the first question regarding rotamers by addressing the following issue: "What is their geometry *in vacuum* (implying *ideal* gas, we can say: *in the gas phase*)?" and "Which among them are more stable under these conditions?" Remember, our starting, although abstract, position is that we have no preliminary structural or energy information about rotamers of organic compounds. In this position, how to analyze molecular geometries of rotamers and estimate their relative stabilities? One can reasonably suppose that, by using a "magic" tool of theoretical calculations, it is possible, without principal difficulties, to predict the molecular geometry of rotamers as well as estimate their relative molecular energy. At this point of our account, the reader should merely believe that calculation methods can model the rotation trajectory quantitatively, i.e., to gradually change the rotational angle and calculate the molecular energy.[a] Trajectory points, which are energy minima, correspond to stable structures, and, thus, the problem of rotamer

---

[a] See Section 4.4.3, Fig. 83, for explanation of this methodology.

location may be solved for each particular case. For instance, according to such calculations,[a] our probe molecule **12** (Fig. 8) corresponds to an energy minimum when adopting the geometry of structure (**a**) shown in Fig. 10, i.e., this geometry belongs to a thermodynamically stable structure. Further computational modeling of the rotation of one of the *iso*-Pr groups (as explained above, by means of gradual change of the rotational angle and calculations of the energy of the molecule for each angle value) leads to a series of rotamers of amine **12** of smoothly changed geometry. These modeled rotamers do not correspond to thermodynamically stable 3D structures. When the rotation angle reaches ∼100°, a new minimum, although of a higher energy, appears (structure (**b**) in Fig. 10).

There is no reason to relate only one of the stable geometries to the molecular geometry of **12** and to reject the other as an irrelevant one. Both of the structures, together with additional rotamers (not shown; those can

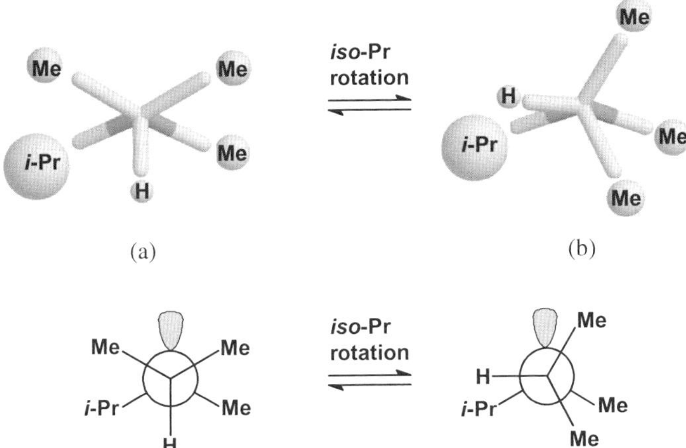

Figure 10. Rotation of a C-centered elementary molecular fragment relatively to the adjacent N-centered one in amine **12**. Structures (a) and (b) of optimized calculated geometry (top; spheres of different radius symbolically indicate different volumes of substituents) and the corresponding schematic representations using Newman projections along the C–N bond (bottom) are shown.

---

[a] By molecular mechanics, using the MM3 force field (Ref. 9). For molecular mechanics and this force field, see Chapter 4.

be located by rotating *iso*-Pr's further), represent the molecular shape of compound **12** at thermodynamic equilibrium. Note that, using theoretical calculations, we can be sure that we have answered the question about the geometry of rotamers of compound **12**.

However, this example would not inspire experimentalists to learn practical computational chemistry. Traditional stereochemistry pushes them to analyze rotational flexibility of molecules in an easier manner. What is the "pocket-size" approach of organic chemists that provides at least verisimilar qualitative results regarding rotamer structures and rotational equilibrium for quite different compounds?

Rotation is a very convenient motion for being considered by non-mathematicians. Since no complex movement law is involved, relative orientation of vicinal substituents is clear at any point of their 3D trajectory. Indeed, a single geometrical parameter, the rotation angle ϕ, determines their mutual location (Fig. 11).

Newman projections, which are ubiquitously used primitive graphics (Figs. 10 and 11), reflect this easiness. For instance, it is shown in Fig. 10 that the calculated geometries and the corresponding Newman projections provide qualitatively the same view.[a]

It is useful to recall here the conventional nomenclature (the Prelog–Klyne nomenclature) of rotation-derived geometries shown in Newman projections (Fig. 11), as well as the classical plot of *rotational angle–molecular energy* for ethane, in a qualitative representation (Fig. 12).

An extended nomenclature for rotational orientations of vicinal substituents in a pair of σ-bonded elementary molecular fragments of tetrahedral geometry (the Michl–West nomenclature) proposes more names associated with certain absolute values of rotational angle ϕ: transoid (T) for $\phi \approx 165°$, deviant (D) for $\phi \approx 150°$, ortho (O) for $\phi \approx 90°$, gauche (G) for $\phi \approx 60°$, cisoid (C) for $\phi \approx 40°$, anti (A) for $\phi \approx 180°$, syn (S) for $\phi \approx 0°$, and E (eclipsed) for $\phi \approx 120°$.[19c] This new trivial nomenclature reflects the current progress in understanding rotational isomerism.

---

[a]Such geometrical simplicity is, of course, not warranted. For instance, in many cases of 3D rearrangements of elementary molecular fragments themselves (Section 1.7), trajectories of substituent motions are described by non-trivial mathematical expressions. Tracking of the relative location of these ligands at any trajectory point is not a simple task for non-computational chemists.

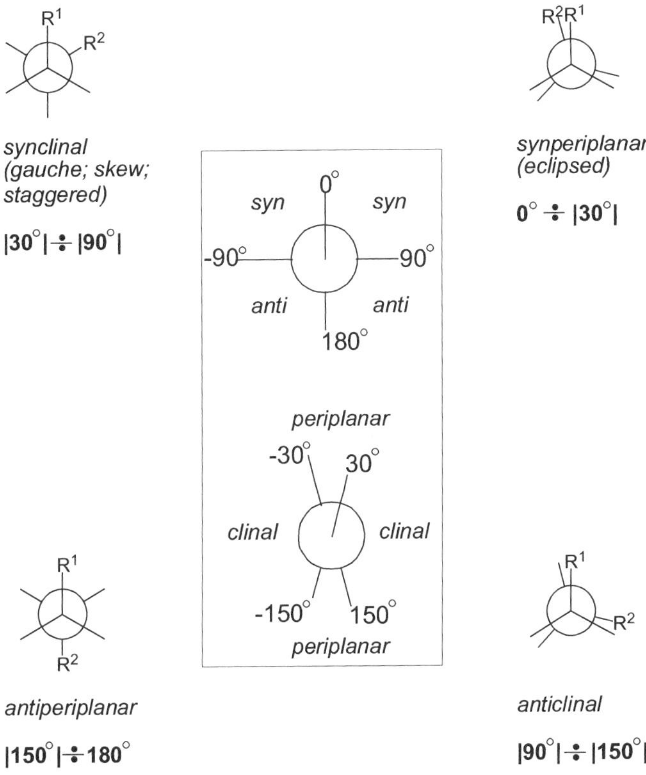

Figure 11. The Prelog–Klyne nomenclature for relative disposition of two vicinal substituents (absolute values for the corresponding torsion angle ϕ are given in bold). The conventional name (placed outside the frame) of a certain orientation of two vicinal substituents (shown by the related Newman projection) is derived from trivial names (placed inside the frame) of the ranges for torsion angles by combining a *syn/anti* term with a *periplanar/clinal* term.

It specifies some rotamers (T, D, O, and C), which normally are thermodynamically stable 3D molecular structures for some milestone compounds (Section 1.6), but are not singlet out in the becoming obsolete Prelog–Klyne nomenclature.

In light of this geometrical easiness, it seems that, if it is possible to supply any pair of vicinal ligands in such a model with relevant energies of ligand–ligand interactions, one could qualitatively estimate attraction/repulsion for different pairs of vicinal substituents and thus identify

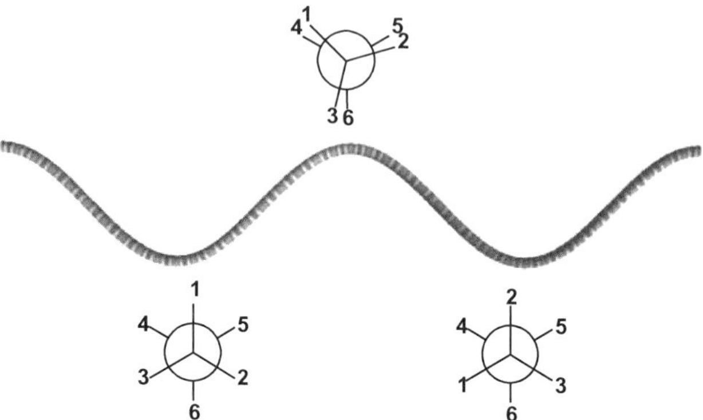

Figure 12.   Change of molecular energy (vertical coordinate) upon rotation (rotation angle ϕ is the horizontal coordinate) around the C–C bond in ethane (numbers depict H atoms). Stripes in the curve indicate structures (rotamers) that are smoothly transformed one into another, when changing the rotation angle ϕ. The continuity of this "sinusoidal" line shows that the number of transient rotamer structures is infinitely large. Stable forms (thermodynamically stable rotamers) correspond to energy minima; the energy maxima are related to transition states of rotation. Neighboring minima or maxima are separated by a 120° turn. Three 120° turns in series lead to the structure of the initial geometry.

the favored mutual rotational orientations of one–bond-separated elementary molecular fragments (i.e., arrange rotamers according to their relative thermodynamic stability). Many organic chemists follow this additive, "pocket-size" approach.[a]

An extremely simple idea is embedded in this visual comparison of vicinal interactions. This qualitative model (mentioned in Section 1.4) equips vicinal substituents with two main "visible," rotation angle-dependent chemical attributes: (*1*) vicinal σ bonds prefer maximal angular remoteness (as dictated by orbital energetics within the molecular orbital model of atom–atom interactions); this factor is favored for non-encumbered ethanes, and (*2*) vicinal substituents tend to occupy maximally distal positions one

---

[a]Due to the obvious simplicity of Newman projections, these pictural 3D structures are most often used in organic laboratories when considering rotational preferences for different interlinked elementary molecular units of tetrahedral geometry.

to another, due to repulsive VDW interactions (this factor predominates for ethanes with bulky substituents). Both factors (*1*) and (*2*) are destabilizing and act in the same direction. As a result, the most known conformational rule in organic practice (let us call it the main conformational rule) is uncompounded and compact. One may phrase it as the following: *for two adjacent elementary molecular fragments of tetrahedral geometry and ligands σ-bonded to the central atom, the staggered orientation of two vicinal substituents is favored, while vicinal bond eclipsing is the least preferable; an increase in the bulkiness of vicinal substituents leads to predomination of their antiperiplanar disposition over the synclinal (gauche) one.* This rule is actually a "simple look" at the relative stability of rotamers; textbooks on stereochemistry equip with it every generation of organic experimentalists, and determination of relative stability of staggered rotational isomers by means of this mechanistic tool is still a traditional "theoretical" conformational analysis in organic laboratories.

Indeed, only molecular geometries that correspond to energy minima may be detected and thus studied experimentally (Section 2.1). According to the above "simple look" at stable rotamers, those are supposed to be structures of staggered geometry. In identifying *favored* stable rotamers, this "theoretical" modeling (most often, by analyzing Newman projections; sometimes, by using mechanical molecular models or on-the-screen molecular graphics) is focused on visual determination which pair of vicinal substituents, among others, has minimal destabilizing interactions in synclinal disposition. Let us inspect this popular pencil-and-paper approach of additive vicinal interactions.

Counting vicinal (actually, synclinal) steric interactions is not problematic for the textbook case of butane. Two vicinal substituents, Me's, are essentially more bulky than other substituents, H's. Taking into account steric contacts of these substituents in both geometries, it is trifling to conclude that an antiperiplanar disposition of Me's is more stable than the synclinal one (Newman projections in Fig. 11 plainly illustrate this). Notably, textbooks compare interactions only between the most bulky groups (Me's) for both rotamers (synclinal and antiperiplanar Me's), and this seems reasonable. Indeed, as the well-proven VDW concept suggests, certain steric volumes, which resist to interpenetration (hard VDW spheres), may be assigned to non-bonded atoms. Their relative size correlates with the energies associated with their interactions, and consideration of a larger or a smaller group means consideration of stronger or weaker repulsion, respectively. The chemical structures of these groups indicate which among them are larger or smaller as, e.g., this is shown schematically for the H, Me, and *iso*-Pr groups in Fig. 10. It is convenient to apply the VDW concept to vicinal substituents. Therefore, the "simple look" at rotamers, in a pictorial language of Newman projections, says that methyl groups in butane (i.e., the most bulky substituents) prefer a distant, antiperiplanar orientation (Fig. 11). Similarly to the VSEPR rules (Section 1.2) that diverge in the molecular space *geminal* substituents as maximally as possible, this simple principle performs the same geometrical arrangement for the most bulky *vicinal* substituents.

One may characterize this quick comparison as "visual" estimation of the energy of vicinal steric interactions for each rotamer. This energy is represented by energy E_{syn}, which actually is the energy of steric interactions E_i of *all* synclinally positioned substituents of a rotamer, i.e., $E_{syn} = \Sigma E_i$, where E_i is estimated in qualitative terms *higher/lower*. The simplicity of this approach consists of the association of these "values" *higher/lower* with "visual" estimates of the relative bulkiness of synclinal substituents. The more bulky two synclinal substituents are, the higher destabilization may be attributed to a related abstract value E_i for this pair of vicinal, synclinally oriented substituents.[a]

It is not difficult to see that the butane case in the textbook version may mislead. "Visual" estimation of rotamer stability compares E_{syn} for alternative rotamers. Let us examine how we can proceed following the butane example, e.g., with more substituted hypothetical structures **A** and **B**. These model systems have three substituents, H, more bulky R^1, and the most bulky R^2; R^1 and R^2 are geminal (Fig. 13). Rotational isomers of **A** are arranged according to their relative stability as (*c*) (one synclinal interaction R^1/R^1) > (*c*) (one synclinal interaction R^1/R^2) > (*b*) (one synclinal interaction R^1/R^1 and one synclinal interaction R^1/R^2). Similarly, rotational isomers are arranged as (*c*) > (*b*) > (*a*) for **B**. As mentioned, proton substituents are actually neglected, i.e., synclinal interactions

---

[a] It is only a formal abstract value. Nevertheless, it is clear for any pair of vicinal substituents, which energy value, E_i or E_j, is higher: the substituent size predetermines their *higher/lower* relationship.

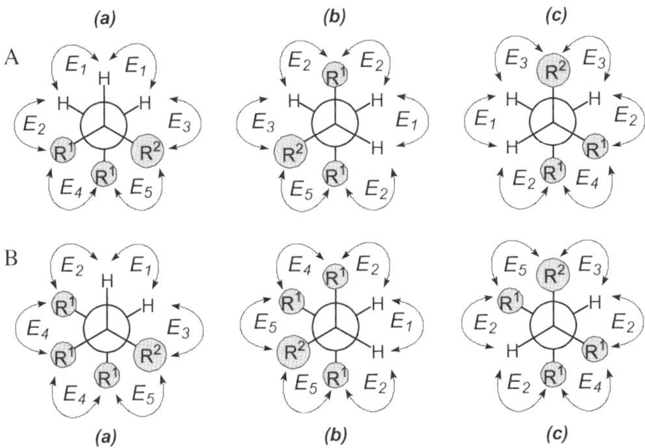

Figure 13.   Determination of relative stability of rotamers by comparing energy of steric repulsion for synclinally positioned substituents in hypothetical molecules **A** and **B**. Bigger circles indicate the most bulky groups R^2. Smaller circles indicate substituents R^1, which are more bulky than protons. Arrows depict steric interactions of vicinal substituents in synclinal orientation.

proton–proton as well as *proton–R* are outside the scope. The problematic point is that there is no indication that steric interactions with protons can be ignored.

   If steric interactions with protons are taken into consideration, this "synclinal" tool unexpectedly becomes useless. Let us include these H-related interactions in the analysis of rotational isomers of **A** and **B**. There are five pairs, H/H, H/R^1, H/R^2, R^1/R^1, and R^1/R^2. Abstract E_i values are E_1, E_2, E_3, E_4, and E_5, respectively, with $E_1 < E_2 < E_3 < E_4 < E_5$ (Fig. 13). Arithmetic summing of E_i for all synclinal pairs gives E_{syn} for each rotational isomer (remember, $E_{syn} = E_i + E_k + \cdots + E_n; n = 6$). It is transparent that the sign of the difference ΔE_{syn} for rotational isomers, e.g., *(a)* or *(b)*, says which among them is more stable.[a] It appears that, albeit the qualitative $E_i, < E_{i+1}$ relationship is known for each pair E_i/E_{i+1}, one cannot determine whether ΔE_{syn} for two rotational isomers is negative or positive. Below are formal energy expressions for rotational isomers of A and

---

[a] $\Delta E_{syn} = (E_{syn})_{\text{rotamer } b} - (E_{syn})_{\text{rotamer } a}$

B (Fig. 13) and the corresponding differences ΔE_{syn}:

$$\textbf{A: isomer } (\textbf{\textit{a}}).\ (E_{syn})_a = 2E_1, +E_2 + E_3 + E_4 + E_5$$
$$\textbf{A: isomer } (\textbf{\textit{b}}).\ (E_{syn})_b = E_1, +3E_2 + E_3 + E_5$$
$$\textbf{A: isomer } (\textbf{\textit{c}}).\ (E_{syn})_c = E_1, +2E_2 + 2E_3 + E_4$$

and

$$\textbf{B: isomer } (\textbf{\textit{a}}).\ (E_{syn})_a = E_1, +E_2 + E_3 + 2E_4 + E_5$$
$$\textbf{B: isomer } (\textbf{\textit{b}}).\ (E_{syn})_b = E_1, +2E_2 + E_4 + 2E_5$$
$$\textbf{B: isomer } (\textbf{\textit{c}}).\ (E_{syn})_c = 3E_2 + E_3 + E_4 + E_5$$

Then, we have for compound **A**

$$(\Delta E_{syn})_{c-b} = -E_2 + E_3 + E_4 - E_5$$
$$(\Delta E_{syn})_{c-a} = -E_1 + E_2 + E_3 - E_5$$
$$(\Delta E_{syn})_{b-a} = -E_1 + E_2 + E_2 - E_5$$

and for compound **B**

$$(\Delta E_{syn})_{c-b} = -E_1 + E_2 + E_3 - E_5$$
$$(\Delta E_{syn})_{c-a} = -E_1 + E_2 + E_2 - E_4$$
$$(\Delta E_{syn})_{a-b} = -E_2 + E_3 + E_4 - E_5$$

None of these differences may be associated with the negative or positive sign. Indeed, four items of the difference expressions may be grouped into two pairs $-E_i + E_{i+1} = E_{S1}$ ($i = 1, 2,$ or 3) and $E_j - E_{j+m} = E_{S2}$ ($j = i + n$; $n = 1$ or 2; $m = 1, 2,$ or 3). Note that E_{S1} is positive while E_{S2} is negative. Then the expression for any difference $(\Delta E_{syn})_{i-j}$ may be written as

$$(\Delta E_{syn})_{i-j} = E_{S1} + E_{S2}$$

The sign of the algebraic sum E_{S1} or E_{S2} for each pair of substituents is predetermined by the substituent's size. Since these signs are opposite for E_{S1} and E_{S2} in each expression for ΔE_{syn}, there is no information concerning which algebraic sum, E_{S1} or E_{S2}, is larger in the absolute value. Therefore, the sign of any difference ΔE_{syn} in this model is unknown. This is an obvious indication of a weakness of this primitive additive approach to rotational mobility. Our consideration shows that this weakness appears even for non-functionalized (alkyl substituted) ethanes.

Of course, one may turn to the literature data and conclude from them that the "synclinal" estimation is valuable, if R^1 and R^2 are bulky ligands relatively to H (e.g., α-branched alkyl groups). If the discrepancy in effective volumes of substituents R^1 and R^2 *vs.* proton is not sufficiently large (e.g., they are halogen atoms, or hydroxy, alkoxy, cyano, or nitroso groups), steric interactions with protons should be taken into account. However, the mechanistic synclinal model cannot estimate relative bulkiness of such unbulky substituents. This means that, surprisingly for many synthetic chemists, *often, two connected elementary molecular fragments cannot be subjected to "visual" estimation of relative stability of rotational isomers.* Favored rotamer geometries are reliably predictable in this way only for textbook examples. Moreover, if rotational orientations of a non-tetrahedral elementary molecular fragment (i.e., of trigonal pyramid geometry) are analyzed by using this primitive technique of additive vicinal interactions, we could not indicate for it any rotational preference at all.

Chemical diversity of vicinal substituents leads to further difficulties with this ill-grounded analysis. For instance, H-bonds between functionalized vicinal substituents or other chemical bridges between them (often formed by coordinated cations, e.g., Li^+) stabilize gauche orientation while vicinal lone pairs, substituents of a small effective volume, destabilize it more than some other larger groups. Hence, the simplest model should be supplemented with additional "visible" interactions that take into account (again qualitatively) these specific gauche interactions. The additivity of vicinal interactions is kept in such "extended" models, but their prediction ability is minimal. It is transparent that any information, which concerns relative contributions of vicinal interactions of different types into rotamer stability, cannot be extracted from such pictorial level estimations.

Not rarely, this simple tool of counting synclinal interaction loses its validity due to other reasons. Those may be different. For instance, the lone electron pair is a peculiar, "Janus-behaving" ligand. On the one hand, not being an atom, it may be characterized as a very non-bulky substituent. On the other hand, it can interfere strongly because of its tendency to energetically interact with valence electrons of other atoms, i.e., it can behave as a bulky group. Therefore, lone electron pairs may introduce uncertainness

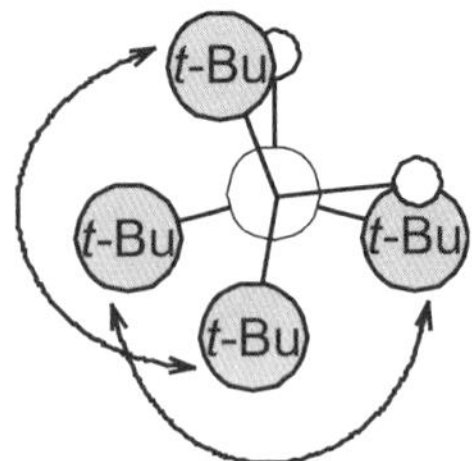

Figure 14. Newman projection for the most stable rotational isomer for symmetrically substituted tetra-*tert*-Bu-ethane (white circles depict vicinal protons). Arrows indicate significant distortion of the tetrahedron of each C–C bond-forming carbon.

into "simple look" predictions. For instance, rotational form (***b***) of amine **12** (Fig. 10) is a stable (minimum energy) form despite an appreciable proximity of vicinal substituents in this geometry.

Sterically overcrowded molecular fragments are another example of the weakness of predictions which are based on "visual" consideration of synclinal interactions. As *calculations* show for 1,1,2,2-tetra-*tert*-Bu substituted ethane, two vicinal *tert*-Bu's almost fully eclipse the corresponding vicinal H's (Fig. 14) in the minimal energy structure.[20a] This eclipsing is because of spatial divergence of two geminal *tert*-Bu groups of each carbon of the C–C bond and essential elongation of this bond. These distortions of the ethane fragment geometry are not self-evident. ■ Consequently, the unusual mutual orientation of vicinal substituents probably is a surprise for researchers only equipped with the "simple look" of the canonical orientations (Fig. 11).[a]

Let us continue our critical analysis of this abused conformational tool of counting "obvious" destabilizing vicinal interactions in alternative rotamers. Different stabilities of non-identical rotamers are not solely because of steric or another "visible" (e.g., H-bonding) interactions of vicinal substituents. Rather often, quantum effects have no classical analogy

---

[a]Some researchers use Newman projections as *proofs* for the proposed geometry. This practice is fallacious such graphics are only a convenient *illustration* of steric repulsion between two vicinal substituents in canonic orientations (Fig. 11). As explained, Newman projections are too primitive to reliably estimate vicinal interactions in many 1,2-disubstituted systems and therefore are not a valid tool for qualitative structural analysis. Nevertheless, they could be useful for an initial or tentative hypothesizing "for internal use."

and, therefore, they are not recognized or estimated by the researcher who does not dispose of tools of theoretical chemistry. For instance, as it follows from many experimental observations, vicinal substituents may perturbate simple steric rules by the ubiquitously known anomeric effect.[20b] This effect consists of preferring a synclinal geometry (Fig. 11) over an antiperiplanar one for vicinal non-bulky R and electronegative X substituents in a saturated fragment R–A–B–X, where A is an atom of virtual tetrahedral geometry (Fig. 1) that bears a lone electron pair, and B is a tetrahedral atom without such pairs. The more electronegative X is, the greater the anomeric effect is. Not going deeper into theoretical analysis of this effect, one can mention that a simple concept of bonding and non-bonding orbitals predicts it by proposing an overlap of these orbitals of 1,2-positioned molecular units of interest. From the perspective of the "non-electronic" concept of hardly deformable VDW spheres (i.e., a mechanistic concept of steric contacts), the anomeric effect does not exist at all. In order to take into account both effects (electronic and steric) in predicting the favored rotational orientation of vicinal groups, one may equip Newman projections with graphical representation of these specifically oriented orbitals. Regrettably, this trick does not supply these images with improving prediction ability: the magnitude of the anomeric effect for different groups X remains obscure in this illustrative modeling. Trying to "rescue" this apparently convenient approach of additive vicinal interactions, one can supply R and X with supplementary X-specific "increments" of the anomeric effect (let us hypothesize that we somehow could extract these qualitative estimates). Recall, steric bulkiness acts in the opposite direction, i.e., it "pushes away" synclinally positioned vicinal groups. Obviously, the balance between substituent bulkiness and the magnitude of the anomeric effect is outside the limits of accuracy of the "improved" visual additive model. Its ability *to predict* rotational preferences in such R–A–B–X systems is worthless.

The same conclusion could be drawn also for the gauche effect systems. Of another orbital origin, this effect means a tendency to adopt a synclinal disposition (Fig. 11) for two vicinal electronegative substituents X^1 and X^2 bearing lone electron pairs (e.g., halogen atoms, OAlk, and SAlk groups). A situation is possible that two other vicinal substituents adopt a synclinal orientation as a result of the gauche orientation of X^1 and X^2. Then the tendency of X^1 and X^2 to have this vicinal geometry competes with that of other substituents to avoid their synclinal orientation. Transparently, there is no simple answer to the question which tendency predominates. This is a common problem with the "simple look" at rotamers: by operating with very qualitative, "binary" estimates *lower/higher* of repulsion for a selected vicinal substituent pair in their different mutual orientations shown in Fig. 11, we are unable to compare contributions of two or more pairs of such substituents into relative thermodynamic stability of alternative rotamers. One should notice that, singling out *effects* (here, geometries disfavored due to increasing steric contacts), stereochemists have actually indicated difficulties in quick prediction of rotational preferences.

Another example deals with eclipsing *vs.* staggering. Unfortunately, many organic experimentalists associate eclipsing with rotamer instability. Surprisingly for them, the above described main conformational rule (predominance of staggering over eclipsing) considers only ethane-like fragments R–A–B–R. For many other compounds, experimental as well as theoretical studies have shown that a single bond and a double bond prefer eclipsing. These observations have established the next conformational generalization: *in fragments* $R^1(R^2)(R^3)C-C(X)R^4$ *of alicyclic olefins, eclipsing prevails over staggering for a pair of vicinal bonds* $C-R^i$ *(σ bonding) and* $C=X$ *(both σ and π bonding).*

Newman projections, which illustrate this statement (let us call it the second conformational rule because it is used as a "simple look" tool), show that two certain geometries — eclipsed and bisecting (Fig. 15) — should correspond to the related thermodynamically stable rotamers of, e.g., alicyclic alkenes, azomethines, and carbonyl compounds.

This complimentary conformational rule implies that vicinal substituents R^i or X should not be too bulky. Otherwise an eclipsed geometry of one of them and group X becomes unfavorable because of increased steric repulsion. The combination of two opposite factors, i.e., steric (a visually estimable factor) and certain orbital (an "invisible" factor) interactions, inserts essential ambiguity in the "no calculations prognosis" of staggering *vs.* eclipsing for chemical structures that contain π-bonded atoms. Indeed, it is difficult to postulate in this visual analysis which groups R are not bulky for R/X eclipsing.

In the commonly known LCAO MO description,[a] the reason for this preference is as follows. A σ bond, when eclipsing the adjacent vicinal π bond, lies in its nodal plane (i.e., in the plane of zero π_z electron density; Fig. 15). Obviously, a decrease of electron repulsion leads to stabilization of this spatial orientation (eclipsing).

In addition, this orbital consideration contains an important message for the user of this conformational rule: the rule does not consider elementary molecular fragments, which have ligands bonded to the central atom by the bonding that has order 2 (or near 2) but does not include a π-bond. In simple words, chemical fragments with the *central atom* **A**–*ligand* **L** bonding of the second order, but without π-bonding **A88L**, are often incorrectly written in chemical notation **A**=**L** and are not distinguished from chemical fragments, which do have "genuine" double bonds (both σ and π bonds in chemically connecting **L** to **A**). Such "no π-bond structures" are not subjects for this conformational rule. For instance, in sharp contrast to C=C and C=O units, two atoms in phosphoryl fragments P=O are not π-bonded (this chemical structure should be written rather as P^+–O^-). Therefore, we are not surprised by the fact that thermodynamically stable C–P rotamers of, e.g. phosphine oxides as well as phosphonic and phosphinic acid esters, have an undistorted staggered (i.e., characterized by the torsion angle $\phi = 60°$, Fig. 11) and not eclipsed angular geometry for vicinal substituents at the C–P bond.

The above "orbital" message also clarifies that none of the common conformational rules can be applied to predicting stable rotamer geometries for a pair of chemically connected elementary molecular fragments until revealing that the bonding between their central atoms and related ligands includes exclusively σ, or both σ and π bonds. Figure 15 (top, right) contains this reminder. For instance, bonding in ylides cannot be described in terms of only sigma and pi bonds. Both the main and second conformational rules are inapplicable to predicting the geometry of thermodynamically stable rotamers of ylides because these empirical generalizations have been made by considering chemical structures that only have such bonds. ■

Similarly to the main conformational rule, this next empirical generalization actually says that there are some canonical geometries for stable rotamers of molecular structures that have certain chemical bonds. It is complementary to the mentioned main rule that considers vicinal σ bonds A–R and B–X in fragments R_3A–BX_3 and, together with it, occupies

---

[a]We can recall from the basic course on theoretical chemistry that, in the LCAO MO approach, MOs (molecular orbitals, i.e., mathematical functions that deal with the energy of individual electrons in the molecule) are built as linear combinations (LC) of the corresponding atomic orbitals (AOs) of individual, i.e., non-interacting, atoms. Familiar bond-characterizing terms σ and π are related to exactly such MOs [symmetrical and antisymmetrical (sign-changing) functions, respectively] combined from a pair either of s or p AOs.

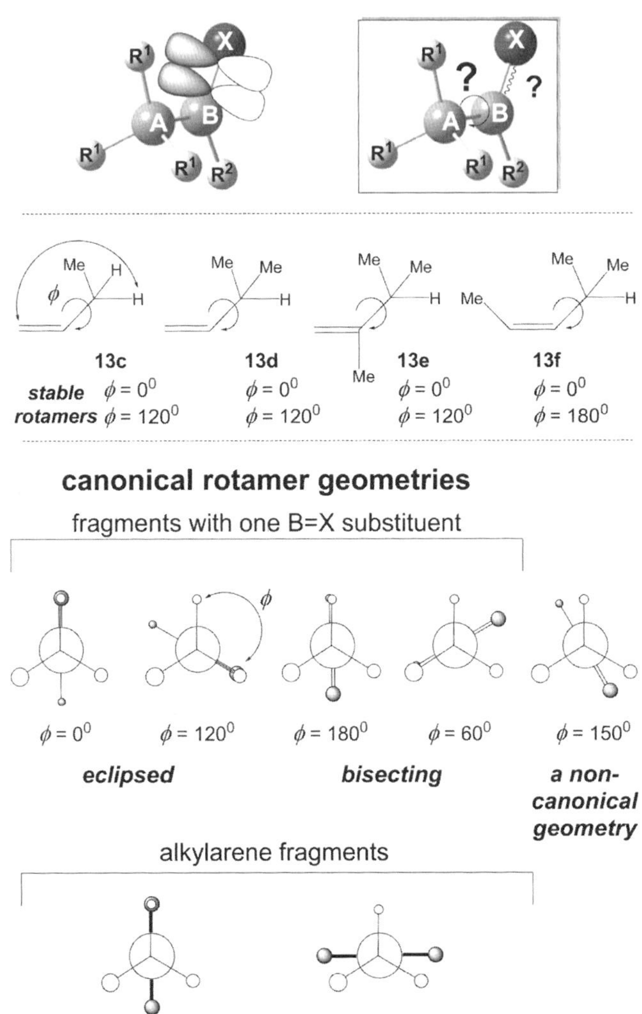

Figure 15. Eclipsing *vs.* staggering in alicyclic structures with π-bonding. Top: The favorable (eclipsed) orientation of substituents R^1 and X in fragment R^1ABX with double bond B=X. If the bonding between atoms B and X is neither σ nor both σ and π, common conformational rules are inapplicable. Mid: Canonical geometries of stable rotamers and geometries of stable rotamers for alkenes **13c–f**; two–head arrow shows vicinal substituents related to the indicated torsion angle ϕ. Bottom: Canonical geometries of rotamers of aryl–alkyl-interlinked molecular fragments.

stereochemistry textbooks. However, the second conformational rule is not more reliable than the previous one; we will see below that the simplicity in predicting rotational geometry for such fragments quickly encounters difficulties.

Let us glance at simple alkenes **13c–f** (Fig. 15). Stable rotamers of compounds **13c**, **13d**, and **13e** have eclipsed geometries while such rotamers for homolog **13f** adopt both certain eclipsed and bisecting geometries, as shown by thorough experimental and theoretical studies. It is clear that the considered conformational rule for alkenes equally approaches all alkenes **13c–f** and, consequently, prescribes both canonical geometries (eclipsed and bisecting) for stable rotamers of all of them. Hence, by being "simply ruled," we are neither able to correctly indicate stable rotamers for each of these compounds nor to reveal the difference in the molecular geometry of thermodynamically stable rotamers for alkenes **13c–e** and **13f**. In the half-chair-shaped cyclohexene (the stable geometry of this ring; Section 3.4, Fig. 61), there are no canonical geometries at all. The torsion angle ϕ for the endocyclic C=C and exocyclic C–H$_{pseq}$ bonds (H$_{pseq}$ is the vicinal pseudoequatorially oriented hydrogen) is 136°. Correspondingly, the angle ϕ for the endocyclic C=C and exocyclic C–H$_{psax}$ bonds (H$_{psax}$ is the vicinal pseudoaxial hydrogen) is −104°. Clearly, our next pencil-and-paper tool for analysis of the molecular shape has not passed even a brief examination.

This unreliability in "visually" predicting rotational geometries increases when considering ketones. According to the second conformational rule, equatorially oriented α-positioned protons H$_{eq}$ in cyclic ketones with a six-membered saturated ring backbone should eclipse the carbonyl oxygen in the chair-shaped ring. In fact, the C=O and C–H$_{eq}$ bonds are not always coplanar in these cycles. For instance, the corresponding torsion angle ϕ is 0° for cyclohexanone while it is 14.4° for the bicyclic azaanalog **13b** (for its chemical structure, see Fig. 9).[9] Thus, we learn that this rule is too rough to accurately predict mutual orientation of vicinal substituents in cyclic ketones. Let us cast a glance at acyclic ketones. While the eclipsed geometry ($\phi = 0°$; Fig. 15) does provide thermodynamic stability to acetone rotamers, the favored stable rotational orientation of the *iso*-Pr group of methylisopropylketone as well as of these groups of diisopropylketone is characterized by $\phi = 150°$ (for the H–C–C=O torsion angle; the torsion angle for the vicinal C–H and C–C bonds is –30.0°), i.e., there is a non-canonical staggered orientation of vicinal substituents (Fig. 15, mid). We see that the second conformational rule may essentially mislead also in the case of acyclic ketones. In attempting to rescue this common rule, one may suppose that α-branching in these chemical backbones leads to stabilization of this incomplete staggering. However, this amendment to the rule is inconsistent with the eclipsing/staggering situation for other α-branched ketones. For instance, angle ϕ is 162.6° for the lowest energy rotamer of methylcyclohexyl ketone (chair-shaped ring, equatorially oriented ring substituent);[a] that is, the vicinal Me and H substituents are rather eclipsed (the corresponding torsion angle is –17.4°), and the rotation-related molecular geometry of this ketone may be characterized as near bisecting. Also the molecular geometry of the thermodynamically stable rotamer of di-*tert*-butyl ketone is unpredictable by manipulating with the deterministic angular geometry of vicinal substituents (i.e., with the related set of canonical geometries shown in Fig. 15). One *tert*-Bu group of this ketone "follows" the second conformational rule (the carbonyl oxygen and one of the Me's of this group eclipse each other), while the other *tert*-Bu "obeys" neither the rule itself nor its amendment (the torsion angles ϕ for the Me's of this group and the O atom are 98.6°, –138.3° and –18.1°).[a] α-Branching further entangles the situation for users of common conformational rules. The thermodynamically predominating rotamer of 4,4-dimethyl-2-pentanone has the carbonyl and *tert*-Bu groups eclipsed (*i.e.*, the H–C–C=O torsion angle ϕ is 120°), as theoretical calculations show.[19d] The other stable, higher energy rotamer has $\phi = 45°$, i.e. it adopts a distorted staggered geometry for vicinal substituents. These conclusions scarcely can be deduced from "visually" balancing canonical geometries dictated by the second conformational rule and the possibility of their distortions caused by steric interactions of vicinal substituents.

The third sufficiently known conformational generalization considers rotamers that arise from rotation around a single bond that connects an aromatic carbon to an aliphatic one. According to it, a stable rotamer in such molecular fragments may adopt both the perpendicular and eclipsed

---

[a]By *ab initio* calculations performed at the MP2/6-31+G(d,p) level[9] (as indicated, these abbreviations are explained in Chapter 5).

geometries (Fig. 15, bottom). For simple alkylbenzenes, the perpendicular rotational orientation is favored, as shown, e.g., for ethylbenzene, isopropylbenzene and neopentylbenzene. For *tert*-butylbenzene, the eclipsed geometry is a single one that delivers thermodynamic stability to rotamers of this compound.[a] Of course, one cannot reach these conclusions by analyzing Newman projections: we cannot "manually" estimate a fine balance of electronic and steric factors. ■ Nevertheless, such structural conclusions obtained for these and other simple alkylbenzenes form a data set that is the empirical basis of the quoted third conformational generalization. Examples of rotamers of many alkylarene and alkylheteroarene compounds, e.g. well-known *syn* and *anti* rotamers of nucleosides and nucleotides,[a] confirm that their geometry corresponds to either expected orientation (i.e., a canonical one; Fig. 15) of their planar aromatic ring relative to the vicinal substituents of the attached tetrahedral atom. It seems that the third conformational generalization definitely may be accepted as a successful conformational rule. However, in view of multiple failures of other conformational rules (see above), could we believe that this rule is sufficiently general to allow predicting the geometry of stable rotamers for any other alkylarene?

Regrettably, this generalized "simple look" at these fragments is not sharp: sometimes rotamers have angular geometries that do not correspond to either canonical form (eclipsed or perpendicular). For instance, the dimethylene chains in paracyclophane **4d** (Fig. 2) are not oriented relative to the phenylene rings in the manner that is dictated by the "third conformational rule." Minimal (by absolute value) torsion angles C_{arom}–C_{arom}–C_{aliph}–C_{aliph} are 69° (and not 60° or 90°, as they are for the eclipsed or perpendicular 3D structures, respectively, indicated in Fig. 15).[19b] One may object that the angular geometry of aliphatic and aromatic vicinal carbons C_{aliph} and C_{arom} would be canonical eclipsed if there was no structure-deforming strain in this molecular backbone. Therefore, an example of the C_{sp2}–C_{sp3} bond rotamers of Δ^9-tetrahydrocannabinol (Δ^9-THC; Fig. 16),

---

[a]Trivial names of stable rotamers that result from rotating the nucleobase around the glycoside bond in molecules of chemically unmodified nucleosides and nucleotides. In both of these rotational isomers, a planar aromatic ring and the H-1′ atom (the hydrogen in the anomeric position of the furanose ring) lie in the same plane, i.e., there is the canonical eclipsed geometry (Fig. 15) for these vicinal substituents. In some nucleosides with a chemically modified sugar backbone,[19e] the nucleobase eclipses the endocyclic C-1′–O-4′ bond.[9]

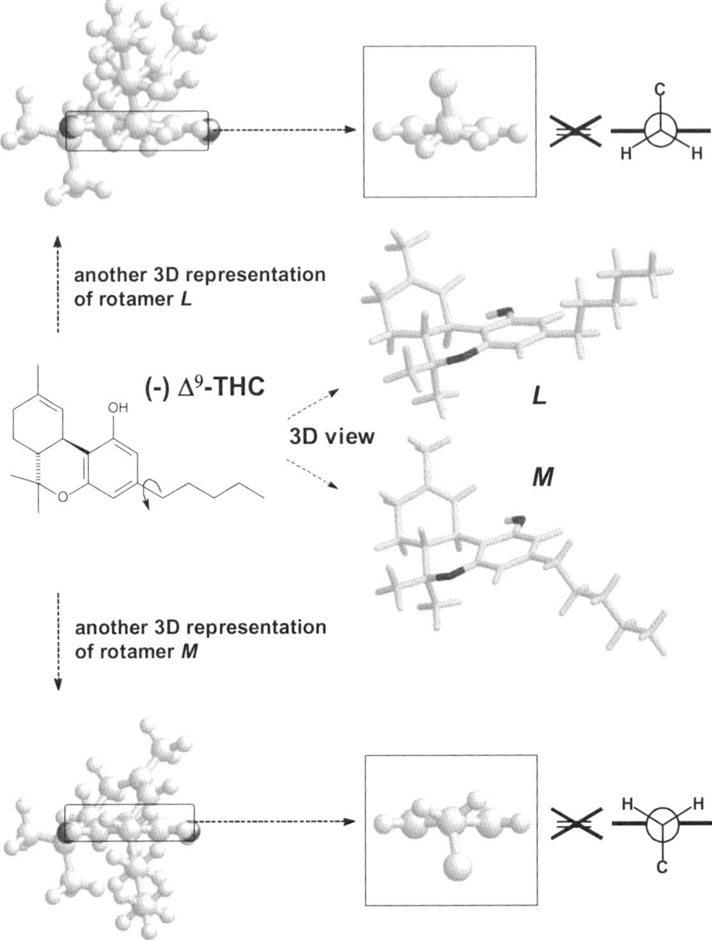

Figure 16. Rotational (angular) orientation of vicinal substituents in low energy C-1′–C-2′ rotamers **L** and **M** of Δ^9-THC (the molecular geometry calculated for the gas phase). The rotation-related tetrahedral and planar molecular fragments connected by the C-1′–C-2′ bond are shown in rectangles.

the main psychoactive component of marijuana, is a better illustration of the gap between "simply predicted" and actual molecular geometries.

Concerning C_{sp2}–C_{sp3} rotation, this molecule has a trivial, unstrained structure, and, according to our conformational rule, expected rotamers certainly are canonical ones, i.e., eclipsed or perpendicular rotamers.

Stereofaces of the phenyl ring in Δ^9-THC are diastereotopic. Therefore, equipped with the information regarding predominance of rotamers of perpendicular geometry for unbranched alkylbenzenes, we may anticipate that two stable rotamers of perpendicular geometry are the most plausible low energy 3D molecular structures for Δ^9-THC (see Newman projections in Fig. 16). Of course, six eclipsed rotamers for this fragment, four with H/C=C eclipsing and two with C/C=C eclipsing, cannot be rejected *a priori*. These 3D structures may be considered as alternative rotamers that have a lower probability of appearing as major rotamers.

However, one should not be satisfied by these predictions when carrying out *research*. "True" rotamers located by means of a "kosher" tool, *ab initio* calculations, have another angular geometry for the considered vicinal substituents. These calculations[9] demonstrate that, among low energy structures of this compound in the gas phase, there are only two rotamers *L* and *M* (Fig. 16).[a] Neither of them has eclipsed geometry. However, the C-1′–C-2′ bond of the alkyl chain is not perpendicular to the plane of the phenol ring in these rotamers; the angle between this bond and the plane is 80° (as we remember, this angle is 90° for the canonical rotamer of perpendicular geometry). That is, the rotation-related geometry of the C_{arom}–C_{arom}–C-1′–C-2′ fragment of Δ^9-THC is not standard (canonical). One could argue that this 10° difference in rotational angles for *L* and *M vs.* the canonical (perpendicular) geometry is not significant from the perspective of the 360° range of torsion angle changes. This is incorrect. The whole range for the change of this rotational angle between canonical eclipsed and perpendicular rotamers is only 30°. The angular deviation from the closest canonical geometry is one third for Δ^9-THC! Obviously, the general description of alkylarene rotamers that we have nearly accepted as a guiding conformational rule is not that canon.

In summary, after considering the three most known guiding descriptions of stable rotamer geometries, we have to conclude that, for non-absolutely-trivial chemical structures, these simple instructions are capable of neither predicting potamers of non-canonical geometries nor reliably indicating

---

[a]Initially located by means of stochastic search combined with geometry optimization by molecular mechanics (see Chapter 4 for this methodology of quantitative molecular modeling). Their geometry has been refined at the B3LYP/6-31+G(d,p) level, for vaccum conditions (see Chapter 5 for explanations of these acronyms). Calculations also show a similar molecular geometry for Δ^9-THC in chloroform.

thermodynamically favored rotamers of concrete canonical geometries. Probably, a heuristic understanding of these empirical generalizations is that they actually describe the *tendency* of vicinal substituents to adopt certain (herein, called canonical) mutual orientations which are sole orientations of such substituents in stable rotamers of structurally simplest C,H-compounds (reference structures). With increasing structural complexity, various molecular factors that disturb this tendency appear. One cannot simply estimate the impact of the disturbing factors on the 3D geometry of a concrete chemical backbone, and frequently we do not know whether the primary structural tendency still prevails. ▮ Therefore, the considered conformational rules are not robust dictums that deliver correct geometries of stable rotamers to any molecular fragment of appropriate chemical structure. It is worth realizing that these and other, similar primitive "theoretical" models for predicting molecular geometry of stable rotamers and estimating their relative energy are not research tools.

For instance, there are chemical structures, where a pair of two elementary molecular units (1) has only ligands σ-bonded to their central atoms, and (2) geometrically is a tetrahedron connected to a triangle, as two such units are interlinked in, e.g., trialkylboranes $R^1(R^2)(R^3)C-B(R^4)_2$ including, of course, 9-alkyl derivatives of 9-BBN (*in situ* generated reagents of Suzuki couplings). An empirical generalization that describes their rotamer space is as follows: the perpendicular rotational orientation (Fig. 15, bottom) in a pair of interlinked elementary molecular fragments of tetrahedral and triangular geometry provides thermodynamic stability to rotamers of this pair. However, as indicated, it would be hard to accept this empirical statement as a rule for predicting the geometry of rotamers for any chemical pair *tetrahedron–triangle* of σ-bonded elementary molecular fragments.

Chemical structure provides some indication about the kinetics in rotational flexibility. This simple look is that an increase of steric crowding (increase of the size of vicinal substituents) correlates with the slowing of internal rotation. In sterically overcrowded systems, the rate of rotation becomes so slow that individual rotamers are NMR-observable even at increased temperatures (a well-documented phenomenon of hindered rotation). Also, the length of the central bond in an ethane-like fragment X–Y is important: the longer the bond, the faster the rotation. E.g., rotation around C–C bonds in alkanes is appreciably slower than rotation around Si–Si bonds in silanes. Thus, the bulkiness of the vicinal substituents and the element type of atoms X and Y in such fragments provide qualitative information, in which systems internal rotation is slower.

As one may expect, these plain tendencies cannot be a basis for structural analysis in *research*. For instance, with only being equipped with the bulkiness "rule," can one indicate that C–N rotation in N-*tert*-Bu amine **5d** is $\sim 5.5 \times 10^4$ times faster than this rotation in a less sterically crowded congener, N-*iso*-Pr amine **5c** (Fig. 2), as theoretical calculations[4a] show? Or, can one predict the result of theoretical modeling demonstrating[4b] that ring inversion[a] in 1,2-disilacyclohexane, a molecule with a Si–Si bond in the ring, is even slightly slower that the ring flipping in 1,3-disilacyclohexane, a molecule with essentially shorter ring bonds? Many structural factors may contribute to the kinetic barrier of internal rotation; in such a case, any simple "rule" cannot take into account all of them. ■

1.6 Rotation Around Several Bonds: Difficulties of Mechanistic Predictions of Molecular Geometry

Usually, in organic molecules, at least several bonds have rotational freedom. By taking a "simple look" at internal rotation (conformational rules), one can roughly estimate the number of rotamers for a rotationally flexible molecule. Let us see how this mechanistic modeling assists at least in general understanding of the variability of molecular shape due to rotational isomerism in such molecules. The three-fold rotation model for two neighboring tetrahedrons (Fig. 11) in the butane "reference" provides a trivial estimate. If the number of bonds is n, the number N of rotational isomers is determined as $N = 3^n$ for an open chain molecule [three isomers generated by rotation around the i-th bond produce nine isomers when considering rotation around the $(i + 1)$-th bond; $i \in n$; that is, for $n = 2$, $N = 9$]. For cyclic compounds, N is obviously lower. However, it cannot be simply estimated for an n-membered ring, because, in contrast to a trivial case of C_n rotation in an open chain, we are unable to reasonably assume what the ranges of energetically tolerable endocyclic torsion angles are for rings of

---

[a]As mentioned, as an intramolecular motion, ring inversion is a restricted concerted rotation coupled with small changes of endocyclic bond angles.

different size and which sets of these angles deliver thermodynamic stability to the ring.

Real molecules are molecular ensembles [large sets of chemically identical molecules obeying the thermodynamic statistics (a certain law of distribution of molecules according to their energy)]. The discussed mechanistic analysis of molecular shape (conformational rules) concerns a single molecule. Therefore, even when it is correct, this simple prediction tool, by indicating the most favored stable rotamer of a *molecule*, is not fully relevant to an ensemble of such molecules (i.e., it does not indicate the rotational geometry of all molecules from the ensemble). For instance, the antiperiplanar orientation of larger substituents *for each bond* in the polymethylene chain (in respect to internal rotation, this chain may be considered as the butane case; Fig. 11) means minimal energy of repulsive interactions for the *entire chain*. Shortly, a regular zigzag (W) form of the backbone should qualitatively represent the molecular geometry of this compound. However, the view is purely internal energy-related and, therefore, it is incorrect. It does not take into account the entropy contribution (to Gibbs energy) which is significant due to a large number of possible rotamers. Unbranched chains in such a molecular ensemble are absolutely not exclusively W-shaped. They exist as a mixture of many interconverting rotational isomers in considerable stationary concentrations that are dictated by Bolzmann statistics [Eq. (6), Section 3.2].

This elementary algebraic expression shows that it is absolutely impossible to analyze all rotational isomers of polymers or even non-polymeric large molecules (e.g., those of phosphotriglycerides). Let us touch a simplified model of a 100-residue protein. If we even restrict its rotational freedom to only three bonds per amino acid residue by neglecting rotation around all side-chain bonds except the C_α–$C_{1'}$ bond in each monomeric unit and rotation around amide bonds in the main chain, the number of distinct rotamers is "astronomical," i.e., $3^{100} \approx 5 \times 10^{47}$. If we meaningfully extend the rotational space of this hypothetical protein by adding the amide bond HN–C(O) rotamers, the corresponding number increases to $2^{99} \times (5 \times 10^{47}) \approx 3 \times 10^{77}$! Even if only one quintillionth (10^{-30}) of these 3D structures do not have intolerable VDW overlapping of non-bonded molecular fragments and, thus, may exist, the remaining number of rotamers, 3×10^{47}, is still untreatably colossal. In this light, oligosaccharides seem to be quite analyzable structures, as polymers with a much lower number n of "rotating" bonds. However, the estimated number of the main chain rotamers for, e.g., a linear 1,4'-linked undecasaccharide, is 3^{10}, i.e., 59049. One scarcely can examine biomolecular VDW or chemical interactions for all of such distinct 3D molecular structures of large size. It is an inestimable fortune for researchers that only a very few rotamers of a biopolymer (they usually are among the lowest free energy rotamers) are biologically significant.

VDW interactions between remote fragments play an important role in rotational freedom of long chain molecular backbones, e.g., polymers. Due to manifold steric contacts in folded chains, changes of their rotational angles may fracture a smooth character of the sinusoidal energy plot of the ethane "reference" (Fig. 12). *Small* changes of these angles may be sufficient to transfer the entire system from an energy minimum to another, very near minimum of energy (in the PES; Section 3.1), because, upon this tiny rotation, VDW contacts of some remote non-bonded atoms are weakened, while VDW contacts between some other distal atoms appear. For such molecules, each energy "minimum" in the sense of the rotation plot in Fig. 12 is actually an area of many local energy minima which are separated by low height energy barriers.[a] These minima correspond to rotamers which insignificantly differ by values of some or all rotational angles. That is, an energy valley appears instead of a single energy minimum, and a set of "subrotamers," i.e., a set of the chains of the same structural motif,[b] appears instead of a single rotamer. From the viewpoint of molecular energy, these chains are almost equal in energy, and, from the viewpoint of molecular geometry, they have almost the same 3D geometry.

Should we consider these subtly different molecular structures as separate thermodynamically stable rotational isomers? Real molecules show quantum behavior and, therefore, geometries of long chain compounds should also be assessed in the light of the energy minimum formalism (Section 2.4, Fig. 42). If "subrotamer" interconversion barriers are of the order of molecular vibration energies for the same temperature, experiments only detect a "vitual" geometry, which corresponds to one energy minimum in the sense of the rotation plot given in Fig. 12. Simply put, there are no real individual "subrotamers" in the case of very low interconversion barriers. If the interconversion barriers are higher, "subrotamers" are individual, thermodynamically stable stereoisomeric structures (rotamers of the same structural motif). Thus, the number of rotamers for a long open chain molecule may be much larger than the three-fold rotation model estimates as 3^n (see below). As real individual structures, "subrotamers" may participate into chemical or biochemical interactions with distinct rates or form complexes with a biopolymer that have distinct thermodynamic stability. Understanding of the molecular mechanism of a biochemical act (e.g., ligand binding) would be essentially complicated in this situation.

Our important methodological question, which closes Section 1.3, is more "small-molecule-oriented." Here, it asks whether *entire* molecular geometries of rotamers can be deduced from the chemical structures in a simple fashion. Indeed, how do synthetic chemists experiment

---

[a] See Section 3.1 for a brief description of the surface of potential energy.

[b] The term motif means here some narrow range of rotation angles.

with their primitive of modeling of separate molecular fragments tools in order to have an idea of what molecular geometries of various organic compounds are?

The deep-rooted practice is trivial. There are several compounds of different chemical structure (reference compounds, e.g., butane, 1,3-butadiene, saturated and some partially saturated cyclic hydrocarbons of small and medium ring size; Fig. 61) that are assumed to represent all possible thermodynamically stable rotamers for related fragments of organic molecules.[a] More accurately, each of these chemical structures is formally associated with a few preferred rotational orientations (i.e., with relevant canonical geometries) that correspond to stable rotamers (Section 1.5); for simple cyclic hydrocarbons, some ring-size-specific geometries are accepted as canonical 3D structures for these rings (since Section 3.4, such 3D templates are called *conformons*; Fig. 61). That is, each reference structure is a source of a few structure-specific canonical molecular geometries (i.e., conformons). One can formally divide the molecule of interest into small fragments that are structural analogs (in the sense of chemical connectivity and chemical bonds) of relevant reference structures. The next assumption is that each such a fragment *in the molecule* may adopt the same canonical geometries that the corresponding reference compound adopts. In other words, the geometry of an entire organic molecule is assessed as a 3D structure assembled from conformons. The ultimate step in this deterministic design consists of arranging conformons according to the conformational rules that establish the preference for either rotation-related geometry (Section 1.5). Indeed, combined in a molecular framework, neighboring 3D "building blocks" (neighboring conformons) may be differently *rotationally* oriented relatively one to another. Therefore, in modeling rotamers of a molecule in this mechanistic manner, the alternative entire molecular geometries are constructed merely as distinct combinations of different staggered or eclipsed mutual rotational orientations of the used 3D "building blocks" (conformons).

---

[a]Cyclic structures are also included into consideration. As indicated in Section 1.4, geometry transformations in saturated or partionally saturated cycles occur as a concerted rotation around ring-forming bonds accompanied by some changes of endocyclic bond angles. Therefore, different 3D forms of a cyclic molecule may be roughly considered as rotamers.

This sole "theoretical" approach exploited in organic laboratories is technically elementary. For instance, the molecular geometries of two formal rotamers of cyclohexane — chair and boat — are merely transferred to the piperidine cycle in azabicycles **13a,b** (Fig. 9). Similarly, all molecular geometries of the pyranose ring are "borrowed" from cyclohexane, rotational orientations in carotenes are supposed to be equal to those orientations in 1,3-butadiene (Section 3.4, Fig. 61), windings of protein chains are approached by considering their (O)C–NH fragments either as the *cis* or as the *trans* (O)C–N rotational isomer of N-methyl acetamide, rotamers in the polysaccharide chain are modeled by orienting the chair-shaped puranose rings so that they have staggered orientations for each glycoside bond, etc.

Nevertheless, one should learn that synthetic experimentalists are equipped with a tool that shows rather a tendency of organic molecules to adopt geometries dictated by conformational rules. As empirical conclusions made from studying some, even many, compounds, these rules cannot reliably predict the molecular geometry of rotamers of any organic molecule. First, association of molecular geometries of some parent reference compounds with those of similar structural fragments in the chemical structure of interest is itself unreliable (Section 1.5). Second, by assembling canonical geometries (conformons) of molecular fragments into an integral 3D framework, we create additional structural factors that appear due to non-bonding (e.g., steric) or bonding (e.g., H-bonding) interactions of these fragments; these interactions may corrupt the standard "3D building blocks."

It is frequently not obvious for a concrete chemical structure whether assembly of conformons into an integral molecular 3D framework disturbs the tendency of vicinal substituents to adopting canonical orientations. One can only indicate the most appreciable structural features that may be signs of existence of "anti-rule" rotamers of an organic molecule. They are (*1*) a cyclic or polycyclic structure of a molecule and (*2*) spatial proximity of molecular fragments which are not neighbors (are not directly connected) in a chain of chemically bonded atoms. These structural features may lead either to stabilization of rotamers with geometries which are thermodynamically *unstable* in reference compounds (i.e., geometries of transition states of rotation, including concerted rotation, or geometries of other transient structures in the pathway of rotation), or to destabilization of *stable* geometries inherent to conformons and, thus, appearance of stable rotamers of non-canonical geometry.

Third, the tendencies to adopting some rotational orientations followed for C,H- or C,H,N,O-containing organic compounds may not correspond to structural tendencies for organic compounds that include other chemical elements. In other words, the textbook-provided geometries of thermodynamically stable rotamers (that we call canonical geometries here) in fact are not general patterns.

There are some chemical structures that illustrate these "anomalies" in rotamer geometry. A trivial example is supplied by the molecular geometry of alicyclic compounds. For them, the quickly recognizable shape of the saturated six-membered ring — the chair (Fig. 17) — is a rare case of molecular geometry that firmly obeys the main conformational rule (Section 1.5). Formally, it is an assembly of three 1,2-disubstituted dimethylene fragments in an ideally staggered geometry (Fig. 11) for a pair of any 1,4-positioned methylene groups. According to the mentioned rule, this ring cannot have another shape. The fact is that the twist geometry (Fig. 17) with four endocyclic torsion angles of absolute value of $30.4°$ is the equally occurring, though usually of higher energy, stable 3D form of the ring for cyclohexane and its heteroanalogs. One may reasonably explain that the structural factor in the twist-shaped ring, which breaks up the tendency of vicinal substituents of being staggered, is the second bridge CH_2CH_2 between vicinal (1,4-positioned) methylene units. However, with assuming the ring closure effect and possessing no information regarding thermodynamic stability of the twist geometry, one cannot rudimentarily indicate this stable form of the ring, i.e., guess the stability of exactly this ring shape. Even having an evidence for thermodynamic stability of a twisted form of the ring, one cannot non-computationally predict its exact geometry. ■ The main conformational rule does not guide us in predicting the geometry of thermodynamically stable ring shapes for other saturated cycles either. For instance, two such torsion angles in the thermodynamically favored geometry of cyclooctane (structure **42a** in Fig. 95) are $100.7°$ (absolute values; a $40.7°$ deviation from the canonical value of $60°$), two others are $44.4°$ and $-44.4°$, and only the remaining four angles have a near-canonical absolute value of $\sim65°$. In *chemical research*, exploration of conformational space for any n-membered ring with $n > 4$ (i.e., location of stable ring forms and, among them, identification of the directly interconverting ones) is a subject of thorough theoretical calculations.

Alkaloid **13a** (Fig. 9) supplies the next example. We have mentioned that canonical ring shapes for cyclohexane are *chair*, and *twist* (stable forms; Section 3.4, Fig. 17); one also can recall that *half-chair*, *boat*, and *sofa*

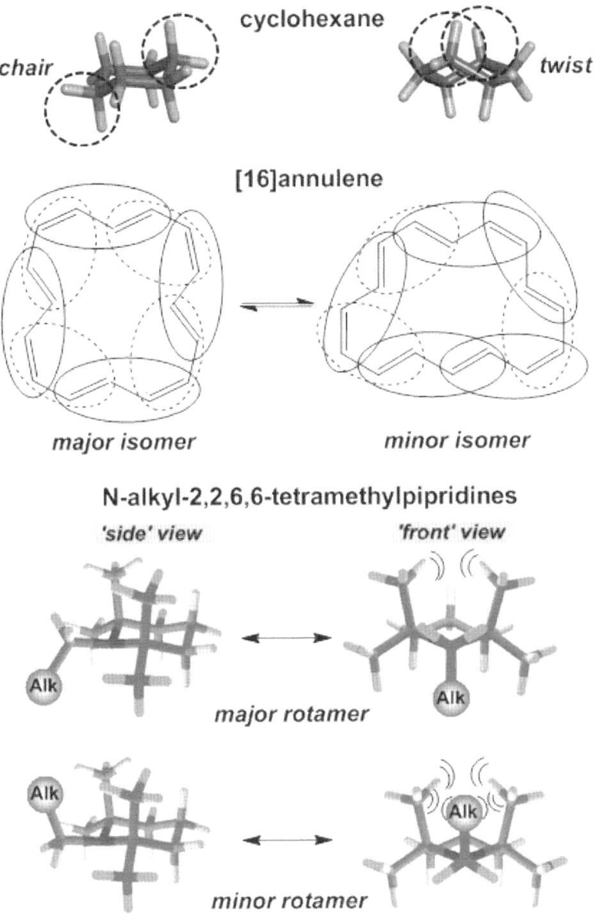

Figure 17. Examples of compounds that have stable rotamers of 3D geometry which does not follow from canonical rotational preferences. Top: the chair-shaped cyclohexane ring [ideal staggering for any 1,4-positioned (indicated by circles) methylene groups] *vs.* the twisted one (highly distorted staggering for two pairs of 1,4-positioned methylene groups; a pair of circles shows such a pair). Mid, left: the major chemical isomer of 16[annulene], which is formally composed from four *s*-cis and four *s*-trans 1,3-butadiene fragments (indicated by dotted and solid ovals, respectively). Mid, right: the isomer, which is formally composed from five *s*-trans and three *s*-cis 1,3-butadiene units. Bottom: the thermodynamically most stable C–N rotamer of sterically crowded piperidines and the less stable C–N rotamer of these compounds. Arc lines indicate destabilizing steric 1,3-diaxial interactions.

are transition states in stereoreorganizations of this molecule (Fig. 53). For bicycle **13a**, a stable stereoisomer has a boat geometry of the piperidine ring. In contrast, this six-membered ring in a twist-shaped form imparts instability to the entire structure.[a]

Antiaromatic annulenes further illustrate limitations of the "simple look" at the ring shape in entire molecules. Formally, 16[annulene] (Fig. 17) may be split into eight "1,3-butadiene" fragments. Thus, if we know the 3D geometries of these fragments, we should be capable of predicting possible ring shapes of this molecule (i.e., the sets of endocyclic torsion angles for eight single bonds of the ring). Theoretical *calculations* have established that 1,3-butadiene has two stable rotational isomers,[b] a planar *s*-trans rotamer and a non-planar, gauche-shaped *s-cis* rotamer (Section 3.4, Fig. 61).[20c] Let us employ these conformons for predicting the ring shape of two low energy structural (chemical) isomers of this annulene.

According to this simple tool, the major isomer is a 3D construction of eight overlapping 1,3-butadiene conformons (four non-overlapping ones of the *s-trans* geometry and four non-overlapping ones of the *s-cis* geometry; Fig. 17). Similarly, the ring shape of the minor isomer is a 3D framework composed from five *s-trans* conformons and three *s-cis* conformons where the all *s-cis* fragments overlap with the *s-trans* fragments while not overlapping one with another, and two *s-trans* units overlap. Theoretical *calculations*[20d] show that the first prediction is correct while none of chemical 1,3-butadiene fragments of the minor isomer has either the gauche *s-trans*- or planar *s-cis*-shape, i.e., this isomer only has non-canonical orientations of all vicinally positioned CH groups. Clearly, without possessing these adamantly based results, it is impossible either to reveal the true geometry of the ring in the minor isomer, or to accept with confidence the "paper-and-pencil-made" geometry of the major isomer. Why is the mechanistic approach so unreliable in predicting the ring shape (estimating endocyclic torsion angles) in annulenes? Fragments CH=CH–CH=CH, which

---

[a]An energy minimum in the surface of potential energy corresponds to a stable structure (Chapter 3). No energy minimum is located by *theoretical calculations* for piperidine **13a** in the region of twist geometries.

[b]As known, they rapidly interconvert at ambient conditions. This multi-step interconversion includes both concerted rotation around ring bonds and double bond migration.[20d] Obviously, one can separately consider molecular shapes of these individual components of the chemical equilibrium.

are combined by means of single bonds into a ring, are not independent 1,3-butadiene structures: their backbones are additionally strained due to the ring closure, and their π-systems interact. In this light, one can single out molecular factors that indicate the energy changes when screwing the ring of annulenes with alternating single and double bonds. They are the degree of delocalization of the π-electrons in the ring and the strain of the latter.[a] If assembling the molecular geometry of the annulene ring from the individual 3D "Lego fragments" of 1,3-butadiene (its conformons; Fig. 61), we merely ignore these theoretically rough, but indicative components of molecular energy.

As mentioned, spatial proximity of skeletally remote molecular fragments may also indicate appearance of rotamers with unexpected geometry since interactions between such fragments may stabilize it. An example of N-alkyl-2,2,6,6-tetramethylpiperidines (Fig. 17) well illustrates this 3D scenario. As molecular mechanics calculations (Chapter 4) show,[20e,f] the substituents at the C–N bond are eclipsed in the minimal energy 3D structure while the molecular geometry with staggered substituents is significantly higher in energy. For instance, for the N-neopentyl compound, the calculated energy difference for such C–N rotamers is 5.5 kcal mol^{-1}; that is, the rotamer with bond staggering is so minor that none of spectral methods could detect it.[b] From the viewpoint of the main conformational rule (Section 1.5), the N-substituent is oriented "incorrectly" in thermodynamically most stable rotamers of these compounds.

Even mechanistically, the reason for stabilization of the eclipsed geometry and destabilization of the staggered geometry is transparent for these molecular structures. Repulsive VDW interactions between two exocyclic Me's and the β-alkyl fragment of the N-substituent are neglectfully weak in the former case while they are weighty in the latter (Fig. 17). However, while satisfying us as an explanation, this "simple look" could not serve as a prediction. Without a "genuine" theoretical modeling, it is impossible to reveal whether these VDW interactions reach the magnitude which permits these molecules to override the tendency of vicinal substituents to

---

[a]Both molecular strain and degree of delocalization of π-electrons are not strict quantum mechanical considerations.

[b]In other terms, it is a hidden partner (Section 3.3) in the rotamer equilibrium.

adopt only certain (canonical) orientations reflected by the main conformational rule.[a]

The next example that shows the impact of interactions of remote molecular fragments on molecular geometry of rotamers is delivered by linear biopolymers. Conformational "rules" (Section 1.5) dictate to a biopolymer chain to adopt rotation-related geometries from some pool of molecular geometries restricted by canonical values for the torsion angle for each four–atom unit of the chain. For instance, twisting of protein chains is described by the α-carbon-framing torsion angles $C_{i+1}-N_i-C_\alpha-C_i$ and $N_i-C_\alpha-C_i-N_{i-1}$ (traditionally called φ and ψ, respectively, for the i-th α-amino acid residue). From this mechanistic perspective, both torsion angles φ and ψ may adopt discrete absolute values of $0°$, or $60°$, or $120°$, or $180°$ that correspond to eclipsed and bisecting geometries of interlinked tetrahedral [here: $N_iC_\alpha(R)C_i$] and trigonal [here: $C_{i+1}N_i(H)C_\alpha$ or $C_\alpha C_i(O)N_{i-1}$] elementary molecular fragemets (Fig. 15). Let us glance at different geometries that protein chains do adopt. On the average, for the so-called α-, 3_{10}- and λ-helix,[b] φ is $-63°$, $-57°$ and $-70°$, respectively, while ψ is $-42°$, $-30°$ and $70°$, respectively. For classical structures of turns of the protein chain, so-called type-I β-turns that include the four–residue fragment $i-i+1-i+2-i+3$ of the chain, $\varphi_{i+1} = -63°$ and $\varphi_{i+2} = -90°$ while $\psi_{i+1} = -30°$ and $\psi_{i+2} = 0°$ (ideal values for these turns) for amino acid residues $i+1$ and $i+2$, respectively. Thus, none of the typical secondary structures of proteins has any canonical rotamer geometry in the chain. Why? The reason is obvious: all these 3D structures in proteins are formed with involving H-bonds $C=O\cdots HN$, where the H-bonded carbonyl oxygen and amino nitrogen belong to different amino acid residues separated in the polyamide chain by one or more amino acid units. By chemically linking remote fragments, this weak bonding nevertheless corrupts the tendency of this linear backbone to orient its vicinally positioned fragments equally to orientations of vicinal substituents in stable rotamers of the reference compounds (Section 1.5, Fig. 15).

---

[a]Routinely used theoretical calculations (Chapters 4 and 5) do not single out the role of either structural (e.g., steric) factor. They merely supply any molecular geometry with the relevant molecular energy and, thus, show which molecular geometry is more stable.

[b]Protein chain helixes differently extended in the longitudinal direction.

Other interactions of remote structural fragments of a twisted main chain of proteins (*e.g.*, π,π- or hydrophobic interactions of spatially proximal side-chains of amino acid residues) lead to additional uncertainty in attempting to plainly predict the molecular geometry of the chain rotamers. The above considered non-canonical rotamers of the polyamide chain together with its other typical secondary structures absolutely do not cover rotational space of proteins, and we realize how ardous the problem of predicting 3D structures of these very large biomolecules is and how troublesome general conformational rules are.

Bioorganic chemists do not need the reminder that there are no standard geometries for entire molecular frameworks. They have become used to assessing biomolecules as systems with individual fine 3D geometries. A very specific 3D structure is necessary in biorecognition, because of high steric requirements of enzymes or biomembrane receptors to their ligands. If we touch upon further structural analysis, one should notice that modeling of ligand uptake by its receptor is in a strong need in an accurate, detailed molecular geometry. A one third inaccuracy for 3D geometry of a ligand fragment (as it is in the case of Δ^9-THC, Fig. 16) probably would deorient in modeling receptor–ligand interactions and, finally, restrain the understanding of the receptor functioning. However, chemical interactions, both in living cells and inside of reaction flasks, require a fine 3D fit for reacting molecules. The need of knowing molecular geometry *accurately* is no less important in organic chemistry than it is in molecular biochemistry.

Our final example is supplied by butane analogs A_4L_{10} (A is a central atom of a elementary molecular unit of tetrahedral geometry, and L is a ligand σ-bonded to A). The butane molecule C_4H_{10} is actually the reference structure of the main conformational rule: only staggered rotational geometries ($\phi \approx |60°|$ and $\phi \approx 180°$; Fig. 11) of butane provide thermodynamical stability to the molecule, and the latter geometry with the spatially more distal 1,4-Me's has the lowest free energy. As we remember, in the Michl–West nomenclature (Section 1.5), the acronyms of these geometries are G and A, respectively.[a] We anticipate that geometries G and A (and only they) are inherent of stable rotamers of so close analogs A_4L_{10}. Quantitative theoretical calculations show that this prediction is essentially inaccurate for many chemical heteroanalogs of butane; among them, only tetrasilane Si_4H_{10} does follow the butane-based main conformational rule.[19f]

---

[a]Thus, there are three stable rotamers in the butane molecule: one rotamer of geometry A, and two enantiomeric rotamers of geometry G ($\phi \approx 60°$ and $\phi \approx -60°$).

Many other chemical structures with the four–atom chain A–A–A–A, e.g. perfluoro butane C_4F_{10}, perchloro tetrasilane Si_4Cl_{10}, 2,2,3,3,4,4,5,5-octamethyl hexane C_4Me_{10}, and 1,1,1,2,2,3,3,4,4,4-decamethyl tetrasilane Si_4Me_{10}, have six stable rotamers: two enantiomeric rotamers of distorted geometry G, two enantiomeric rotamers of distorted geometry A, and two enantiomeric rotamers of geometry O (for the O form, $\phi \approx 90°$ or $\phi \approx -90°$, according to the Michl–West nomenclature). That is, for these compounds, each stable rotational geometry of butane is "split" into two enantiomeric geometries. Also, two additional enantiomeric structures with rotational geometry O are thermodynamically stable. One can certainly speak about an explicit structural tendency that differs from the tendency demonstrated by the butane pattern and superficially accepted as a main rule for rotational preferences in open chain aliphatic molecules without examining for its generality.

The difficult cases for the primitive modeling of rotamer geometries, which are discussed in this and previous Sections, show that canonical rotational orientations are sufficiently often corrupted even in small and medium size organic molecules. With realizing this fact, we get a correct idea about the practice to view the geometry of rotamers of an entire 3D molecule as an assembly of rotamers of basic reference systems. Canonical rotamer geometries and their combinations in entire molecular backbones are important topics only in basic chemical education. In chemical *research*, geometries of stable rotamers of reference systems are definitely inadequate models of rotational preferences in other organic molecules. Molecular geometrical variability due to internal rotation is not determined by only a few (though main) structural tendencies.

What is the methodology that should guide us in considering rotational isomers in organic research? An obvious simplicity of the ubiquitously used mechanistic models (conformational rules) makes them indispensable in preliminary, light, considerations of possible molecular geometries of concrete chemical structures or in general understanding of rotational freedom for classes of chemical compounds, when possessing no additional data in hand. However, as we learnt, the relevance of canonical or other geometries of rotamers to a "geometrically unexplored" molecule is not warranted for elaborating weighty conclusions (most often, we need exactly them). That is, in answering our question regarding a simple tool for designing molecular geometries of rotamers of organic molecules, one should make certain that there is no such a reliable tool available. Our final methodological conclusion regarding the "simple look" at rotation-generated stereoisomers is therefore uncompromised: one should exclude the textbook-introduced conformational rules from the research arsenal of organic chemists. Refusal from using them supposes that researchers should turn to another, reliable methodology of prediction of rotamer geometries. The reader can guess what such a convenient alternative is: *in considering the rotamer geometry of a compound of interest, calculations based on a quantitative chemical theory are the required practical tool* (see Chapters 4 and 5).

At this point, the discussion of internal rotation and the modeling of this intramolecular motion is temporally interrupted (and continued in Section 4.5). We have come back to the starting position, i.e., that the shapes of a majority of organic molecules are not rigid because of internal rotation occurring in their backbones. There is only one difference — now we know that there is no reliable methodology that can locate stable rotational isomers in a simple way. Nevertheless, suppose that all rotational isomers of an organic molecule are identified, without concern for the level of the trustworthiness, by means of Newman projections, theoretical calculations, or through experimentation. Would this mean that its 3D structure is fully described? More concretely, are all possible molecular geometries generated by rotating elementary molecular fragments around the bonds that connect them?

1.7 Internal Mobility of Elementary Molecular Fragments and Interconversion of Abstract Geometrical Polyhedrons

At first glance, for changing the molecular shape while not disrupting the chemical connectivity, there are no alternatives to rotation of interlinked elementary molecular fragments relative one to another. However, this is an arbitrary self-limitation by supposing absolute rigidity of elementary molecular fragments. In fact, they are not rigid. Not going into the details, one may be reminded that flexibility of many of these fragments has been detected in hundreds experimental studies of various compounds.

The concept of positional exchange by geminal ligands explains these observations well, and quantitative theoretical modeling reproduces the experimental structural and kinetic data. Thus, the 3D structure of elementary molecular fragments (relative spatial disposition of geminal ligands) possesses a freedom of internal changes. The second factor of the variability of molecular shape does take place. Thus, internal rotation does not exhaust all possibilities of a molecule to undergo geometrical changes. The next component of molecular flexibility is positional exchange of two or more ligands $\mathbf{L}_1,\ldots\mathbf{L}_i$ in many chemically and geometrically different elementary molecular units $\mathbf{A}(\mathbf{L}_i)_n$ (one may term it geminal stereorearrangement). How does it occur?

In Euclidean space, one may convert a polyhedron into a polyhedron of the same shape by moving its apexes synchronously, each apex in a certain trajectory, in order to prevent the intersection of the edges. Then some apexes exchange their spatial positions. Figure 18 shows how tetrahedral, trigonal bipyramidal, and octahedral polyhedrons can interconvert through square, square pyramidal, and trigonal prism geometries, respectively. Along the interconversion coordinate, the transient structure is geometrically most different from the initial (or resulting[a]) polyhedron. The indicated pathway of interconversion of each polyhedron is associated with minimal alterations of its geometrical parameters including the angles formed by the polyhedron symmetry center (the angle vertex) and any two of polyhedron apexes (Fig. 18).

As discussed in Section 1.2, exactly these polyhedrons represent the geometry of elementary molecular fragments. Thus, one can speak about *chemical polyhedrons* [elementary molecular fragments $\mathbf{A}(\mathbf{L}_i)_n$ of certain geometry]. For them, geometrical transformations, which are shown in Fig. 18 for abstract polyhedrons, mean positional exchange of some ligands $\mathbf{L}_1,\ldots\mathbf{L}_i$. The analogy between geminal stereorearrangements in geometrical and chemical polyhedrons is sufficiently fruitful. As in the case of abstract polyhedrons, the overall geometry of the elementary molecular fragment eventually is not changed; the geometrical transformation of the fragment only results in ligand permutation. Concerted displacements of polyhedron apexes schematically indicated in Fig. 18 reflect simultaneous

---

[a]The initial and final geometries are identical for regular polyhedrons.

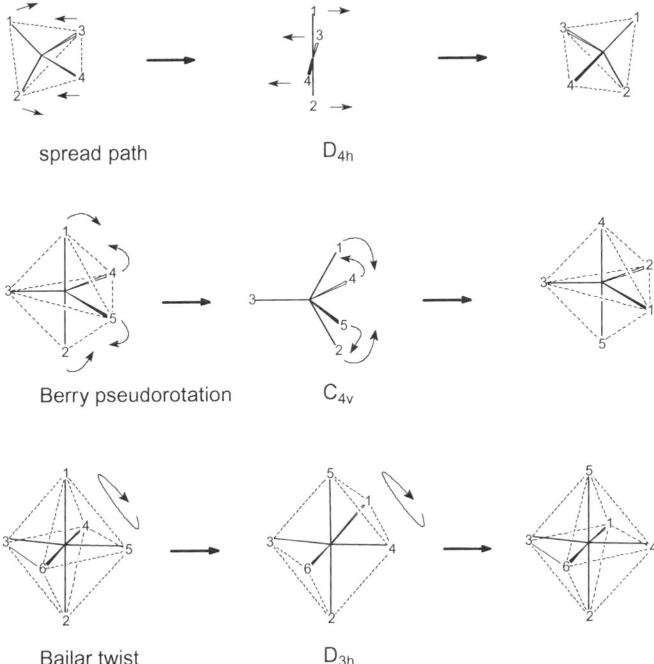

Figure 18. Minimal distortion pathways of interconversion of chemically significant regular polyhedrons. Bold lines connect the apexes (depicted by numbers) with the polyhedron symmetry center, dotted lines depict the polyhedron shape, and arrows show the direction of apex motion in rearranging the polyhedron shape. If considering elementary molecular fragments $A(L_i)_n$, bold lines indicate chemical bonds, the polyhedron symmetry centers correspond to central atoms A, and numbers depict ligands $L_1,...L_n$. Symmetry point groups are indicated for transition states.

motions of ligands in spatially rearranging chemical polyhedrons, and the transient 3D structures shown in this Figure correspond to the geometries of transition states for 3D rearrangements of many elementary molecular fragments. For instance, in such a rearrangement of a chemical trigonal bipyramid, equatorial ligands 4 and 5 become apical, while apical ligands 1 and 2 adopt the equatorial orientation. Formally, this positional interchange of geminal ligands (known as the Berry pseudorotation) corresponds to the rotation of four ligands 1, 4, 5, and 2 around the axis which is specified by equatorial ligand 3 and the bipyramid symmetry center (the central atom). In fact, there is no such a rotation, and the four ligands move in non-linear trajectories, bending around each other, as indicated in Fig. 18 for the abstract geometrical bipyramid. The transient structure shown there for the Berry pseudorotation is a square pyramid of C_{4v} symmetry. This is exactly the geometry of the transition state (probably, of the lowest energy) for the *apical ligand–equatorial ligand* positional interchange occurring in chemical trigonal bipyramids: ligand *3* is the pivot, four ligands *1*, *4*, *5* and *2* (basal ligands) lie in the plane (basal plane) that is perpendicular to the bond between the central atom and ligand *3*. In one Berry pseudorotation (in one concerted motion of all ligands), two apical ligands (ligands *1* and *2* in Fig. 18) become equatorial, and two equatorial ones (ligands *4* and *5* in Fig. 18) become apical.

An essential feature of *chemical* polyhedrons is stereoisomerism; the relevant classical example is a chiral tetrahedral carbon. Does the *flexibility* of a chemical polyhedron mean that all its stereoisomers interconvert? For instance, let us look from this perspective at an elementary molecular fragment of trigonal bipyramid geometry. One Berry pseudorotation can lead to three different stereoisomers because all three axes *equatorial ligand–central atom* are equivalent, and formal rotation can be performed around each of them (the corresponding equatorial ligand is the *pivot*; in Fig. 18, the pivot is ligand *3*). In language of stereochemists, the term *connectivity* is sometimes used for characterizing the mechanism of a stereoisomerization. Connectivity means the number of distinct stereoisomers can be formed from a stereoisomer in one step transformation by a given mechanism. Hence, connectivity 3 characterizes the Berry pseudorotation. If all five ligands of a trigonal-bipyramid-shaped elementary molecular fragment are different, there are 20 stereoisomeric structures that differ by combinations of five ligands $L_1, \ldots L_5$ which can occupy five positions [3 (equatorial) + 2 (apical)].[a] Connectivity 3 of the Berry pseudorotation permits to any two from all these stereoisomers interconvert via not more than three sequential Berry pseudorotations.

This example well illustrates what the flexibility of elementary molecular fragments is: *in non-rigid chemical polyhedrons, all ligands in turn occupy all polyhedron apexes*. Experiments cannot reveal current positions of the ligands, if the latter interchange their positions very rapidly in the timescale of the observation method (Section 3.3).

Nevertheless, abstract geometrical polyhedrons are not perfect models of chemical polyhedrons. The latter have apexes (ligands $L_1, \ldots L_n$), but

---

[a] Among these structures, each stereoisomer has an optical antipode. That is, this pool of stereoisomers consists of 10 pairs of enantiomers.

they do not have edges. Therefore, the *no edge intersection* condition for interconversion of geometrical polyhedrons is annulled for interconversion of elementary molecular fragments. In principle, there is an infinite set of interconversion pathways for chemical polyhedrons since we can exchange two or more apexes (atoms) by moving the atoms in any arbitrary trajectories, making sure to avoid their intersection. On the other hand, chemical connectivity puts strong restrictions on the atom motion upon 3D rearrangements of elementary molecular fragments, fashioning chemical polyhedrons on some imbalance with geometrical ones. One implication of this inadequacy is that geometrical pathways (mechanisms) of interconversions of real chemical polyhedrons, as a rule, cannot be deduced from considering geometry transformations of abstract 3D frameworks. One can only claim that there are a few general mechanisms of stereoreorganizations for each geometrical type of chemical polyhedrons. For instance, for the trigonal bipyramid, including the Berry pseudorotation, there are six other interconversion mechanisms (called turnstile rotation and mechanisms M1-M5; M1 corresponds to the Berry mechanism).[21a,b] They differ from each other by the number and positions of permutating ligands as well as the geometry of the transition state (the streochemical outcome by these mechanisms is different stereoisomers of the trigonal bipyramid) or only by the geometry of the transition state if the number and the type of orientation of the position-exchanging ligands are the same in these alternative mechanisms (the streochemical outcome by these mechanisms is bipyramids of equal stereochemistry).

The turnstile rotation mechanism (Fig. 19) is the most often discussed alternative to the Berry mechanism. Its connectivity is 3, the same as for the Berry pseudorotation. Both permutations lead to the same stereoisomers, and these mechanisms can be distinguished by comparing initial and final structures. However, the turnstile rotation is distinct from the Berry mechanism because it passes the transition state of C_S symmetry (Fig. 19).

As mentioned, the interconversion pathways shown in Fig. 18 are distinctive in that they are accompanied by minimal distortions of the shape of the initial polyhedrons including, of course, the trigonal bipyramid. Therefore, in turnstile rotation, the 3D structure is distorted to a greater extent than it is during the Berry pseudorotation. However, this does not mean that

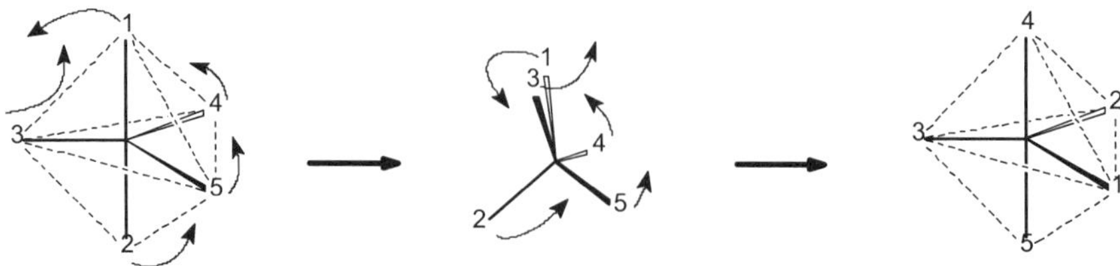

Figure 19. Turnstile rotation in trigonal bipyramid. Bold lines and numbers indicate chemical bonds and ligands $L_1, \ldots L_5$, respectively, dotted lines depict the polyhedron shape, and arrows show the direction of atom motion.

this mechanism[a] is less preferable than the Berry pseudorotation for any chemical elementary molecular fragment of trigonal bipyramid geometry. Considerations of abstract geometry of molecular structures cannot assist in identifying the actual mechanism of a stereorearrangement of any chemical polyhedron (see, e.g., the example of the inverting tetrahedral fragment of methane, Section 1.8, Fig. 26); why? At first glance, minimal distortion pathways would be a convenient *simple tool* for identification of favored mechanisms in such stereomutations of elementary units, disregarding their chemical composition.

Indeed, pathways of minimal distortions for common nuclear frameworks (Fig. 1) are derived from geometrical considerations by introducing a certain quantitative measure for estimation of an averaged change of the geometry for any regular polyhedron (a parameter called shape measure).[22a,b] Unfortunately, identification of the mechanisms of geminal stereorearrangements is not reliable when applying this attractively plain deterministic tool to real chemical polyhedrons. The approach of minimal deformation of the molecular framework brings to mind the mechanical theory of elasticity of macroobjects with its idea of a continuously positive relationship between energy and deformation. This relationship may be expressed by a single-valued continuous function that is defined in the continuum (the set of all real numbers) and often may be written in an algebraic form. The preference of the pathway of minimal geometrical changes is

---

[a]Or another mechanism (e.g., cuneal inversion, Section 1.8).

exactly what the elasticity theory requires. However, there is no such explicit function for *deformation–molecular energy*. The wave function, and only it, predetermines any physical property of a microsystem (e.g., a molecule), including its energy; because of the cryptic character of this unique molecular descriptor,[a] values of molecular physical quantities, which satisfy the wave function for the given geometry of the backbone (eigenvalues; e.g., molecular energy values for various electron configurations), are "invisible" for classical physics. In spite of this uncertainty, one can, nevertheless, claim that the wave function does not require the relationship *the more deformation, the higher molecular energy* for any electron configuration of a molecule. Molecular energy is associated both with nuclei and electrons. Thus, molecular energy, including potential energy, depends on spatial location of each electron. It is impossible to both know the trajectory of even one electron in a molecule and to follow the changes of electron trajectories upon distorting the nuclear backbone. In simple words, we cannot know even in principle what occurs in the space with the cloud of interacting electrons when changing coordinates of nuclei. It would be unphysical to speak about a deformation or elasticity of this cloud and to suggest to the wave function what the relationship *deformation–molecular energy* should be. In this light, one should not be astonished that spatial rearrangements of the framework of nuclei fastened by the united electron "glue"[b] do not obey any primitive law or an analogy to classical physical models. The pathway of minimal geometrical changes of the backbone does not necessarily correlate with minimal changes of molecular energy. Contrary to the classical theory of deformation of macroobjects, there are no physical requirements of minimal deformations in spatially rearranging molecules.

Note that, in considering *interfragmental* and *intrafragmental* intramolecular motions [rotation of elementary molecular fragments around the bonds that connect them (Sections 1.5 and 1.6) and positional exchange of geminal ligands in such a fragment (this section), respectively],[c] we are able to distinguish between them. Molecular backbones

---

[a]The mathematical form of the fundamental quantum mechanical wave function is unknown for any molecule, excluding H_2^+.

[b]Molecular 3D rearrangements are visualized as trajectories of the movement of sole nuclei (see, e.g. Fig. 18) because of a principal uncertainty of the exact locations of electrons.

[c]Not relating yet whether these intramolecular stereoreorganizations are abstract or real intramolecular processes.

(nuclei-in-molecules) are very massive objects in molecular scale, and can be represented as geometrical frameworks (in contrast to the "electronic glue" of molecules). In other words, only because of the massivity of atom nuclei, one can speak about certain geometries of organic molecules and their geometrical changes (stereorearrangements).[a]

Our metric for the two types of spatial reorganizations in organic molecules (rotational stereorearrangements and geminal stereorearrangements) that result from the two considered distinct intramolecular motions is commonly known parameters of molecular geometry, i.e., interatomic distances, bond angles, and torsional angles. Rotational stereoisomerization is often defined as a stereoreorganization that changes exclusively torsional angles (including their signs) between vicinal substituents in a molecular structure neglecting small alterations of bond angles if present. Upon such a stereoreorganization, the bond angles and bond lengths are essentially the same in the starting and final structures as well as in the intermediate ones; rotation, as an interfragmental motion, is not supposed to affect the 3D geometry of elementary molecular fragments (both involved and not involved in the rotation). Interconversion of chemical polyhedrons is different in this aspect. By this "measuring" definition, geminal stereorearrangement is associated with a change of some or all bond angles (including their signs; Section 2.1, Fig. 31) in an elementary molecular fragment. If geminal ligands occupy spatially non-equivalent positions (e.g., apical and equatorial positions in trigonal bipyramids), their positional exchange also leads to different bond lengths **A–L** in the initial and resulting stereoisomeric elementary molecular units.

These definitions of the two considered types of intramolecular stereoreorganizations have no generality because they consider interconverting stereoisomeric structures that have any symmetry relationship. The weak point here is that one cannot say whether a pair of stereoisomeric molecular fragments, which may be formally transformed one into another by symmetry operation S_1,[b] are rotational isomers [3D structures related by changing relative orientations of *vicinal* substituents, i.e., via rotation around a bond (or several bonds)] or configurational isomers (3D structures related by

---

[a]For limitations of this understanding, see Section 2.4.

[b]In simple words, elementary molecular fragments of planar geometry. Thus, there is a geometrical transformation of polygons; they may be considered as degenerated polyhedrons.

exchanging spatial positions of *geminal* ligands, i.e., via their concerted bending around motions), when comparing these geometrical parameters, i.e., bond and torsion angles, for them (see Section 1.8).

Methodologically, in order to distinguish these 3D transformations for such two "inconvenient" stereoisomeric molecular structures, one should consider an additional structure that lies between the initial and resulting stereoisomers along the transformation coordinate, i.e., consider related stereoisomers that do not have S_1 symmetry relations.[a] If torsion angles for the same vicinal substituents are different for this non-planar and the initial (or the resulting) 3D structures while *absolute* values of all bond angles are essentially the same, we deal with internal rotation; if bond angles of an elementary molecular fragment in such two structures are different, the considered stereorearrangement is positional exchange of geminal ligands. In other words, one should "see" (by means of theoretical modeling) the molecular structure in moving.

Sometimes intramolecular stereochemical interconversions are considered in more common terms of polytopal rearrangements. Polytopal rearrangements are intramolecular transformations that convert a molecular geometry to another (non-identical) one; thus, this terminology does not specify the rotational or configurational origin of intramolecular stereoreorganizations. There are three types of polytopal rearrangements: (1) non-identical ligands exchange their non-equivalent positions in the molecular skeleton, (2) a 3D molecular structure changes its overall shape, and (3) the initial and resulting structures have chiral skeletons that are mirror images of each other. One may add that, if molecular fragments or whole molecular structures do not fall under none of categories (1)–(3), one can enumerate ligands in the relevant elementary molecular fragments (as, e.g., in Fig. 18), and then their geometry changes can be referred to as formal polyptopal rearrangements. Otherwise, symmetry relations can obscure the type of intramolecular motion occurred. For instance, inversion of the nitrogen geometry in i-Pr(Et)NMe or iso-Pr rotation in amine **12** (Fig. 8) is a polytopal rearrangement. The corresponding intramolecular processes, inversion in, e.g., Me_3N, or $tert$-Bu rotation in, e.g., t-BuNMe$_2$, respectively, are not polytopal rearrangements (as interconversions of invertomers or rotamers of equal geometry). If ligands $\mathbf{L_1}$, $\mathbf{L_2}$, and $\mathbf{L_3}$ of the amino fragment are formally enumerated, inversion in this elementary molecular unit in both i-Pr(Et)NMe and Me_3N or rotation of both the N-iso-Pr and the N-$tert$-Bu substituent in **12** and t-BuNMe$_2$, respectively, may be viewed as a polytopal rearrangement.

---

[a] If performing theoretical modeling of a direct interconversion of two stereoisomeric molecular structures which have a symmetry plane (symmetry C_S), the geometry of the located transition state, i.e., of the mentioned transient 3D structure, is the most convenient indicator of the type of intramolecular stereoisomerization of interest.

Thus, the spatial structure of organic molecules seems to be more flexible than it would seem at first glance, even taking into account rotational freedom. The complexity of analysis of molecular shape, however, does not increase drastically with supplying chemical polyhedrons with flexibility. The number of possible stable 3D geometries is, of course, increased, but it remains finite, and, as a rule, no forms of an unexpected stable geometry appear when ligands exchange their spatial positions. That is, each thermodynamically stable chemical polyhedron, which has the overall shape of its abstract geometrical prototype (Fig. 18), usually keeps it if transformed into another stable stereoisomeric structure.

1.8 Internal Mobility of Real Elementary Molecular Fragments is Unpredictable by the Unarmed Eye

At this point, we stop to assess the permutation of geminal ligands only as an abstract geometrical transformation of molecular elementary fragments and start to consider stereoreorganizations of chemically different elementary molecular fragments. Our first question is: "Are all elementary molecular units flexible, disregarding their chemical composition?" In principle, the answer is yes, i.e., each elementary molecular fragment may interconvert its stereoisomeric geometries.[a] For instance, there are no principal physical restrictions for positional exchange of substituents R^1 and R^2 in ketones $R^1R^2C{=}O$ if these groups are moved along non-intersecting trajectories. The question is only about the rate of such an interconversion which may be extremely slow. However, the kinetics of many intrafragmental stereomutations is immeasurably slow, practically of the zero rate. For instance, derivatives of four-coordinated phosphorus (e.g., phosphate, phosphonate, phosphinate esters as well as phosphinoxides) do not invert the P configuration at any physical conditions either in the gas or in the condensed phase if chemical processes are not involved. Experimentalists

---

[a] It should not embarrass us that the initial and resulting structures may be experimentally undistinguishable. For instance, interconversion of molecule Me_3N via pyramidal N-inversion also occurs between undistinguishable (3D identical) structures. Nonetheless, one can "discern" inversion of the nitrogen configuration (addressed in Section 2.1) by formally labeling, e.g., enumerating, the N substituents. Then the initial and the resulting geometry formally are enantiomeric. Similarly, formal invertomers of a planar structure $R^1R^2C{=}O$ (addressed below in the main text) may be differentiated if prohiral faces of the trigonal carbon in these structures are labeled in either manner.

could recall an incomparable difference between geminal flexibility of the carbon tetrahedron in alkanes and in the corresponding carbanions. The former is rigid at any pre-decomposition temperatures and the latter are usually flexible even at low temperatures. Therefore, it is reasonable to view chemical fragments of totally "frozen" kinetics as rigid. Fortunately for researchers, those predominate in organic frameworks: carbon-centered elementary molecular fragments are their major building blocks.

In some cases, elementary molecular fragments do have a zero rate of positional exchange of geminal ligands. These structural units completely lose their geminal flexibility if they are anchored in a backbone, which does not permit their ligands to move, keeping its chemical connectivity. For instance, N inversion does not occur in urotropine (Fig. 20), an amination reagent for preparation of primary amines. Similarly, in contrast to pyramidalized carbons in corannulene **4a** (Figs. 2 and 68), such carbons in C_{60} fullerene do not undergo C inversion. One can speak herein about absolute rigidity of elementary molecular fragments of pyramidal geometry.

However, there are some chemically and geometrically different elementary molecular fragments that possess a high intrafragmental flexibility at room and even lower temperatures. Can one recognize them by considering the chemical structure? Note that our knowledge is less about the internal mobility of elementary molecular fragments than it is about the rotational mobility of such two neighboring fragments. This is mainly due to the circumstance that geminal stereorearrangements are a much more complicated target for both experimental and theoretical studies, and stereodynamics of scarce chemical polyhedrons has been thoroughly examined yet. Therefore, the answer to this practically important question is that one cannot *a priori* estimate the flexibility of an elementary molecular fragment of concrete chemical structure, but one can sometimes predict which chemical polyhedron has higher kinetic stability (interconverts with a lower rate) among two flexible ones of the same overall molecular geometry and the same chemical functionality.

Steric bulkiness is worth being considered as a simple indicator of such a relative flexibility. A general concept that changes in the ligand size have a predictable impact on the kinetics of intramolecular stereoreorganizations[a]

---

[a]The idea of the rate-decreasing role of steric bulkiness is more or less fruitful in estimating the kinetics of internal rotation in very qualitative terms faster/slower (Section 1.5). Of course, this mechanistic principle is far from being a universal rule of relative kinetic stability of rotamers, as stressed in this Section.

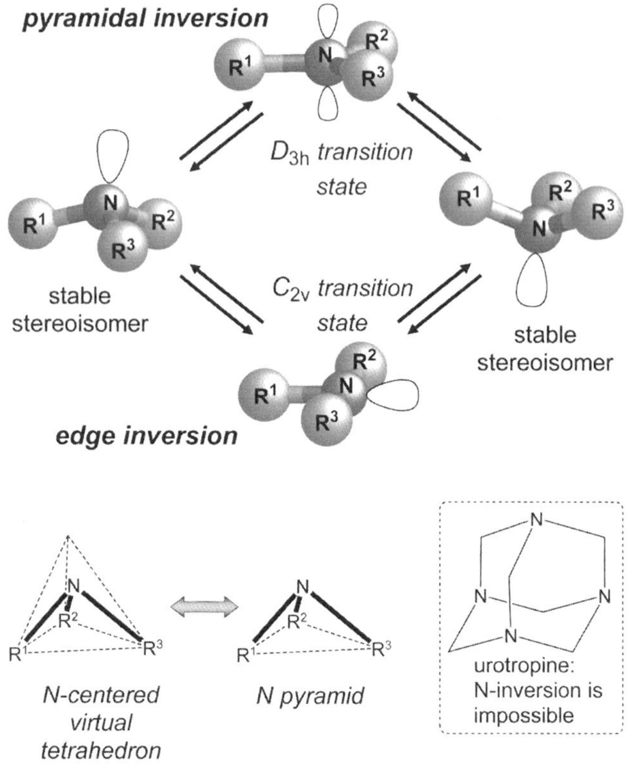

Figure 20. Geminal stereorearrangement of the elementary molecular N-centered unit in alkylamines through a trigonal transition state (pyramidal inversion) and a T-shaped transition state (edge inversion). Formal enumeration of N substituents R^1, R^2, and R^3 helps to figure out that this 3D rearrangement stereochemically is enantiomerization (inversion of the configuration of the nitrogen; in stereochemical jargon, nitrogen inversion).

may be, with caution, applied to presuming a higher rate of a geminal stereorearrangement for a chemical polyhedron relative to another one of the same geometry and the same chemical functionality. According to this simple look, the ligand bulkiness may affect the kinetics of geminal stereorearrangements in two ways. If ligands $L_1,...L_n$ are more distanced in the initial geometry than in the geometry of the transition state, molecular units with the same number of ligands and their same chemical functionality undergo the rearrangement slower as the ligand size increases. If the situation with $L_i \cdots L_j$ distances (distances between non-bonded ligands L_i and L_j) in the stable and transition states is reversed, the rate of a geminal stereorearrangement diminishes in parallel with increasing the ligand size.

The reason for this *rate–bulkiness* dependence is obvious. The kinetic barrier in such a series of tightening steric contacts in the transition state is increased because, with enlarging the ligand size, steric interactions in the transition states increase more than they increase in the corresponding stable structures. That is, the energy gap between the initial and transition states (i.e., the kinetic barrier) rises because of a faster increase of the transition state energy in the series. If, when increasing the ligand size, the energy of the thermodynamically stable structure rises faster than that of the transition state, the kinetic barrier diminishes in the order from lesser steric bulkiness to larger one.

This "quick prediction" approach has a limited significance since, as mentioned, it can supply with relative estimates less flexible/more flexible only chemical polyhedrons of the same chemical functionality. Nevertheless, it is practical due to its simplicity. It is not difficult to compare $L_i \cdots L_j$ distances by considering the relevant interconverting abstract geometrical polyhedron in the stable and the transition form. For instance, for polyhedrons shown in Fig. 18, position-interchanging ligands $L_i \cdots L_j$ are spatially more proximal in the transitional geometry than in the initial one. This means that sterically crowded elementary molecular fragments that are flexible have lower rates for their geminal stereorearrangements than their analogs with less bulky ligands. However, we cannot make ourselves comfortable with this "simple look" because, as the reader may guess, it does not take other structural factors into account.

Tertiary alkylamines are a clear illustration of both the success and the failure of this mechanistic idea. As we know from textbooks, the N fragment is configurationally unstable in these compounds, i.e., their N pyramid (Fig. 20) rapidly inverts under ambient conditions. The geometry consideration is not difficult: the N substituents are more proximal in the stable structure, the N pyramid, than they are the planar transition state of N inversion, where all three bond angles CNC are larger. In other words, these substituents diverge in the molecular space more and more when moving the amino system along the transformation coordinate from the initial structure to the transition state. Hence, with the bulkiness rule in hand, we can conclude that barriers of this geminal stereorearrangement should be lower when the size of N alkyl groups increases. Experiments and theoretical calculations have shown for a large set of different tertiary alkylamines that the barrier of this stereorearrangement usually does decrease in parallel with an increase of the bulkiness of the N substituents. Does this mean that we have a simple guiding principle for qualitatively comparing the kinetic flexibility of the nitrogen fragment in different alkylamines? For answering this question, let us return to internal rotation.

As indicated in Section 1.5, the general tendency for the kinetics of rotation is that the more bulky the vicinal groups are, the slower the rotation is (i.e., the rotation barrier rises). Thus, if a 3D rearrangement of geminal ligands is coupled with ligand rotation, the increase of steric crowding has the opposite impact on the rate of internal rotation *vs.* the rate of geminal stereomutations that follow the kinetic scenario *the larger the bulkiness, the lower the barrier*. That is, there is a mechanistic idea that declares the opposite tendency *the larger the bulkiness, the higher the barrier* for rotation-including concerted intramolecular stereo-rearrangements. There are two simple conflicting ideas of equal quality. Without clearance by means of an accurate tool of theoretical calculations, it is vague which tendency prevails in a concrete chemical polyhedron. Remarkably, a conflict of primitive structural concepts is a dead-locked situation in stereochemical evaluations which do not resort to genuine (see Chapters 4 and 5) molecular modeling. This artificial conflict demonstrates an additional weak point of "no calculations" tools.

N inversion and C–N rotation occur for a vast majority of alkylamines as a synchronous (coupled) intramolecular motion (abbreviated here as NIR, see Fig. 21 and the related explanation in the text). Stereodynamics of these compounds may be considered in light of both mechanistic ideas discussed above, and one cannot indicate whether or not an unordinary sterically hindered trialkylamine follows the general kinetic scenario: *the larger the bulkiness, the lower the barrier* for these chemical molecular pyramids. Expectably, theoretical calculations have identified alkylamines, which have higher kinetic barriers of NIR than their less hindered analogs.[23a]

Also an empirical approach of structural analogies is often not successful in assigning a certain flexibility to a chemical polyhedron. For instance, the fact that the N unit of aniline rapidly interconverts even at very low temper-atures 40–50 K may only misleads in attempting to estimate the flexibility of this fragment in N,N-difluoroaniline: this molecule undergoes N inver-sion $\sim 10^9$ times slower. Alkylamines provide an additional example. An absolute majority of trialkylamines N(Alk)$_3$ invert their N configuration at room and somewhat lower temperatures with the rates that even cannot be measured by NMR at these ambient conditions. Close molecular similarity of trialkylamines and trialkylphosphines does not suggest whether the rate of P inversion in the latter should be higher or lower than the rate of N inversion in the former [as known from *studying* alkylphosphines P(Alk)$_3$, they possess an unchanging spatial geometry of the P-centered elementary molecular unit at room and even increased temperatures]. If, possessing

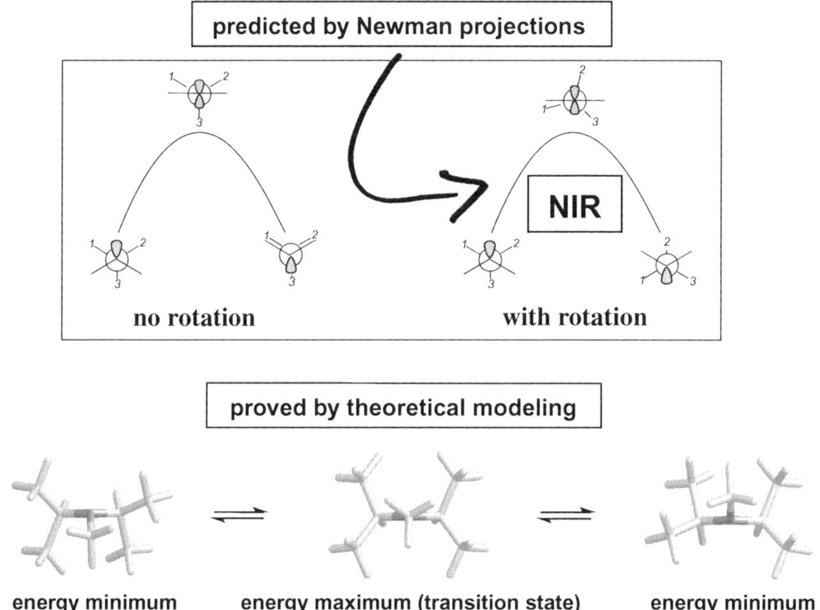

Figure 21. Nitrogen inversion in alkylamines as a concerted process of *nitrogen inversion–rotation* of N-substituents (NIR). Top: schematic representation of NIR. Bottom: evidence of NIR in amine **12** provided by molecular mechanics calculations. The *initial, transition state, and final* structures of the optimized, i.e., calculated, geometry show that each N substituent rotates with moving the molecule along the coordinate of N inversion.

this information regarding the relative flexibility of $N(Alk)_3$ and $P(Alk)_3$, we decide that the decrease of the electronegativity of the central atom **A** diminishes the flexibility of the unit $A(L_i)_n$, the next example disturbs this superficial conclusion. Positional exchange of apical and equatorial ligands in the trigonal bipyramid (Fig. 18) is appreciably slower for phosphoranes than for the related arsoranes. If, from learning that the measured rates of N inversion are significantly lower for haloamines and alkoxyamines than for alkylamines, we conclude that the increase of ligand electronegativity makes elementary molecular fragments more rigid, we overinterpret this fact. Our conjecture is incorrect: although the Si-centered tetrahedron does not interconvert its geometry even at high temperatures, theoretical calculations show that it is much more rigid (less prone to Si inversion) in SiH_4

than in SiF_4; values of the calculated kinetic barriers of pseudorotation are very close for cyclic oxy- and thiophosphoranes;[24] etc.

We see that the situation with internal mobility of elementary molecular fragments is entangled for any "simple look." The mobility of geminal ligands in elementary molecular unit $A(L_i)_n$ is obviously determined by chemical elements that form this fragment, the number of coordinated ligands $L_1, \ldots L_n$, and the chemical bonding $A - L_i$. However, this trivial understanding is not a telling one. These general descriptors of molecular structure only say that molecule-specific electron "clouds" determine the stability–instability balance for the shape of molecular backbones. However, molecular electronic structure is too often essentially dissimilar even for close structural analogs and, remarkably, absolutely unreadable without undertaking a solid theoretical analysis. Thus, without experimenting or computationally modeling one cannot indicate which chemical structures impart to elementary molecular units, e.g., to 4–6-coordinated central atoms in chemical moieties, a detectable non-rigidity in a "regular" range of temperatures in chemical organic experiments, say, from 100 K to 600 K.

Bersuker's rationalization[3d] for solving the stability/instability problem for molecular geometry reasonably foresees that any "simple look" ultimately fails in guessing flexibility of stereodynamically unexplored chemical polyhedrons. A non-negligible vibronic coupling, i.e., coupling of the ground and excited electronic states, does not rarely change the shape of the potential energy surfaces (Section 3.1) both for the ground and excited states of molecules (usually, in the form of the pseudo-Jahn-Teller effect), and, thus, dictates to either 3D structure to be thermodynamically stable and convert to another stereoisomer with passing over a certain energy barrier. Vibronic coupling is highly chemical structure-specific, i.e., a significant change in this interaction between electron states usually accompanies a change of the chemical structure. Oversimplified, pencil-and-paper generalizations for predicting the molecular shape are always restricted, exactly because they "have no idea" that any molecule has different electron states and these states may depend on each other (may be coupled). Examination of electronic states of molecules is a prerogative of quantum mechanical calculations.

In order to understand customary stereomutations in elementary molecular units more deeply as well as to analyze other "desk-tools," we probably need to familiarize ourselves with concrete chemical examples. Below are three examples (*1*)–(*3*) of geminal stereoreorganizations arranged according to their fame for organic chemists.

(*1*) **Pyramidal inversion.** Nitrogen inversion in amines is, of course, the most recognizable stereorearrangement of geminal ligands because of its ubiquitous primitive description as pyramidal nitrogen inversion[23b,c] (sometimes termed vertex inversion) in stereochemistry textbooks.[a] From the geometrical viewpoint, it is an interconversion of an N-centered quasitetrahedron (pyramid) into an enantiomeric masked quasitetrahedron (pyramid; Fig. 20). Why is this 3D transformation treated as the inversion of a *pyramid*?

In principle, there are two alternative pathways (mechanisms) of N inversion in amines. The pathway, which occurs through the transition state of D_{3h} symmetry (an sp^2-hybridized nitrogen), is called pyramidal or vertex inversion (Fig. 20, top). The second pathway corresponds to the transition state of C_{2v} symmetry (a T-shaped transition state with equal positions of substituents R^2 and R^3); it is called edge inversion.[23d]

In the first pathway, a trigonal N pyramid is reflected in a virtual mirror, which is actually the plane where a planar D_{3h} transition state lies (Fig. 20). Inversion of the N unit corresponds to vertex inversion in a flattened trigonal pyramid with the nitrogen atom in the inverting apical vertex. From the viewpoint of molecular geometry, only N alkyl groups are the ligands of the N unit while the unshared electron lone pair of the N atom is not a ligand (because one cannot localize electrons in the space). In other words, only the backbone of the elementary molecular fragment is considered. Therefore, the trivial name *pyramidal inversion* has been attributed to this mechanism of inversion. Pyramidal inversion of N geometry is, of course, associated with a specific change of the density of the 3D distribution of the electron lone pair of the nitrogen. One could say that this VSEPR ligand is gradually "decreased" on the outer side of the nitrogen umbrella (actually, a decrease of local electron density) and is synchronously "increased" (i.e., a local increase of electron density) on the opposite side when going over the first half of the coordinate of the N pyramid inversion (in terms of physical chemistry, the reaction coordinate). This N "substituent" is equally exposed in both stereofaces in the transition state of a trigonal (sp^2-related) geometry. Similarly, one may say that this unusual ligand "restores" its size on the

---

[a] Such a stereorearrangement (inversion of a masked tetrahedron) in three-coordinated phosphorous compounds, sulfoxides, carbanions, silaanions, onium cations, etc., is stereochemically equal to that in amines.

outer side of the reverting N umbrella, finishing the second half of the transformation. Formally, the N pyramid is reflected in the σ_v symmetry plane of the transition state.

A quasitetrahedron (and not a pyramid) formed by four VSEPR ligands represents the aminic nitrogen under a more relevant consideration of this molecular unit (Fig. 1; the lone electron pair is not ignored). In accordance with this physically reasonable formal geometry, edge inversion in amines (Fig. 20) corresponds to the canonical mechanism of interconversion of geometrical tetrahedrons, i.e., the spread pathway through a planar tetragonal transition state that, in the amine case, has symmetry C_{2v} (Fig. 20) instead of D_{3h} shown in Fig. 18 for an abstract tetrahedron of symmetry T_d. In this pathway, the nitrogen electron pair is formally a "full and equal member" of the motion of geminal ligands along the coordinate of this 3D rearrangement.

Which mechanism of inversion predominates for alkylamines? As shown by theoretical calculations for different amines, this ligand stereorearrangement occurs through the transition state of D_{3h} symmetry and not in the pathway with the transition state of symmetry C_{2v} (Fig. 20). Our probe amine **12** is also not an exception. MM calculations[9] locate for it a planar trigonal transition state for the N unit (Fig. 21). Thus, the nitrogen unit in alkylamines does not behave in the manner which can be expected from a canonical masked VSEPR tetrahedron shown in Figs. 1 and 20. Edge inversion is not realized, while the "abnormal" pyramidal inversion is a single occurring mechanism. What is the reason for this "ignoring" the physically relevant chemical N-centered tetrahedron?

A lone electron pair is an unordinary ligand in 3D interconversions of polyhedrons. As mentioned, upon edge inversion, the N lone pair "behaves" as a regular ligand that does not change its size when stereorearrangement occurs. This "normal size" of the peculiar ligand (the lone pair) requires a "normal" spread pathway of minimal distortions (Fig. 18) that, as we know for amines, is the pathway with the transition state of a T-shape (symmetry C_{2v}; Fig. 20). In pyramidal N inversion in amines NR_3 (the transition state of symmetry D_{3h}), the "flowing" of this pair from one side of the symmetry plane to another one (Fig. 20) is accompanied with less structural distortions of the backbone than edge inversion requires. The decrease of local electron density of unshared electrons upon pyramidal inversion actually provides a new minimal distortion pathway. One could say that, in amines, the interconversion with minimal distortions of the *backbone* (i.e., pyramidal inversion) predominates over interconversion with minimal distortions of the VSEPR quasitetrahedron composed from the nuclear backbone and the N lone electron pair (edge inversion).

If this interpretation does not satisfy the reader who has learned that, in essence, mechanistic, non-quantum mechanical descriptions of molecular shape are inadequate, one can recall Bersuker's concept that *multitudinous* molecular systems tend to avoid degeneracy by lowering molecular symmetry.[3d] In less general terms, higher symmetry molecular geometries very often are thermodynamically unstable and correspond to transition states of intramolecular stereorearrangements because a severe vibronic coupling between non-degenerate excited electronic and ground states leads to stabilization of lower symmetry geometries in the ground state (a strong pseudo-Jahn-Teller effect), i.e., the latter geometries correspond to energy minima. The symmetry of the masked N tetrahedron is C_{3v} in amines NR_3 while the symmetry of the transition state is D_{3h} and C_{2v} for the pyramidal and edge inversion of this chemical tetrahedron, respectively (Fig. 20). Thus, pyramidal inversion is associated with a decrease in symmetry of the N unit when going from the transition state to the stable 3D structure, while edge inversion increases the symmetry in this path *transition state → stable structure*. It is obvious that, among these mechanisms for alkylamines, the former (pyramidal inversion) formally corresponds to Bersuker's concept, and the latter contradicts it.

One cannot claim that the pathway of minimal distortions of the *backbone* is preferred in the 3D rearrangements of any other chemical tricoordinated fragments AL_3 with the electron lone pair at central atom A. Let us glance at phosphines Me_3P and PF_3. The molecule of Me_3P is inverted only via pyramidal inversion (the D_{3h} transition state); there is no alternative transition state of a T-shaped geometry (the C_{2v} transition state) for the P inversion. In PF_3, edge inversion is a highly predominated mechanism for P inversion in this molecule (the related transition state of symmetry C_{2v} has significantly lower energy), although pyramidal inversion also occurs (the related transition state of symmetry D_{3h} does exist).[23e] Thus, the course of minimal distortions may fail in its prediction of the favored mechanism for positional interchange of geminal substituents [an additional illustration is example (3) discussed below]. The occurrence of edge inversion in PF_3 becomes understandable when recruiting the above mentioned concept of vibronic coupling. In this molecule, vibronic coupling leads a more symmetrical D_{3h} transition state of pyramidal inversion to have higher energy than a lower symmetry (C_{2v}) transition state of edge inversion. In the potential energy surface (Section 3.1) of PF_3, the decrease in the energy of the D_{3h} transition state include pathways of descending both to the stable enantiomers of this phosphine (pyramidal molecules) and the C_{2v} transition state. This manifestation of the pseudo-Jahn-Teller effect is, of course, revealed by quantum mechanical modeling of stereodynamics of this molecule.[23e]

We again observe that a "simple look" — the use of structural analogies for molecules or mechanistic parallels between real molecules and abstract elastic polygons or polyhedrons — may mislead in identifying flexible or rigid elementary molecular fragments. A practical solution suggests itself: in examining a geometrically simple geminal stereorearrangement, inversion of unit **AL_3**, theoretical modeling (Chapters 4 and 5) is the required research tool.

When considering stereorearrangements in alkylamines, it is useful to know that inversion of their nitrogen-centered units is a more complex intramolecular motion than the flipping "umbrella" of the nitrogen pyramid from stereochemistry textbooks. One can see from Newman projections (Fig. 21, top), a sole pyramidal inversion, without rotation around C–N bonds, results in an eclipsed geometry of vicinal substituents. In other words, the interconversion of a non-rotating N unit terminates in the energy maximum of rotation (see for the rotational energy plot in Fig. 12). To escape the eclipsing, the N substituent undergoes a 60° turn and, hence, the classical 3D rearrangement of the nitrogen unit in amines is actually a concerted process of *nitrogen inversion–rotation* (NIR).[a] Remarkably, the three N-alkyl groups are equivalent in considering NIR. This eclipsing-related reasoning (Fig. 21) may be given for any one of them. Hence, this mechanistic consideration shows that all three N-substituents are synchronously rotated during NIR.

However, we cannot be based on Newman projections. NIR is a hypothesis until it is proven by using a trusted model. Molecular mechanics calculations (Chapter 4) performed for different alkylamines[4,23a,25a,b] confirm the occurrence of the predicted concerted 3D stereorearrangement by demonstrating that two intramolecular motions — N inversion and C–N rotation — are concerted (i.e., that they have a common transition state), i.e., they do not occur sequentially. To illustrate this, the initial geometry, the transition state geometry and the final geometry[9] of amine **12** are shown in Fig. 21. Comparing these geometries, one can easily recognize that each alkyl substituent has undergone a rotation of ~60° synchronously with completing the inversion of the N pyramid.

Now, on the basis of a reliable model, we are able to conclude that pyramidal N-inversion in alkylamines is usually coupled with rotation of N-substituents. The example of NIR shows

---

[a]Organic chemists have become accustomed to understanding intramolecular dynamics in alkylamines as sole nitrogen inversion. This obsolete view is, at best, oversimplified.

that 3D rearrangements of vicinal and geminal substituents may be coupled in polytopal rearrangements. It is meaningless to indicate which intramolecular motion is predominant in a concerted 3D rearrangement; all participating motions are equally involved. Therefore, coupling of intrafragmental and interfragmental spatial reorganizations does not permit linking the flexibility of a fragment solely with any one of them. This is a remarkable circumstance. It appears that sometimes it is impossible to separate internal mobility of elementary fragments and their rotational freedom as entire units, at any point of the transformation coordinate. As we will see later (Chapter 2), this inference is important for the *conformer* definition.

Let us deal with an easier question on whether the relative stability of diastereoisomers, that interconvert exchanging positions of geminal substituents, (here: N-invertomers; Fig. 22), can be established at least qualitatively without undertaking experiments or computational modeling. Discussing internal rotation (Sections 1.5 and 1.6), we have reached the conclusion that often such a comparison of two rotamers of distinct geometry is scarcely reliable if trustworthy theoretical calculations are not conducted. To a greater extent, this conclusion may be related to stereoisomers

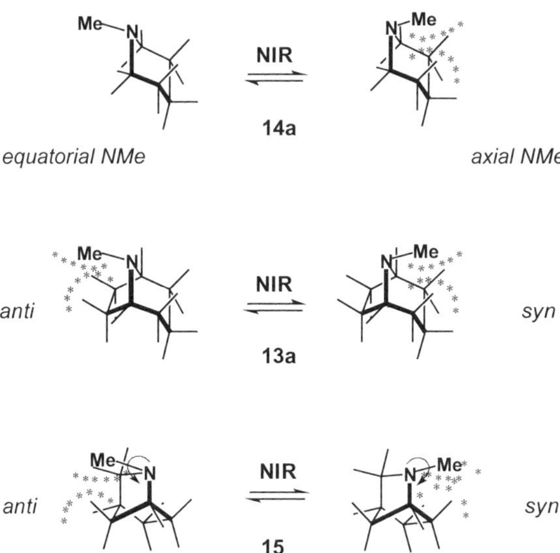

Figure 22.   NIR in azamonocycle **14a** as well as azabicycles **13a** and **15**. Asterisks depict steric contacts of the N-Me substituent with the bicyclic backbone, and arrows indicate rotation of the N-Me group in **15**.

that result from permutating geminal ligands. The energy of interactions of geminal substituents is not changed upon rotation of elementary molecular fragments since these substituents keep their relative spatial disposition in the molecular skeleton, and only vicinal interactions are dissimilar in different rotamers. In contrast, in exchanging positions of geminal ligands, more interactions are different for the initial and resulting structures because changes of vicinal interactions are additional to changes of geminal interactions. Therefore, it is difficult to estimate the relative stability of NIR-generated diastereoisomers, using "lab-made" tools. ■

Let us illustrate this thesis. The equatorial N-invertomer of N-Me piperidine (**14a**; Fig. 22) is more stable than the axial invertomer. By a "simple look," one can indicate the reason for this predominance. If the N-Me group is oriented axially, destabilizing steric 1,3 diaxial interactions appear. Thus, a geometrically rough molecular model (Fig. 22), with its main idea of hard VDW spheres of atoms, is sufficient to identify a "visible" factor of the preference of an equatorial orientation of a ring substituent *vs.* its axial orientation in saturated six-membered cycles. However, steric hindrances, as a rule, do not seem explicitly dissimilar for distinct orientations of ring substituents in other systems. In addition, as we know, "invisible" electronic effects essentially contribute to molecular energetics. Then, only theoretical computations for accurately modeled geometries are capable to "catch" the desired energy difference for N-invertomers.

Therefore, the mentioned simplicity in "extracting" the *correct* estimate for diastereoisomeric N-invertomers of amine **14a** is rather accidental. For instance, there is no obvious choice of which stereoisomer, *syn* or *anti*, is favored in bicyclic N-Me amines **13a** and **15** (Fig. 22). As in the case of amine **14a**, NIR in these amines translocates the N substituent from some steric surrounding to another one. The difference even in sole steric interactions for either of the two geometries is tiny, and the primitive, VDW-radii-based, modeling is absolutely unfruitful.[a] Such "research tools" do not even manipulate with accurate molecular geometries since they cannot model them!

---

[a]As one can guess, the correct choice is not problematic for theoretical calculations. They have demonstrated a predomination for the *anti* form for azabicyclooctane **13a** and the *syn* form for azabicycloheptane **15**.[26,27]

(*2*) The second example of interconverting chemical polyhedrons is the **trigonal bipyramid**. This subject is stereochemically more intricate due to a larger number of ligands as well as their spatial non-equivalence (two apical and three equatorial positions). As we remember, there are twenty distinct arrangements of five non-identical ligands in this pyramid, i.e., 10 pairs of enantiomers. Thus, the first problem in understanding flexibility of such chemical polyhedrons consists of determining favored relative arrangements of five ligands $L_1,\ldots L_5$ in the trigonal bipyramid of a concrete chemical composition, i.e., identifying favored stereoisomeric structures. Due to the reasons explained in example (*1*), there are no reliable empirical or mechanistic models that permit us to predict the most stable arrangement of different organic ligands in the bipyramid apexes. Nevertheless, there is at least one empirical rule which is more or less general for trigonal bipyramids. According to it, lone electron pairs of the central atom **A** rigorously prefer equatorial orientations; probably, this is due to the energy gain in forming a 3c-4e bond in triad L_{apical}–**A**–L_{apical}. For instance, in selenurane **16** (Fig. 23), the selenium lone pair is oriented equatorially, while four aromatic ligands occupy four remaining apexes, i.e., two apical positions

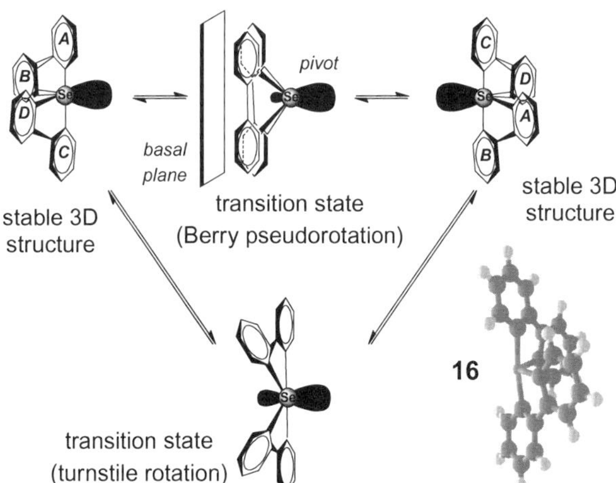

Figure 23.   Berry pseudorotation *vs.* turnstile rotation in selenurane **16** (phenylene rings are labeled by letters **A, B, C**, and **D**). Bottom, right: stable 3D structure of geometry optimized[9] at the MP2/6-31+G(d,p) level.

and two equatorial positions.[28] However, it is a light case because these four ligands are chemically equal. In general, information about orientation of one or two ligands among five chemically different ones does not determine occupational preferences of other three ligands: the apicophility "rule" explained in Section 1.2 is not general. ■

The problem of identification of the actual (favored) mechanism of ligand stereomutation in trigonal bipyramids is much more complicated. Note that mechanisms of polytopal rearrangements, which lead to the same stereoisomers, are indistinguishable in today's standard experimental techniques (i.e., not using ultrafast spectroscopies; see notes below). For instance, traditional spectral experiments cannot establish whether interconversion of the synclinal and antiperiplanar forms of ethane derivatives (Fig. 11) occurs as rotation around the C–C bond or a 3D rearrangement of all three *geminal* ligands in one of the C-centered elementary units (of course, the latter mechanism only is hypothetical).[a] To the same extent, no NMR experiment[b] could determine whether the rearrangement of the amino fragment in amines occurs as pyramidal inversion or edge inversion. Notice that tetrahedrons are stereochemically simpler than trigonal bipyramids. From the point of view of stereodynamics, each case of a chemical trigonal bipyramid is a difficult challenge. Six alternative mechanisms of relevant geminal stereorearrangements are mentioned in the previous section; for the seventh (cuneal inversion), see below. In principle, all six mechanisms should be included into consideration, if there are at least two identical ligands in a chemical trigonal bipyramid. The favored mechanism of 3D reorganizations in such a compound definitely cannot be identified by means of any "simple look."■

Experimentalists monitor only stable structures, e.g., the initial and the final stereoisomers, i.e., structures of a finite lifetime (see also Section 2.5). A transient structure, i.e., the structure that has an unlimitedly short time period of existence, is not traced by any experimental method. Intermediate stereoisomeric structures in the trajectory of the ligand motions upon a 3D rearrangement are transient structures. However, this does not mean that ligand motions are in principle undetectable. Spectral methods of the femtosecond (fs; 1 fs

---

[a]It is worth mentioning that these alternative mechanisms may be examined by means of even not laborious theoretical calculations.

[b]In respect to many other spectroscopies, NMR is not a fast method (see Section 3.2, Table 1).

$= 10^{-15}$ s) time window may follow in-time-evolution of a transient molecular structure including its *geometrical* changes since this process of further transformation takes a short, but finite, time period. Alternative motions of ligands are actually characterized by alternative mechanisms of a 3D rearrangement. Thus, in principle, alternative mechanisms of stereomutations can be distinguished experimentally. However, such spectral experiments, due to their complexity, are far from being routinely conducted.

In order to understand the difficulties, let us look at a concrete example plainly. In accordance with VSEPR estimates, experimental[29a] and theoretical[9] examination of selenurane **16** show that the molecule possesses the geometry of a masked trigonal bipyramid (Fig. 23). This chemical polyhedron is not ideal: the two apical liganding carbons and atom Se do not form a straight line (here, we recall the geometry-distorting impact of the pseudo-Jahn-Teller effect[3d]). The Se configuration may be interconverted via the Berry pseudorotation; the turnstile rotation is an explicit alternative. Let us try to indicate the favored mechanism from these two. In order to identify the preferable one, one must compare transition states for these alternatives by estimating their relative energy. That is, we need to estimate "visible" additive contributions to molecular energy. However, one may quickly find out that we are incapable of qualitatively analyzing the energetics involved.

First, we have no idea regarding the accuracy of the general geometries of these transition states (Fig. 23). Sufficiently accurate molecular geometries are required to permit us to compare geometrical distortions (and, thus, the strain) in this molecule in the alternative pathways of the considered enantiomerization. Clearly, we do not have such information without undertaking a relevant theoretical modeling. Even obvious alterations of the lengths of bonds Se–C cannot be compared for the alternative transition states when using a "simple look" of general (Figs. 18 and 19) geometries. In addition, the strain-developing locking of the four phenylene rings into two biphenylene fragments may unpredictably distort the considered general molecular geometries of the alternative transition states for this molecule.

Second, in the absence of accurate data for molecular geometry, one can only speculate regarding occurrence of "visible" electronic interactions in the alternative transition states. Are the selenophene rings of **16** to some extent aromatic in the transition state of turnstile rotation? Is the Se lone electron pair involved into conjugation with the "apical" phenylene rings in this transition state? This questioning might be continued, and each answer would be arbitrary.[a]

Third, when assessing transition states for selenurane **16** with some general geometries in hand, we do not know whether or not these transition states (or any among them) exist for this molecule at all and, thus, other alternatives should be put into the focus. This irrefragable argument actually terminates our discussion of two mechanisms of positional ligand exchange in selenurane **16**. We are forced to conclude that there is no reasonable *simple* choice between the Berry pseudorotation and the turnstile alternative for this enantiomerization.

Moreover, one can suggest for this molecule an additional stereodynamic alternative called cuneal inversion. Cuneal inversion is a particular case of through-plane inversions that is related to through-plane inversion of a trigonal bipyramidal geometry. In principle, through-plane inversion may occur in a chemical polyhedron, the central atom of which has a lone electron pair(s). It is the mechanism that leads to the mirror image structure of a molecular elementary fragment by inverting the initial geometry relative to the plane (reflection plane), which contains the central atom and is perpendicular to the C_∞ symmetry axis of the p orbital occupied by the lone electron pair of this atom. The orbital has symmetry S_1 in the transition state; when reaching it, the other ligands lie in the reflection plane, i.e., the inverting elementary molecular fragment (the central atom with the liganding atoms) is planar in the transition state.

Obviously, through-plane inversion interconverts enantiomeric geometries of the fragment (the entire molecular structures that contain such interconverting fragments not necessarily are enantiomers). In-plane inversion that occurs in chemical fragments of trigonal geometry (e.g., N-inversion in imines; Section 2.2) is actually a through-plane inversion, where the inversion plane is perpendicular to the plane formed by the fragment atoms; in general, the interconverted molecular structures are diastereomers.

---

[a] Such questions are attributes of a "simple look." As a rule, non-empirical (quantum mechanical calculation) methods do not deal with questions that are based on primitive structural concepts, e.g. strain or conjugation. Being based on quantum mechanics, i.e., on a much higher theory, *ab initio* calculations "extract" molecular geometry and the corresponding molecular energy directly, using fundamental physical descriptions of the interactions of electrons and nuclei (Chapter 5).

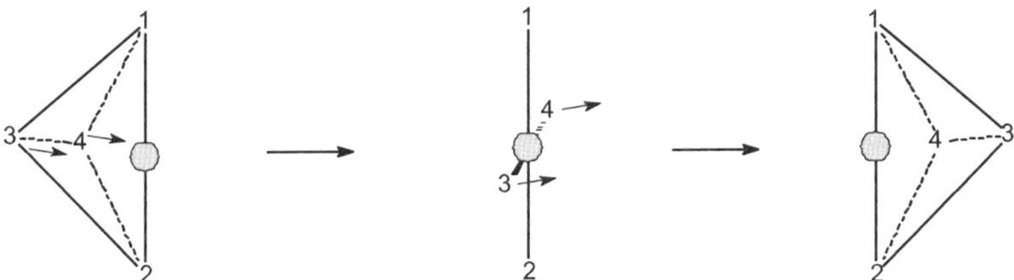

Figure 24.   Geometrical changes of the wedge-shaped polyhedron (trigonal pyramid; lines show the edges, and numbers indicate the apexes) upon cuneal inversion. In considering this trigonal pyramid as a chemical polyhedron, the gray circle indicates the central atom, numbers depict the ligands, and arrows show the direction of ligand motion.

Then, what is cuneal inversion of molecular trigonal bipyramids? Nuclear frameworks of elementary molecular fragments with the central atom that bears four substituents and one lone electron pair are wedge-shaped, or, in other words, are trigonal pyramids (Fig. 24). From the viewpoint of canonical geometries of chemical polyhedrons (Section 1.2, Fig. 1), they are "materialized rests" of a fantomic trigonal bipyramid that does represent their "true" molecular geometry.[a]

As mentioned, lone electron pairs are peculiar ligands. For positional exchange of geminal ligands in real molecules, this also means that, in contrast to atoms, these unordinary ligands do not require a localized 3D domain in the space in the molecular backbone changing its 3D geometry. If a wedge-shaped central atom bears a lone electron pair (as, e.g. in sulfurane **16**), its configuration may be inverted through a planar transition state of a square geometry (Fig. 24), i.e., via cuneal inversion. Thus, it is a through-plane inversion. Unequally to Berry pseudorotation and turnstile rotation that orient two apical ligands equatorially and two equatorial ones axially, the ligands of chemical polyhedrons of wedge shape do not exchange their apical and equatorial positions upon occurring cuneal inversion. What indeed occurs is that equatorial ligands exchange their positions relative to one another; formally, two equatorial ligands permutate, i.e.,

---

[a]The meaning *molecular geometry* is exclusively related to the mutual spatial disposition of nuclei in a molecule. As explained, one can more or less accurately determine location of any nucleus in the molecular space (and, thus, speak about the geometry of the molecule), while no electron can be supplied with some intramolecular 3D domain of its location.

cuneal inversion interconverts enantiomeric geometries of such chemical polyhedrons.

This difference in positional exchange of geminal ligands in Berry pseudorotation *vs.* cuneal inversion often provides the possibility to distinguish these mechanisms experimentally. For instance, spatial equivalence of all four phenylene rings demonstrated by NMR for selenurane **16** at room temperature[28] indicates occurrence of a fast Berry pseudorotation [Fig. 23; indicating a slow (in the NMR timescale) Berry pseudorotation, only diphynelene fragments of **16** are equivalent at low temperatures[28]] if we assume that turnstile rotation or a sequence of other mechanisms is not involved.

The S and Te units of sulfurane **17a** and tellurane **17b** (Fig. 25) geometrically are masked trigonal bipyramids. A fast diastereomerization observed by NMR for **17a** at increased temperatures almost evidently shows occurrence of cuneal inversion of this Se-centered elementary molecular unit (a hypothetical alternative is a chemical ligand dissociation — re-association).[29a] Indeed, only an inversional process can interchange the *endo* and *exo* orientations (relative to the equatorial ligands, i.e., the phenylene rings) of the Et group.

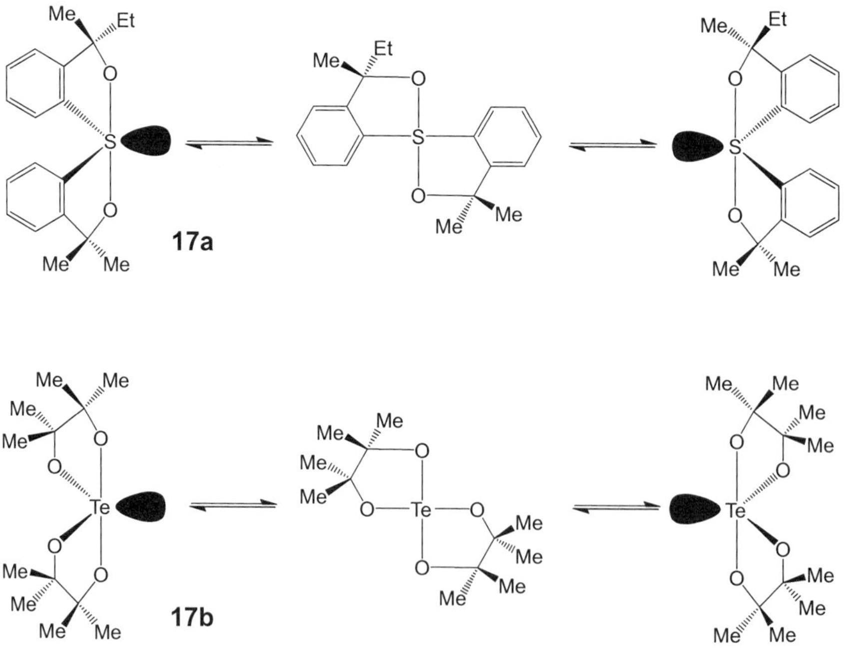

Figure 25. Cuneal inversion in sulfurane **17a** and tellurane **17b**. In the middle: a square-shaped transition state [the *p* orbital (not shown) occupied by the S lone electron pair is normal to the plane of the chart].

Tellurane **17b** is an example of occuring two separate stereoreorganizations (in other words, two different individual intramolecular motions) for a chemical trigonal bipyramid. All four Me groups of **17b** are not equivalent in NMR spectra registered at low temperatures.[29b] Equivalence of *pairs* of Me groups, which is manifested in room temperature NMR spectra of **17b**, shows that the apical and equatorial ligands of the Te-centered elementary molecular unit are isochronous, indicating by this a fast exchange of spatial orientations for them. Assuming no turnstile rotation or a few step-by-step geminal rearrangements by another mechanism, a hypothesis of occurrence of Berry pseudorotation well interprets these temperature-variable NMR spectra. Equivalence of all four Me groups detected by NMR at increased temperatures[29b] is a firm evidence for cuneal inversion that inverts the configuration of the central atom (Te) in this molecular fragment of **17b** much slower than Berry pseudorotation re-orients its apical and equatorial ligands.

A thorough theoretical modeling of mechanisms of geminal stereorearrangements in simple molecules of trigonal pyramid geometry (i.e., location of the transition states that correspond to distinct pathways of interconversion of this geometry), e.g. SF_4, SeF_4, TeF_4, and PF_4^-, has shown that cuneal inversion is associated with the transition state (the latter, as expected, has symmetry D_{4h}) of essentially higher energy than the transition states related to other intrafragmental motions of ligands $L_1, \ldots L_4$.[29c] Nevertheless, one should not view cuneal inversion as an absolutely irrelevant stereodynamic hypothesis, when assessing the flexibility of an organic chemical polyhedron AL_4 of trigonal bipyramid geometry with the central atom **A** on its longest edge. In the transition state of cuneal inversion, four ligands $L_1, \ldots L_4$ may have VDW contacts that are sufficiently tolerable to enable this intramolecular motion to be unexpectedly fast, if bonds $A–L_i$ are very long.

These examples show that trigonal bipyramids are a chemical polyhedron too complicated for a primitive mechanistic analysis of their stereodynamics. Most of known "no calculation" assignments of either mechanism (usually, Berry pseudorotation) to the observed geminal stereorearrangement actually should be supplied with a comment: *assuming no alternative mechanisms*. If attempting to reach more certain conclusions, the alternative pathways (transition states) should be modeled by using reliable theoretical calculations. The necessity of locating many transition states (well illustrated in Ref. 29c) for almost any concrete chemical polyhedron of trigonal bipyramidal geometry suggests to organic experimentalists to leave modeling of these molecular objects to computational chemists.

(*3*) As the third example, we take the **classical carbon tetrahedron** (Fig. 6), now considering its rigidity in methane. The minimal distortion pathway for the hypothetical configurational inversion of CH_4 is the spread path through a planar structure, a square (D_{4h} symmetry, Fig. 18). It would

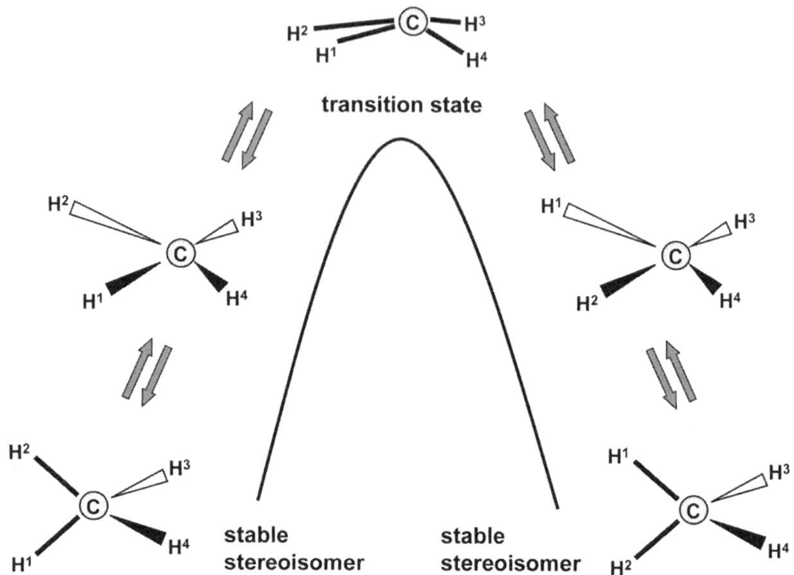

Figure 26.   The lowest energy pathway for C inversion in methane (modeled by *ab initio* calculations, Refs. 30a,b).

not be too presumptuous to say that, for the "simple look," this pathway would seem to be favored.

Theoretical calculations show that the "true" transition state for methane is of C_S symmetry (Fig. 26; protons 1 and 4 "eclipse" protons 2 and 3, respectively).[30a,b] In progressing along the interconversion coordinate in the path *starting structure–transition state*, substituent 2 moves a longer distance in the molecular space than substituent 1. Then these substituents interchange their roles, and, in the path *transition state–final structure*, substituent 1 moves a longer distance in the molecular space than substituent 2. The energy barrier is significantly lower in this pathway than in the alternative pathway with the transition state of D_{4h} symmetry. The change of total molecular energy, surprisingly from our perspective, does not correlate severely with the degree of the geometry changes. To state this issue more explicitly, the minimal distortion pathway for this polytopal rearrangement, which *a priori* seems preferable, is not like that. This example is a good illustration of the correct understanding that the "quantum mechanical" electron cloud and not a mechanistic law of elasticity dictates the geometry of the molecular backbone and its internal mobility.

Interestingly, this "message" regarding stereodynamics of organic molecules is from the apparently well-understood tetrahedron from textbooks. Until solid computations were performed for the methane molecule,[30a,b] one might have thought that no stereochemical news should be expected from this milestone 3D structure. The surprise of the tetrahedral carbon certainly reminds us that the internal mobility of elementary molecular units, in contrast to internal rotation, is yet rather poorly explored.

For instance, the ylidic carbon still remains tetrahedral in spite of the absence of four explicit VSEPR bonds (Section 1.2); the extent of non-planarity depends on the chemical structure of the concrete ylide. Is this carbon-centered elementary molecular unit configurationally stable at decreased temperatures? In other words, can "tetrahedrally the best" ylidic carbons, which bear different substituents, provide chirality to ylide molecules in the absence of chiral fragments?

A similar question may be addressed to familiar reagents, dialkyl phosphite salts $[(AlkO)_2P^+-O^-]^- Cat^+$ which are phosphorylating species in the Michaelis–Becker, Todd–Atherton, Abramov, and Pudovik reactions. Their anions have a pyramidal molecular geometry (a masked VSEPR tetrahedron). In Fig. 27, this pyramid is shown for the dimethyl derivative, which structure has been optimized by quantum mechanical calculations.[9] According to these calculations, such species are delocalized anions, where the anionic O partially donates one of its electron lone pairs to the P atom that, in its turn, shifts the ester bonds P–O (i.e., the σ-electron pairs) to the related O's (Fig. 27). As a result, these P–O bonds are longer than ester bonds P–O in other organophosphorous structures, e.g. in dimethyl phosphate anion (0.158–0.168 nm), i.e., they are weakened. The parent dialkyl phosphites $(R^1O)HP(O)(OR^2)$ (often called H-phosphonates) are configurationally stable at room temperature. Does the P pyramid of the considered P,O-anions slowly invert at this temperature? In terms of synthetic chemistry, are phosphite salts $[(R^1O)P(O)(OR^2)]^- Na^+$ configurationally stable in solution so that one can speak about racemic mixtures and separate the phosphite enantiomers, e.g., by using chiral resolving amines?

The main pool of chiral reagents currently consists of compounds which molecules include either a tetrasubstituted carbon, or trisubstituted phosphinic phosphorous, or sulfoxidic sulfur. Obviously, exploration of internal mobility for as much as possible elementary molecular fragments with a heteroatom as a central structural element may facilitate the further development of stereoselective synthesis by adding to this well-explored pool of new, highly effective reagents that have atoms of other chemical elements as centers of chirality.

One may summarize from our discussion of polytopal rearrangements in elementary molecular fragments that these 3D reorganizations resist to

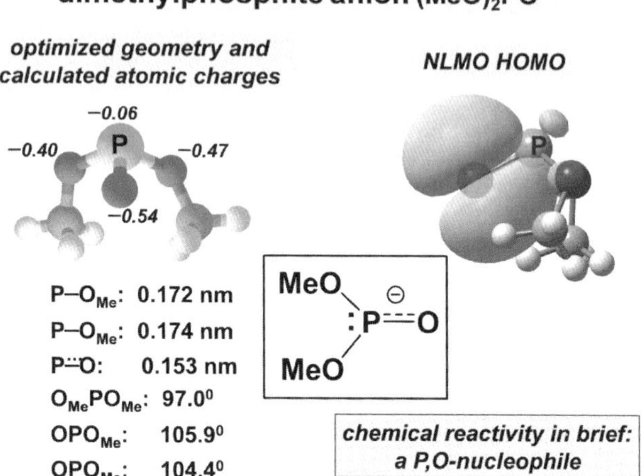

Figure 27. The lowest energy P–O rotamer of dimethylphosphite anion in vacuum: the molecular geometry [calculated at the MP2/aug-cc-pVTZ level (see Sections 5.2.3 and 5.2.4)], the HOMO (among NLMOs), and atomic charges (shown by numbers in italics) calculated by using the CHELPG methodology.[33b]

primitive mechanistic analysis much more than internal rotation. There is no simple non-empirical way to identify energetically preferable stereoisomers from the structures that result from permutations of geminal ligands. To a greater extent, there is no "pocket" approach to estimate flexibility of elementary molecular fragments; simplified approaches to the stereodynamics of these fragments are rather deorienting. In-time stability of elementary molecular units is usually known to organic chemists from textbooks as a larger set of configurationally stable fragments and a smaller set of configurationally flexible moieties (e.g. alkylamino and five-coordinated phosphorane molecular fragments). Both sets include only elementary fragments of trivial chemical structures. The rigidity/mobility of elementary 3D units of chemical structures, which do not fall within these sets from textbooks, e.g., the I-centered units of hypervalent organic compounds or P-centered fragments in aromatic phosphines, is obscure for many specialists in organic synthesis. One may claim that, unfortunately, the problem of internal mobility of elementary molecular fragments is often shadowed in analysis of reaction stereospecificity.[a] Besides, such molecular units of the same chemical functionality may have different flexibility, depending on the concrete chemical structure of their ligands.

1.9 Conformational Concept Is a Description of a Non-Static Molecular Shape

Our final, very general notes regarding spatial structure of organic molecules (next paragraphs) are actually of heuristic character. These notes indicate why conformational studies are possible in principle and what their principal difficulties are. Before analyzing the essence of the conformational matter in the next chapter, we need to know what researchers look for in routinely exploring molecular shapes of organic molecules.

Molecular shape is variable. However, this variability is not infinite, and molecular shape may be described by means of primary geometrical meanings; the set of typical 3D geometries for elementary molecular fragments is not even large. The static 3D geometry of these fragments is in general readily available in a simple way (i.e., by VSEPR). At room temperature, many elementary molecular fragments of different chemical structure keep their geometries during a long time period. At the same time, some chemical units are configurationally unstable, and there is no simple tool for identifying them. Assembly of elementary molecular fragments in an entire molecular framework essentially blurs its "3D image," because of interfragmental rotational freedom of many fragments and intrafragmental mobility of some of them. Nevertheless, for such (i.e., non-absolutely rigid) molecules, the number of resulting stereoisomeric structures of short, but non-zero time of existence (short-living thermodynamically stable stereoisomers, i.e., *conformers*) is still finite and, hence, these stereoisomers can be, in principle, detected. In fact, the threshold of the analytical sensitivity for the

---

[a]This "texbook dependence" leads to conceptual restrictions of hypotheses in studies of concrete compounds, if non-customary (in context of geminal flexibility) species are assessed. For instance, in rationalizing the stereochemical outcome, many synthetic chemists do not take into account that pyramidal carbanions are usually configurationally unstable. Clearly, consideration of only one invertomer is insufficient when analyzing stereoselectivity of reactions of carbanions, which, e.g., have three different C-substituents or bear a chiral auxiliary.

used spectral method predetermines the lower limit in detecting molecular species present at low concentrations.

Conformers interconvert with a different rate under different physical (mainly, temperature and pressure) and physical chemical (medium) conditions. That is, molecular flexibility is not unchangeable: a certain flexibility of a molecular system is firmly associated with certain external conditions. Thus, the possibility to "catch" individual 3D forms of molecules experimentally is limited by the capability of the used spectroscopy in time resolution.

The number of conformers essentially increases as the size of the molecule increases; this number is of an "astronomic" scale for large molecules, e.g., biopolymers. An entire set of conformers is practically inaccessible for exploring it in detail, and researchers assess "important" (low energy) conformers. Nevertheless, large molecules are absolutely "indigestible" objects for non-specialists in structural chemistry.

This resulting complexity of the molecular shape — many conformers, their distinct in-time-stabilities and the dependence of this flexibility on external conditions — requires a special consideration from researchers; this consideration is what we call the conformational concept. Thus, the 3D structures of rapidly interconverting stereoisomers (conformers), their thermodynamic as well as kinetic stabilities at different external conditions are the focus of conformational studies. Methodologically, conformational studies include either experiments of different complexity, or routine theoretical modeling, or both of them. In Chapters 2 and 3, we will learn conformational meanings and understand conformational ideas at the level required in organic *research*.

Finally, one should notice that only molecular systems, which are integral chemical structures, fall within the scope of the conformational concept. Changes of molecular shape due to the relative motion of large molecular fragments, which are not connected chemically but are geared topologically (e.g., rings passed through each other), are not considered here. Examples of compounds possessing this unusual, but easily understandable intramolecular mobility are catenanes and rotaxanes. To some extent, intramolecular motion of chemically non-bonded structural components in these systems is similar to the free motion of individual molecules.

2

Conformer and Conformation

2.1 Conformation: The Traditional Stereochemical Understanding

To this point, the usage of conformational terminology has been avoided, and different non-specified terms, e.g. *3D form, rotational geometry* and *invertomer*, have been used instead in formally describing molecular geometry and its changes. These common meanings have provided a general understanding of 3D polymorphism of molecular shape and its intrinsic mobility. Based on this background, we will define in this chapter a specific usable vocabulary in order to describe the phenomenon precisely.

If one looks back, it can be found that the description of 3D structures is embedded into the firmly established concept of stereoisomerism. It specifies different spatial, but constitutionally equivalent, structures (stereoisomers). Rotamers exemplify that 3D structures (they are called conformers in stereochemistry textbooks) come within the stereoisomerism concept. According to this description of molecular geometry, stereoisomers, which are not enantiomers, are diastereomers. Then, antiperiplanar and synclinal geometries of 1,2-disubstituted ethane derivatives (Fig. 11) are diastereomeric, and their rotational interconversion should be considered as diastereomerization. Correspondingly, the two synclinal rotational forms are enantiomeric and their equilibrium is enantiomerization. Obviously, one may see herein a reasonable description of molecular geometry and its changes.

However, definitions of diastereomers and enantiomers are focused on symmetry relationships between 3D structures, disregarding chemical specificity of different molecules. The stereoisomerism concept may successfully consider mechanical ball-and-stick molecular models instead of the corresponding molecules. In other words, structural specificity and molecular flexibility are ignored in that stereochemistry which considers the molecular geometry of stereoisomers separately from chemical (electron) molecular structure and time.

For instance, from the viewpoint of the stereoisomerism concept, NIR in open-chain amine **12** (Fig. 21, bottom) and ring inversion (RI) of cyclohexane **18a** (Fig. 28) provide similar transformations; for both compounds, they are an enantiomerization. Also, ignoring the circumstance that the same motion, NIR, changes the molecular geometry in amines **12** (Fig. 21) and **14a** (Fig. 28), the molecular symmetry concept considers these stereo-rearrangements as dissimilar; for **14a** it is a diastereomerization. Azacycle

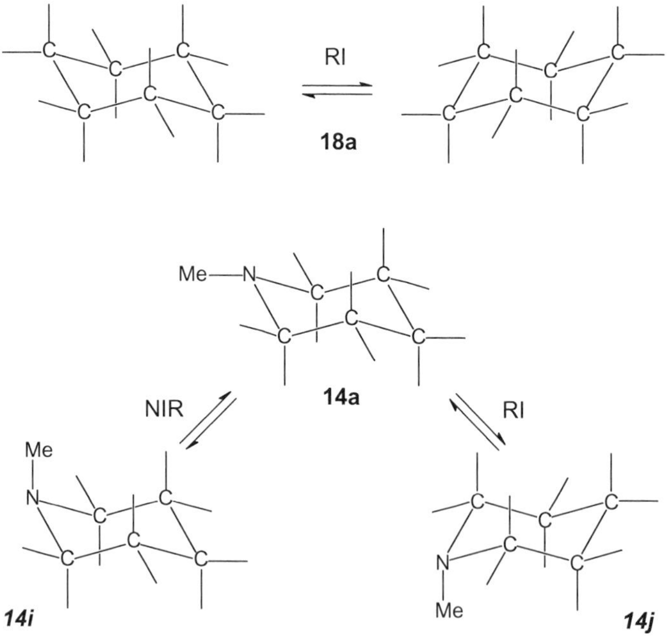

Figure 28. RI (ring inversion) in cyclohexane **18a** as well as RI and NIR in piperidine **14a**.

14a provides an additional example. Two *different* intramolecular motions in this molecule, RI and NIR (Fig. 28) that also differ in their rates, lead to structures *14i* and *14j* of identical geometry. It is obvious that, stating enantiomerization or diastereomerization for these interconversions, we do not make up our mind to understand, e.g., the striking kinetic instability of the masked N tetrahedron in amines *vs.* the incomparably higher stability of the C tetrahedron in hydrocarbons. This formal description of stereoisomer interconversions does not intend to distinguish between internal flexibility of an elementary molecular fragment (here: the N pyramid in **12**) and intramolecular rotational flexibility of interlinked rigid elementary molecular fragments (here: carbon tetrahedrons of the methylene units in **18a**); simply put, it does not characterize molecular flexibility at all. One may say that considerations in terms of classical stereochemistry are rather mathematical (a simple application of group theory) and not chemical: they do not contain any information that could be used to establish the relationship *chemical structure–molecular geometry*, which intrigues us. Thus, for understanding molecular shape and its changes, it is useless to consider molecular symmetry.[a] The phenomenon of variability of molecular geometry — existence of stereoisomers and stereoisomer interconversions — requires a separate formalism to be described in a more informative, "chemical" manner.

At this point, for the first time in this account, we specify the commonly used term *conformational*. This term is a common name of stereoisomers that have a high molecular flexibility (conformational stereoisomers, i.e., conformers) as well as their interconversions (e.g. conformational transformation, conformational dynamics).

Featured molecular properties or chemical behavior may be named *conformational* if tightly associated with the in-time-lability of stereoisomers; e.g. conformational control is a commonly known term. Consequently, this name carries any description or consideration that deals with the in-time-lability of molecular shape, or, more accurately, with such stereoisomers.

Thus, first of all, conformers are stereoisomers. The point is which criterion should be chosen in order to formally single out a subset of conformers from the stereoisomer set. The understanding of the term *conformational*

---

[a]Molecular symmetry is a chemically telling matter if it is assessed in connection with molecular electronic structure (see for Bersuker's rationalization of the pseudo-Jahn-Teller effect[3d]).

presented above is most universal (Sections 2.4 and 2.5), but it is not a single one. Before overviewing all of them (see below), let us touch on the conformational idea by glancing at stereoisomers from the perspective which is unusual for many organic chemists. *Most organic compounds have many stereoisomers that are short-lived molecular structures in the liquid or gas phase at ambient conditions and, therefore, cannot be isolated (and often cannot be detected) at room temperature.* A simple example of cyclohexane (**18a**, Fig. 28) illustrates this. X-ray diffraction demonstrates that this six-membered cycle adopts a chair geometry (Figs. 17 and 28). Axial and equatorial protons of this carbocycle should not be chemically equivalent in ^{1}H NMR spectra. Why do the room temperature ^{1}H NMR spectra of this compound registered by NMR spectrometers of different frequencies show only one resonance signal? The first suggestion is that there is an occasional absolutely precise overlap of the signals of axial and equatorial H's; however, one can reject this hypothesis because of its low probability. The next interpretation is that the ring of **18a** is planar, and all the protons are spatially equivalent. It explains these NMR spectra. It appears that the molecular 3D geometries supplied by X-ray crystallography and NMR spectroscopy for compound **18a** are absolutely dissimilar. However, this contradiction is readily annulled if we assume that the molecule is not static in solution, and two "enantiomeric" chair forms (Fig. 28) rapidly interconvert at room temperature. Then, all the experimental data are in accordance. ^{1}H NMR detects only one signal that belongs to an individual resonance neither of axial and equatorial protons, nor protons of another spatial disposition. It is an averaged signal that results from in-time "mixing" of their two individual resonances, i.e., signals of axial and equatorial protons. Concerning the X-ray results, molecular packing is very tight in the crystal lattice and does not allow chair-shaped molecules of **18a** to flip.

What is interesting in this brief discussion of spectral data for the textbook compound **18a** is that we, not even suspecting *a priori* the existence of its short-lived stereoisomers, "meet" some of them in routine NMR spectra. What is more important in this example is that it introduces to the reader a very rational idea: at ambient conditions, thermodynamically stable organic molecules frequently are a finite set of short-lived interconverting stereoisomers. It is what we call conformational idea.

Identification of molecular structures (geometries) and relative stabilities of stereoisomers from this set has become known as conformational analysis. The consideration of both a finite set of thermodynamically stable stereoisomers (structures with some, even very short, lifetime) and an infinite set of intermediate stereoisomeric structures (thermodynamically unstable structures; they include transition states of conformational interconversions) has become known as conformational theory. Neither of these terms is accurate. Conformational theory does not contain original explanations of intramolecular mobility, which would not follow from contemporary molecular physics or theoretical chemistry. Conformational analysis does not have its own research methods; it borrows methods from physical and theoretical chemistry. Probably, a less pretentious term conformational concept is preferable. Nevertheless, the conformational concept has appeared to be a very convenient integrative description of intramolecular flexibility, and, importantly for organic experimentalists, this simple approach is not "too theoretical." It is indispensable in synthetic organic chemistry in explaining or predicting the regio- or stereochemical outcome of quite different reactions by means of primitive models. Besides, spectral methods usually do not detect short-lived stereoisomeric forms in the gas or liquid phase at room temperature. Therefore, one has to realize that spectral data for substances at these experimental conditions give information not about a single 3D structure. Because of fast stereoisomerizations, they always "present" a set of interconverting stereoisomers[a] (conformers) as an averaged virtual molecular structure. Organic experimentalists exploit various spectroscopes, and, among them, NMR is in daily use. Due to this circumstance, they feel a permanent "presence" of the conformational concept in the laboratory. Even ordinary NMR spectra as, e.g. the above spectral example of cyclohexane, could not be correctly interpreted if ignoring a fast interconversion of conformers.

Probably because of the diversity of their sources, conformational ideas have met "assembling" difficulties during their development. Ever since conformational theory was introduced to organic chemists, its key meanings, *conformer* and *conformation*, have been ill-defined. Notably, because stereochemistry textbooks still present this concept as if it was a finished consideration, this unpleasant situation remains unknown for most non-computational chemists. In fact, one may differentiate three distinct approaches to *conformer* and *conformation* that co-exist in chemical practice. They are: (*1*) the approach that puts focus on internal rotation; (*2*) a purely kinetic approach; and (*3*) a broad approach based on considering molecular potential energy. All three of these approaches are problematic, although to quite different extent. The rotation-grounded view currently rules in the environment of synthetic organic chemists. It has

---

[a] Even if masked by a total predominance of one conformer.

been reasonably criticized by many chemists, but, it is still "alive" and, unfortunately (we will see why), has migrated from old textbooks on stereochemistry to the newest ones as well as to monographs on conformational analysis, chemical dictionaries, and encyclopedias. In the following (Sections 2.1–2.3), we shall consider these three definitions of *conformer* or *conformation*. Each definition is discussed in the related Section, and, there, terms *conformer* or *conformation* carry the sense of this definition.

Let us start from the textbook-supplied approach that regards internal rotation to be of paramount indicativity; for many years, this view is also recommended by IUPAC. "*A conformation is the spatial arrangement of the atoms affording distinction between stereoisomers that can be interconverted by rotations around formally single bonds.*"[31a] Three aspects of this definition, one formal and two "kernel" ones, are worth stressing. First, the term *conformation* means some molecular geometry (the molecular geometry that is related to a rotational stereoisomer) and not the molecular backbone itself. Consequently, the term *conformer* implies a rotational stereoisomer itself. Second, this IUPAC convention deals exclusively with intramolecular rotation. Geometries of other stereoisomeric structures, i.e., the structures that arise from positional exchange of geminal ligands in elementary molecular fragments (Sections 1.7–1.8), are not covered by this definition of *conformation*. Why this limitation contradicts the definition itself as well as the conformational idea is discussed below in this Section.

These restrictions have appeared due to historical reasons. Molecular flexibility was observed for the first time when stable rotamers were detected in ethane derivatives. Further progress in intramolecular dynamics has shown that sufficiently fast rotation may occur not only for fragments connected by a single bond, but also in push–pull olefins and imines, for example. Detection of pyramidal inversion in ammonia and amines as well as pseudorotation in simple phosphoranes were the first indication that rotation is not a single fast intramolecular motion. Nevertheless, most stereochemistry textbooks continue to present conformational theory as was done in the 60s, i.e., as an internal rotation-based concept for single bonds with one formal exception of atropisomerism considered as configurational isomerism. And vice versa, N-inversion in amines is inconsistently introduced in these textbooks as conformational transformation. Fortunately, the fundamental *Stereochemistry of Organic Compounds* by Eliel and co-authors (Ref. 33a) has indicated for a broad readership the problem of this assigning atropisomers (they actually belong to hindered rotation systems) to configurational isomers as well as a more general problem of the IUPAC *conformation–configuration* relationship (see below).

Third, rotation only around formally single bonds is indicated in this understanding of *conformation* and *conformer*. The use of the term "formally single bonds" inserts an obvious uncertainness in conformational concept since the meaning of the bond multiplicity is taken into consideration. This characterization of chemical bonds, originating from some level of chemical bonding theory, has its own limitations, and, here, they migrate to the conformational formalism. For instance, single or double bond character may be arbitrarily attributed to phosphoylides since these compounds actually do not obey the schematic Lewis description of chemical structures. According to the molecular geometry of some ylides,[32a] the ylidic bond may be approximated well by a double bond representing their structure as ylenes (Fig. 29). For other ylides, the single bond approximation is suitable for the ylide bonding. Should the real molecular structure be ignored and such bonding in ylides always be assigned to a "formally single bond"?

At first glance, the answer should be yes. Formally, one can accept all ylides as "ylide" structures; however, this certainly would be misleading as a general approach. If a real chemical structure (i.e., the 3D geometry of the framework of nuclei-in-molecule and the related electron density) is not important for such a consideration, we could also represent the $C=C$ bond in alkenes or the $Si=Si$ bond in disilynes as a ylide structure C^+-C^- or Si^+-Si^-, respectively. It is clear that double bonds completely disappear under manipulation with Lewis structures in this manner; *formally*, one can represent any double bond as a single bond of ylide X^+-Y^-. Obviously, the term "formal single bonds" is meaningless. Although the conformational definition of IUPAC requires distinguishing between simple and double bonds,[a] it actually deprives of this ability.

If the answer is negative, i.e. the "true" chemical structure should be taken into account, other difficulties are unavoidable. As mentioned, the problems come from another source, i.e., the chemical bond formalism. Single and double bonds are qualitative descriptors of chemical bonding and, in many cases, bonds cannot be characterized as either single or double. A broad, continuous variability of chemical bonding in ylides of phosphorous (see above) is the first mentioned illustration of the "chemical" inaccuracy of this

---

[a] For instance, according to this definition, alkenes do not fall under the category of compounds that have rotational freedom.

definition of *conformation*. Rotational stereoisomers of purely ylidic structure P^+-C^- are conformers, according to the rotation-based criterion in defining *conformation*. Should one study every phosphoylide for reaching the ultimate verdict regarding the P–C bond multiplicity in order to reveal whether its rotational stereoisomers are conformers from the viewpoint of the IUPAC nomenclature regulations?

Let us suppose that chemical bonds in question might be formally reduced either to single or to double bonds by means of a hypothetical sophisticated criterion. Even in this case, the situation could occur in which one would have to classify rotational isomers differently for compounds that, in fact, have the same chemical functionality (e.g., such as the above phosphoylides that actually only differ by the relative weight of the ylide and ylene resonance forms; Fig. 29). One group would fall under the rotational definition of *conformation* and the second group would not.

It remains absolutely unclear which molecular structural criterion should separate "formally single bonds" and "true" double bonds. Chemical structures with highly delocalized charge show how arbitrary such a choice may be. The dissipation of electron density in allyl cation (Fig. 29) prescribes no single/double representation of chemical bonding; by "summing" its resonance structures, we obtain a value of 1.5 for the order of each carbon–carbon bond. This middle value may be equally referred either to a formal single or formal double bond.

In this connection, an interesting example is anion $(NO_2)_3C^-$ of nitroform (trinitromethane). At first sight, one may assign the bonds, which bind the nitro groups with the central carbon atom, to formally single ones. The C-associated π bonding is electron-deficient since the lone electron pair of the carbon atom seems dissipated in all three nitro groups (the equalization of three radially positioned bonds proposed with name Y-aromaticity). Then, the order of each C–N bond should not exceed 1.33, i.e., these bonds are near-single. However, the C–N bonds are different in this anion, as theoretical calculations[9,32b] (Fig. 29) and X-ray analysis[32c] show; this desymmetrization is in line with Bersucker's rationalization[3d] of the pseudo-Jahn-Teller effect. The length of one bond is very close to that of single C–N bonds (0.147 nm), while two others have the same length which value is rather near the value of the double C=N bond length (0.132 nm).[a] Hence, there are two different chemical bondings between the carbon and the nitrogens in $(NO_2)_3C^-$, and the order of each shorter bond carbon–nitrogen is near 1.5. If we ignore this essential bond shortening and accept the shorter bonds as formally single, we make an illegitimate assumption. The NMR-detected chemical non-equivalency

---

[a]Remarkably, one can withdraw referring to the Y-aromaticity concept when considering this anion.

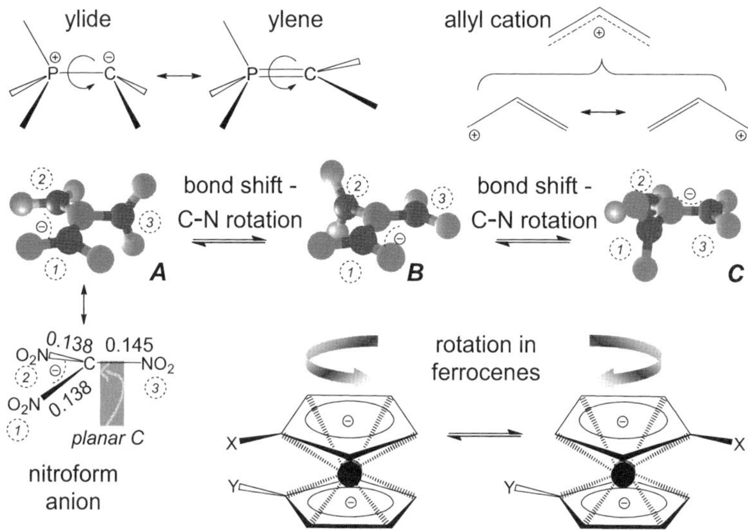

Figure 29. Examples of chemical structures (phosphoylides, allyl cation, nitroform anion, and 1,1′-disubstituted ferrocenes) that cannot be adequately represented by the single/double bond formalism. Arrows indicate rotation around the axes of considered bonds. Theoretically derived geometry and calculated bond lengths[a] (nm) are shown for anion $(NO_2)_3C^-$; numbers in circles are formal labels of the nitro groups. For ferrocenes, the gray ball depicts the Fe^{2+} ion. For each ferrocene ring, the related dotted lines indicate one 6e-6c bond.

of the nitro groups in this anion indicates that identical structures **A–C** (Fig. 29) are chemical isomers (i.e., not stereoisomers), which may be classified as degenerate positional isomers. Note that two nitro groups of $(NO_2)_3C^-$ lie in the CNNN plane of the molecule, while the third NO_2 group is oriented perpendicularly to this plane. Interconversion of structures **A**, **B**, and **C** consists of reorganization of chemical bonding in the Y-shaped fragment CNNN; this bond isomerization is coupled with concerted rotation around C–N bonds.[b] More concretely, this π-tropic isomerization gears π bond shifting and concerted internal rotation. As known, if interconverting thermodynamically stable rotational isomers, internal rotation does not change their chemical structure. Thus, our association of all C–N bonds in nitroform anion with rotational, i.e., purely stereoisomeric, freedom, appears absolutely irrelevant.

---

[a]Calculated at the MP2/aug-cc-pVTZ level;[9] for explanation of acronyms of widely used methods of in computational chemistry, see Chapter 5.

[b]In solution, ^{14}N NMR registers one resonance signal of the nitrogen nuclei of this anion.[32c] Thus, isomers **A–C** rapidly interconvert.

In its exact form, the discussed definition also does not include molecular structures in which internal rotation is not around a bond. For instance, rotation of cyclopentadienyl rings in metallocenes is such a case. Donor-acceptor bonding, e.g. in ferrocenes, is between the sandwiched Fe^{2+} ion and all of the aromatic carbons; six delocalized π electrons of each Cp ring provide it. This bonding obviously cannot be described using the formalism of isolated covalent bonds. One can say that there are two equal multi-electron multi-center (i.e., 6e-6c; Fig. 29) bonds in ferrocenes $(\eta^5\text{-Cp})_2\text{Fe}$. It is meaningless to define rotation around a multi-center bond since the bonded atoms do not lie in a straight line. Ring rotation in $(\eta^5\text{-Cp})_2\text{Fe}$ is around the C_5 symmetry axis, which goes through the centers of the Cp rings as well as the iron atom; it is not around a chemical bond. Nevertheless, the "core" of the traditional definition of *conformer* and *conformation* is internal rotation in molecular backbones. That is, if a stereoisomer may be converted into another one by rotating a molecular fragment around any axis with keeping chemical connectivity for any pair of bonded atoms of the molecule upon this rotation, these isomers informally satisfy the traditional definition of *conformers*.

The problematic condition of single bonds for the conformer-determining rotation has been discerned. Later analogs of the IUPAC definition, which currently predominate in the organic world, exclude any connection with the chemical character of bonds and are more focused on molecular geometry. Probably, the most widespread form has been given by the above mentioned influential *Stereochemistry of Organic Compounds*,[33a] which states *"By conformation of a molecule of given constitution and configuration is meant the rotational arrangement about all the bonds as defined by the magnitude and sign of all pertinent torsion angles"* (see also Ref. 34a for an analogous definition. This definition has given a more strict form to basically the same description of *conformation* that is scattered in different related monographs and textbooks; see, e.g., Refs. 19,34b–g). Rephrasing, we could say that a certain conformation of the molecule is its certain rotational geometry.

A ubiquitously accepted name for thermodynamically stable rotamers is *conformers* (or, as a rarely used synonym, conformational isomers). These 3D structures correspond to the energy minima on the rotational pathway (Fig. 12), and, therefore, their accurate name should be *stable conformers*. Consequently, thermodynamically unstable rotational isomers, e.g. the transition state of rotation, should be called unstable conformers. Stable conformers are a finite subset of an infinite set of rotation-produced stereoisomers, as the rotational plot in Fig. 12 exemplifies. Indeed, the rotational plot is continuous (does not have "lacunas"), and an infinite number of points there indicates that the number of conformers (i.e., stable and unstable together) is infinite. The number of energy minima points (obviously, they correspond to stable rotamers) is finite and even very small when changing the rotation angle from $0°$ to $360°$.

Let us discuss the meanings used in this definition that are essential in the context of *conformer/conformation*. The first is *constitution*. Literally, Eliel's definition (let us call it the traditional definition) only considers the molecular backbone, i.e., nuclei-in-molecule, by requiring its connectivity of being unchanged for conformers. One should realize that this definition, in fact, requires the *chemical structure*, i.e., the chemical bonding (molecule-specific electron–electron and electron–nucleus interactions), of being the same for molecular geometries that are considered as conformations. For instance, by the traditional definition, rotation around a bond is supposed not to affect multiplicity of this bond or interconvert molecular geometries that are related to distinct electronic states. This condition of keeping chemical structure intact is self-obvious: conformers are stereoisomers and, therefore, are chemically identical molecular species.

Nevertheless, this trivial condition in defining *conformation* leads to difficulties when assigning some rotamers to conformers. For instance, E and Z diastereoisomers of olefins satisfy this definition since their rotational interconversion keeps the chemical structure of the molecule unchanged. However, the double bond character is fully lost in the transition state of rotation. By representing this bonding in terms of σ and π orbitals, the carbons are connected in the transition state only by a single bond (a σ bond) because their p orbitals, being orthogonal, do not overlap. We find out that chemical bonding is altered upon rotation around C=C bonds. According to the traditional definition of *conformation*, the geometry of the transition state of rotation is not a conformation of an *alkene*, and the transition state itself is not its unstable conformer, although the chemical connectivity is not lost in this transient molecular structure. Moreover, the double bond character of the C=C bond is smoothly diminished in parallel with moving along the rotation coordinate from the stable rotamer to the transition state. One cannot put a border between rotamers with the "still doubly bonded" carbons and rotamers with these carbons that are "mainly σ-bonded." Should we accept that only the stable rotamers, i.e., E and Z diastereomers, are conformers, while all intermediate 3D structures, which lie between these stable rotamers, are not conformers (i.e., thermodynamically unstable conformers)?

The alkene C=C bond is a simple case; one can discern straightaway which rotamers are problematic for this definition of conformer. However, it exposes a weak point of conformational considerations: should intimate chemical bond concepts be the instrument for establishing conformational relationship of interconverting isomers of non-trivial organic compounds? The above discussed example of nitroform anion (Fig. 29) shows that the condition of unchanging chemical structure which is an obvious prerequisite in conformational concept which requires examination of chemical bonding in rotamers in assigning them to conformers.[a]

In respect to rotation around double bonds, this extended analog of the IUPAC definition is not problem-free. The circumstance that ordinary E- and Z-olefins do not interconvert to any appreciable extent without exciting them photochemically or thermally quickly indicates the problem. In considering intramolecular stereorearrangements, the conformational idea singles out rapidly occurring stereodynamic processes, and organic chemists reasonably cannot accept that, e.g. E- and Z-alkene diastereoisomers should be approached as conformers. Indeed, we do not feel a discomfort when characterizing a fast E/Z diastereoisomerization in many push-pull ethylenes as conformational transformation while consideration of maleates and fumarates as conformers would cause no enthusiasm. It is quite understandable, therefore, why the traditional requirement of rotation around only single bonds has been maintained in organic practice, in spite of the definition of *conformation* given by the mentioned milestone monograph on stereochemistry.[33a] *Stereochemistry* states regarding the *cis*- and *trans*-butene "conformers" that "such a view has not become popular." With some irony, this worldwide known textbook notifies that "the question as to the delineation of single and double bonds is swept under the rug."[33a]

Eliel's traditional definition of *conformation* sometimes adopts an equivalent form, which has recourse to parameters of molecular geometry. Then the usage of diffuse descriptors "arrangement in the space," "rotational

---

[a]In daily research routine, the situation of meeting this "chemical" condition is implied when considering any stereoisomers. As we see, such a practice may lead to naming isomers incorrectly.

arrangement," "rotation around bonds," etc. is avoided. *Conformation* of molecular fragment $(\mathbf{L}_i)_n\mathbf{A}_1-\mathbf{A}_2(\mathbf{L}_j)_m$ is represented by a set of all torsion angles $\mathbf{L}_i\mathbf{A}_1\mathbf{A}_2\mathbf{L}_j$ for the given configurations of elementary units $\mathbf{A}_1(\mathbf{L}_i)_n$ and $\mathbf{A}_2(\mathbf{L}_j)_m$. A set of conformations of all fragments $(\mathbf{L}_i)_n\mathbf{A}_1-\mathbf{A}_2(\mathbf{L}_j)_m$ of a molecular backbone is its conformation. According to this definition, stereoisomeric structures with the same configurations for all elementary fragments have distinct conformations, if the corresponding torsion angles for at least one fragment $(\mathbf{L}_i)_n\mathbf{A}_1-\mathbf{A}_2(\mathbf{L}_j)_m$ are not equal (including the angle signs).[33a,34g] In other words, only a change of torsional angles including inversion of only their signs indicates conformational transformation. By this quantitative description, a conformation of a system is a set of all torsional angles of the structure with a fixed configuration for each elementary molecular fragment.

The traditional and related conformational definitions contain the condition of keeping *configuration* of elementary molecular fragments unchanged; thus, the next (to *constitution*) essential meaning in the traditional definition of conformation is *configuration*. Under *configuration*, the discussed Eliel's definition of *conformation* implies "*the arrangement of the atoms in space of a molecule of defined constitution without regard to arrangements that differ only by rotation around one or more single bonds, provided that such rotation is so fast as not to allow isolation of the species so differing.*"[33a] We need some comments on this statement. First, *configuration* herein is a synonym of a certain rotation-invariant geometry of a molecule or a molecular fragment. Second, this definition, nevertheless, firmly links *configuration of the molecule* (i.e., some molecular geometry) with the kinetics of rotation by requiring this motion to be fast. That is, *fast rotation* is an additional condition in defining *configuration* of a *molecule* according to Eliel: only stereoisomers produced by fast internal rotation have the same configuration, while slow rotation affords *configurational* isomers (*atropisomers*) of the molecule. Remarkably, this configurational approach considers internal rotation as a really occurring intramolecular motion. In contrast, many chemists, in accepting rotation as the criterion for distinguishing configurational and traditionally understood conformational (rotation-provided) stereoisomerizations, understand under this term a certain symmetry operation and not as an intramolecular rotational motion of either kinetics (see Section 2.2 for the discussion followed). In other words, they consistently consider *configuration* without mixing geometrical and kinetic determinants.

Concerning *configuration of an elementary molecular fragment*, Eliel's approach, of course, has no rotation to consider, and it returns to operating with only geometrical parameters regardless of the kinetics of the fragment stereoisomerization. Herein, signs of bond angles combined into a set serve as a configuration determinant:[33a,34g] a set of signs of bond angles $L_i AL_j$ determines configuration of ligands $L_1, L_2, \ldots, L_n$ in monocentric unit $A(L_i)_N$. Or, simply speaking, a set of positive and negative signs of the bond angles of an elementary molecular fragment describes its configuration.

For instance, any ligand L_c may be selected as a pivot, if all other ligands $L_1, \ldots, L_i, \ldots, L_N$ ($n \neq c$) lie in the plane, which is perpendicular to the pivot bond $A_1 - L_c$ and includes the central atom A, and/or beyond this plane. These ligands $L_1, \ldots, L_i, \ldots, L_N$ ($n \neq c$) may be supplied with formal ranks in respect to the arbitrary chosen ligand L_i of highest rank (Fig. 30). The choice of the highest rank ligand is arbitrary because ranking of geminal ligands according to any structural criterion may appear impossible — the ligands may be identical, e.g., $L_i = L_j = H$. The rank of spatially neighboring ligand L_j, which lies in the clockwise order of ligands L_i and L_j, is defined as being lower; the rank of the next ligand is lower than the rank of L_j, and so on. The clockwise order for any two neighboring ligands (e.g., L_j and L_k) in the *initial* geometry may be associated with the positive sign of angle Ω_i between them (bond angle Ω_i; $0 < \Omega_i < 180°$); for the counterclockwise order, the sign is negative.[a] In order to "+/—label" *all* bond angles of the fragment and rank them (i.e., to include bond angles $L_c AL_i$ and $L_d AL_i$, where L_c is a pivot ligand, and bonds $A - L_c$ and $A_1 - L_d$ lie in the same axis), one should perform such an assignment by selecting an additional pivot ligand. The chosen ranking of all ligands is unchangeable for the considered elementary molecular fragment, i.e., the ligands keep their ranks in stereorearrangements of the fragment. In this way, we actually supply a certain relative arrangement of all ligands $L_1, \ldots, L_i, \ldots, L_N$ (a certain configuration) in elementary molecular fragment $A(L_i)_N$ with a $N!/2(N-2)!$-membered set of all "+/—labeled" bond angles Ω_i.

---

[a] Also, the choice of the angle sign is arbirtrary. With the same success, one may associate the clockwise round direction with the negative sign.

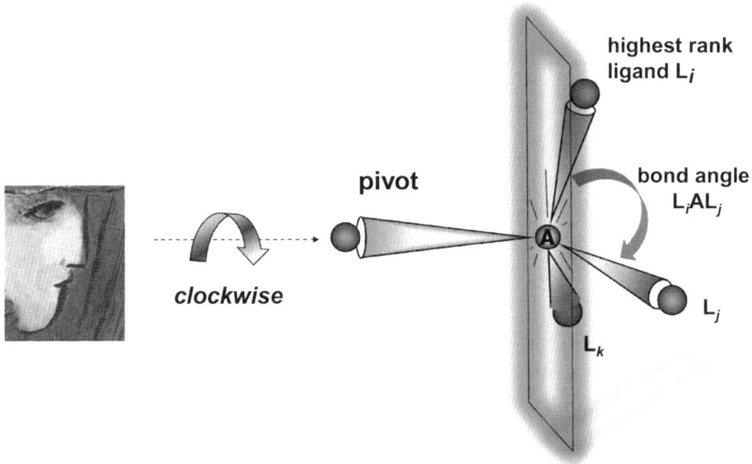

**one may associate going from atoms
L$_i$ and L$_j$ around the pivot clockwise
with the positive sign of angle L$_i$AL$_j$**

Figure 30. Establishment of the sign for a bond angle. Ligands L_i, L_j and L_k are beyond the plane which passes through atom **A** and is perpendicular to the pivot bond (i.e., are to the rear). For simplicity, a tetrahedral elementary molecular fragment is shown.

This description of configuration is impractical. First of all, the sign of a bond angle L_iAL_j in elementary molecular fragment $A(L_i)_N$ is not predetermined by its chemical or spatial structure. It may be defined following any reasonable rule that unambiguously relates to the bond angle either the positive or the negative sign. However, without employing a computerized algorithm, it is somewhat burdensome to supply all 10 or 15 bond angles with positive or negative signs in a stereoisomer of a five– or six–ligand elementary molecular unit, respectively, and to perform the same procedure for the stereoisomer with configuration in question (note that stereochemical nomenclature reasonably does not provide any recommendation for such a formal assignment).

A stereomutation of ligands inverts signs for several or all bond angles Ω_i. Let us suppose that ligands L_i and L_j (Fig. 30) have exchanged their positions. It is transparent that, keeping the same pivot, one should invert the round direction in order to keep the established ranks of these ligands. This means the inversion of the signs of angles L_iAL_j, L_jAL_k, and L_kAL_i, i.e., the initial configuration is reversed.

These somewhat ponderous definitions differ from the alone standing accurate definition that exclusively deals with individual elementary molecular fragments and considers them in clear terms of ligand permutation. *Configuration* of elementary molecular fragment $A(L_i)_n$ is a permutation-variant spatial arrangement of geminal ligands $L_1, L_2, \ldots, L_n$ in this molecular unit, where A is the central atom, $L_1, L_2, \ldots, L_n$ are fragments directly bonded to A, n is an integer ($n > 1$); ligands $L_1, L_2, \ldots, L_n$ may be different or identical. Note that the overall shape of a chemical polyhedron or polygon (Figs. 1 and 18) is supposed to be unchanged after positionally permuting the ligands. Thus, a certain 3D arrangement of ligands relative one to the other in a chemical polyhedron or polygon is what is understood as *stereochemical configuration*. The mechanism of ligand permutation (geminal stereorearrangement or internal rotation) is irrelevant when defining *configuration* in this manner; in other words, this definition permits an elementary molecular unit to alter its configuration (to permute the ligands) by any intramolecular motion(s). The permutation-based definition probably supplies organic chemists with the most convenient and strict description of stereoisomerism of elementary molecular fragments; it is further called the stereochemical definition of configuration.

According to the stereochemical definition of configuration, any chemical polyhedrons and polygons possess configuration; it may be changed via permutation of two ligands. This is illustrated by N-inversion in trigonal N-centered units, e.g., in imines $R^1(R^2)C=NR$ or azo compounds $RN=NR$. For these structures, the permutation of the N substituent and the N lone electron pair is "visible" even to the untrained eye since this stereorearrangement is a diastereomerization (Fig. 32).[a] In organic practice, chemists sometimes lend stereochemical configuration to trigonal units (Ref. 34f), and, sometimes they do not (Ref. 34a), specifying nonetheless heterotopic stereofaces (*Re* and *Si*) in a prochiral trigonal unit.

Stereochemical configuration is not a synonym to the 3D geometry of an elementary molecular fragment. If no ligand permutation occurs in such a fragment, but its geometry is changed as a result of some stereoreorganization of the molecule, this change does not affect the configuration of this

---

[a]Configurational inversion is NMR-undetectable if it interconverts elementary molecular fragments which are not diastereomeric as, e.g., invertomers of amines $MeN(Cl)Me$, *tert*-$BuN(Cl)Me$, and NMe_3 (Fig. 27) or telluranes R_2TePh_2.

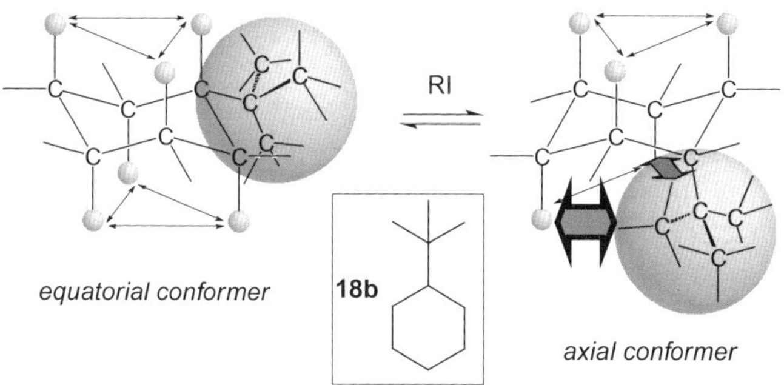

Figure 31. 1,3-Diaxial steric interactions in the equatorial and axial chair-shaped conformers of cyclohexane **18b** (the axial chair is shown of being turned by 120° relative to the equatorial structure). Small grey spheres indicate axial protons. Bold arrows depict steric interactions with the *tert*-Bu group (shown as a large grey sphere), while thin arrows indicate steric interactions between 1,3-positioned axial H substituents. Due to 1,3-diaxial steric interactions in the axial conformer, this bulky substituent somewhat deviates out of the ring in this conformation.

fragment. A proper illustration is *tert*-Bu cyclohexane **18b** (Fig. 31). While 1,3-diaxial steric interactions are not essential in the equatorial chair of **18b**, these interactions of the *tert*-Bu group with other ring substituents (protons) are significant in the axial conformer of chair geometry because of a large bulkiness of this ring substituent. These steric interactions force the exocyclic C–C bond to deviate from a "perfectly axial" orientation. As a result, the C-1 carbon tetrahedron is distorted in the axial conformer, in contrast to its shape in the equatorial conformer. In other words, the $C_{ring}C_{ring}C_{tert-Bu}$ bond angles are not equal for the equatorial and axial conformer. According to the stereochemical definition of *configuration*, changes in angle values are ignored, and we conclude that there is no configurational change upon RI in cycle **18b**.

Also, internal rotation in trivial bicentric molecular fragments $(\mathbf{L}_i)_N\mathbf{A}_1$–$\mathbf{A}_2(\mathbf{L}_j)_M$ is geometrically characterized solely by a change of torsion angles $\mathbf{L}_i\mathbf{A}_1\mathbf{A}_2\mathbf{L}_j$. Bond angles $\mathbf{L}_i\mathbf{A}_1\mathbf{L}_j$ and $\mathbf{L}_i\mathbf{A}_2\mathbf{L}_j$ in rotating molecular units $\mathbf{A}_1(\mathbf{L}_i)_N$ and $\mathbf{A}_2(\mathbf{L}_j)_M$, respectively, are altered to some extent because vicinal ligands $\mathbf{L}_i$ and $\mathbf{L}_j$ "repel/attract" each other differently at different distances. Nonetheless, configuration of such fragments is considered unchanged because ligands, which belong to the same unit $A_1(L_i)_N$ or $A_2(L_j)_M$ do not interchange their positions.

Another general interpretation for *configuration* (also recommended by IUPAC), in parallel to the *Stereochemistry* definition,[33a] has penetrated into the stereochemical literature (see, e.g., Ref. 34a). This IUPAC-recommended definition also tries to separate rotational and non-rotational stereoisomerism. It may be formulated as following: configuration is a 3D arrangement of atoms in stereoisomers that are not conformationally (i.e., rotationally) related. Similarly to the Eliel's definition, this *configuration* is related to the molecule or its fragment; however, herein, configuration is secondary to conformation. As we remember, the conformational statement in *Stereochemistry*[33a] is for "a molecule of given constitution and configuration..." It implies that the meaning *configuration* has been settled before, or, in other words, the term *conformation* is secondary to *configuration*. Hence, from the viewpoint of Eliel's *configuration*, the IUPAC's *configuration* is meaningless.[a] From the viewpoint of the stereochemical definition of configuration, both of these two definitions are restricted due to consideration of only a certain molecular flexibility (geminal flexibility of elementary molecular fragments).

A separate and to some extent mathematical concept of *configuration* has been proposed by frontier Moscow chemists[35a−c] in the end of the research prosperity period in the USSR. As basic meanings, it uses *conformation* in the sense of the broad approach (Section 2.4) as well as chirality. The main qualitative feature of this approach is that it relates configuration to only chiral systems. In this aspect, this *configuration* is similar to the canonic *absolute configuration* of chiral molecular fragments that is described by the Cahn-Ingold-Prelog rules (Ref. 33a and references therein). One may say that this concept is "too mathematical" for employing it in today's organic laboratories; in any case, it has remained unused.

A quite irrelevant usage of the term *configuration* may be encountered in publications on physical or structural chemistry. By a configuration, a given 3D arrangement of atoms in a molecule or a structural molecular fragment is meant there (see, e.g. Ref. 3d). This *configuration* is actually synonymous to *molecular geometry*. This interpretation does not have any relationship to *configuration* as it is understood in stereochemistry. As explained, stereochemical *configuration* means a certain distribution of ligands in the apexes of a chemical polygon or polyhedron (Figs. 1 and 18).

---

[a]The IUPAC definition of *configuration* is, of course, in line with the its own approach to *conformation* (see above), which does not mention configuration at all; rotation is the only point.

After systemizing our knowledge of *configuration*, we can analyze the traditional definition of *conformation* in more detail.

2.2 Why Rotation-Associated Definitions of Conformer and Conformation are Unsuccessful

We remember that there are two rotation-focused definitions of *conformation*. The traditional definition of *conformation* resorts to meaning *configuration,* and the IUPAC definition does not mention configuration. It is reasonable to start our analysis of approaches to *conformation* from the former definition, a somewhat obsolete convention that is presented by Eliel's formulary[33a] of stereochemistry basics. As we have learned in the previous Section, Eliel's *"Stereochemistry"*[33a] considers *configuration* of a molecule as a distinct rotation-invariant molecular geometry if rotation is fast, and any distinct, rotation-variant spatial molecular geometry if rotation is slow. The IUPAC approach to *configuration* is appreciably different and is incompatible with Eliel's *configuration*. Therefore, it is reasonable to consider Eliel's *conformation* (i.e., the traditional conformational definition) as coupled with Eliel's *configuration*. In considering them, we only refer to the main part (that does not include atropisomerism) of this meaning of *configuration* — the rotation-invariant, but ligand permutation-variant geometry of elementary molecular fragments (what we call stereochemical configuration).

At first glance, the traditional conformational definition provides an accurate criterion for assignment of stereoisomers to configurational or conformational isomers. In stereoisomeric structures in question, one should compare the corresponding bond angles (i.e., their signs) and torsional angles. As stated above, opposite signs of the same bond angles in stereoisomers indicate a configurational relationship between their fragments. Dissimilar values of torsion angles for the same molecular fragments together with retention of configuration of all elementary molecular fragments show a conformational relationship between stereoisomers. Thus, there is an unconditional separation between conformational and configurational transformations. Conformational transformations (rotations around chemical bonds) do not change configuration, and alteration of configuration (changes of bond angle signs) does not generate a new conformation. Does

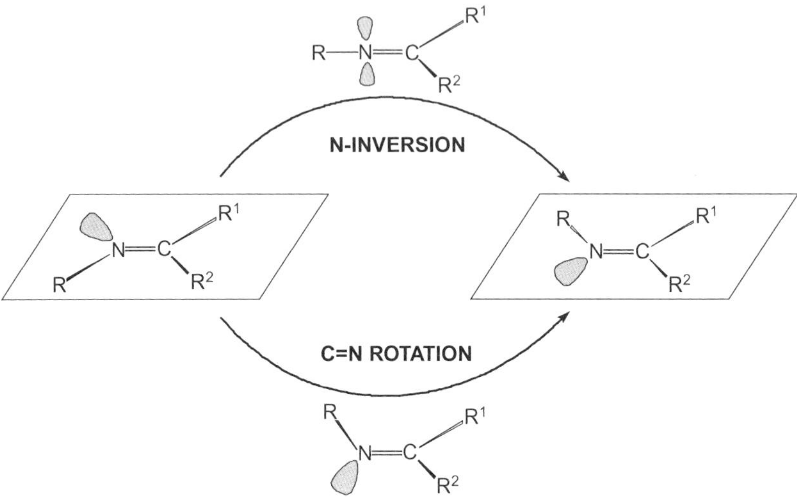

Figure 32. Two pathways of diastereomerization of imines. Arrows indicate two alternative intramolecular motions.

the reader suspect that there is a striking conflict between these definitions? An unexpected obstacle is that, for some structures, internal rotation leads to the 3D geometry that also results from a configurational change.

Imine derivatives R(R)C=N–X are a vivid example. For instance, N alkyl imines $R^1R^2C=N–Alk$ undergo diastereomerization changing the orientation of the N substituent in the plane of diagonal (virtual trigonal) N-centered fragment C=N–Alk (Fig. 32).[36a–c] Geometrically, this stereomutation may be provided by reflection through the plane that is perpendicular to the plane where vicinal substituents lie or by rotation around the C=N bond by 180°.[a] That is, both the geometrical transformations lead to the same structure. Are the initial and final structures conformers or configurational isomers?

We have an obvious contradiction here. As explained, a formal positive or negative sign may be attributed to bond angles in order to distinguish related configurational isomers.[33a,34g] From this point of view,

---

[a] If considering real molecules, this diastereomerization may occur as inversion of the nitrogen (in-plane rearrangement through the transition state of the C_S symmetry; Fig. 32) or as rotational rearrangement of vicinal substituents through a non-planar transition state.

configuration of the imine nitrogen is changed due to inversion of the sign of the CNC angle (in this consideration, the olefinic carbon is the pivot). Or, in equivalent terms, configuration of the nitrogen in the initial and the final imine structures (Fig. 32) is different because two ligands (ligand R and the N lone pair) exchange their positions in the trigonal N-centered elementary molecular unit. Thus, we can reasonably conclude that these stereoisomers should be approached as configurational isomers (here: invertomers) and not as conformers. On the other hand, rotation around the C=N bond by 180° interconverts the considered diastereoisomers. Thus, these rotamers, according to the Eliel's conformational definition, are conformers (they have different $RNCR^1$ and $RNCR^2$ torsional angles) because internal rotation does not change the Eliel's *configuration*, according to the definition. But it appears that the N configuration is rotation-variant in imines!

It appears that our question, whether we deal with conformational or configurational isomers, has no answer. The 3D structure of these interconverting stereoisomers indicates that a configurational relationship exists between them, while the possibility of interconverting them by means of rotation demonstrates conformational relationship (in the sense of the traditional conformational definition). One cannot eliminate this contradiction by arguing that the relationship is determined by the interconversion pathway, i.e., by answering the question how the real diastereomers of N-alkyl imines are interconverted, by N-inversion or by C=N rotation. First, the traditional definitions of *conformation* and *configuration* only consider geometries of stereoisomers themselves and are not concerned with any transition state (i.e., do not consider both the pathway of geometrical changes and the kinetics in any pathway) of stereoisomer interconversion. Second, both intramolecular dynamic processes, though of quite distinct rates, take place in imines.[a] This means that the same real diastereomer appears at one moment as a rotamer and the next moment as an N-invertomer. It is transparent that neither conformational definition would suggest classifying a stereoisomer sometimes as a conformer and

---

[a] In-plane N-inversion is a fast intramolecular motion in imines and may be detected by spectral methods. The related C=N rotation has a much higher kinetic barrier.

sometimes as a configurational isomer. Similarly, rotation in C=C compounds affords the structure that is equivalent to a diastereomer with an inverted olefinic carbon.

Here, we make a few brief remarks regarding methodologies to track the real movement of nuclei-in-molecules upon stereomutations of vicinal or geminal ligands. It is obvious that stable stereoisomers can be detected and studied experimentally. And what about stereoisomeric structures of unlimitedly short live time, e.g., stereoisomerization transition states? Experiments cannot "display" the nuclei-in-molecule framework at any moment of the stereoisomer interconversion. On the other hand, 3D rearrangements take a finite, although a very short time period, time to occur, say, in the range of 1–10 picoseconds[36e] (ps; 1ps = 10^{-12} s). This circumstance enables ultrafast, time-resolved spectroscopes of femtosecond time resolution to indicate the trajectories of ligand movements.[36d,e] How is this possible? Transient structures of unlimitedly short time of existence (transition states, i.e., structures that correspond to a saddle point in the PES; Fig. 43) cannot be detected, this is indeed impossible. However, transformation of any transition state takes some finite time, and ultra-fast laser spectroscopies may track in-time-evolution of such a molecular structure by using two short electromagnetic pulses separated by a 3–5 femtosecond time period and monitoring, e.g., the absorption of the sample. Then, dependence *absorption–time (fs)* reflects dependence *geometrical parameter (selected bond angle or interatomic distance)–time (fs)*. The latter dependence actually represents the trajectory of intramolecular motion of nuclei that occurs during several picoseconds. Such experiments are not routine, and interpretation of experimental data is often supported by theoretical calculations.

Therefore, when identifying the intermolecular motion that provides the 3D rearrangement of interest, it is much easier to undertake theoretical modeling (Chapters 4 and 5). There, the starting point is that any two 3D frameworks of a molecule in principle may be interconverted via different arbitrary trajectories of atom movements; these movements may be of geminal or vicinal substituents. Geometrically, we could interchange ligands, moving them in arbitrary non-intersecting trajectories that only deform the framework of chemical bonds and do not elongate them to unacceptable lengths. These geometrical trajectories of the ligand movement can be modeled by energy calculations of the corresponding transient structures. Trajectories, which are represented by modeled structures of low energy, indicate real (high probability) intramolecular motions.

Thus, experimental and theoretical studies are capable of identification of the ligands (vicinal or geminal) that change their relative disposition in the interconversion of concrete stereoisomers. Such identifications do not establish or disestablish any conformational formalism.

This *conformation/configuration* mess is not limited to compounds containing double bonds. Even if we follow the definition of *conformation*, which only considers rotation around single bonds, we cannot

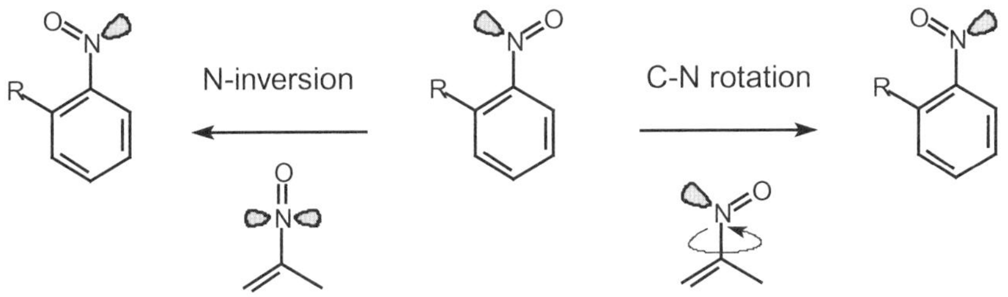

N-inversion

C-N rotation

Figure 33. C–N Rotation and N-inversion in nitrosobenzenes.

unequivocally relate a pair of stereoisomers of some compounds to conformers or configurational isomers. An easy-to-understand example of nitrosobenzenes illustrates this (Fig. 33). Rotation around the C–N bond and in-plane N-inversion afford the same stereoisomeric structure leading to the same dilemma — are the starting and final structures conformers or configurational isomers? Internal rotation around a *single* bond inverts the sign of bond angle CNO. Also in this case, the traditional criterion that separates conformers and configurational isomers — changes of torsion angles or inversion of the sign of bond angles, respectively — is controversial. Changes of torsion angles (values) and bond angles (the sign) may appear geared.

The systems shown in Figs. 32 and 33 actually represent trigonal or masked trigonal elementary molecular fragments (i.e., these fragments are planar; Fig. 1). However, the traditional *conformation* may expose its inconsistence when considering some masked chemical polyhedrons. The hydroxyl group is a O-centered virtual tetrahedron that does not have lone pairs in the plane of the C, O and H nuclei (Fig. 34). It is clear that the O–H bond may be oriented distinctly relative to the rest of the molecule. There are two pathways for the reorientation of an O–H bond in alcohol ROH without dissociating the bond (Fig. 34). The stereoisomeric structures may be interconverted by means of trivial rotation around the C–O bond. The second possibility consists of inverting the O-centered virtual tetrahedron, i.e., of occurring cuneal inversion of the O-centered, elbow-shaped backbone. The initial and resulting structures have two alternative configurations since the formal sign of the COH angle is different in these structures (ligand R may be considered as a pivot). On the other hand, rotation around the C–O bond provides interconversion of these stereoisomers.

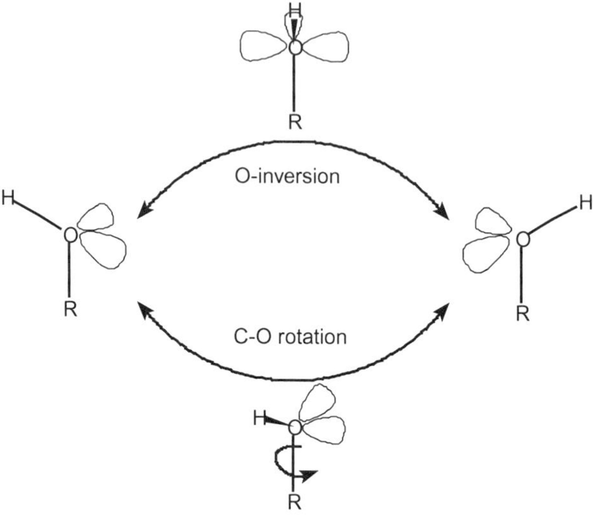

Figure 34. C–O rotation and O-inversion in alcohols.

Again, there is a definition conflict: are they conformers or configurational isomers?

Difficulties in the usage of the terms *conformation* and *configuration* have been discussed by many organic experts. Not vainly, the stereochemical "mentor" of organic chemists, Eliel-Wilen-Mander's *Stereochemistry of Organic Compounds*,[33a] which is cited multiple times in this chapter, characterizes the difference between these stereochemical meanings as subtle. Sometimes the feebleness of the traditional definition of conformation is explicitly exposed. For instance, ignoring its own definition, *Stereochemistry* suggests that pyramidal N-inversion be considered arbitrarily either as a configurational or as a conformational transformation. A recent didactics of stereochemical basics[31b] reports that IUPAC has reached some consensus regarding *conformation* by broadening the related definition as follows: "Some authorities extend the term (conformation — *A. M. B.*) to include inversion at trigonal pyramidal centers and other polytopal rearrangements." As we will see (Sections 2.3 and 2.4), this acknowledgment is a step forward from exclusively considering rotamers as conformers. Prompting the next step, the authors refer[31b] us to their own, well-known, understanding of conformers (see Section 2.3); this alternative approach actually is an endorsed conformational idea.

What makes *configuration* and *conformation* (understood in the traditional sense) indistinguishable in some cases? *The main problem of the traditional conformational definition consists of selecting a certain symmetry operation (proper rotation) as a criterion for conformation.* Proper rotation and some other symmetry operations are equivalent (lead to the same structure), if applied to structures of some symmetry. This is the reason why rotation cannot be a universal criterion in the classification of intramolecular stereoreorganizations. For instance, equilateral triangles with apexes A, B and C, in addition to a proper C_3 axis, have three proper C_2 axes as well as three S_1 improper axes; the C_2 and S_1 axes are mutually orthogonal. Proper rotation for 180° (symmetry operation C_2^1) and improper rotation for 360° (symmetry operation S_1^1) are equivalent for this simplest polygon, i.e., they lead to the equal relative disposition of A, B and C in the triangle plane.[a] This means that, if these symmetry operations are performed for a planar trigonal N-centered fragment of imines, where the C_2 axis is coaxial with the C=N bond, and the S_1 axis is perpendicular to the plane of this double bond, the same geometry of fragment $RN=C(R^1)R^2$ is the stereochemical result. *Stereochemistry*[33a] indicates exactly this definition problem when characterizing the distinction between conformation and conformation as subtle.

Mathematical meanings of symmetry operations are actually the basis of the traditional definition of *conformation*. The reader probably remembers that, applied to molecules, symmetry operation is a formal transformation of molecular geometry that leads to the same shape and the same orientation of a molecular system. Considering symmetry operations for elementary molecular units, we mean herein the positions of ligands in a chemical polygon or polyhedron (Fig. 1) and not ligand themselves, since these geminal substituents can be different. We can recall that there are five symmetry operations for chemical polyhedrons and polygons: (1) reflection with respect to a plane; (2) inversion with respect to an inversion center, i.e., coordinates x_i, y_i, z_i of all N atoms are transformed into coordinates $-x_i$, $-y_i$, $-z_i$; (3) $n-$fold rotation around a proper axis $2\pi/n$ ($n = 1, 2, 3, \ldots$); (4) improper rotation, i.e., n-fold rotation around a proper axis $2\pi/n$ ($n = 1, 2, 3, \ldots$) followed by reflection through a plane that is perpendicular to the rotation axis; and (5) identity operation $\mathbf{E}$, which

---

[a]For generalization, ligands A, B and C are considered as chemically or stereochemically different substituents. Therefore, in applying a symmetry operation here, it is not suggested that it transforms the molecular geometry into an identical one.

transforms the structure as a whole in the structure of the same geometry and orientation. This formalism is important for organic chemists since it describes symmetry relations in all polytopal rearrangements of chemical polygons or polyhedrons, and, thus, provides practical understanding which nuclei are chemically equivalent in NMR spectra of flexible molecular systems.

Of course, symmetry operations are formal manipulations; when applied to molecular structures, they do not describe trajectories of the ligand movement. Nevertheless, they are indicative. In formally establishing conformational relationship in the sense of the rotation-focused definition, we compare molecular geometries only for the stereoisomers in question and ignore the mechanism of how ligands interchange their relative spatial positions in the real molecule. As explained, certain n-fold rotations and other certain symmetry operations may appear equivalent for polygons as, e.g., they are in the case of trigonal N fragments of imine derivatives. Therefore, for intramolecular 3D rearrangements, the type of position-interchanging ligands, vicinal or geminal, cannot be always identified from when considering molecular geometries of stereoisomers. That is, molecular geometries of interconverting stereoisomers do not always indicate the type of ligands (vicinal or geminal) that interchange their spatial positions. This inference is actually fatal for the traditional definition of *conformation* where the rotation condition (the requirement of permuting *vicinal* substituents) plays a key role. This definition has no generality! Applying it, one evidently cannot draw a conclusion about the conformational relationship between stereoisomers of compounds from many chemical classes: any common molecular geometry (Fig. 1) is spread to structures of different elemental composition as well as diverse chemical functionality.

Referring to the related definition to *stereochemical configuration* (permutation of ligands in the elementary molecular unit), one should notice that it presupposes rotational freedom. Not specifying *conformation* in its own definition, this *configuration* is primary to the traditional conformational definition. Therefore, the failure of the latter does not affect this understanding of *configuration*. Configurational difference for stereoisomers or molecular fragments is established ignoring whether or not formal internal rotation interconverts them. For instance, from the viewpoint both of Eliel's *configuration* and the *stereochemical configuration*, rotation around the C=N bond inverts the nitrogen configuration in imines (Fig. 32).

Note that, starting from Section 2.3, only the stereochemical meaning (i.e., ligand permutation in the elementary molecular fragment) is implied for *configurational* changes.

And what about the IUPAC definition of *conformation*? It has no configuration/conformation conflict of definitions: stereoisomeric structures, which are not conformers, are called configurational isomers. It is difficult to consider this approach to *conformation* as more successful. Selection of internal rotation as a criterion of conformational relationship is burdened with the same problem — some stereoisomers may be "connected" via both rotation of an elementary molecular fragment as a whole and ligand permutation in this fragment.

What is not surprising is that the rotation-based approach to *conformation* has survived mostly in textbooks; they, of course, have not criticized it on many counts. Let us indicate some of these drawbacks.

(1) This approach is not adequate for describing molecular flexibility. If we recall the conformational idea only which states that most organic molecules are flexible at ambient conditions, we realize how accidental this association *fast stereorearrangements–internal rotation* is. Internal rotation is not a unique intramolecular process that lets the shape of an organic molecules vary with time (Section 1.7). There is no reason why an intermolecular stereorearrangement associated with positional exchange of geminal ligands cannot occur rapidly. Indeed, geminal stereorearrangements of many chemical polygons and polyhedrons, e.g. positional interchange between equatorial and apical ligands via Berry pseudorotation in tellurane **17b** (Fig. 25) or flipping of the molecular bowls of corannulene **4f** via synchronous inversion of pyramidalized carbons (Fig. 68), are fast at ambient conditions: their room temperature NMR spectra do not show resonance signals of individual stereoisomers. Why should this non-rotational flexibility (i.e., internal mobility of elementary molecular units) have a status that differs from the traditionally defined conformational flexibility?

(2) The traditional definition of conformation does not treat rotation-including concerted motions rationally. Relative movement of geminal substituents and rotational movement of vicinal substituents may be concerted (i.e., occur synchronously) during an intramolecular stereorearrangement. Let us have a quick look at simple examples of geared intramolecular dynamic processes. X-ray diffraction and molecular

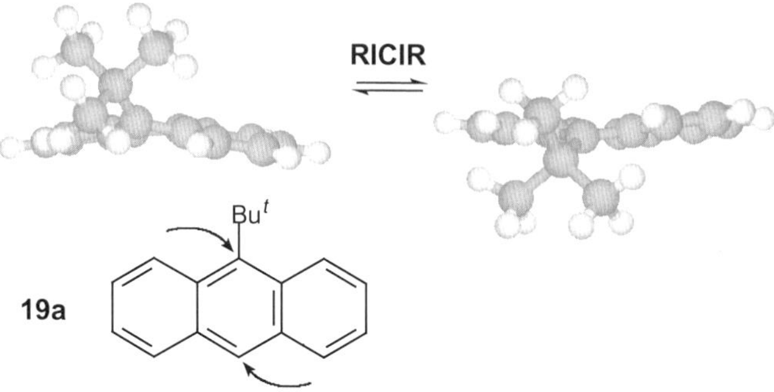

Figure 35. Stereoisomers of antracene **19a** are interconverted by gearing three intramolecular motions: rotation of the *tert*-Bu group, inversion of the central aromatic ring, and inversion of the configuration of two carbons (indicated by arrows). The descriptive name of the integral motion *ring inversion–carbon inversion–rotation* may be abbreviated as RICIR.

modeling have shown that the molecule of antracene **19a** (Fig. 35) is not planar, and the two central carbons are pyramidalized.[36f] Thus, two identical stereoisomers of **19a** may interconvert (Fig. 35).

If we formally (i.e., geometrically) perform either rotation of the *tert*-Bu group, or inversion of the central, non-planar, ring, or inversion of the configuration of the two pyramidalized carbons for one of the stereoisomeric structures of **19a**, none of these geometrical manipulations provides the stereoisomeric counterpart. If we "merge" these three mechanistic motions into one concerted motion or execute them one after the other, we succeed to "interconvert" these stereoisomers. As shown, these intramolecular motions in **19a** are concerted in solution.[36f]

According to the stereochemical definition of *configuration*, these stereoisomers of compound **19a** fall under the *configurational isomers* category. There is a gap between the conformational idea in considering internal rotation and the rotation-based definitions of *conformation*. The idea consists of a firm association of fast internal rotation (i.e., a certain real intramolecular motion) with meaning of *conformational*. The definition consists of a firm association of a C_n symmetry operation (i.e., a certain formal geometrical transformation) with meaning of *conformational transformation* under the condition of unchanging stereochemical configuration. However, although neither C_n symmetry operation "connects" the considered enantiomers of **19a**,[a] the C_3 rotation of the *tert*-Bu group is involved in the stereoisomer-interconverting concerted intramolecular motion, equally to the configurational inversion. The traditional definition of *conformation* excludes from conformational consideration the rotational component (i.e., the really occurring internal rotation by 120°) of the *concerted* intramolecular motion. That is, this formal viewpoint that considers only the initial and ultimate molecular geometries ignores the occurrence of a fast internal rotation which, in pair with configurational inversion, directly interconvert these geometries. The idea that links any *fast internal rotation* with *conformational transformation* obviously is more insightful than the idea (realized in the traditional definitions of *conformation*) that suggests an arbitrarily restricted bundle *symmetry operation (rotation)–conformational transformation*.

Diastereomerization *cis–trans* in qunolizidine (**19b**, Fig. 36) is a similar example. Both structural parameters, the bond and torsion angles, are changed significantly by NIR in alkylamines (Fig. 21, top). Preserving the integrity of the molecular skeleton, inversion of the nitrogen configuration changes *all torsion angles* of one of the rings of bicycle **19b** by 120°. In terms of dynamic stereochemistry, this reorganization of molecular geometry occurs via a coupled intramolecular motion,[36g] a concerted ring inversion–N-inversion (RINI).

As in the previous example, the *cis* and *trans* stereoisomers of **19b** should be related to configurational isomers. On the other hand, both individual

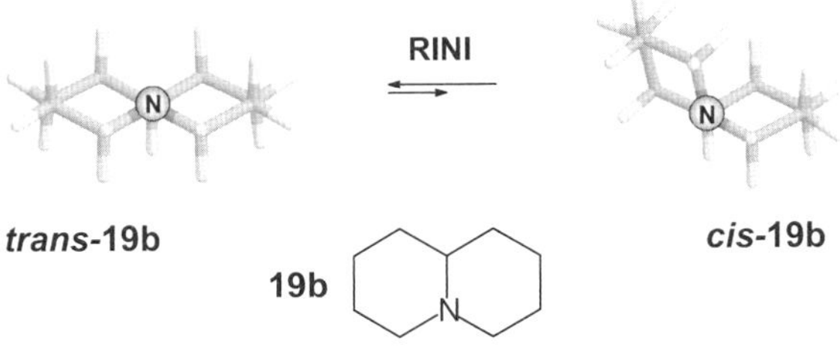

Figure 36. *Cis–trans* transformation in qunolizidine **19b** occurs via a concerted process *ring inverison–nitrogen inversion* (abbreviated here as RINI).

---

[a] Simply put, they are not rotamers.

motions are *equal* components of the resulting motion since the inversion of the N fragment and the inversion of one of the piperidine rings occur synchronously; stereoisomers *cis*-**19b** and *trans*-**19b** cannot be interconverted by only inverting the configuration of the N fragment and then inverting the ring: the chemical connectivity would not allow this stereoisomerization to occur stepwise. The traditional conformational approach adamantly links meaning of *conformational transformation* to formal rotation of molecular fragments around the bonds that connect them. A concerted restricted rotation around ring bonds is a real component of the stereodynamics in **19b**. It would not be rational to ignore occurrence of this rotation and not relate interconversion of *cis*-**19b** and *trans*-**19b** to conformational transformation, relying on the reason that these stereoisomers have opposite configurations for the N unit.

Concerted intramolecular motions are not news in stereochemistry. However, as we see, the rotation-associated definitions do not comprise these non-rare stereoisomerizations. Indeed, "internal rotation has haunted conformational analysis as an ill fate."[13]

(3) The rotation-associated approach discomforts the use of the terms conformer and conformation. Suppose there is a compound that is capable of configurational changes. For each configuration, internal rotation generates a set of stable conformers. In other words, each configuration is "surrounded" by a set of related conformers. As the definitions require, conformers from different sets are configurational isomers, i.e., they do not have a conformational relationship.

How uncomfortable this hierarchy is may be seen from the example of piperidine **14b**.[a] In Fig. 37, a part of stereodynamics of this azacycle is shown; this scheme[37] represents interconversions of all twist-shaped conformers of **14b**.[b] The intramolecular motion involved in this pseudorotation is a restricted concerted rotation around endocyclic bonds.

Note that this pseudorotation cycle is not an adopted copy of the pseudorotation model for cyclohexane. The latter is the fairly known basic model of interconversion of twist forms in saturated six-membered cycles (Fig. 53; this conformational model can be found

---

[a]In synthetic research, compound **14b** is well-known as a non-nucleophilic proton scavenger.

[b]For rings, such a cyclic sequence of conformational transformations is called pseudorotation (do not confuse this term with pseudorotation in trigonal bipyramids, Section 1.7)

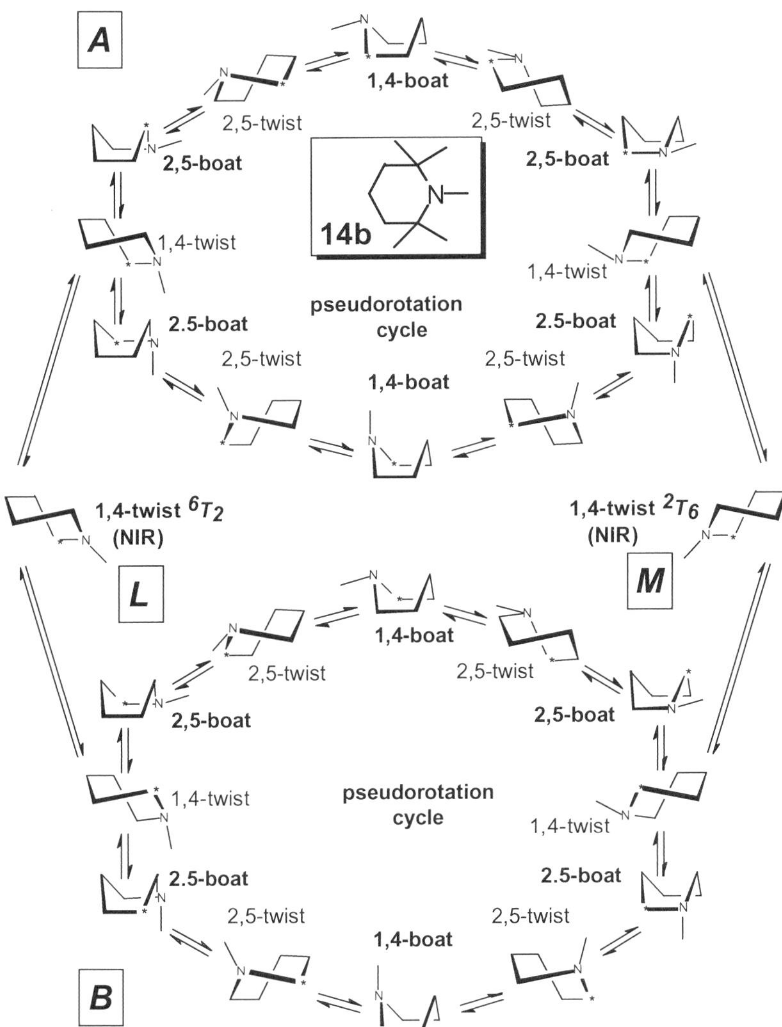

Figure 37. Thermodynamically stable and unstable conformers of piperidine **14b** (α-methyl groups are not shown, the asterisks are formal labels). Trivial names of conformers are given in bold for transition states; *A*, *B*, *L* and *M* depict distinct N configurations. The pseudorotation cycle represents only a part of piperidine stereodynamics.

in monographs on stereodynamics[38]). The structures of stereoisomers of six-membered ring **14b** have been modeled by means of theoretical calculations.[37] Situated so successfully, we can discuss these 3D structures with confidence.

The scheme shows that there are two formal sets, A and B, of structures of opposite configuration of the nitrogen. Each set is formed by structures of the same configuration and includes twist (stable stereoisomers) as well as boat (intermediate, unstable stereoisomers[a]) forms. That is, in the frame of the rotational definitions of *conformation* and *configuration*, the members of set A are in conformational relationship. Similarly, set B comprises conformers (twist forms) and thermodynamically unstable conformers (boat forms). As structures of opposite configuration, conformers of the first set and conformers of the second set do not have a conformational relationship. Sets A and B are "connected" via two NIR transition states $^2T_6$ and $^6T_2$ of identical geometry (two thermodynamically unstable 1,4-twist conformers with a planar nitrogen, i.e., two NIR transition states).

Thus, there are conformers from set A and conformers from set B. Remarkably, transition states $^2T_6$ and $^6T_2$ belong neither to A nor to B. The signs of their bond angles CNC in a planar N fragment cannot be defined meaningfully since its geometry in $^2T_6$ and $^6T_2$ is exactly "in between" the geometries of this fragment in the related interconverting stereoisomers of *opposite* signs of the CNC angles. One may suggest that configurations of $^2T_6$ and $^6T_2$ and configurations of stereoisomeric structures from set A or B are distinct, and structures $^2T_6$ and $^6T_2$ form two one-member sets L and M of conformers that are not conformationally related to conformers from A or B. Thus, we have accepted that there are four sets of conformers (in the sense of the traditional conformational definition), which are in a conformational relationship only inside of their own configurational set. This conclusion is not changed if *conformer* and *configurational isomer* are understood in the light of the IUPAC recommendations. Note that piperidines are not conformationally or configurationally muddled up objects; they have only one elementary molecular fragment (the N-centered unit) that is capable of non-hypothetically inverting its configuration. If there were n $(n > 1)$ centers that changed their configurations, the number of distinct sets of conformations, would be increased as 2^n. Conformers from such a set are

---

[a]They are transition states in the shown twist–twist interconversions.

not conformers for conformers from any set among the other $2^n - 1$ sets. Does not such a picture look terrible?

(4) The rotation-associated approach is problematic from the viewpoint of kinetics. As we remember, internal rotation in different chemical compounds is characterized by kinetic barriers of essentially different height. One could recall toluene as well as ferrocene (Fig. 29; X = Y = H): internal rotation in these compounds is almost absolutely free. The kinetic barriers are lower than 1 kcal mol^{-1}, and this value is of the order of the energy of molecular vibrations. On the other hand, it is worth casting a glance at tetra-*ortho*-phenylene (a [8]annulene, Fig. 38), a cyclic system with alternate single and aromatic bonds. The barrier of ring inversion (RI) is estimated[36h] to be 75.8 kcal mol^{-1}; this value is only 9–10 kcal mol^{-1} lower than the typical value of dissociation energy of the C–C bond (~85 kcal mol^{-1}). Clearly, the range of rotation barriers is enormous.

The barrier value for this [8]annulene is supplied by theoretical calculations;[36h] even at very high temperatures the rate of RI is so slow that it cannot be measured. Geometrically, concerted rotation around single bonds of the central eight-membered ring interconverts stereoisomers of this cycle. The rotation-based definitions permit consideration

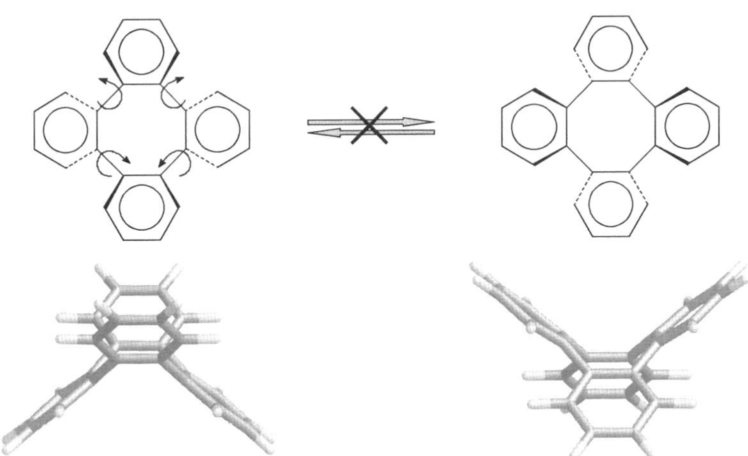

Figure 38. Tetra-*ortho*-phenylene: RI as a concerted disrotatory rotation (top); 3D structures of stereoisomers (bottom).

of single bonds in molecular frameworks of any steric crowding, of any chemical composition. Should the undetectably slowly interconverting stereoisomers of this compound be considered as conformers, just as formal rotamers of toluene were? Eliminating the main conformational idea of rapidly interconverting stereoisomers, the rotation-based definitions of *conformer/conformation* give an absurd answer that the rotamers are conformers in both cases. This equalization means that *fast* stereoisomerizations are not classified as a separate group, and, thus, we do not need some specific term *conformer*. Stereochemical term *rotational stereoisomer (rotamer)*, which exclusively concerns molecular geometry and does not pertain to its kinetic stability, appears sufficient within the limits of the rotation-based definitions.

We certainly need a description that includes the conformational idea of molecular shape that rapidly changes in time. From analyzing the rotation-related conformational definitions, it is clear that such a description should not be based on considering molecular geometry. The next few sections discuss these alternative approaches to delineating flexible molecular shape.

2.3 An Alternative Approach to Conformers: Kinetic Barriers

The elusiveness of conformers for organic chemists during the first years of the development of conformational analysis adversely affected their understanding. It was an intuitive view of many experimentalists that if stereoisomers cannot be isolated because of their fast interconversion, they are conformers.

This diffuse understanding has been transformed into accurate terms of kinetic barriers. If a low kinetic barrier or a set of only low barriers separate two stereoisomers (Fig. 39), they are conformers;[39a,b] consequently, stereoisomers with a high interconversion barrier are classified as rigid stereoisomers. The criterion that singles out *conformers* from *stereoisomers* obviously is purely kinetic. As we know, the rotation-related approach to *conformer/conformation* ignores the rate of stereoisomer interconversion (Section 2.2). This "insusceptibility" to fast stereoisomerizations is removed in the kinetic barrier approach.

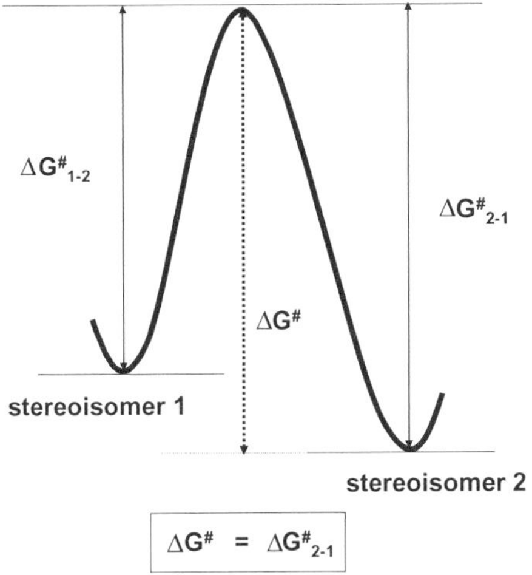

Figure 39. Kinetic barrier $\Delta G^{\#}$, which separates two stereoisomers at equilibrium. The barrier height $\Delta G^{\#}$, which characterizes the interconversion rate, is the highest from the two half-barriers.

However, this description of molecular flexibility has its own weakness; it is an arbitrariness in deciding which value(s) of the kinetic barrier should separate fast and slow stereoisomerizations. For instance, different observation methods, e.g., IR, NMR, ESR are characterized by distinct timescales (Section 3.3, Table 1), which, of course, are much shorter than that of chemical experiment time. A spectral method cannot register individual molecular species, if their conversions occur in shorter time periods than the method timescale is. Therefore, by "viewing" individual, but rapidly interconverting stereoisomers, it would be an absolutely arbitrary choice to select the timescale for defining *conformers.*

The stereoisomerization barrier of 80 kJ mol^{-1} ($\sim$19.4 kcal mol^{-1}) with small, arbitrary chosen, deviations was proposed as a border between conformers and rigid stereoisomers.[39a] At room temperature, compounds with this and lower barriers of stereoisomerizations are actually mixtures of stereoisomers in the equilibrium that cannot be separated because they too rapidly interconvert at these conditions. A minimal barrier for stereoisomer interconversion should be, say, of 22–23 kcal mol^{-1} at room temperature, in order to provide a possibility of isolating them at this temperature as individual isomers. However, the time of existence τ (lifetime[a]) of an individual equilibrating component depends not only on the kinetic barrier, it is also determined by ambient physical conditions (in the first instance, by the temperature; Section 3.5.3, Eq. 14) and the medium (solvent). Thus, stereoisomerization barriers are not an absolute criterion for *conformer*. It appears that assignment of a stereoisomer to the conformer or to the rigid stereoisomer depends on physical conditions in the environmental medium.

This additional circumstance diminishes the convenience of the kinetic barrier as a single conformational descriptor. For instance, the lifetime τ is 1.1×10^9, 667 and 0.01 s for each of two stereoisomers of equal molecular geometry that interconvert passing the 20 kcal mol^{-1} barrier (83.6 kJ mol^{-1}; for simplicity, we suppose that this barrier is temperature-independent) at 200, 300 and 400 K, respectively. Should these stereoisomers be classified as conformers or not? Or, among *ortho*-substituted diphenyls, should chiral binaphtyl ligands for asymmetric metalorganic synthesis, e.g., $M-$ and P-BINAP (Fig. 40), be considered as conformers at highly increased temperatures and as rigid stereoisomers under ambient conditions? The kinetic and the related (Section 2.4) approaches to *conformers* provide no answers; therefore, a voluntary limit for τ is introduced in Section 2.5. Besides, the kinetic approach considers only stable stereoisomeric structures relating them to either conformers or rigid stereoisomers,

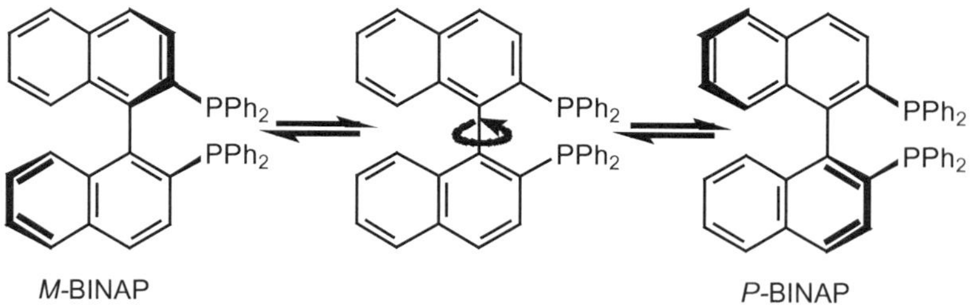

PPh$_2$
PPh$_2$
PPh$_2$
PPh$_2$
PPh$_2$
PPh$_2$

M-BINAP

P-BINAP

Figure 40.   Interconversion of enantiomeric BINAP ligands.

---

[a]The lifetime for a state, τ (s), which is interconverted with kinetic rate constant k (s^{-1}), is defined as k^{-1}. Note that while kinetic rates are significantly changed with the temperature (Eq. 14), kinetic barriers (term $\Delta G^{\#}$ in Eq. 14), as a rule, are weakly temperature-dependent.

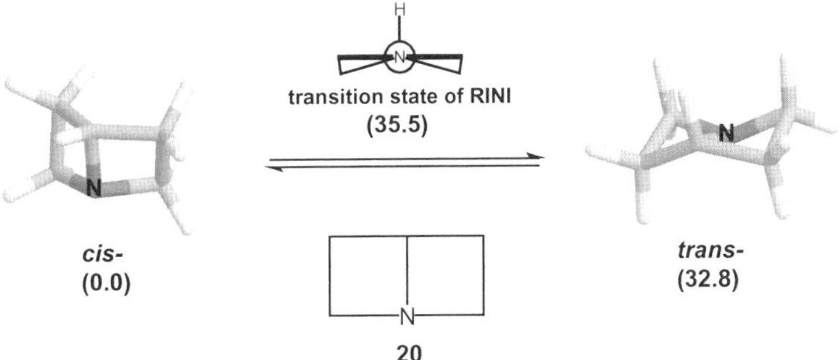

Figure 41. RINI in 1-azabicyclo[2.2.0]hexane **20** (*ab initio* optimized geometries are shown). Calculated relative energies (kcal mol^{-1}) for these structures under normal conditions are given in parentheses. We could not conclude without an experiment (in this case, a computational one) that the *cis* and *trans* stereoisomers are not conformers from the viewpoint of the kinetic approach.

while transient 3D structures of unlimitedly short lifetime (in the first instance, transition states of stereoisomerizations) are in no way classified.

Note that, because of ignoring molecular geometry in assigning stereoisomers to conformers, the kinetic barrier approach cannot characterize stereoisomeric structures as conformers or as rigid stereoisomers. Until the interconversion barrier is determined or reasonably estimated, the conformational relationship between stereoisomers will remain unknown.

For instance, a non-trivial structure (N-fusion of two small rings) of azabicycle **20** (Fig. 41) does not permit, even roughly, prediction of the barrier of RINI in this amine. Therefore, within the limits of this conformational approach, *cis* and *trans* isomers of bicycle **20** cannot be treated as rigid stereoisomers or as conformers if only considering their molecular structure. Only since a 35.5 kcal mol^{-1} barrier was provided by *ab initio* calculations for the *cis/trans* transformation of this azabicycle,[36f] the *cis* isomer may be supplied with rank *rigid stereoisomers* while the *trans* stereoisomer may be classified as a conformer.

Nevertheless, this kinetic description of *conformers* is essentially more successful than the rotation-considering one. Not looking for any characteristic geometrical feature in various 3D structures of stereoisomers (that, of course, does not exist), the kinetic barrier approach lends a universal character to describing non-static molecular shape: it focuses on molecular flexibility itself. For instance, all stable twist-shaped stereoisomers of polymethylated piperidine **14b** (Section 2.2, Fig. 37) are conformationally related (are conformers) since the barriers of their interconversion are low. Considering in light of the kinetic barrier approach, this example well illustrates the conformational idea: many rapidly interconverting stereoisomers in the thermodynamic equilibrium — this is what most organic molecules are in non-solid phases at ambient conditions.

2.4 The Third Alternative: The Broad Conformational Approach

Unfortunately, this purely kinetic approach has not been widely distributed; nevertheless, it can be viewed as the first, though unpolished, version of a solid approach to *conformer*. The next version represents the meaning of *conformer* as it is understood by theoretically-oriented chemists. They also have tried to reflect the 3D mobility of chemical structures very explicitly, without separating rotational and non-rotational 3D rearrangements. In this formalism, only relationship *molecular geometry–molecular energy* is assessed, and, therefore, this description may be called the broad conformational approach. It operates with energy terms and, linking molecular geometry and molecular energy, specifies stereoisomeric structures either as conformers or as rigid stereoisomers.

We know that any molecular geometry is characterized by a point in the potential energy surface (PES; Section 3.1). Due to this deep link, the broad conformational understanding[13,40a,b] is very simple but pithy. *Conformer is the molecular structure that corresponds to a point in the PES.* This definition actually contains an additional, "refining" condition. A single PES characterizes a certain electron state of a molecule; that is, molecular geometry is electron configuration-variant. Therefore, when considering molecular geometries related to a certain PES, we consider the molecule in a certain electron configuration. When discussing a PES in this account, we will imply that it is of the molecule in the electron configuration of the non-degenerate ground state.[a]

---

[a]Strictly speaking, it is insufficient to include into any chemical consideration only the ground state of the molecular system. Molecules react not only of being in this lowest energy state. Because many

Probably, the reader remembers that n electrons may be distributed in the same set of orbitals in different ways, and that each distribution of these n electrons is termed electron configuration. Since different orbitals usually differ in energy, the total molecular energy depends on this arrangement of electrons, among other factors. The ground state does not exclusively represent a molecule, because a set of different electron configurations must be attributed to this molecular system. In this light, it is obvious why any physically meaningful definition of *conformer* must take electron configuration into account: a certain electron configuration is characterized by a certain PES. Unequal energies correspond to identical geometries of a molecule of different electron configurations. This means that the molecular geometries of conformers (energy minima) or transition states (energy maxima) in the ground state are not necessarily the molecular geometries of conformers or transition states, respectively, in excited states with either electron configuration (i.e., a singlet or triplet exited state).

Consequently, structures that correspond to energy minima are stable *conformers*. Following this approach, we should relate, e.g., *cis* and *trans* stereoisomers of amine **20** (Fig. 41) to *conformers*. The circumstance that, due to a very high barrier of their interconversion, the *cis* isomer is utterly kinetically stable at room temperature, does not affect the assignment. Thus, by the broad conformational approach, an infinite set of geometries, which are permissible by chemical connectivity of bicycle **20**, i.e., conformations, represents the molecular shape for its predetermined electronic state.

At the beginning of conformational analysis, Barton proposed that *conformations* are 3D arrangements of atoms in a flexible molecule that cannot be superimposed. From the geometrical point of view, this definition seems to be an analog of the above approach. However, it disregards electron configuration as well as distribution of molecules into vibrational levels in each point of their PESs (see below). It is more suitable for abstract 3D frameworks than for real molecules.

At present, *conformation* is often understood as a 3D geometry of a molecule in an unlimitedly short time moment; a predetermined electron configuration is implied (see, e.g., Ref. 34g). Metaphorically, this description likens *conformation* to a "flash hologram" of a changing molecular skeleton, with unlimitedly short duration of the "flash." *Conformer* is defined similarly, under the condition that its 3D structure is unchanged in a finite time period. Clearly, these descriptions do not take into account the vibration level-related distribution that blurs the 3D structure in the "flash hologram" (see below). Nevertheless, this understanding of *conformer* and *conformation* is very near the broad conformational approach, and takes more and more room in research practice of organic chemists.

---

reactions (e.g., carbene-involving, photochemical) occur via excited states, molecular geometry in these states is important for fruitfully assessing chemical reactivity.

This description of flexible molecular shape covers all possible stereorearrangements in organic molecules. In defining the conformational relationship, it does not require auxiliary stereochemical descriptors (symmetry operations) for 3D structures or any limits for kinetic barriers of their interconversion. As indicated, it does not specify certain intramolecular motions, conferring upon them the status of conformational transformations. For example, similarly to the kinetic barrier approach, all stereoisomeric structures of piperidine **14b** shown in Fig. 37 are considered as conformationally related. Opposite configurations of the nitrogen do not disturb this convenient ranking of all stable stereoisomers of cycle **14b** as conformers.

Organic chemists are used to assessing *conformer* (also stereoisomer) as a single, case-specific, 3D molecular backbone. However, many molecules (understood as a statistical molecular ensemble) of the *same* stereoisomer cannot be characterized by a single 3D geometry. Molecular vibrational energy has a noticeable impact on molecular geometry, and this energy is different for molecules of the ensemble. Note that the vibrational energy cannot be fully taken out of molecules even at 0 K.

In order to understand this non-trivial stereochemical aspect, we *should* consider an abstract adiabatic minimum of energy for the ground electronic state of a molecule. Recall the "topography" of this minimum. For real systems, the potential well is non-symmetrical, but, for the sake of simplicity, we can approximate it with a symmetrical well of parabolic form (Fig. 42); the abscissa (coordinate r) indicates the molecular geometry, and the ordinate (coordinate *energy*) shows potential (i.e., non-kinetic) energy of the molecule. The r_0 value corresponds to the so-called equilibrium geometry, i.e., the geometry that is characterized by the minimal energy value of the energy well. Any energy minimum in the PES (Section 3.1) has vibrational and rotational levels of energy; their number is infinite. Only the levels can be populated, i.e., energy cannot have a value that is "between levels."[a] The allowed values for vibration energy are $v_0, v_1, v_2, \ldots, v_i, \ldots$ (shown as frequency values). Quantized values of potential energy that appear due

---

[a]This a manifestation of a fundamental physical principle of energy quantization. Even the values that can adopt the energy of rotation for a molecule rotating as a whole are split to rotational levels.

to rotation of the molecule as a whole (in spectroscopic terms, rotational levels) are inherent to each vibrational level.

This description presents an "empty" abstract potential energy well (an unpopulated energy minimum in the PES) illustrated by the energy diagram in Fig. 42. It shows which energy can have a molecule if it has the geometry (described by generalized geometrical parameter r in this figure) that corresponds to the location r_0 (in the PES) and width $r_0 \pm \delta r$ of the potential energy well. In other words, there is no single molecular geometry inherent to the energy minimum; instead, there is some range for varying molecular geometry without leaving the minimum (the potential well).

Now, let us "fill" an abstract minimum of energy with many molecules, or, more strictly, suppose that an ensemble of chemically identical molecules has adopted the geometries required by an energy minimum. An obvious assumption is that the molecules will occupy different "floors" (i.e., energy levels) when appearing in this energy minimum (Fig. 42, top). Which "floors" will be more populated? Concretely, what is the law that distributes molecules into energy levels v_0, v_1, v_2, ..., v_i, ... in the well of potential energy?

When recalling the basics of chemical thermodynamics, we can guess the answer: in a statistical ensemble, molecules, which occupy identical energy minima, are distributed into vibration energy levels v_0, v_1, v_2, ..., v_i, ... according to Boltzmann statistics. This means that statistical weights of unexcited molecules, which occupy the *i-th* and *j-th* levels, are described by Eq. 1,

$$p_j/p_i = (g_j/g_i) \times \exp(-\Delta E_{j-i}/k_B T) \tag{1}$$

where p_i and p_j represent the occupancy of the *i-th* and *j-th* levels, respectively, g_i and g_j represent the level degeneracy, ΔE_{j-i} is the energy difference for these two levels, T is absolute temperature (K) and k_B is the Boltzmann constant (3.298×10^{-24} cal/K $\times$ mol). Transparently, for vibration energy levels of low energy, $g_i = g_j = 1$ (with increasing the numbers i and j for vibrational levels v_i and v_j, the latter get closer to each other, i.e., these energy levels become near-degenerate; see Fig. 42 for a typical potential energy well of real molecules). Note that another partition function (another statistical distribution) may be only applicable to excited states of molecules.

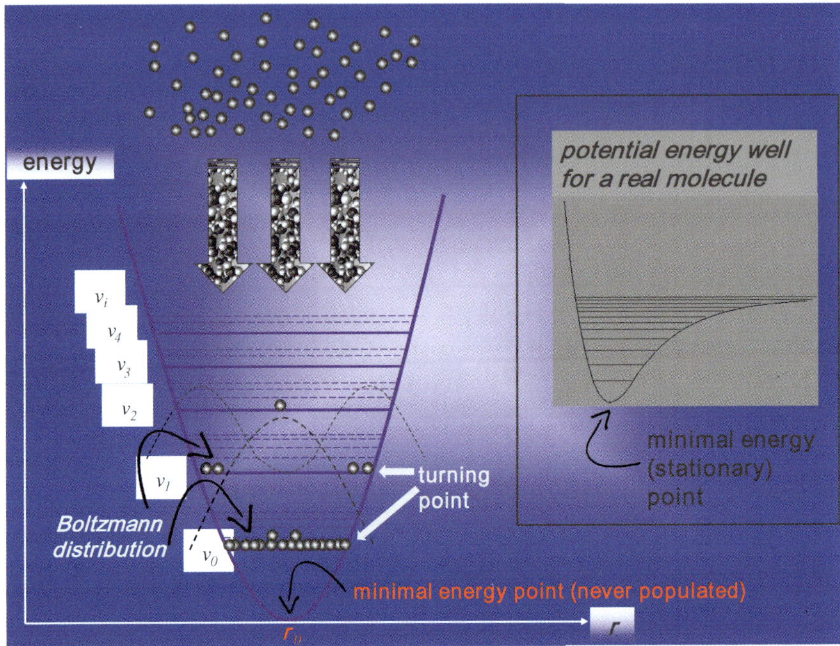

Figure 42.   Schematic representation of a minimum of molecular potential energy (left; r is a generalized geometrical variable; the r_0 value corresponds to the equilibrium geometry) for a hypothetical molecule. Solid horizontal lines indicate levels of vibration energy v_i, dashed horizontal lines show rotational levels (note that the number of vibrational and rotational levels is infinite for real systems). Dashed curves indicate (probability)$^{1/2}$ for different geometries inside the potential energy well for the molecule at the v_0 and v_1 levels. Molecules (shown as small balls) are distributed into energy levels according to Boltzmann statistics (Eq. 1).

This is a very useful equation for molecular research. Spectral experiments supply values ΔE_{j-i}. Using Eq. 1, we can reveal relative populations of molecules that occupy the *i-th* and *j-th* levels of energy. For instance, the occupancy of the first vibrational level v_1 (i.e., the second in energy) of organic molecules is $\sim$1% at 298 K. It also indicates that molecules do not fully lose their energy of vibrations at any low temperature.

This partition function demonstrates that molecules always occupy all energy levels, and that the relative occupancy of the levels is a function of the temperature. It shows that the minimal energy point (Fig. 42) is not populated at all; the first populated level is ν_0, and it is highly occupied ($100\% > p_0 > 98\%$) at room and lower temperatures. Nevertheless, some occupancy takes place for all other vibrational and rotational levels: as Eq. 1 shows, the higher the temperature, the greater the occupancy of the higher energy levels. In other words, there is some residual occupancy of the higher levels. This occupancy leads to the increase of amplitudes as well as frequencies of backbone oscillations in the classical model of molecular oscillations. This provides an opportunity to cite a prominent Cramer's monograph[33b]: "... one may think of the system as a cloud hovering over the PES, with the density of the cloud thinning as it rises according to Boltzmann statistics." Thus, molecules, which populate energy minima as a statistical ensemble (i.e., not falling to the lowest energy level ν_0 of these wells all together), actually cannot be characterized by a single energy value or, consequently, by a single 3D structure. What actually is *molecular geometry* of such a cloud of molecules (Boltzmann molecular ensemble)? Certainly, one should make this important point clear.

First, under *static molecular geometry*, one should understand some virtual geometry that is attributed to molecules statistically distributed into vibration and rotation energy levels of a minimum of potential energy. This virtual geometry is a molecular 3D structure that arises from averaging the geometries of molecules occupying these vibration and rotation energy levels, proportionally to statistical weights in the occupancy of these levels of potential energy. Notably, experiments deal with molecular ensembles and, for an individual stereoisomer, supply us with exactly such a statistically averaged geometry (see below).

Second, one cannot associate a certain molecular geometry with a vibration or rotation energy level of the energy minimum of a molecule. It is clear from Fig. 42 that, at the i-th level of vibration energy, the molecule may adopt any geometry, which is within the range of the width $(r_0 \pm \delta r)_i$ of the potential well for this level.[a] According to the Heisenberg uncertainty principle, in no way we are able to reveal which geometry the molecule adopts when possessing a certain vibration energy. Otherwise, one would know both spatial coordinates for the nuclei of this molecule and its energy;

---

[a]There is some low probability for a molecule that has the i-*th* vibration energy to adopt the geometry which is outside the geometry range $(r_0 \pm \delta r)_i$ of the potential energy well for this i-*th* level (Fig. 42).

realization of such a possibility contradicts this fundamental physical principle. One has to discuss the geometry of molecules, which populate an individual level, only in terms of probability. For instance, the probability for adopting either geometries at the zeroth vibrational level v_0 as well as the first vibrational level v_1 is schematically shown in Fig. 42 for a symmetrical well. For molecules occupying the v_0 level, the equilibrium geometry r_0 has the maximal probability. In contrast, for molecules that populate the v_1 level, this geometry has the minimal probability, and the geometries that correspond to the turning points, have the maximal probabilities. In considering real molecules, the description of the relationship *maximal/minimal probability–molecular geometry* is less concrete. The distribution of this probability is asymmetric for any vibrational level in wells of significantly distorted symmetry that describe potential energy of these molecular structures (Fig. 42), and its exact form is difficult to derive.

So, static molecular geometry cannot be mechanistically understood as a geometric construction of chemically bonded atoms. Concerning stereoisomers, such a structure — a 3D structure resulted from averaging 3D geometries of molecules distributed into all energy levels of the energy minimum — is a *stable conformer*, in the sense of the broad conformational approach. This "phantom" geometry (*conformation*) is the geometry of *stable conformer*. Thus, by this approach, *thermodynamically stable conformer is a statistical ensemble of molecules, which are distributed into vibrational and rotational levels of a minimum of the potential energy surface*. A similar statement would be problematic if considering one or a few molecules. One cannot assign a certain geometry to a non-statistical set of molecules that occupy an energy minimum. The probability of possessing any allowed energy within this energy minimum and, thus, adopting any corresponding geometry is uncertain for one or a few molecules.

Let us ascertain which molecular geometry is provided by experiments. The physical tools of organic chemistry (e.g., NMR, ESR, X-ray diffraction, conventional IR, or UV spectroscopies) never register the spectral response of a population of molecules that is associated with only one level of vibrational energy, even if the level occupancy is changed by selectively exciting selected level(s). Experiments sample molecular ensembles in the solely possible way — they supply chemists with a virtual 3D structure, which results from averaging molecular structures that are statistically distributed into all vibrational and rotational levels. It is obvious that molecules from different vibration and rotation levels contribute to the averaged (virtual) structure according to the weighting in level occupation. Thus, when dealing with a stereoisomer that is rigid during the observation (e.g., during the time period of an external electromagnetic pulse and followed relaxation of molecules), the experiment provides information exactly regarding the molecular structure that is *stable conformer* from the viewpoint of the broad conformational description.

Many organic chemists have not realized that any structural study provides exactly a virtual (averaged) structure. Probably, they even would prefer not to use this "inconvenient" term. There is no reason to escape the term *virtual structure* in descriptions of molecular geometry. Time-averaged structures of *other types*, which are also supplied by instrumental methods, are well known by organic chemists and, of course, are not viewed as non-sensical. Virtual structures that are most familiar to synthetic experimentalists result from averaging of structures of conformers, which rapidly interconvert in the NMR timescale (Section 3.3). A simple example is 1,1,1-trisubstituted ethanes $CH_3–CR^1(R^2)R^3$. For such a compound, the geminal H's are non-equivalent in each of three identical rotamers. However, their ^1H magnetic resonance is manifested as a sharp singlet even at low temperatures. The geometry of the methyl fragment (three *equivalent* protons) revealed by NMR is *virtual*. This non-existent structure "appears" in ^1H NMR spectra because of the averaging of the signals of these actually *non-equivalent* protons that one after another rapidly occupy three non-equivalent spatial positions, due to Me rotation.

Customary thermal ellipsoids, which are provided by X-ray analysis and show protons in a 3D molecular backbone obtained by this method, are another easily recognizable example. The obtained molecular structure is virtual, i.e., protons themselves do not occupy such a space. These 3D regions show the averaged area of location for each proton, which results from their thermal, large amplitude (i.e., non-harmonic) oscillations. 3D structures provided by X-ray analysis, with protons blurred in some space, do not disturb us.

An unexpected conclusion is that one cannot absolutely exactly indicate geometrical parameters (a value of r in Fig. 42) for the virtual averaged molecular structure which is "extracted" from experiments. In simple words, where in the energy minimum (i.e., in coordinate r) is the structure that corresponds to the experimental structure? For any vibrational level of an asymmetric well of potential energy of a real molecule, one cannot accurately determine the distribution of the probability for molecular geometry or, at least, precisely indicate the most probable geometry. This means that, without succeeding in spectral experiments, we do not know what the exact geometry of the vibrationally and rotationally averaged virtual structure is.

After realizing this irremovable uncertainty, let us comprehend which *theoretically considered* geometry is assigned to molecules in routine organic research and what its relationship with the experimental molecular geometry is. First of all, the theoretically modeled molecular geometry is an approximate geometry. This approximate molecular geometry of a thermodynamically stable stereoisomer (or a stable conformer) is the geometry of the molecular structure that corresponds to the minimal energy point (Fig. 42) of the related energy minimum, i.e., it is the equilibrium geometry r_0. The latter is an approximation of the experimental molecular geometry. The bottom point of any minimum in the PES is not populated at all, while there is an uncertainty for the geometry of molecules that occupy any vibrational level of energy (see above). However, as mentioned, unexcited molecules predominantly occupy the zeroth level v_0. One can accept an approximation that the population of molecules that occupy this level represents a whole molecular ensemble, and, hence, they provide a whole unexcited 3D molecular structure "extracted" from experiments. The next approximation is that the probability of adopting the equilibrium geometry r_0 is maximal for molecules that populate this level (Fig. 42). This is reasonable since the energy well is near-symmetrical at the depth of the zeroth vibrational level.[a] The geometry, which corresponds to this probability maximum and the minimal energy point of the symmetrically shaped energy well, is the same, r_0. Therefore, for the ground state of the conformer, this 3D structure "from" the minimal energy point of the energy minimum successfully represents the conformer geometry. That is, the experimentally detected molecular geometry[b] of a stereoisomer (a conformer, if short-lived) of a non-excited organic compound is satisfactory

---

[a] If the energy minimum is not broad. Chemical structures with weak (essentially elongated) bonds or with multiple VDW contacts of spatially proximal molecular fragments (e.g., individual and associated polymers, including those of biological origin) may have broad energy minima. Obviously, the molecular geometry that corresponds to such a minimum is less certain. For instance, if assessing a hypothetical unrealistically broad minimum for a biopolymer [e.g., with the zeroth vibrational level width $r_0 \pm \delta r$ of 50 nm for a distance between some non-bonded fragments or of 180° for a torsion angle], one cannot speak about a certain molecular geometry at all, although relevant spectral experiments would supply some 3D structure for this conformer of the biomolecule.

[b] I.e., vibrationally averaged geometry; called effective molecular geometry (Fig. 92).

approximated by the equilibrium geometry. The equilibrium geometry is a "niche" that researchers usually allocate for the experimentally obtained molecular geometry in the energy minimum.

There is an additional circumstance that also justifies this approximation of the molecular geometry of *stable conformer*. If the energy minimum is narrow, the geometry variations are subtle for molecules that occupy low energy levels of their energy minima. Hence, the minimal energy point is a satisfactory "supplier" of molecular geometry if a conformer is equilibrated at moderate and even increased temperatures. High energy levels become appreciably populated only at "non-organic" temperatures, say, higher than 700 K. Obviously, the equilibrium geometry of *conformer* is valid for interpreting molecular geometry "obtained" in a vast majority organic experiments, since the latter are conducted at much lower temperatures.

And what about *unstable conformers*? In the broad conformational approach, if strictly considered, they are only transition states and energy maxima, i.e., structures that correspond to saddle points (Fig. 43) and peak tops, respectively, of the PES (Section 3.1). All other points of the PES are associated with energy minima. Even high energy points are "structural components" of potential energy wells, and the related molecular geometries are associated with high levels of vibration energy. If considered loosely, an unstable conformer is a stereoisomeric structure which corresponds to a PES point that is not a minimal energy point.

Regarding the geometry of the strictly considered *unstable conformer*, it is usually understood as the geometry of the structure that corresponds to the related minimal energy point in the PES (saddle point; Fig. 43). One has to realize that, as in the case of *stable conformer*, it is an approximation. The actual geometry of *unstable conformer* is an averaged geometry of statistically distributed structures of the mentioned Cramer's cloud for this PES point. However, due to the same reasons as for *stable conformer* (see above), the geometry, which is associated with the saddle point of the PES, is a good approximation for the conformation of the transition state[a] at moderate temperatures.

---

[a]It is worth reminding that *conformation* is a synonym of molecular geometry applied to a *conformer* (stable or unstable thermodynamically). For instance, chair and twist are trivial names for certain geometries of some saturated rings that, being structural fragments in many mono- and polycyclic compounds, rapidly interconvert these geometries at room temperature. Terminologically, these ring

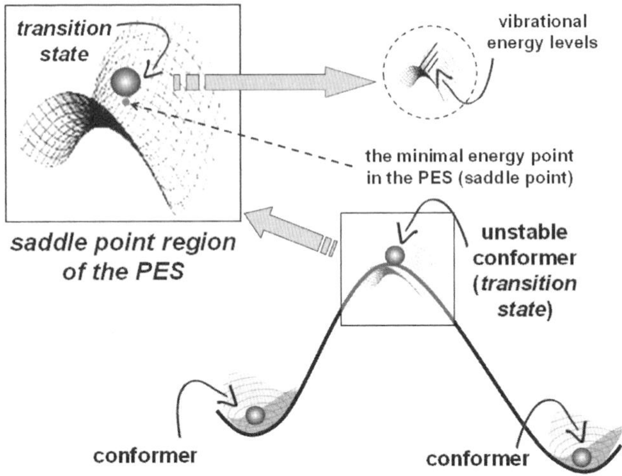

Figure 43. Schematic representation of *conformers* (stable stereoisomeric structures) and *unstable conformers* as they are defined by the broad conformational approach. The bold curve shows the MEP for interconversion of two conformers (for MEP, see Fig. 58 in Section 3.2), and the ball depicts the virtual averaged molecular structure (the mentioned Cramer's cloud of molecules distributed into vibrational and rotational energy levels).

At this point, it probably seems that, due to the well-based formalism of PES, the broad approach to *conformers* is perfect. However, some inaccuracies have escaped our attention. Suppose that the barriers to the interconversion of stereoisomers are very low, i.e., energy minima are separated by the barrier that is lower than the energy of the zeroth vibrational level (Fig. 44). The essential consequence is that any molecule from the molecular ensemble does not occupy a certain minimum from these two energy minima; as indicated in this Section, allowed values of vibrational energy start from the zeroth vibrational level v_0. In other words, there are two minima 1 and 2, but no real molecular structure corresponds to either of

---

shapes are *conformations*. This term (conformation) is frequently used in recognizing typical conformations (e.g., chair, staggered, and eclipsed; they are called conformons in this account; Sections 1.6 and 3.4) of flexible molecular fragments of "conformational reference compounds" (Section 3.4) in geometrically similar molecular fragments in other compounds as, e.g., cyclohexane chairs are in the urotropine skeleton (Fig. 20).

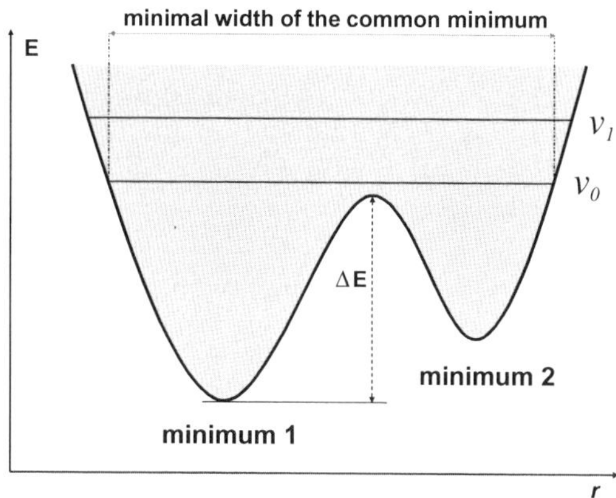

Figure 44. Schematic representation of two energy minima separated by a low barrier. These minima form one common energy minimum (shown in gray).

them. The molecules actually "lie" in a common minimum that incorporates minima 1 and 2 as a bottom of its potential well. The minimal width $r_0 \pm \delta r$ of the common minimum is the interval of geometrical parameter changes that corresponds to the vibrational level v_0. Of course, experiments would provide information about a structure with energetic and geometrical parameters that are averaged over vibrational-rotational levels of this broad common minimum.

The broad definition of *conformer* does not comprise this situation. How this problem should be treated was proposed by Eliel, and the treatment has become a ubiquitously accepted criterion. In the context of conformational definitions, one may say that this suggestion is an amendment to the broad definition of *conformer*.

Accordingly to this criterion for a statistical ensemble of molecules, two energy minima correspond to two conformers if the value of the separating barrier (ΔE in Fig. 44) is larger than the mathematical product $k_B N_A T = RT$, where k_B is the Boltzmann constant; N_A is Avogadro's number; R is the universal gas constant, i.e., $R \approx 1.985878 \, \text{cal}^{-1} \times \text{K}^{-1} \times \text{mol}^{-1}$; and T is absolute temperature. For instance, $RT \approx 0.6 \, \text{kcal mol}^{-1}$ for 300 K. The value RT shows the thermal energy of one mode vibration for one mole of an individual compound at temperature T in the gas phase. It is obvious that, if the barrier is lower than the value RT, all vibrational levels, including level

ν_0, "cross" the energy well over the barrier. This is exactly the situation of the one common energy minimum (Fig. 44), and we now understand why the value RT has been chosen as a numerical criterion that discriminates between one and two conformers. The value RT for the minimal height of the kinetic barrier that still separates an individual conformer from another individual conformer means that its minimal lifetime τ (Section 2.5) is $1.31 \times 10^{-10}/T$ seconds [i.e., the minimal value of the lifetime that a conformer (a statistical ensemble of molecules occupying different energy levels of the same potential well) may have], according to the Eyring equation (Section 3.5.3).[a]

Probably, the reader would agree that the broad conformational approach is more solidly grounded that the rotation-based and kinetic barrier approaches to *conformer* and *conformation*. At first thought, it might seem universal and blameless, but, unfortunately, it is not. The disappointing aspect is that the broad conformational approach actually includes all kinds of stereoisomerism. By the above definition, any stable stereoisomers of a compound are conformers if the condition of intact chemical connectivity is maintained in an isomerization. Then, molecular rearrangements, which have very high kinetic barriers and are not accompanied by bond migration or rupture, fall under the category of conformational transformations. For instance, the experimental barrier of interconversion of M- and P-enantiomers of nonahelicene (Fig. 45) is of a significant height of 43.5 kcal mol^{-1}.[41a] According to the broad conformational approach, we have to accept this 3D rearrangement as a conformational transformation. Racemization of M- and P-BINAP ligands (Fig. 40) is even slower. In this interpretation of *conformer*, these non-racemic ligands, which are ubiquitously used in stereoselective synthesis, should be considered as conformers. There is no doubt how contrary this statement is to the organic practice. The equalization of meanings *stable conformer* and *stable stereoisomer* actually annuls the conformational idea. Obviously, the third attempt to approach *conformer* requires some correction.

---

[a]Of course, the tunnel effect (also called tunneling; transfer from an energy minimum to another energy one without possessing the energy sufficient to cross the barrier) is neglected here. According to the quantum mechanical description of molecules, they have some non-zero probability of tunneling through low energy barriers. Tunneling strongly decreases with increasing barrier height and width.

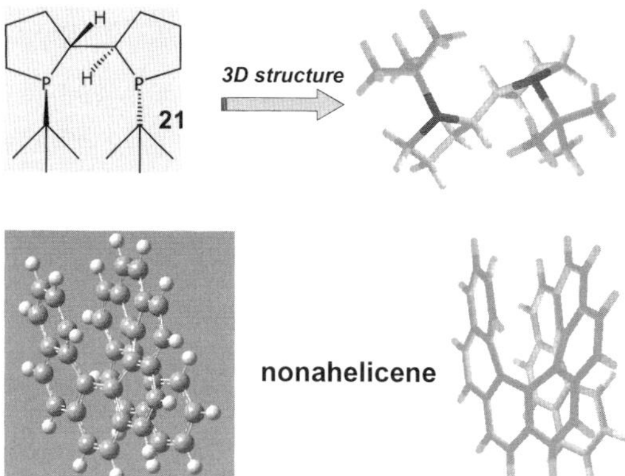

Figure 45. Examples of molecules with very high kinetic stability. Top: a stereoisomer of chiral phosphine ligand **21** with two stereogenic carbon atoms and two such phosphorous atoms. Both the rings and the exocyclic bonds have conformational freedom while all the elementary molecular fragments are deprived of it. Bottom: 3D structure of nonahelicene (for clarity, two graphical representations are shown). Inversion of the molecular helix occurrs via a restricted concerted rotation around aromatic bonds that causes additional, high energy cost, distortions of the rings.

2.5 Amendments: The Universal Conformational Approach

Summarizing this short review of approaches to *conformer* and *conformation*, we can conclude that the current situation in stereochemistry is paradoxical. The conformational description of molecular flexibility is extensively used even though there is no successful definition of *conformer*. Three different *conformational* descriptions co-exist in chemical publications as quite legitimized, all-comprising approaches while none of them is, in fact, universal. Of course, their true legitimacy is different. The rotation-based definition is inconsistent. Besides, its discrimination of non-rotational stereomutations is vacuous, since molecular shape is changed via both rotational and geminal flexibility (Section 2.2). Therefore, one may express a hard critical opinion. *The rotation-associated approach to conformation should be abandoned.*[a] The two other approaches do not contain

---

[a]One may realize that, in practice, it is not easy to withdraw this understanding which is deep-rooted in organic literature. At least, serious chemical accounts[31b] have become mentioning alternative

irremovable conflicts but they have some limitations. Are they a basis for a correct conformational language?

One may try to tolerate this uncomfortable situation with distinct co-existing conformational definitions by accepting any "difficult" case of interconverting stereoisomers as an exception from the applied conformational classification. For instance, when using the rotation-focused definition of conformers, atropisomers may be accepted as configurational isomers. We quickly meet a difficulty there: this exception ignores the absence of principal discrepancy between bisaryls with hindered rotation and bisaryls with free internal rotation. This difficulty of the traditional conformational approach remains in considering many structural congeners.

If following the broad conformational approach, too many exceptions also have to be introduced. Phosphines, e.g., diphosphine **21** (Fig. 45),[a] are kinetically stable even at increased temperatures. Should such species with a pyramidal phosphorous be considered by the broad conformational approach as an exception (i.e., as rigid stereoisomers)? In the end of Section 2.4, nonahelicene (Fig. 45) is mentioned as a system with an extremely high kinetic barrier [43.5 kcal mol^{-1}; in terms of stereoisomer lifetime τ (see below), $\tau \approx 1.6 \times 10^{14}$ min at 330 K]. Such enantiomers of pentahelicene are interconverted with an essentially lower barrier of 24.4 kcal mol^{-1} ($\tau \approx 63$ min at 330 K).[41b] Should the *M*- and *P*-helical enantiomers of nonahelicene categorized as exceptions (i.e., rigid stereoisomers), in contrast to the *M*- and *P*-enantiomers of pentahelicene (i.e., conformers)?

Obviously, in such a way the assignment of many structures to *conformer* becomes voluntary, established by a convention. This situation is very undesirable. Selection of *conformers* from the set *stereoisomers* without applying a strict criterion would lead to additional confusions in conformational considerations. Also, as time passes, more and more new chemical structures (sometimes of new chemical functionality) would be a matter of discussion, and it is likely that the list of exceptions would be extensively expanded.

This situation is avoidable, if the somewhat limited kinetic and broad definitions of *conformer* (Sections 2.3 and 2.4) could be modified to eliminate their disadvantages. We remember that these kinetic and broad approaches at least do not contain an inconsistent basic principle, and, therefore, they seem to be proper starting concepts.

Let us recollect the primary conformational idea. It consists of two propositions: (1) molecules possess a finite number of 3D structures of short lifetimes (conformers) and (2) these structures are capable of fast

---

approaches, and this is a promising sign. One may also notice that theoretically-oriented organic chemists do not use the traditional approach to *conformer*, preferring to describe molecular flexibility in terms of potential energy surface, i.e., to understand molecular flexibility in the sense of the broad conformational approach.

[a] This chiral compound (commercial name: TangPhos) and related diphosphines are widely used in stereoselective synthesis, e.g., in Rh-catalyzed asymmetric hydrogenations of functionalized alkenes.

interconversions without losing chemical connectivity. Therefore, formulating basic conformational notions, one should keep a specificity that reflects *fast* intramolecular 3D reorganizations.

Now, let us revisit the broad approach to *conformer*. Its problem may be characterized as the problem of the upper kinetic limit for *conformer*. As we have seen in the previous Section, there is no principal discrepancy between the meanings of *conformer* and *stable steroisomer* within the limits of the broad conformational approach. As we remember, in order to define what *conformer* means for the case of stereoisomers that are separated by a very low kinetic barrier (Fig. 44), Eliel introduced the "artificial," but meaningful, $k_B N_A T$ criterion (Section 2.4). It is an essential improvement because it defines the kinetic lower limit for *conformers* and defines it so that this lower limit is temperature-dependent. Then, we know for any temperature when rapidly interconverting stereoisomers "are" conformers.

However, why not introduce a similar, but upper limit that could cancel equalization of *conformer* and *stereoisomer* in the case of slow 3D rearrangements of molecules? Then the improper totality of the broad conformational approach would be avoided.

Obviously, the upper limit for *conformers* should be a certain kinetic stability of stereoisomers that, for different temperatures, distinguishes between *stable conformer* and *stable stereoisomer*. What is not obvious in defining the upper limit is this "certain." Kinetic stability is not a molecular physical quantity, it is a qualitative, very subjective, characteristic. For instance, molecular structures, which only exist for pico- and even femtoseconds, are sufficiently stable from the point of view of laser spectroscopists. Such structures would be considered as labile by NMR professionals since NMR only tracks individual chemical species that have longer lifetimes. For synthetic organic chemists, stereoisomers, which do not change their molecular geometry at least during a chemical experiment, are stable. For these professionals, stability also implies principal isolability. Therefore, similar to the criterion of the barrier height in the kinetic approach to *conformer* (Section 2.3), this upper kinetic limit may only be postulated. A diffuse chemical understanding of stability is the only criterion.

Often, it is sufficient to have 24 hr or even less at one's disposal when separating chemical species that are stable during this time period. The isolability criterion for stability of stereoisomers has been mildly characterized as imperfect[33a] because the possibility of being isolated is not a chemical system property. Nonetheless, isolation of unchanged species within a reasonable time period of 24 hr does not seem to be an unsolvable problem to organic experimentalists. The success in this separation tells them that they are dealing with stable structures. And vice versa, if a stereoisomer is transformed into another stereoisomer during this time period, its molecular 3D structure seems to them to be escaping observation, i.e., flexible.

Here, before we set the upper kinetic limit for definition of *conformer*, it is probably useful to refresh our knowledge for a very few routines in chemical kinetics. Lifetime τ (s, seconds) represents the mean life expectancy for a statistical ensemble of molecules.[a] As often explained, τ shows the averaged time that molecules spend in the ensemble. This statement is not quite accurate. In first-order decay processes with kinetic constant k (s^{-1}), τ is the time required for a concentration to decrease to $1/e$ (e is the transcendental number; numerically, $e = 2.71828\ldots$) of its initial value; τ is the reciprocal value of k, i.e., $\tau = 1/k$. Thus, lifetime τ shows the time period during which a population of molecules loses slightly less than one-third of its original quantity. As we see, the lifetime does not depend on the initial concentration for the first-order kinetic processes. Non-chemical conversion of a stereoisomer into another stereoisomer obeys first-order kinetics. Then, we can use this simple, concentration-invariant characteristic τ in setting up the upper kinetic limit that separates *conformers* and rigid stereoisomers.

Consider two interconverting stereoisomers that correspond to two energy minima in the PES, i.e. two vibrationally-rotationally averaged 3D structures (stereoisomers P and Q in Fig. 46). Let us suggest the time period to 24 hr (8.64×10^4 s) for a stereoisomer (stereoisomer P from Fig. 46) as the upper limit of its lifetime τ for assigning P to a conformer. Consequently, if the lifetime τ for the stereoisomer P is of a higher value (i.e., $\tau > 24$ hr), P is a rigid stereoisomer. Since the equivalent of stereoisomer lifetime τ is the kinetic constant k ($k = 1/\tau$) of the rate of stereoisomerization, the kinetic

---

[a] In the stereoisomer case, this statistical ensemble is an ensemble of molecules with geometries that correspond to an energy minimum in the PES.

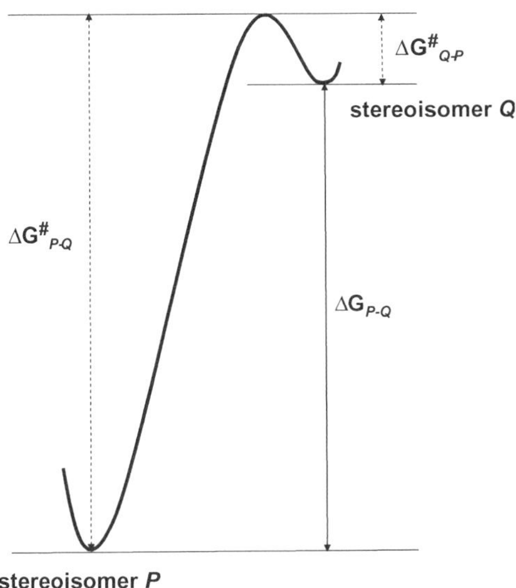

Figure 46. The case of two stereoisomers of very different thermodynamic stability. Kinetic half-barriers are indicated by dotted arrows.

border that separates molecular flexibility and rigidity corresponds to the kinetic constant $k = 1.16 \times 10^{-5}$ s^{-1}. This means that, if stereoisomer P from a pair of stereoisomers is interconverted into its partner Q with the rate that is characterized by a k value higher than 1.16×10^{-5} s^{-1}, P is a conformer. In the opposite case ($k < 1.16 \times 10^{-5}$ s^{-1}), it is a rigid stereoisomer. The term *rigid* does not mean that all molecules from the ensemble are rigid. For instance, if a stereoisomer is characterized by the lifetime which value is slightly larger than 24 hr (i.e., is rigid from the viewpoint of our suggestion), this means that two-thirds of the molecules have not been converted to the related stereoisomer by the end of this time period.

The lifetime of a stereoisomer is a universal criterion of classifying the stereoisomer as a conformer or as a rigid stereoisomer. It implies any temperature, in contrast to the kinetic barrier approach to *conformers* (Section 2.3) that neglects the dependence of kinetic barriers on temperature. Note that the half-barrier reflects the rate of conversion of a stereoisomer at a given temperature (see for the Eyring equation in Section 3.5.2). Nevertheless, one can use half-barrier values (in other words, the rate of conversion of the stereoisomer of interest[a]) when assigning the stereoisomer to *conformer.* The conformational criterion — the stereoisomer lifetime τ — permits, for different temperatures, to obtain values of the τ-associated kinetic barrier by deriving these values from the Eyring equation (Section 2.6). Let us call the broad definition of *conformer* (Section 2.4) augmented with the introduced upper kinetic limit (as the stereoisomer lifetime of 24 hr) *the universal definition of conformer.*

A similar, brief form of the universal approach to *conformer*, which can be encountered in the physical chemical literature, describes thermodynamically stable conformers as rapidly interconverting stereoisomers that may, *in principle*, be detected as individual structures at given ambient conditions. Indeed, the first requirement — detectability — implies that a stereoisomer is thermodynamically stable, i.e., has some non-infinitely short lifetime. This means that the molecular geometry (the statistically averaged, virtual 3D molecular geometry; Section 2.4) of the stereoisomer (a molecular ensemble of molecules distributed into vibrational and rotational levels of potential energy) corresponds to an energy minimum in the PES. The second requirement — *rapidly* — sets up some upper limit for the stereoisomer lifetime. Thus, the universal definition of conformer may be rephrased in the macrocharacteristic language. *Stable conformer is a stereoisomer that has a finite lifetime τ bounded both from above and from below; the boundary limits may be defined as* $1.31 \times 10^{-10}/T\ s <$ $\tau < 24 \times 3600\ s$ (s, seconds). Note that τ, being linearly associated with the kinetic constant k, is strongly temperature-dependent (Section 3.5.3, Eq. 14). Thus, according to this conformational description, a stereoisomer may be rigid above some temperature, while it should be considered as a flexible molecular structure, i.e., a conformer, below this temperature (Section 2.6). A more accurate conformational terminology would use terms *conformational state* and *rigid state* (formal, not physical ones) for a stereoisomer instead of *conformer* and *rigid stereoisomer* for the same stereoisomer.

For a primary orientation of non-physical organic chemists, let us encounter τ values for some familiar molecular fragments. Lifetimes of regular amide fragments (O)C–NH in

---

[a]This quantity is the macrocharacteristic of dynamic molecular processes that is measured in kinetic experiments.

proteins lie in the range 0.01–1 s. More concretely, amide fragments in proteins "spend" this time period when being *trans*-shaped before C–N rotation occurs.[a] Lifetimes are significantly longer for amide fragments in *acylamino acid–prolyl* domains of protein chains. At room temperature, they are 10–100 s. At ambient conditions, the lifetime of a chair-shaped cyclohexane ring (see Fig. 28) is about 10^{-4}–10^{-5} s, then the chair interconverts in several picoseconds. The inverting molecular bowl of corannulene **4a** (Fig. 68) has a similar lifetime of $\sim 5 \times 10^{-5}$ s at room temperature. Lifetimes of *anti* C–N rotamers[b] of nucleotide residues incorporated into DNA duplexes are $(1-5) \times 10^{-3}$ s and $(1-5) \times 10^{-2}$ s for pairs A:T and G:C, respectively. The lifetime of the C pyramid of *sec*-BuLi is ~ 1 s at 298 K before the carbon configuration is inverted. At the same conditions, the lifetime of the *cis* fragment CN=NC in azobenzenes ArN=NAr lies in the range $(1-5) \times 10^{6}$ s.[c] According to our classifying border $\tau = 24$ hr, any amide fragments in proteins, the cyclohexane ring, the corannulene bowl, *syn* and *anti* rotamers in DNA, and the anionic carbon-centered fragment in lithium alkyls are conformationally mobile at T = 298 K, while unit *cis*-CN=NC is rigid.

What is interesting is the situation where stereoisomer P has lifetime $\tau > 24$ hr, while a very short lifetime $\tau \ll 24$ hr characterizes stereoisomer Q, i.e., there is a large energy gap $\Delta G_{P\text{-}Q}$ between P and Q, and it is essentially larger than the half-barrier $\Delta G^{\#}_{Q\text{-}P}$ (Fig. 46). This means that the barrier for conversion $P \to Q$ (i.e., the half-barrier $\Delta G^{\#}_{P\text{-}Q}$) is very high, while the related half barrier $\Delta G^{\#}_{Q\text{-}P}$ is low. For instance, compound **20** (Fig. 41) is exactly such a case. Equipped with required energy values, we can see that the *cis* stereoisomer totally predominates over the *trans* isomer (they may be associated with stereoisomers P and Q, respectively, indicated in Fig. 46). The barrier (i.e., the half-barrier $\Delta G^{\#}_{P\text{-}Q}$) of the *trans* $\to$ *cis* stereoisomerization is very high; i.e., its lifetime τ is a large value. Thus, the *cis* stereoisomer is a rigid stereoisomer. The *trans* isomer is a short-lived structure at ambient conditions (the related half-barrier $\Delta G^{\#}_{Q\text{-}P}$ is low). Thus, this stereoisomer should be treated as a conformer. However,

---

[a]Amide fragments are capable of internal (O)C–N rotation; *cis-trans* isomerization is its trivial, non-systematic name. Here, lifetimes are of the predominated *trans* conformers; the related kinetic barriers are half-barriers $\Delta G^{\#}_{trans-cis}$ of this isomerization.

[b]In double stranded DNA molecules, the so-called *anti* conformation for the glycosidic bond is highly favored over the opposite conformation (*syn*) since rotational orientation *anti* of the nucleobase relative to the furanose ring provides spatial proximity to complementary nucleobases from the anti-parallel strands. This proximity is absolutely necessary for chemoselectively H-bonding these aminobases (i.e., for forming Watson–Crick pairs A:T and G:C) and, thus, for binding the two DNA strands together.

[c]This fragment undergoes *cis-trans* isomerization via double in-plane N inversion. Here, lifetimes are of the *cis* stereoisomers; they correspond to half-barriers $\Delta G^{\#}_{cis-trans}$ of this stereorearrangement.

how should we term the equilibrium between the *cis* and *trans* structures of bicycle **20**? It is not a conformational transformation because one of the equilibrium components is not a conformer. On the other hand, it is not a stereoisomerization of rigid stereoisomers because one of the stereoisomers (the *trans* isomer) is a conformer. We have to be satisfied by stating that, from the perspective of the universal description of *conformers*, this equilibrium is an equilibrium *conformer–rigid stereoisomer*.[a]

Within the limits of the traditional, rotation-focused, definition of *conformer* (Section 2.2), only two or more rotamers of a molecule can be considered as conformers or as rigid stereoisomers. That is, in assigning a molecular 3D structure to *conformer* in this manner, the structure is another stereoisomer-dependent. If considered separately, one molecular 3D structure, e.g. that of paracyclophane **4d** shown in Fig. 2, is not in the focus of this definition since rotation means a certain change of molecular geometry. The universal conformational approach abolishes the dependence of a stereoisomer from another one when assigning it to a conformer or to a rigid stereoisomer. Indicating the kinetic stability of a stereoisomer, its lifetime (in kinetic terms, the half-barrier that corresponds to this lifetime at the given temperature) characterizes the stereoisomer itself. That is, the lifetime of stereoisomer Q (its half-barrier at the corresponding temperature) has no relation to the conformational mobility or molecular geometry of its stereoisomeric counterpart. If Eliel's condition for the lower limit for *conformer* is fulfilled ($\Delta G^{\#} > k_B N_A T$), the upper kinetic limit allows an unambiguous assignment of an individual molecular 3D structure either to a conformer or to a rigid stereoisomer, according to its half-barrier value. Note that the lowest half-barrier should be taken into account, if the stereoisomer undergoes alternative stereoisomerizations (e.g., see Fig. 68 for conformational transformations of tricorannulene **4e**).

2.6 Kinetic Barriers that Discriminate Between *Conformers* and *Rigid Stereoisomers*

We have learned that the stereoisomer lifetime τ is the principal distinctive characteristic of a stereoisomer in the universal approach to *conformer*. As explained, kinetic barrier $\Delta G^{\#}$ (more accurately, the half-barrier; Section 2.5, Fig. 46) is an equivalent of the lifetime τ for the given temperature, in formally distinguishing *stable conformer* and *rigid stereoisomer*. Note that the rates of stereoisomerizations are what is measured in related kinetic

---

[a] In order to characterize the kinetics for pair Q and P by means of one kinetic characteristic, sometimes the meaning effective lifetime (τ_{eff}) is used. It is defined as $\tau_{eff} = p_Q \times \tau_P = p_P \times \tau_Q$ (p_Q and p_P are fractions of Q and P, respectively, in their equilibrium; $p_Q + p_P = 1$). Values of τ_{eff} do not indicate whether a stereoisomer is flexible or rigid.

experiments at different temperatures. The rate values provide kinetic constants k, and kinetic barriers $\Delta G^{\#}$ are derived from the k values by employing the Eyring equation (Section 3.5.3, Eq. 14). As a convenient, weakly temperature-dependent measure in molecular kinetics, kinetic barriers are exactly the kinetic data favored in organic considerations. For easily assigning relevant stereoisomers to *conformers*, let us turn to evaluating the relationship between kinetic barriers, which correspond to our upper kinetic limit for *conformer* ($\tau = 24\,\mathrm{hr}$), and temperature.

In the Eyring equation, barrier $\Delta G^{\#}$ is determined by two independent variables. They are temperature T and kinetic constant k of the stereoisomerization (or, equivalently, lifetime τ of the stereoisomer[a]). Since our upper limit $\tau = 24\,\mathrm{hr}$ (or $8.4 \times 10^{4}\,\mathrm{s}$) is a constant value that formally separates between flexibility and rigidity of stereoisomers, the related value of k is also a numerical parameter ($k = 1.16 \times 10^{-5}\,\mathrm{s}^{-1}$). Then, we can simplify the Eyring equation so that the barrier depends only on the temperature variable. When inserting the above numerical k value into the equation, the barrier may be expressed as

$$\Delta G^{\#} = 4.576\mathrm{T}(15.26 + \log \mathrm{T}) \tag{2}$$

where absolute temperature T is in degrees Kelvin (K) and the Gibbs energy of activation $\Delta G^{\#}$ is in calories per mole ($\mathrm{cal\ mol}^{-1}$).

Treating lifetime τ implicitly, Eq. 2 represents the direct relationship *flexibility-rigidity border–kinetic barrier* for any temperature. This near-linear dependence $\Delta G^{\#}(\mathrm{T})$, i.e., Eq. 2, shows the maximal value of the barrier of stereoisomerization of a *conformer* at a given temperature. Let us call such barriers *border barriers*. Higher barrier values for a given temperature point out stereoisomerization of a *rigid stereoisomer*, while lower values indicate stereoisomerization of a *conformer*. Thus, Eq. 2 provides the desired upper limit for *conformer* as border barriers, each of which is

---

[a]When considering two interconverting stereoisomers *P* and *Q* (Fig. 46), the related kinetic constant k_P or k_Q (equivalently, lifetime τ_P or τ_Q), respectively, is implied. Note that, for the equilibrium $P \rightleftharpoons Q$, kinetic *experiments* measure the slowest rate, i.e., the rate of stereoisomerization of the lower energy stereoisomer *P* and, thus, provide the higher half-barrier $\Delta G^{\#}_{P\text{-}Q}$. The lower half-barrier $\Delta G^{\#}_{Q\text{-}P}$ is obtained as the difference $\Delta G^{\#}_{Q\text{-}P} = \Delta G^{\#}_{P\text{-}Q} - \Delta G_{Q\text{-}P}$, where the energy difference $\Delta G_{Q\text{-}P}$ for stereoisomers *P* and *Q* is derived from the ratio *Q*:*P* determined for their equilibrium.

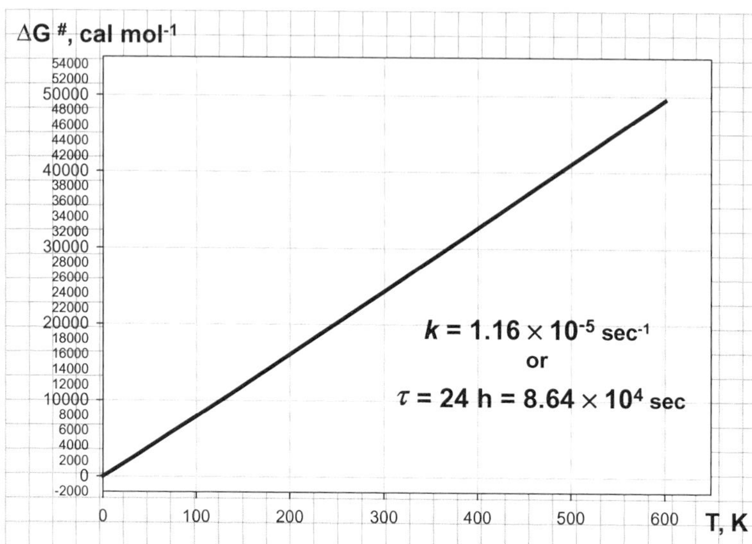

Figure 47. The border line (shown for the temperature range $0-600\,\mathrm{K}$), which separates between *conformers* ($\tau < 24\,\mathrm{hr}$) and *rigid stereoisomers* ($\tau > 24\,\mathrm{hr}$), corresponds to the separating lifetime $\tau = 24\,\mathrm{hr}$ ($k = 1.16 \times 10^{-5}\,\mathrm{s}^{-1}$). The area below the border line indicates *conformers*, while the area over the border line is associated with *rigid steroisomers*.

associated with a certain temperature. The border barriers for the temperature range of 0 to 600 K are shown as a plot (border line) that graphically represents Eq. 2 (Fig. 47).

Using this graphical representation (border line), one can quickly classify any stereoisomer either as a flexible stereoisomer (conformer) or as a rigid stereoisomer, according to the kinetic barrier of this stereoisomer and temperature. For instance, for 298 K, the border barrier is 24.2 kcal mol^{-1}. For this temperature, higher barriers of stereoisomer stereorearrangements indicate the rigidity of the stereoisomer. This barrier value straightaway supplies us with a general understanding of how flexible a majority of organic compounds are. Kinetic experiments for thousands organic compounds of different chemical structure have shown that stereorearrangements of their molecular skeletons, i.e., permutations of geminal ligands in elementary molecular fragments or rotation of neighboring molecular fragments relative one to the other, are characterized at this temperature by barriers lower than this value of 24.2 kcal mol^{-1}. This means that, in a rough generalization, *at room temperature, organic compounds in the liquid as well as the gas phases usually are conformers in the equilibrium.*

Why do we find different organic molecules, as a rule, to be flexible? The answer is trivial. Organic backbones usually contain at least one single bond that connects elementary molecular fragments. Rather rarely, these fragments are covalently locked with forming cyclic structures that prevent rotation around endocyclic single bonds (as, e.g., in cyclopropane, adamantane, and twistane). Also rarely, rotation in C–C or in similar fragments is hindered due to steric factors. Therefore, while geminal ligands do not permutate their positions in many chemical elementary fragments, free intramolecular rotation is an attribute of many organic compounds.

It is difficult to overestimate the significance of this notion for synthetic chemists. Mixing reactants in a reaction vessel at 250–300 K, we almost surely deal with dozens of conformers for each component of the reaction mixture. Location of conformers as well as determination of their relative stability (Chapters 4 and 5) is ubiquitously known as conformational analysis; an entire set of conformers is often called conformational space. Metaphorically speaking, organic researchers operate in the conformational space. With realizing this situation, many synthetic chemists are challenged to depart from the common practice to consider only some "obvious" 3D geometries for reactants (e.g., only a chair shape for the saturated six-membered ring when solving the long staying problem of stereoselective synthesis — to accurately predict the stereochemical outcome in 1,2-additions to substituted cyclohexanones). One should not approach an organic system at "organic reaction temperatures" as a single 3D structure; for kinetically controlled reactions, it is methodologically correct to locate all the conformers and explore involvement of each of them, being equipped with the commonly known Curtin–Hammett principle.[33c]

The universal approach to *conformer* has some terminological consequences. For an organic compound, several stable stereoisomers of non-equal stability usually interconvert with different rates, at "organic reaction temperatures" of 200–500 K. Relative energy of stereoisomers may lie in a very wide range, and the height of kinetic barriers is not determined by thermodynamic stability of 3D forms. This means that any stereoisomers of a molecule, either of low or high energy, may be separated from its counterpart (the neighboring energy minimum) by low or high barriers (Fig. 48). That is, molecular shape is represented by a set of different stereoisomers that have dissimilar lifetimes at a given temperature. For

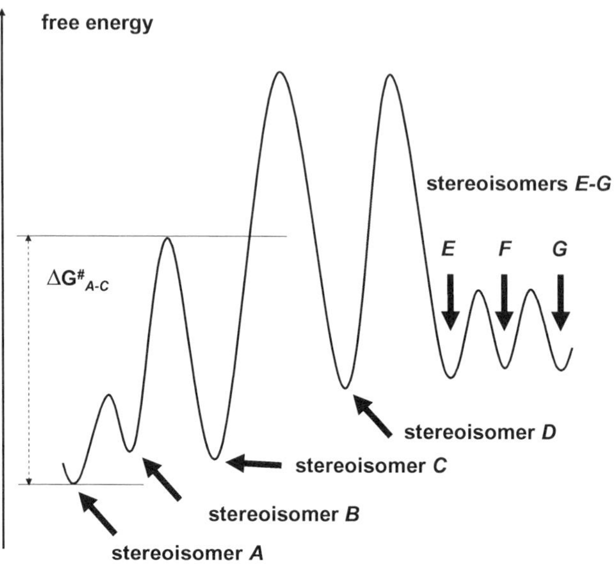

Figure 48.   A typical case of stereodynamics in organic compounds (T = *const*). Several stereoisomers with different thermodynamic stabilities are separated by kinetic barriers of different heights. The barrier of the *A*→*C* conversion for this hypothetical compound is indicated by the dotted line.

some molecular systems, stereoisomers with $\tau < 24$ hr and $\tau > 24$ hr co-exist at the same conditions. Should we consider such molecules as flexible or as rigid?

It would be unreasonable to give priority to any lifetime. For instance, one could suggest that an entire molecular structure is accepted being flexible if at least one stereoisomer has $\tau < 24$ hr. However, this suggestion is unsuccessful. One may see this for an example of bicycle **20** (Fig. 41). An extremely high barrier for disappearance of the totally predominant *cis* isomer means that practically any experimental observation of this compound would show that the molecular shape does not change. Thus, it is incorrect to view the backbone of **20** as being flexible.

The opposite suggestion is that rigidity should be ascribed to the molecule with one or more stereoisomers of $\tau > 24$ hr. Unfortunately, this view is also unsuccessful. Suppose that for a compound with stereodynamics shown in Fig. 48, stereoisomers *A–C* and *E–G* are characterized by $\tau < 24$ hr, while stereoisomer *D* is long-lived ($\tau > 24$ hr). Thermodynamically, *D* is significantly less stable than the low energy stereoisomers *B* and *C* of shorter lifetimes. The lowest energy stereoisomers are usually the structures that are monitored by spectral methods, in particular by NMR. If revealing interconversion $A \rightleftharpoons C$ of these lowest energy stereoisomers, NMR experiments would show us that this molecular system is labile in time; they would even fail to detect a very minor form *D*. However, following our primary condition for conformational mobility of *a molecule*, we would be obligated to claim that this molecular system is rigid because stereoisomer D is a long-lived form. This situation does not seem attractive.[a]

Therefore, *it is preferable not to characterize entire organic molecules in terms of flexibility*. This term does describe individual stereoisomers, but not a set of them (a molecular structure "split" into stereoisomers). In contrast, characterization of relevant molecular fragments as flexible or as rigid is valid and, of course, convenient. Only if all stereoisomers of a molecule are short-lived ($\tau_i < 24$ hr) or all stereoisomers have $\tau_i > 24$ hr at given conditions, one can say that the molecule is flexible or rigid, respectively. In order to determine all τ_i for the molecule of interest, one should perform laborious theoretical calculations. Of course, a formal statement of whether a molecular system should be considered as conformationally mobile or rigid is not worth extensive efforts.

Another terminological nuance is related to *stereoisomer*. For many compounds, an individual stereoisomer converts to different stereoisomers with the distinct rate (conformers of the three-bowl molecular structure **4e** are an example; Fig. 68). Clearly, the assignment of such a stereoisomer to *conformer* or *rigid stereoisomer* is determined by the lowest kinetic barrier among the barriers for the alternative transformations.

Let us briefly reiterate what the universal approach to *conformer* is. Description of an individual, stable, 3D molecular structure as belonging to an energy minimum in the PES is a valid approximation of quantum mechanical theory applied to organic molecules. Individual stereoisomers correspond to individual energy minima, and a "time-resolved" geometry of an organic molecule at given conditions may be adequately represented as a set of stereoisomers of different lifetimes. The lifetime of a stereoisomer is the principal characteristic that semantically relates it either to *conformer*, or to *rigid stereoisomer*. The lower (Section 2.4) and upper (Sections 2.5 and 2.6) kinetic limits for stereoisomerizations "extract" *conformers*[b] from a common set of thermodynamically stable stereoisomers of an organic system; in other words, they select certain energy minima from all minima in the related PES. Any stereoisomer is assessed separately, either as a conformer or as a rigid stereoisomer; only its own lifetime under given physical and chemical conditions of the stereoisomer equilibrium (temperature,

---

[a]Stereoisomer A has a shorter lifetime relative to, e.g., a higher energy stereoisomer C; as shown in Fig. 48, the barrier of the conversion $A \rightarrow B$ is appreciably lower than the lowest kinetic barrier for disappearance of C (conversion $C \rightarrow B$). Since T $=$ *const*, the barrier ratio qualitatively indicates that the kinetic constant $k_{A \rightarrow B}$ is larger than the constant $k_{C \rightarrow B}$ (see Section 3.5.3 for the Eyring equation) and, thus, $\tau_A < \tau_C$.

[b]Further, the term *conformer* is only used in the sense of the universal conformational definition.

pressure, and medium) is taken into account in assigning the stereoisomer to either category. In contrast to the rotation-based definitions of *conformer* (Section 2.1), relationships between the geometry of a stereoisomer and the geometries of related stereoisomers are irrelevant in classifying whether this stereoisomer is a conformer or not. Thus, this approach to *conformer* may be called universal for good reasons.

2.7 Conformers and H-Bonding

The last passage of our discussion of the universal approach to *conformer* could be a final concluding remark that this definition of *conformer* is general as well as accurate; expressing the specificity of the phenomenon of rapidly interconverting stereoisomers, it is not at variance with other stereochemical meanings. However, an additional feature of molecular systems is important for defining *conformer*, and it has not been reflected yet. It concerns H-bonding.

An indispensible attribute of all definitions of *conformers* is the requirement of preserving the chemical structure intact in stereoisomer interconversions. One could note that textbooks and monographs on stereochemistry, as well as organic research articles, unexpectedly discuss some chemical isomers as conformers. The first example of this terminological inconsistence is supplied by stereochemical considerations of ethylene glycol in textbooks. The synclinal (favored) structure has an intramolecular H-bond (vicinal OH groups are H-bonded), the antiperiplanar (disfavored) isomer has no H-bonding. Ignoring the contradiction with all conformational definitions, these *chemical* isomers are termed there conformers. This practice is absolutely prevalent; organic chemists consider interconversions, which are accompanied by disruption or formation of intramolecular H-bond(s), as conformational transformations. Pseudotropine (**22a**), biotine (**22b**), and calix[4]arenes **23** (Fig. 49) are simple demonstrative examples of numerous organic compounds, for which changes of molecular geometry are classified as conformational transformations, although these changes include formation or dissociation of H-bond(s). Intramolecular H bonds in peptides, proteins, oligosaccharides and nucleic acids are considered as a structural attribute of certain conformers or a family of similar conformers (in terms of bioorganic chemists, structures of the same conformational

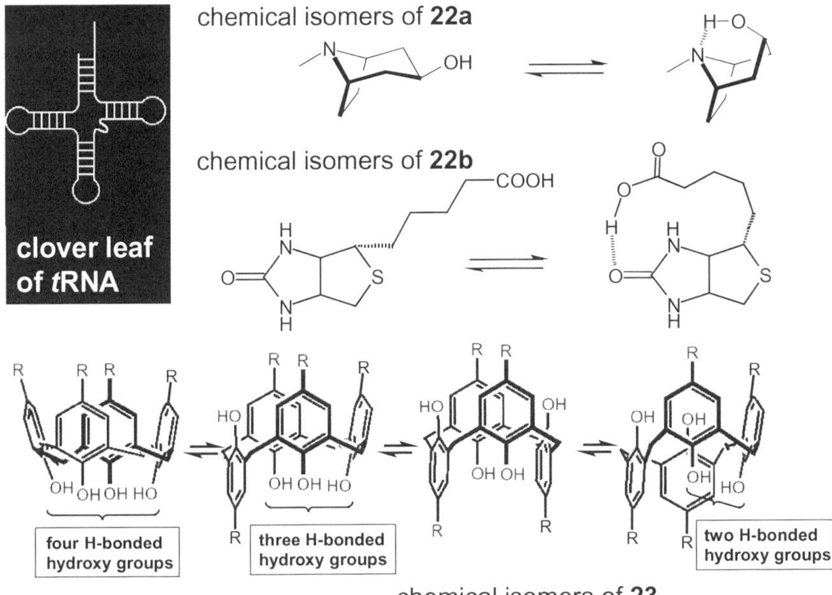

Figure 49. Chemical isomers of aminoalcohol **22a**, biotine **22b**, and calix[4] arenes **23** that chemically differ one from another by the presence (absence) of intramolecular H-bonding. Top left: a schematic secondary structure of tRNAs (short lines indicate H bonds between complementary nucleotides).

motif). For instance, intramolecular assembling or "melting" of two-strand domains (so-called hairpins) in single stranded RNAs or DNAs is caused by formation or dissociation, respectively, of many intramolecular H bonds between complementary nucleotide sequences that are fragments of the nucleic acid chain. The remarkable clover leaf structure of tRNAs (transfer RNAs; Fig. 49) is an example of such chemical structures. Nevertheless, it is frequently called conformation, in parallel to the correct term *the secondary structure* that means the biopolymer-specific chemical structure that results from chemically bonding molecular fragments separated by other fragments in the polymer chain. The mentioned flipping of the nucleobases in DNA duplexes (i.e., rotation around the glycoside bond in nucleotides that interconverts their *anti* and *syn* rotamers; Section 2.5) is also associated with dissociation or formation of H-bonds. As the reader may guess, biochemical literature uses the term *conformational* in characterizing this fast intramolecular transformation. Almost always, the disruption of the H-bond framework in biopolymers, which is accompanied with 3D rearrangement of their chains, is discussed in terms of conformational reorganization. In fact, it is a chemical transformation.

H-bonds are peculiar chemical bonds but their specificity does not annul *bonding* between proton-donating (D) and proton-accepting (A) organic functions. Similarly to covalent bonds, H bonds are characterized by some range of energies of formation, some range of interatomic distances $A \cdots H \cdots D$, and preferable values for bond angles AHD. Isomers with and without intramolecular H-bonding, e.g. those shown in Fig. 49, are chemical isomers and not stereoisomers and, hence, not conformers.

How can we finally assess H-bonding and conformational transformations? It would be absolutely senseless to revise all chemical literature that does not scrutinize dissociation of an intramolecular H bond from the viewpoint of conformational definitions. Probably, it is unavoidable to continue to follow the current, though inconsistent, terminological practice and supply the meaning *conformer* with understanding that H bonds are not implied by the requirement of intactness of chemical structure in conformational transformations.

2.8 Conformational Schemes: A Convenient Representation of Molecular Flexibility

Having learned what intramolecular stereorearrangements are (Sections 1.4 and 1.7) and having defined *conformer* (Sections 2.4–2.7), it is possible to start thinking "conformationally." We know that molecules of an organic system in the gas or liquid phase are actually a set of conformers in the equilibrium, i.e., interconverting short-lived stereoisomers. How could we portray such discretely flexible molecular shape?

A primitive description consists of orienting the molecular backbone of a conformer arbitrarily and superimposing the centers of mass of these differently oriented 3D structures. Imitating rotation of the conformer as a whole, we have obtained the time-averaged molecular shape for this conformer — a ball of a certain, conformer-specific size, e.g., as it is shown in Fig. 50 for the conformer of amine **12** shown in Fig. 21; the ball radius approximates the effective dimensions of the conformer. In this manner, one can produce

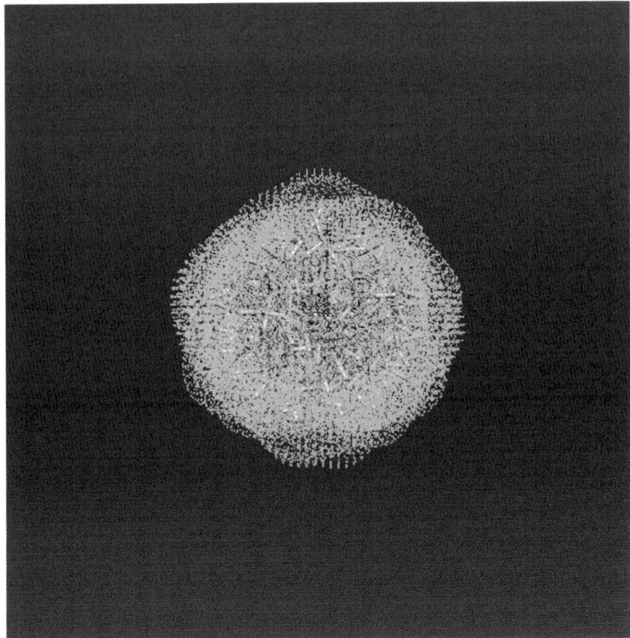

Figure 50. Schematic superposition of different orientations of a conformer of amine **12**. Atoms and bonds are shown in the middle, and the overlapping white clouds represent VDW dimensions of this conformer. Although an infinite number of different orientations of these structures cannot be presented, this image of the molecule gives a correct idea about in-time-averaged molecular size of this conformer.

a set of "molecular balls" of somewhat different size that represent all the conformers of the system of interest. The relative content of conformers at the equilibrium (say, determined experimentally) may impart Boltzmann verisimilarity to the set of "molecular balls." Thus, such a set of statistically weighted (Section 3.2, Eq. 6) "molecular balls" very roughly models the molecular structure that is "split" into conformers.

Such a set of "conformers" may be utilized in mathematical models of macroprocesses, e.g., diffusion, membrane permeability, absorption in pores, surface coating, crystal growth, electrode polarization, chemical kinetics in bulk; however, such molecular sets would scarcely satisfy organic chemists. It is clear that, even correctly predicting quantitative

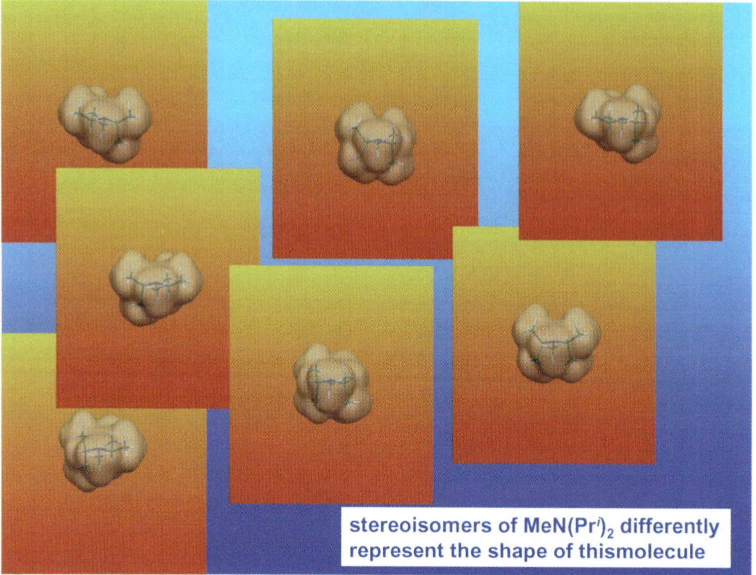

Figure 51. Conformers of amine **12** (MM3-optimized geometries; for MM3, see Chapter 4) shown as structures of VDW size.

or qualitative macrocharacteristics (e.g., diffusion rate, rates of interphase mass or energy transfer, pore selectivity), these "molecular balls" with no chemical structure are absolutely useless for understanding these processes. Molecular interactions rule them.

For gaining deep insight into chemical, physical chemical and biochemical interactions of molecules, researchers, of course, require structured molecular models. An unarranged set of stable conformers, e.g., as shown in Fig. 51 for amine **12**, although showing variability of the molecular shape, does not reflect its dynamics. In other words, in this description, molecular shape is actually represented by a set of rigid stereoisomeric structures. It is clear that the scenario of an interaction with a mixture of rigid stereoisomers differs from the scenario of the same interaction with rapidly interconverting conformers.[a] Thus, in rationally describing conformational space, one should (*1*) arrange conformers, i.e., reveal which of them directly convert one to another (have a common transition state); and (*2*) determine the rate

---

[a]These scenarios are quantitatively considered by the Curtin–Hammettt principle (Ref. 33c).

(in equivalent terms, the kinetic barrier) of interconversion for each pair of such conformers for given ambient conditions. Let us call conformers, which are transformed one to another through the common transition state, *related conformers*.

Thus, in addition to information regarding geometries and relative energies of conformers of a compound, we need information that indicates all pairs of related conformers. In other words, in addition to conformational analysis, we need a description of conformational dynamics.[a] At first glance, it seems that it is not embarrassing to indicate which two conformers are "genetically" related. For instance, those are gauche and antiperiplanar conformers in ethane type compounds (Fig. 11). They are interconverted via a 120° rotation, and no other stable conformer is in this pathway. However, if there are more than two conformers in the equilibrium, determination of pairs of related conformers is usually not a task that can be executed using improvised tools. One should locate all transition states of conformational transformations in the system, and establish for each transition state which conformers interconvert through it [Section 4.4.3, subsection (*C*)].
For instance, are the two enantiomeric conformers of lowest energy of 1,1′-corannulene dimer **41** (Section 6.1.3, Fig. 96) related conformers? The reader should not even try to figure out the correct reasoned answer to this question.[b] Molecular geometries of stable rotamers that result from rotation around the central C–C bond in this diaryl system with non-planar C-substituents obviously cannot be associated with those of stable rotamers of biphenyls, molecular structures with planar C-substituents. Stereodynamics of conformers is complicated in this molecular system, and extensive theoretical calculations — location of several conformers and several conformational transition states with determining related conformers (all of them of *a priori* unknown geometry) — are required for cracking the "conformationally hard nut" **41**.

---

[a] In its primary meaning, conformational analysis is a description of stable conformers that assesses their molecular geometry and relative thermodynamic stability. Thus, in the original understanding, this term reflects a static (time-independent) molecular shape; usually, it is not used when considering conformational dynamics. The latter term is related to conformational interconversions, i.e., it indicates assessment of kinetic stability of conformers as well as structural changes in their interconversions.

[b] For the answer, see Section 6.1.3, the caption of Fig. 96.

Establishment of all related conformers for an organic molecular system introduces the desired order into their set. We know which conformational pathways "connect" any two conformers for this system, or, in other words, which conformer geometries, one after another, are adopted by the system, starting from any conformer in order to reach any other conformer. The "genetic" relationship between two conformers shows which molecular fragment changes the geometry in their interconversion; establishment of all pairs of related conformers is equivalent to identification of all possible geometrical changes of the molecule. Thus, we have reached our goal in describing molecular shape in the gas or liquid phase. *Molecular shape is conformational space where all the elements (stable conformers) appear as pairs of related conformers interconverting with certain rates dependent on ambient conditions*, i.e., it is an arranged set of stable conformers with related lifetimes. When saying *conformational dynamics* (or, more generally, *stereodynamics*) of an organic molecular system, we imply the time-dependent molecular shape, i.e., exactly this arranged set of its stereoisomers supplied with characteristics of their thermodynamic and kinetic stabilities for given ambient conditions. Note that this description of molecular shape only includes thermodynamically stable molecular 3D geometries and does not include transient molecular geometries. This "thermodynamic" flexible molecular shape is the molecular shape which interacts: only molecular structures of non-infinitely short lifetime (i.e., stable stereoisomers) may interact with an "interventionist" [another molecule, electromagnetic quanta, or a supramolecular structure (a nano object)].

It is convenient to represent conformational dynamics in an organic compound in the form of a graph of conformational transformations (called conformational scheme or conformational map). These graphs (conformational schemes) are easily readable by synthetic chemists' eyes, because, showing the transformations of 3D structures, they are a formal analog of the familiar schemes of *chemical* transformations. A restricted conformational scheme is shown in Fig. 52 for piperidine **14b**. It comprises low energy conformers of this amine, while not including highly disfavored conformers.[a] Note that this scheme does reflect the real conformational

---

[a]They are enantiomeric chair-shapred conformers with an axially oriented N substituent.

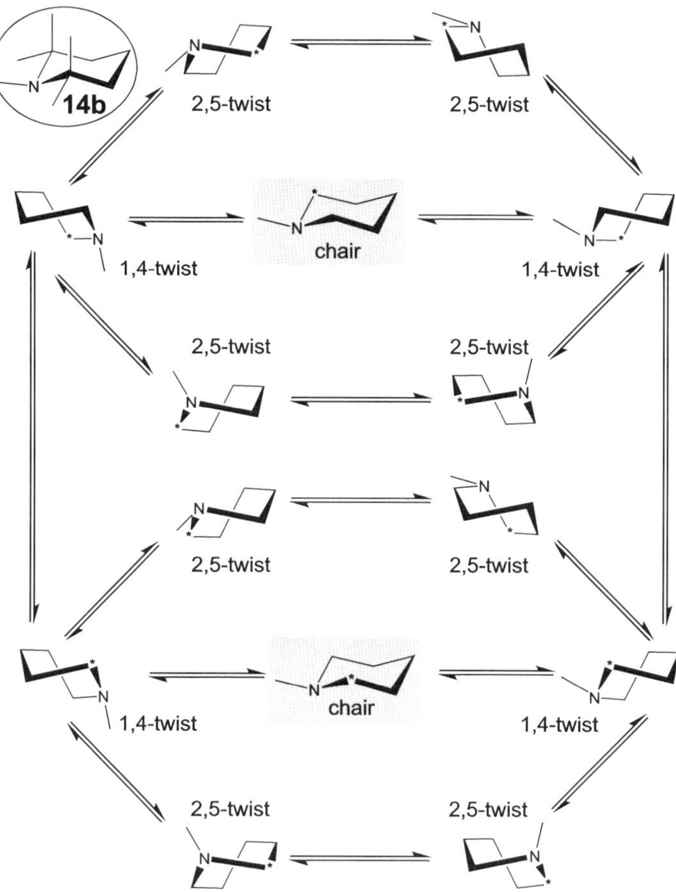

**An arranged set of conformers
represents non-static molecular geometry**

Figure 52. A fragment of conformational scheme for piperidine **14b** that shows interconversions of the twist conformers at room temperature [the related conformers (see the main text) have been located by theoretical calculations]. For clarity, α-methyl groups are not shown; the asterisk is a formal label. Gray background indicates the lowest energy conformers.

dynamics: conformers of **14b** have been located by means of theoretical calculations[37] that have also established the conformer "genesis" as well as relative stabilities of the conformers. Identical structures, which appear due to Me rotation, are not shown in the scheme.

By this representation, it is noticeably simple to conceive how molecular shape of this azacycle is changed. There are only three geometries of the piperidine ring for stable conformers of this compound, chair, 2,5-twist and 1,4-twist. The two chair-shaped conformers of lowest energy have opposite configurations of the N fragment. There are two groups of twist conformers (6 + 6); conformers from the different groups have opposite N configurations. Inside each group, twist forms interconvert, forming a closed cycle of sequential conversions (pseudorotation cycle; see Fig. 37 for more details). The chair conformers do not interconvert directly; this transformation occurs by passing two 1,4-twist conformers of opposite N configurations. The latter two are interconverted via NIR (for NIR, see Section 1.7). Thus, we have here a qualitative description of changing molecular geometry of amine **14b**. Using two calculations-provided "parameters," the geometry of conformers and pairs of related conformers, we easily "monitor" geometrical changes (i.e., 3D rearrangements) of thermodynamically stable stereoisomers (here: conformers) in this organic molecule. If we had spoken about chemical interactions of this amine, we would imply that this entire set of conformers is exactly the reactant.

Note that only stable conformers have been considered to this point. It is legitimately to ask about "fluxing" molecular shape in the path between any two related conformers, i.e., about unstable conformers (transition states). How does the molecular shape look if they were included?

Let us discuss a transition state-including conformational scheme. Textbooks have made the phenomenon of RI in cyclohexane (**18a**, Fig. 28) "illustrious," but a whole conformational scheme for this basic cyclic system (Fig. 53) is probably far less known.

As for cycle **14b** (Fig. 52), we can quickly read the map of essential changes of the molecular shape for this basic cyclic system **18a**. One can readily perceive that the most stable conformers of **18a**, commonly known as chair conformers, are interconverted in two dissimilar ways: (*1*) passing itinerary *half-chair–twist–half-chair* or (*2*) passing itinerary *sofa–boat–sofa*. Twist conformers are interconverted through low energy

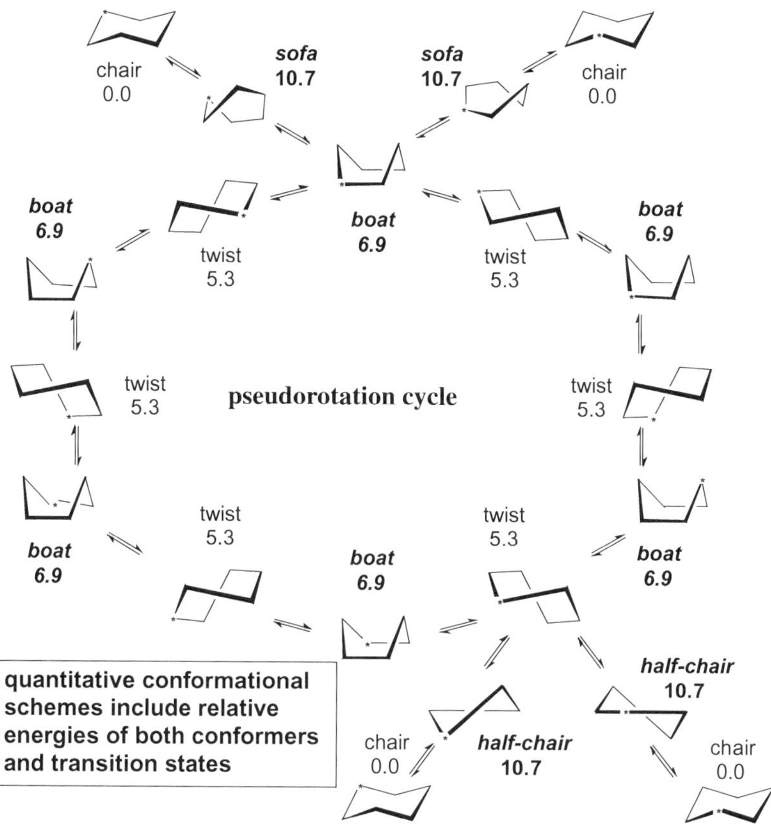

Figure 53. A quantitative conformational scheme for stereodynamics of monocycle **18a** at room temperature (asterisk is a formal label). Trivial names for confomers as well as transition states (in bold italics) are given. Numbers show calculated relative energies (kcal mol^{-1}) for these stereoisomeric structures (in bold for transition states). For clarity, pathway *chair–half-chair–twist–half-chair–chair* as well as pathway *chair–sofa–boat–sofa–chair* are shown only for one twist and one boat structure from the pseudorotation cycle.

transition states of boat geometry, forming a cycle of six conversions (the cyclohexane pseudorotation cycle). Hence, twist conformers are more flexible (kinetically more easily interconvertible) than chair conformers. Each chair conformer changes its geometry through two different transition states (half-chair and sofa), while each twist conformer converts it through four transition states (two boat- and two half-chair-shaped ones). Having been accustomed to in-plane representations of conformational pathways and transition states (Figs. 39, 46, and 48), non-specialists could learn from the conformational scheme for **18a** that a transition state may be transformed to another transition state directly, without passing through an energy minimum (Section 3.1, Fig. 57). Indeed, the sofa and boat stereoisomeric structures of this six-membered ring are transition states, and, nevertheless, no stable conformer separates them. While easily tracking changes of molecular geometry when reading such conformational schemes, one should remember that they show molecular shape that contains "chemically non-functional" elements, transition states: as mentioned, only thermodynamically stable conformers may be reactive species.

In the present account, the reader has met a similar scheme earlier. In Section 2.2, Fig. 37, interconversion of twist conformers for piperidine **14b** is given as a conformational scheme with conformers and the corresponding transition states. The description of a flexible molecular geometry as a set of conformer and conformations is so transparent, as well as convenient, that the conformational scheme in Fig. 37 is clear without the need for explaining it. From the conformational schemes in Figs. 37 and 53, we can quickly "catch" the major discrepancy between conformational dynamics of twist forms in carbocycle **18a** and azaanalog **14b**: N inversion doubles the number of twist and sofa conformers in the azacycle relative to the carbocycle (conformational dynamics of the amino compound includes two pseudorotation cycles; Fig. 52).

Let us assume that we have determined the relative stability of conformers and their lifetimes at a given temperature. Supplying conformers as well as transition states with these values (i.e., with relative energies of conformers as well as transition states), we transform a qualitative description of molecular flexibility (as it is, e.g., in Figs. 37 and 52 for azacycle **14b**) into a *quantitative* description (as, it is, e.g., in Fig. 53 for cycle **18a**). This supplement "returns" the timing to molecular shape, i.e., this "energy language" may be easily translated into relative stationary fractions of conformers in the equilibrium and the rate constants for their interconversions at this temperature (see Eq. 12, Section 3.3, and Eq. 14, Section 3.5.3, respectively).

Thus, trying to describe flexible molecular geometry as accurate as possible, we have traversed the path from a non-informative molecular ball (Fig. 50) to *quantitative conformational schemes* (i.e., the schemes with structures of known relative energies[a] as, e.g., the scheme in Fig. 53), which well represent flexible molecules. Nevertheless, as emphasized above, conformational transition states are not participants of any interactions which chemically change molecular structure (e.g., of chemical reactions). Do synthetic chemists require such detailed conformational delineations (conformational schemes) that include transition states of conformer interconversions?

At first glance, quantitative conformational schemes seem belongings of sole conformational considerations. However, the conformational concept does not solely consists of a description of molecular shape. It also includes the indispensible Curtin-Hammett principle that links *conformers* to *selectivity* of chemical reactions.[b] The prerequisites for using this quantitative prediction of the chemical product ratio (the relative content of reaction products) are the known (determined) values of relative thermodynamic stabilities of conformers and kinetic rates of their interconversions, i.e., exactly the data for the conformational equilibrium that we find in quantitative conformational schemes with included transition states of conformer stereoisomerizations. Therefore, quantitative descriptions of time-dependent molecular shape that contain both relative energies of conformers and kinetic barriers for conformer interconversions (one may read this as quantitative conformational schemes) are of solid benefit for synthetic chemistry.

In simple words, quantitative conformational schemes are a precise tool for *synthetic* research. As stressed above, only in-time stable stereoisomeric structures (i.e., stable conformers) of the compound interact with molecules of the reagent. Why are conformational transition states so important in the analysis of, e.g., stereoselectivity of organic reactions? Conformers are interconverting species. The kinetics of conformational transformations may be competitive with the kinetics of chemical reactions and, then, relative energy of

---

[a] In practice, these energies are supplied by theoretical calculations (Chapters 4 and 5).

[b] It is explained in many stereochemistry textbooks. Nevertheless, a review written by J. I. Seeman[33c] probably is the most useful account on this quantitative principle that, by predicting the ratio of the products of a chemical reaction for different conformational equilibria of an edduct (initial compound), makes conformational studies necessary both in rationally approaching to organic synthesis of desired compounds and understanding the regio- or stereochemical outcome of chemical interaction of concrete compounds.

conformational transitions states (relative height of conformational barriers) becomes crucial for stereoselectivity of the reaction. The Curtin-Hammett principle solves this kinetic problem by considering all the involved kinetic rates together, of both chemical and stereochemical transformations. In terms of kinetic barriers, the selection principle is *the lower the more favored*.

Incorporation of "energy-equipped" conformational transition states into conformational schemes supplies the required kinetic data to the Curtin-Hammett "selection machine" *the lower the more favored* and, thus, provides analysis of *selectivity* in chemical interactions. Thus, if one needs to analyze or, more importantly, predict stereochemical outcomes in chemical interactions of concrete organic compounds in solution or gas, one should seek (i.e., model) exactly this set of arranged conformers and conformational transition states with relative energy "labels."

Conformational analysis (identification of conformers and determination of their relative stabilities) alone is a "lower level" methodology because, in principle, it cannot provide an unshakable basis for stereochemical predictions in synthetic chemistry. However, the modeling of transition states is time-consuming or, sometimes, problematic, and conformational studies are often restricted to only performing conformational analysis. One should remember that a widespread practice to explain stereochemical results by assessing exclusively conformers ignores conformer interconversions and, therefore, may mislead.

Methodologically, a quantitative delineation *conformational dynamics in conformational space* is the ultimate result of conformational studies that are oriented to the needs of organic synthesis. Quantitative conformational schemes are a convenient form of this delineation, and it is worth targeting them. Nevertheless, conformational analysis should not always be discriminated as an absolutely inappropriate platform for stereochemical explanations. If the minimal energy conformer is highly predominant (say, more than $5\,\text{kcal mol}^{-1}$ lower relative to the "nearest" higher energy conformer), other conformers scarcely are more reactive. An appreciably higher chemical reactivity of the less favored conformer means that the kinetic barrier for the related chemical reaction of this conformer is more than $\sim 5.5\,\text{kcal mol}^{-1}$ lower than the kinetic barrier for the most stable conformer. For conformers, which geometries are not very dissimilar, differences of the reaction barriers usually are less than $5\,\text{kcal mol}^{-1}$. Therefore, if relevant, the regio- or stereoselectivity is provided by mainly involving this minimal energy conformer into the reaction.

2.9 Conformer "Freezing" in the Solid State

Our last comment with respect to flexible molecular shape concerns its rigidity in molecules-in-crystal. In the solid phase, *intermolecular* forces are sufficiently strong to hold molecules so tightly that any intramolecular motion is prevented. Only small amplitude motions are present, and those are oscillations of the backbone and not its 3D rearrangements. Thus, no conformers exist in the solid state. In terms of the universal conformational approach, these structures are rigid stereoisomers because their lifetime is much more than 24 hr (these lifetimes may be considered as long as the lifetime of the crystal lattice). Thus, structural studies of compounds in the solid state are not conformational. Nevertheless, molecular geometry supplied by crystallographic researchers is often used as a complementary, supporting or directing information when determining molecular geometry of stable conformers or rigid stereoisomers in the liquid or the gas phase. One should note that molecular geometries extracted from X-ray diffraction analysis require a deliberate focus on their transferability to 3D geometries of molecules in solution.

What is the geometry that is "chosen" by a molecule embedded in a crystal lattice, among its many geometrical alternatives? Sufficiently often, the 3D structure of such molecules is similar to the geometry of the stereoisomer that is the most stable 3D structure in vacuum, and no other stereoisomers are "frozen." However, fixation in the solid state of the most stable stereoisomer in solution or in vacuum is absolutely not a rule. Spatial structure in the solid phase is not determined exclusively by *intramolecular* interactions; *intermolecular* forces may "force" molecules to adopt the geometry of a high energy stereoisomer (in vacuum), or an appreciably distorted geometry of any stereoisomer, or the geometries of two or more stereoisomers. For instance, for dimer **41** (Section 6.1.3, Fig. 94), crystallization "captures" four stereoisomers (two geometrically different pairs of enantiomeric structures) from twelve ones in solution.

For the same stereoisomer, geometrical parameters of the structure "from the crystal" and the structure "from the solution" are always somewhat different. If two or more stereoisomers form the crystal lattice, they are not a statistical ensemble of independent (non-interacting) molecules, and their relative populations absolutely do not follow Boltzmann statistics[a] that describes conformational equilibria in the gas or the liquid phase (Section 3.2, Eq. 6).

The possibility that molecules of a compound have different favored geometries in the solid state *vs.* the liquid or gas phase should be taken into account by organic experimentalists who think that the "freezing" of the most stable stereoisomer upon crystallization ubiquitously occurs.[b] Crystallographically determined molecular 3D structure, in principle, neither supports nor demolishes any structural hypothesis for the favored conformer in solution.

To a greater extent, this methodological remark should be related to large multi-conformer systems, where proteins, oligosaccharides, and lipids are remarkable examples. Small energy differences for many conformers in such compounds or conformer-specific solvation effects permit a minor conformer in solution to become a favored structure when such a compound is crystallized. Therefore, bioorganic specialists consider the 3D structures of biopolymers in solution with an unremitting caution when they only have data for X-ray-resolved geometry. The role of these data is often auxiliary, and they are used as a blank for planning the strategy of the series of further NMR experiments for solution or as an initial structure in computational refining of the geometry determined in the solid state.

An example of even conformationally non-intricate compounds illustrates this need in precautions. The 1,3-dioxophosphorinane ring of xylofuranosides **24a,b** (Fig. 54) is flexible in solution at the room temperature.[42] Compound **24a**, according to X-ray diffraction analysis, has a chair geometry for the six-membered ring in the crystalline state. NMR shows that its preferable geometry in solution is twist. Clearly, the data of X-ray analysis for the solid state do not overthrow the NMR-based assignment for solution; simply, relative energies of the same stereoisomers of this compound are dissimilar for these different condensed phases.

The case of analog **24b** (a compound of an opposite P configuration relatively to **24a**; Fig. 54) should be similarly treated. Two stereoisomers of phosphate **24b** are co-crystallized as 1:1 components of the crystal lattice.[42]

---

[a]Or any other statistics.

[b]The statement *the most stable structure* implies *at given conditions*. Without revealing the molecular geometry of an organic compound in both the solid state and solution, one should not attribute the molecular geometry that has been determined for the solid state to the most stable conformer in solution.

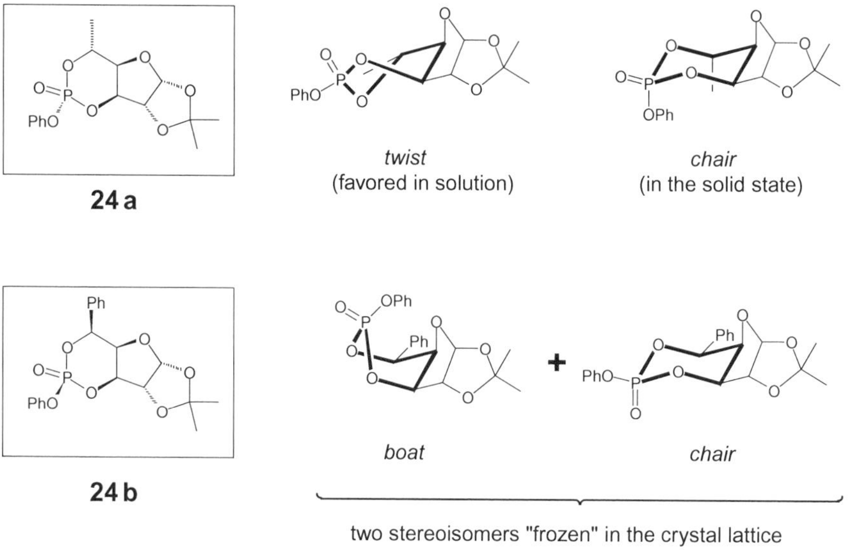

Figure 54. "Frozen" conformers of xylofuranose phosphate diesters **24a,b** in the solid state (Ref. 42).

Does this mean that these stereoisomers are equally predominant conformers in the equilibrium in solution, while other conformers are very minor forms? At first glance, the boat and chair stereoisomers are at least of comparable stability; their 1:1 ratio in the crystalline state seems an obvious indication. However, this conclusion is ungrounded and rather incorrect. Their equal content is due to their successful co-packing of these stereoisomers in the crystal lattice, it does not show the relative stability of isolated 3D structures. Only experimental data for solution (or reliable theoretical calculations, of course) can provide valid information about the preferences for molecular geometry and energy in conformational equilibrium in solution that may appear similar or dissimilar to the preferences in the solid state.

Our ultimate goal in describing molecular shape (Chapters 1 and 2) has been reached. Using the familiar language of interconverting structures, we can easily describe flexible organic molecules as arranged sets of stereoisomers that possess different lifetimes, and transition states that separate the related stable stereoisomeric structures. Such sets in a clear form of quantitative conformational schemes are starting data, which are necessary for a valid understanding or rational planning of stereochemical results in organic synthesis. However, we absolutely do not know yet how to locate such a goal set of arranged conformers and conformational transition states for a concrete compound. We can only guess that this methodology may be theoretical modeling. Therefore, the next topic in our account is the PES (potential energy surface, Chapter 3). This important concept will further assist in the explanation of how to model arranged sets of conformers and transition states (Chapters 4 and 5), with understanding the limitations of the theoretical methods and computational resources that can be exploited in the organic laboratory. Thus, starting from Chapter 3, our goal is mastering in practical realistic modeling of molecular shape.

3

Quantitative Description
of Conformational Space

3.1 Potential Energy Surface

In previous chapters in considering *conformers* of organic molecules, more than once we mentioned the term PES (potential energy surface). Synthetic chemists, of course, freely use or at least often encounter energy-related terms, such as *molecular energy, potential energy, molecular electron energy, steric energy, strain energy (strain), energy minimum/maximum,* and *energy barrier,* in characterizing molecules. However, as a rule, specialists in synthetic methods are not able to explain exactly what these terms mean. Indeed, which energy category do we mean, when associating the energy minimum with either molecular geometry? Which theoretical calculations energy differences that may be associated with Gibbs energy differences given by experiments with non-single molecules? Synthetic chemists usually are unsure of their answers to such questions. Nevertheless, as mentioned, organic and bioorganic experimentalists have become accustomed to operating with energy terms. The principle *the lower the better,* which relates energy and thermodynamic stability, is a convenient postulate for non-theoreticians since it supplies the easily digestible information regarding the relative stability of molecular systems (e.g., conformers). Thus, an accurate understanding of molecular energetics is also required in the organic laboratory. Let us recapitulate very qualitatively some basic energy terms that are relevant to conformational analysis.

The fundamental notion of quantum mechanics is that a microsystem may be fully described by function Ψ (the wave function). How does this unique Ψ describe such systems? Appropriate operators A (mathematical actions), when applied to Ψ, afford physical quantities of a molecular system (e.g., energy, dipole moment, polarizability, and magnetization):

$$A\Psi = a\Psi \qquad (3)$$

where a is the numerical value (eigenvalue) of the corresponding molecular property of the system. In general, independent variables for Ψ are geometry parameters (for simplicity, generalized parameter r) and time t. This means that this $\Psi(r, t)$ "displays" quantitative (measurable or computible) molecular properties for any moment in time, for any geometry of the system, and for its different electron configurations. One could scarcely desire a better delineation of molecules. As Cramer's *"Essentials"* notes[33b], "...perhaps the best description of Ψ at this point is that it is an oracle — when queried with questions by an operator, it returns answers." On the other hand, even postulating the common mathematical requirements to Ψ (Section 1.3), we cannot indicate any physical principle that would provide an explicit, numerically computible mathematical form of the wave function for a given molecular structure. This circumstance seriously obscures our understanding of the wave function; we have no "classical physical" explanation of what Ψ is. The quantum mechanical interpretation of $\Psi(r, t)$ is deep but not instructing: the square of the modulus of Ψ, i.e. $|\Psi(r, t)|^2 = \Psi(r, t)\Psi^*(r, t)$ (where Ψ^* is the complex conjugate of Ψ), shows the probability of locating the system in geometry r at time moment t.

If Eq. 3 is related to the energy of the system, operator A is the commonly known Hamiltonian operator H.[a] Probably, even non-specialists easily recognize Eq. 4 (below) as the Schrödinger equation.

$$H\Psi = E\Psi \qquad (4)$$

For a molecule, the values of E for different geometries characterize its total energy (molecular energy) for these stereoisomeric structures. In the

---

[a] $H = \frac{\partial}{\partial x}\mathbf{i} + \frac{\partial}{\partial y}\mathbf{j} + \frac{\partial}{\partial z}\mathbf{k}$. It is not our intent to present detailed mathematical evaluations from quantum mechanics. We only show the principal way to acquire the energy values for molecules. The curious synthetic chemist could refer to numerous monographs on the basics of computational chemistry (see, e.g., Refs. 14, 33b, and 62a,b).

absence of an external field (electric or magnetic) and appreciable relativistic effects, *H* deals with five contributions to the total energy, i.e., the kinetic energy of electrons, the kinetic energy of nuclei, the energy of attraction of electrons and nuclei, the energy of interelectron repulsion, and the energy of internuclear repulsion. We can see that Eq. 4 provides the relationship between molecular energy and molecular geometry; it is exactly that accurate quantitative relationship which intrigues us. Unfortunately, theoretical evaluation of Eq. 4 is very complex for polyatomic molecules, because the motions of electrons and nuclei are correlated (are not independent of each other). This also means that numerical computations which are the ultimate step in modeling concrete organic structures would consume unbelievably abnormal computer resources and time. However, theoreticians have found a way out of this situation. It appears that very often one can include into consideration not all the mentioned energy contributions and still have a sufficiently accurate relationship *molecular energy–molecular geometry*.

In molecules, electrons are changing their trajectories much more rapidly than the nuclei, which are very massive particles, if compared with electrons. Therefore, one can accept that the 3D geometry of the nuclear backbone has no time to change when electrons are relocated. In other words, electron rearrangement in the space is approximately instantaneous relative to the displacement of nuclei, and one can consider electrons and nuclei in a molecule separately. This deliberate assumption is called the Born–Oppenheimer (or adiabatic) approximation. It is sufficiently accurate for ground states and often for excited states of organic molecules; it becomes invalid for degenerated (equal in energy) or almost degenerated electronic states. Thus, the Born–Oppenheimer approximation is suitable for calculations of energy of the vast majority of organic compounds. Of course, computational chemists have not missed the opportunity to get an adequate quantitative characterization of molecules using a manageable quantum mechanical model. An overwhelming majority of computational studies of organic compounds is performed within the limits of adiabatic approximation, which is such an obvious prerequisite that it is not even indicated in the related publications. Further, we will consider molecules in light of this generally used convention.

Why is the Born–Oppenheimer model so convenient? Fixing positions of nuclei upon "free" relocation of electrons, we perform three essential simplifying actions: (*1*) we set the kinetic energy of nuclei to be independent of the motion of electrons; (*2*) we eliminate the term that describes the correlation of the energy of attraction of electrons and nuclei; and (*3*) we transform the contribution of the repulsion energy of nuclei into a constant value (V_{nucl}) for given coordinates of the nuclei. These changes give the time-invariant electronic Schrödinger equation for electronic state p (Eq. 5)

$$(H_{el} + V_{nucl})_p \Psi_{el}(r_{el}; R_{nucl})_p = E_{el(p)} \Psi_{el}(r_{el}; r_{nucl})_p \qquad (5)$$

where subscripts "el" and "nucl" show that the terms are related to electrons and nuclei, respectively; V_{nucl} is the energy of internuclear repulsion; r_{el} designates the coordinates of electrons, r_{nucl} designates the coordinates of nuclei and $E_{el(p)}$ is the eigenvalue of Ψ_{el} for electronic state p. This energy $E_{el(p)}$ is termed *potential energy*. Further, index p is omitted; it is implied that E_{el} is related to a distinct electronic state. Notably, r_{el} are independent variables, while the backbone geometry r_{nucl} (a set of coordinates of all nuclei) is a set of numerical parameters. Different constant values of V_{nucl} correspond to different sets of fixed coordinates r_{nucl}. Pure electronic energy, which is "supplied" by operator H_{el}, is also dissimilar for non-identical geometries of the nuclear backbone. Therefore, in general, distinct sets of nuclear coordinates (i.e., different geometries of the molecular backbone) provide different values of E_{el}.

Thus, potential energy E_{el} is a numerical solution of the electronic Schrödinger equation in the Born–Oppenheimer approximation, i.e., Eq. 5, for any set of nuclear coordinates (i.e., any fixed set of parameters r_{nucl}). With recalling the above indicated contributions to molecular energy, one can say that adiabatic potential energy E_{el} is the sum of the electron energy and the energy of nuclear-nuclear repulsion. The total energy of a molecular system of fixed 3D geometry, in the adiabatic model, is the sum of potential energy E_{el} and the kinetic energy of the nuclei. Or, it is the sum of the energy of electrons and the energy of nuclei (no contribution of correlation energy *nuclei–electrons*). Certainly, non-theoreticians can readily "digest" these energy meanings.

The solution of Eq. 5 for given coordinates r_{nucl} (i.e., the E_{el} value for the "frozen" nuclear backbone) supplies us with one point in the space *nuclear coordinates* r_{nucl} — *potential energy* E_{el}.[a] For different r_{nucl} values, i.e., for non-identical 3D geometries of the molecular backbone, Eq. 5 provides different values of E_{el}. That is, $E_{el}(r_{nucl})$ is a single-valued function of nuclear coordinates r_{nucl} (independent variables r_{nucl}) and potential energy (dependent variable E_{el}). This function $E_{el}(r_{nucl})$ actually is the PES. In other terms, PES is a surface in the multidimensional space *nuclear coordinates* r_{nucl} — *potential energy* E_{el}. Notice that PES is a function that does not include the temperature variable, i.e., PES "is" temperature-independent.[b]

A hypothetical PES with two geometrical coordinates and the energy coordinate is shown in Fig. 55. This example clearly illustrates that the adiabatic model supplies us with a *quantitative* relationship *molecular*

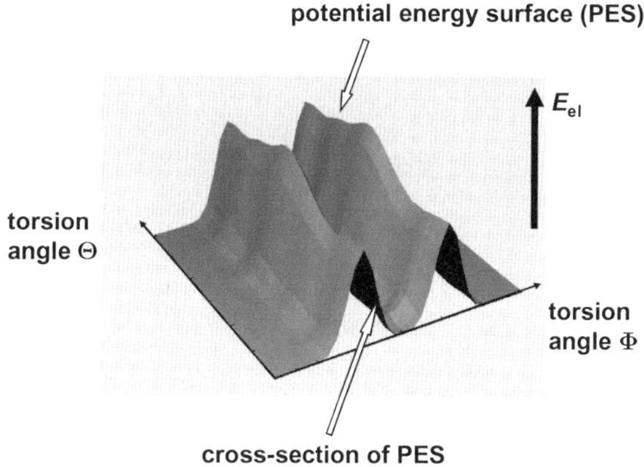

Figure 55. A fragment of the PES of a hypothetical molecule (here, independent variables x and y are two torsion angles Φ and Θ). From this energy landscape, one can learn straight away that internal rotation around a certain bond (changes of angle Φ) significantly alters potential energy of the system, while rotation around another bond (changes of angle Θ) has little effect.

---

[a]Since addition of a constant to the Hamiltonian operator does not change Ψ, Eq. 5 may be simplified by excluding V_{nucl}. To obtain E_{el}, the value of V_{nucl} is simply added to the solution (the energy of electrons) of the simplified equation. Again, no description of how to solve Eq. 5 is given here; this equation is presented only as an indication of the principal way to obtain energy values E_{el}. For further evaluation of Eq. 5, see Section 5.1.

[b]For taking into account the temperature impact on the energy landscape, we would need a function *nuclear coordinates* r_{nucl} — *free (Gibbs) energy* [see Section 5.3.4, subsection (3)].

geometry–energy. In contrast to different primitive approaches to molecular geometry (Chapter 1), this relationship is physically well-based.

Polyatomic molecules contain at least three nuclei, and their nuclear coordinates r_{nucl} are a set of geometrical parameters and not a single parameter. Therefore, the PES for such a molecule is actually a hypersurface (multidimensional surface) in hyperspace (multidimensional space) *nuclear coordinates–potential energy coordinate*. An accurate term would be potential energy hypersurface. However, it is a rarely used term in chemical literature. Parameters of molecular geometry (interatomic distances, bond angles, torsion angles) may be used instead of Cartesian coordinates of nuclei for constructing this hyperspace. The number of these parameters (trivial name in computational chemistry: internal coordinates) is less than the number of Cartesian coordinates for all nuclei of the molecule. Besides, in many cases only local changes of molecular geometry (recall, in Chapter 1, we also discussed *molecular fragments*) are important in stereochemical considerations. Therefore, it is convenient to analyze molecular geometry with using these molecular backbone-oriented descriptors.

Two parameters of molecular geometry (e.g., two torsion angles as shown in Fig. 55) are often sufficient to describe changes of 3D geometry because these changes more frequently are only local. Then, we have a 3D (Euclidean) space and the term PES is indeed accurate. Chemists, of course, have been accustomed to analyze modeled 3D PESs (also called landscapes of potential energy) and locate minima, maxima and saddle points (see below) visually. If more than two geometrical parameters are subjected to changes, the easiness of the minima/maxima/saddle point location disappears, and n-dimensional landscapes with $n > 3$ can only be "viewed" (i.e., analyzed for locating molecular geometries that correspond to these points) by means of computational algorithms.

As indicated, every electronic state of a molecule is characterized by its own PES. Sometimes PESs for distinct electron states (e.g. the ground state singlet $^1S_0$ and the first excited singlet state $^1S_1$) or for electron states of different multiplicity (e.g., singlet $^1S_0$ and triplet $^1T_0$ for the ground state) of a molecule have a common point (points). Such PESs are called degenerate.

It is apparent that familiar 2D energy plots in the coordinate system *geometrical transformation–energy* (e.g., Figs. 12, 39, 43, 44, 46, and 48) are cross-sections of such PESs in the plane given by two coordinate axes, the energy coordinate axis and the axis of one geometrical parameter coordinate. The cross-section of the PES shown in Fig. 55 illustrates the "genesis" of such 2D plots. For this hypothetical PES, the cross-section in plane *rotation coordinate–potential energy coordinate* is a plot that displays the changes in potential energy E_{el} upon rotation around a bond (a continuous change of one torsion angle). It is not difficult to discern that the classical energy plot from Fig. 12 is a similar cross-section of the PES for ethane. Thus, the PES concept is actually known to organic experimentalists in the truncated form of illustrative 2D contours (PES cross-sections) which show MEPs for conformational or chemical transformations (for MEP, see Fig. 58 below). Let us survey briefly some qualitative features of the parent surface, i.e., PES.

In chemical publications, one may encounter other meanings of potential energy of molecules, e.g., the well-known 1D Morse potential, parametric functions used in molecular mechanics (Section 4.1), or other similar semi-empirical functions. Note that this classical physical understanding is limited when applied to molecular objects, while the quantum mechanical meaning of PES is "molecules-targeted." Therefore, when potential energy is mentioned in competent publications related to conformers, the reference is usually to the potential energy of the Born–Oppenheimer approximation.[a]

As we remember, by the universal conformational approach, energy minima correspond to conformers, while energy maxima in 2D energy contours (e.g., shown in Fig. 48) correspond to conformational transition states. Thus, the related PES regions are of interest to us. One could easily recognize in the energy minima from such contours the energy minima of the PES. However, which PES regions are related to transition states?

We need to recall some mathematical meanings before answering this question. Stationary (critical) points in a 2D PES are points in which both first derivatives $\partial E/\partial x$ and $\partial E/\partial y$ are zero; saying metaphorically, at stationary points, energy E_{el} has no inertia to increase or decrease. These points are either minima, or maxima, or inflection points (Fig. 56, top). We know

---

[a]Excluding publications related to molecular mechanics calculations. There, a differently defined energy quantity — steric energy (E_{ster}, Chapter 4) — is in the focus, and the term *potential energy* is applied to this molecular quantity. This potential energy E_{ster} may only be considered as a vague analog of potential energy E_{el}.

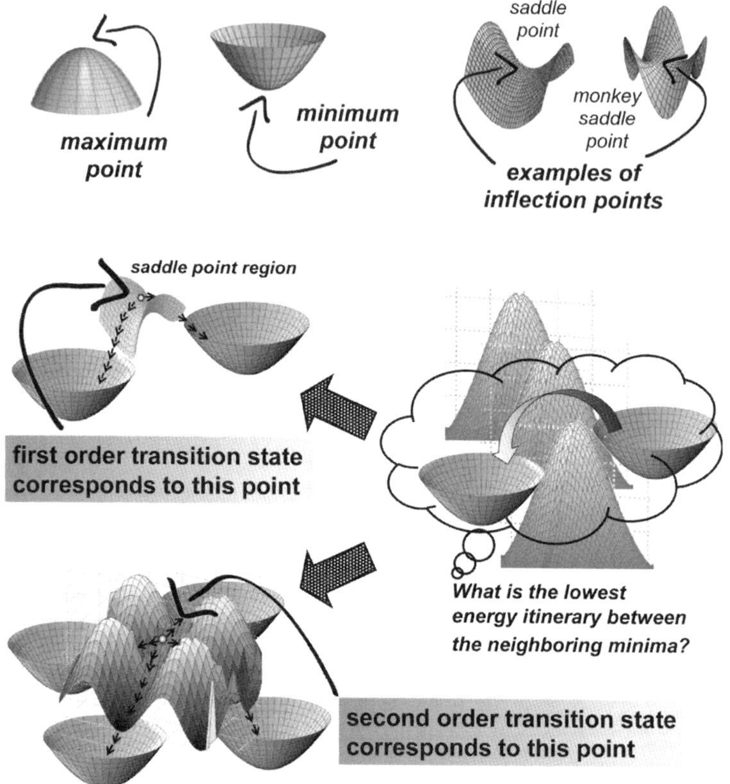

Figure 56. Stationary points for a 2D surface (top) and PES regions (below) which include inflection points that correspond to the first- and second-order transition states (shematically indicated by small balls). Small arrows show departure paths from these inflection points to the related energy minima.

that a maximum in the PES is a point which has the maximal value of the dependent variable (here: potential energy E_{el}) for some continuous interval of both independent variables x and y (here: two independent parameters of molecular geometry). Similarly, a local minimum in the PES is characterized by the minimal value of E_{el} for some continuous interval of both x and y. A point of inflection in a 2D surface is a stationary point which "combines" a minimum and a maximum in the manner as follows: it is a maximum (i.e., a maximal E_{el} value) for a set of values x, y of these independent variables in some their continuous intervals and a minimum for such another set of values x and y. In the simplest case, it is a maximum for a continuous interval of one independent variable (say, x) with keeping the other independent variable y constant ($y = const_1$) and a minimum for an interval of y values with keeping variable x constant ($x = const_2$); such an inflection point is called saddle point (Fig. 56). Clearly, inflection points are neither minima nor maxima in a surface.

A 2D surface is actually a function z of two independent variables x and y [in mathematical notation, $z = f(x,y)$]. Thus, a 2D PES formally is a function $E = f(x,y)$, where x and y are independent parameters of molecular geometry. For the minima, (*1*) both second derivatives $\partial^2 E/\partial x^2 > 0$ and $\partial^2 E/\partial y^2 > 0$, and (2) $\partial^2 E/\partial x^2 \times \partial^2 E/\partial y^2 - \partial^2 E/\partial x \partial y > 0$; for the maxima, (*1*) both $\partial^2 E/\partial x^2 < 0$ and $\partial^2 E/\partial y^2 < 0$, and (2) $\partial^2 E/\partial x^2 \times \partial^2 E/\partial y^2 - \partial^2 E/\partial x \partial y > 0$. For inflection points, $\partial^2 E/\partial x^2 \times \partial^2 E/\partial y^2 - \partial^2 E/\partial x \partial y < 0$. At inflection points, the value of the first derivative $\partial E/\partial x$ is either a local minimum or a local maximum (for the $\partial E/\partial x$ values in some continuous interval of x values), while the value of the other first derivative $\partial E/\partial y$ is either a local maximum or a local minimum in some continuous interval of y values, respectively. When numerically calculated for concrete molecular structures, these derivatives indicate whether the located stationary point is an energy minimum or an inflection point (Chapter 5). That is, such calculated values show whether the located molecular geometry belongs to a stable stereoisomer or to a transition state (see below).

Any two neighboring minima in the energy landscape are separated by a ridge of energy maxima (Fig. 56), and one should cross this ridge if traveling from one of the minima to the other one. Our question regarding transition states and the PES is reduced to indicating the points in the ridge that form the path (in the PES) of passing over the ridge. The energy principle *the lower the better* again appears valid. The favored path for molecular geometry to pass the energy ridge lies through the lowest point there. Such a point is an inflection point in the PES; it corresponds to a conformational transition state (Fig. 56). The transition state that corresponds to a saddle point is called the first-order transition state. Any saddle point has only two descents (valleys in the PES landscape) towards the energy minima. That is, each first-order transition state "connects" only two conformers.

Thus, from the PES perspective, transition states are molecular structures with the geometry that corresponds to PES inflection points. Inflection points in the PES do not have to be exclusively saddle points. In more compound energy landscapes, minima may be isolated from each other by a conglomerate of energy hills that have no energy minimum between them (Fig. 56). The stationary point, which lies in this region and is not a maximum, corresponds to a transition state. In the simple case of a two ridge conglomerate of separating maxima, this point is the common maximum for two sets of values x_1, y_1 and x_2, y_2 of independent variables x and y as well as the common minimum for two other similar sets x_3, y_3 and x_4, y_4, where x_i and y_i are sets of values that correspond to some continuous intervals for independent variables x and y, respectively. This means that there are four descending paths from this point to points of lower energy and, ultimately, to four energy minima. Transition states that are related to such stationary points are called second-order transition states.

Geometrically, multi–descent inflection points may be generated from the first-order saddle point region by raising a maximum in a descent (valley) near the saddle point. For instance, a three–descent saddle point (so-called monkey saddle point; Fig. 56, top) is produced by erecting a maximum in one of the descents of the saddle point. If similarly transforming energy valleys in the region of the second-order transition state, a saddle point with more descents is formed, and one can speak about third-order transition states. Clearly, such formal evaluations of the PES may be continued *ad infinitum*. They reflect real pathways in changing molecular shape: high order transition states do exist for molecules. However, it is worth noting that first-order transition states for intramolecular stereorearrangements of organic systems have lower energy than transition states of higher orders, and, hence, there is practical interest in preferably locating and analyzing exactly the first-order ones.

One should stress that transition states do not correspond to maxima of potential energy; recall that inflection points are not maxima. Therefore, familiar 2D contours of pathways *conformer–transition state–conformer* (exemplified by Figs. 39, 43, 44, 46, and 48) may be misunderstood. Their 2D barriers foster an illusion that transition states are located at energy maxima. A glance at any PES (see, e.g., Fig. 56) corrects this 2D view-inspired inaccuracy. As indicated, molecular geometry "travels" from a minimum to a minimum of the PES surmounting the hindrance of energy hills (maxima) *between* them, i.e., at saddle points.

Interestingly, in contrast to the above described situation with inter-conversion of stable stereoisomers (transfer from an energy minimum to another minimum with inescapably "visiting" an inflection point in the PES), a *direct* transformation of a transition state into another transition state is quite possible. Two "non-parallel" regions of different energy saddle points may merge into a T-shaped one so that there is no energy minimum

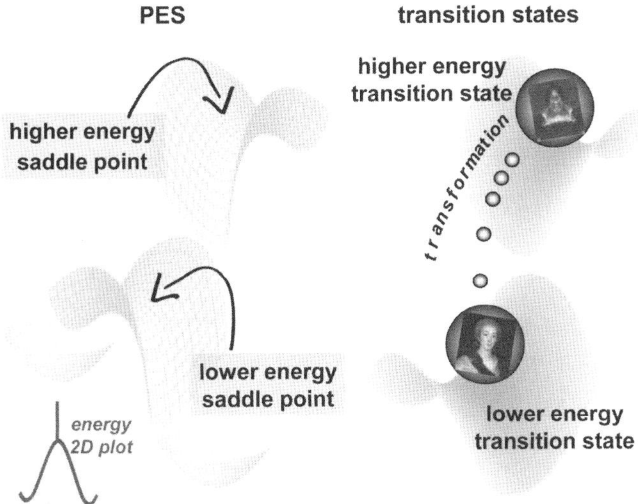

Figure 57. A PES region with two adjacent saddle points of different energy and mutually orthogonal orientation of energy valleys (vertically diverged and horisontally T-positioned "saddles"; left). This shape of PES provides direct transformation of a transition state into another transition state (right).

between them (Fig. 57). Then, the higher energy transition state (related to the higher energy saddle point) converts into the lower energy transition state (related to the lower energy saddle point) without sliding into an energy minimum.

Again, we see that the ubiquitously used 2D energy contours[a] may deorient. This truncated representation of the PES indicates that two neighboring energy "maxima" are necessarily separated by an energy minimum, or, equivalently, two transition states require a thermodynamically stable intermediate when transformed one into another. However, these high energy points in 2D contours actually are inflections points in the PES, and, hence, they correspond to transition states. As we learnt, transition states can transform directly into other transition states, and there is no necessary condition of an intermediate energy minimum. In order to describe the pathway of a direct *transition state–transition state* transformation by means of energy contours, one should superimpose two orthogonal cross sections of the PES in the T-shaped region with two saddle points. The resulting energy 2D contour (Fig. 57) does reflect this pathway in the PES. Nevertheless, in contrast to representations in PES terms, traditionally used energy 2D plots are an insufficient description of the *molecular*

---

[a]As mentioned, Figs. 39, 43, 44, 46, and 48 show examples of such contours.

geometry–energy relationship. We would be unable to interpret, e.g., the energy plot from Fig. 57, if we had considered it alone, i.e., without information about the PES shape in the related region. The graphical "language" of 3D surfaces is exhausting and accurate, while energy 2D contours are only conditionally adequate.

When encountering such geometrical constructions, i.e., different hypothetical PESs, non-computational chemists may ask whether they are a purely explanatory illustration of the relationship *molecular geometry–potential energy* or such an energy surface may be built for a concrete chemical system and, thus, the PES concept has a practical significance for conformational studies. Indeed, the wave function is cryptic for any polyelectron molecular system, and, therefore, at first sight, landscapes of potential energy of real organic molecules seem to be hidden from our eyes. Fortunately, this uncertainty is not absolute. Strict theoretical calculations are capable to mimic this "mysterious" function in one way or another (QM calculations, Chapter 5) or avoid "contacts" with it by using rational equations that include many empirical parameters (MM calculations, Chapter 4). Thus, one can *model* the desired relationship *molecular geometry–potential energy* by means of theoretical calculations. In other words, for any compound, the energy landscape may be reconstituted, and, "looking" at it, we can identify stable stereoisomers and the corresponding transition states as well as supply these 3D structures with related energies. Clearly, any conformational scheme (Section 2.8) actually is a "summary" of such a landscape scrutiny.

3.2 Conformer Traveling Along the PES

The discussion above is more focused on the shape of the PES. In other words, we have "viewed" an unpopulated landscape, i.e., the energies, which a molecule may have, when adopting certain geometries. Let us consider populated PESs. A single molecule is characterized by its own PES; an incorrect understanding would be that a single PES is populated by many molecules. Nonetheless, for convenience, it is legitimate to assess one PES for an *ensemble* of chemically identical molecules, implying that this PES is a superposition of individual *identical* PESs of these molecules. Since the resulting PES is identical to the single PES, let us continue to use the term PES also for it.

Let us neglect the entropy factor for a while (till Eq. 12 from Section 3.3). At equilibrium conditions, conformers are distributed in N energy minima of the PES according to Boltzmann statistics (Eq. 6):

$$p_i = \exp(-\Delta E_i/k_B T)/\sum_i^N \exp(-\Delta E_i/k_B T) \tag{6}$$

In Eq. 6, p_i is the population (fraction) of the i-th conformer, where the total population is numerically regarded as unity (i.e., $p_1 + p_2 + \ldots p_i + \ldots p_N = 1$); ΔE_i is the energy difference, ΔE_{el} for the i-th conformer and the lowest energy conformer. Consequently, the relative energy ΔE_i for the latter ($i = 1$) is zero. Other symbols have the same definitions as those used in Eq. 1 (Section 2.4). Equation 6 indicates that population p_i of the i-th conformer for a given temperature T depends on the number N of conformers and stabilities ΔE_i of all N conformers relative to the lowest energy. This is obvious: an additional conformer decreases the fraction of each conformer by taking a "portion" from the total population of 1. In contrast, relative population p_i/p_j of any two conformers i and j at temperature T is determined by only their relative energy ΔE_{j-i} ($\Delta E_{j-i} = E_j - E_i$), as Eq. 1 (derived from Eq. 6) explicitly shows. The dependence of p_j/p_i on ΔE_{j-i} is appreciably strong. For instance, if $\Delta E_{j-i} = -0.97\,\text{kcal mol}^{-1}$ for conformers j and i at room temperature, ratio p_j/p_i approximately is 9:1. When increasing this energy gap between the conformers by three times ($\Delta E_{j-i} = -2.91\,\text{kcal mol}^{-1}$), ratio p_j/p_i increases by eleven times ($p_j/p_i \approx 99$:1).

Equation 6 "explains" that at any temperature T $\neq$ 0, an ensemble of molecules populates all energy minima of the PES (recall that PES itself is temperature-invariant). Relative occupation p_j/p_i of energy minima by molecules changes with changing the temperature, and an increase of temperature leads to an increase of the content of higher energy conformers. However, this effect is considerable for T $>$ 400 K; high energy minima are almost "empty" at room and lower temperatures.

The bottoms of energy wells in the PES are not populated at all because these bottoms are minimal energy points (Fig. 42). A certain averaged 3D geometry corresponds to each minimum since, in occupying the energy well, molecules are distributed into vibrational-rotational levels there, as explained in Section 2.4. Modifying Cramer's cloud analogy, one may say

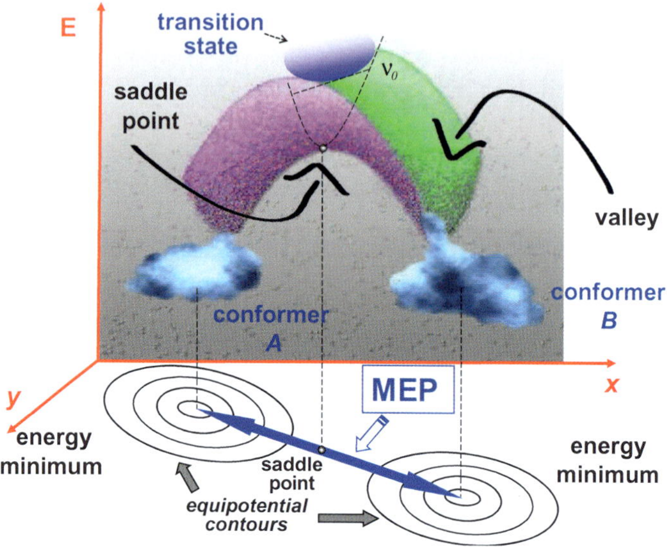

Figure 58. Schematic representation of the MEP in the PES for interconversion of conformers **A** and **B**. Equipotential contours are cross sections of the PES that are parallel to plane *xy*.

that a fog (Boltzmann-distributed molecules) fills minima of the energy landscape, not touching the bottoms of these wells, while other regions of the landscape are fog-free. The fog has a high density in low energy minima, while it is almost indiscernible in higher-lying energy minima.

What is the itinerary of this molecular fog in the PES when molecules are relocated from a minimum to another minimum? In "traveling" in the PES, i.e., in changing their shape, molecules prefer the path of minimal energy. In terms of the PES topography, it is the path that (*1*) passes over saddle points; and (*2*) has the trajectory, which is orthogonal to all equipotential contours (lines formed by equal energy points of the PES; Fig. 58) that it crosses. In simple words, it is the path that leads downwards from each saddle point[a] in two maximally steepest opposite descents, reaching both of the related energy minima. This trajectory along the PES is called the minimal energy path (MEP). One can see that, "connecting" two neighboring minima, the MEP lies at the bottom of two descending energy valleys

---

[a]The first-order inflection point is implied here.

(Fig. 58). Obviously, the saddle point is the highest energy point in the MEP. On the other hand, the saddle point lies lower than the highest energy point of any other itinerary between these energy minima in the PES.

We remember that a conformer is not represented by a point of the PES; what about transition states and saddle points? Similarly to a conformer, the transition state may be seen as a rotationally-vibrationally distributed molecular ensemble (Cramer's thinning cloud), which corresponds to an inflection point (e.g., to a saddle point; Fig. 58) in the PES. Since minimal energy points are not populated, one can say that the transition state lies "slightly above" the saddle point.

The only *evolution* of molecular structure (geometrical or other changes in an ultrashort time period) and not the structure at time moment is detected in ultrafast time-resolved spectral experiments. Although descriptions of such measurements often recruit the term transition state, these experiments actually track a molecular ensemble that moves along the MEP in a PES domain which incorporates a saddle point and some continuous PES area around this point (this domain may be called the transition state region). In contrast, theoretical calculations (Chapter 5) can model the transition state itself [the structure with molecular geometry that corresponds to one point (inflection point) in the PES and with vibrational energy levels that correspond to this critical point].

If there are several saddle points of different energy in the ridge that separates two energy minima, the MEP lies through the lowest energy saddle point, as shown in Fig. 59. We will see (Section 3.5.1) that this principle is important, in particular, for identifying NMR-detected conformational transformations.[a]

A simple analogy for the MEP in the PES with several saddle points is the pathway of a heavy fog through a chain of hills. Being in a potential field (the gravitational field of the Earth), the fog concentrates near the bottom of low places. When moving, the fog moves through the lowest pass, i.e., it follows the pathway with minimal heights of hindrances.

At this point, we have finished our qualitative description of the PES. It is legitimate to ask again why synthetic chemists need to know the PES

---

[a]When detecting occurrence of a fast stereoisomerization, any NMR (including dynamic NMR) experiment alone does not establish which 3D transformation is detected among alternative or sequential ones (Section 3.5.1). As explained there, an external (improved or theoretically based) conformational hypothesis is, in fact, always involved in interpreting the obtained NMR data.

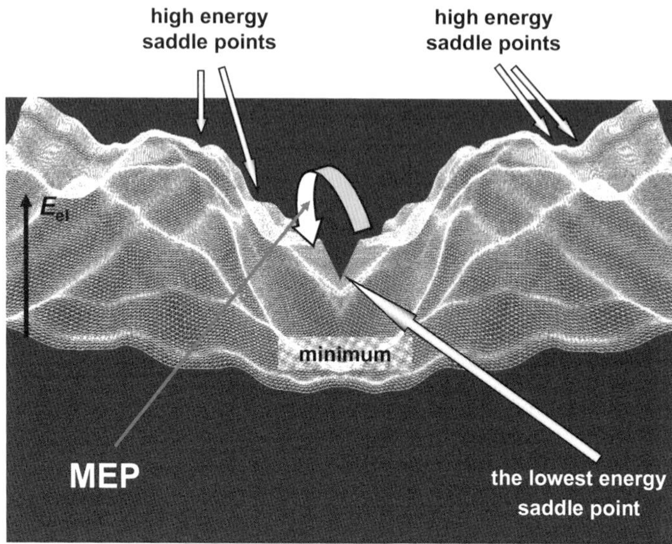

Figure 59. *The lower the better* for the MEP: a ridge of energy hills in the PES is passed through the lowest energy saddle point.

formalism. The answer is as following. Explicitly reflecting the relationship *potential energy–molecular geometry*, the modeled PES of an organic molecule actually supplies us with an arranged set of stable stereoisomers and transition states of their interconversions. These 3D structures "from" the PES are associated with energy values. Then, with the quantitatively modeled PES in hand, we know:

(*1*) which stable stereoisomers are conformers at the given temperature, or, in other terms, are flexible at this temperature (Section 2.5) since we possess information about interconversion barriers (saddle point energies) for each stable stereoisomer,

(*2*) which conformer is thermodynamically more/less stable than any other conformer since the modeled PES provides quantitative information about relative thermodynamic stabilities of stereoisomeric structures,

(*3*) which stable stereoisomers are interconverted *directly* (through a common transition state) since these stereoisomeric structures correspond to the neighboring energy minima in the PES,

(*4*) what the kinetic barrier of interconversion of a pair of any stable stereoisomers (not necessarily neighbors in the PES) is since it is trivial to indicate the MEP for them after modeling the PES (Section 3.5.2).

Thus, the PES modeled for a molecule supplies us with all information relevant to static and dynamic stereochemistry of this molecular structure. In other words, the modeled PES, by indicating stationary concentrations of conformers that populate its minima according to Boltzmann statistics and supplying conformer interconversions with related kinetic barriers, quantitatively describes the conformational equilibrium. This comprehensive information about molecular shape [i.e., answered questions (*1*)–(*4*)] is the ultimate goal of purely conformational studies.

Such studies, if considered separately, have very limited significance. However, they often become absolutely necessary in analyzing molecular phenomena which are more complex than stereoisomerism. The first thought that comes to mind in this connection is chemical and biochemical behavior of flexible molecules. For rational organic synthesis, understanding of biochemical reactions and, of course, mechanism-based drug design, results of conformational explorations are indispensible *starting* data. Chapters 4, 5 and 6 explain how organic experimentalists can obtain them without spending significant efforts. Nevertheless, before introducing this "magic" research tool to the reader, next Sections in this Chapter reiterate essentials of traditional conformational methodologies used in the organic laboratory and critically analyze them.

3.3 Conformational Equilibrium and Conformational Analysis

We have learned that mixtures of rapidly interconverting conformers are what is subjected to chemical interactions in solution. For organic experimentalists, it is important to understand what spectral methods used in organic research report about individual conformers under these conditions of conformer "appearance-disappearance." For instance, we certainly yearn to know from NMR or IR experiments what the number of major conformers in the mixture is and what these conformers are. Unexpectedly, conclusions derived from interpreting obtained spectra are not what one would expect from precise instrumental work. Different physical methods may supply researchers with distinct numbers of detected conformers for the compound of interest and, as a consequence, inconsistent information about the molecular flexibility, for the same experimental conditions.

In order to manage this seemingly contradictory situation in conformational analysis, let us learn qualitatively what the time resolution of conventional spectral methods means. NMR, mass-spectrometry, and sometimes IR and UV spectroscopies accompany synthetic organic work. Other spectroscopic methods, e.g., fluorescent, Raman, electron spin resonance (ESR), photoelectron spectroscopy, and X-ray diffraction are not routine tools in organic laboratories; electron diffraction and neutron scattering are only used in special structural studies. All these methods have different timescales for the observation of molecular structures; these time limits may be different by several orders of magnitude (Table 1). The timescale for a method is a criterion of the applicability of the physical method in kinetic studies. The timescale indicates how fast a structural rearrangement can be for still allowing a concrete spectroscopy to detect individual molecular structures at their rapid equilibrium. Using an analogy with photography, one may say that spectral methods take a "picture" of interconverting molecules with the exposition time of the method timescale. If the exposition time is shorter than the lifetime of a certain molecular structure, we obtain a clear, detailed "picture" of the latter. In the opposite case, a diffuse "image" is the only result. Thus, for detecting an individual stereoisomer, it is required that the timescale of the detection method should be shorter than the lifetime of the stereoisomer. The same interconverting stereoisomers, at the same ambient conditions, may appear as a mixture of separate species if short timescale methods are used or as a single virtual structure of in-time-averaged molecular geometry if longer timescale spectroscopies examine the equilibrium. Therefore, the observation result should be interpreted in connection with the timescale of the employed physical method of detection.

These limitations of the ability of spectral methods in respect to time resolution appear due to the Heisenberg uncertainty principle. Any spectral line cannot have an infinitely narrow width (to include a single frequency); otherwise, for the observer, a certain energy (the single registered frequency) would absolutely precisely associate with a certain time period (the time

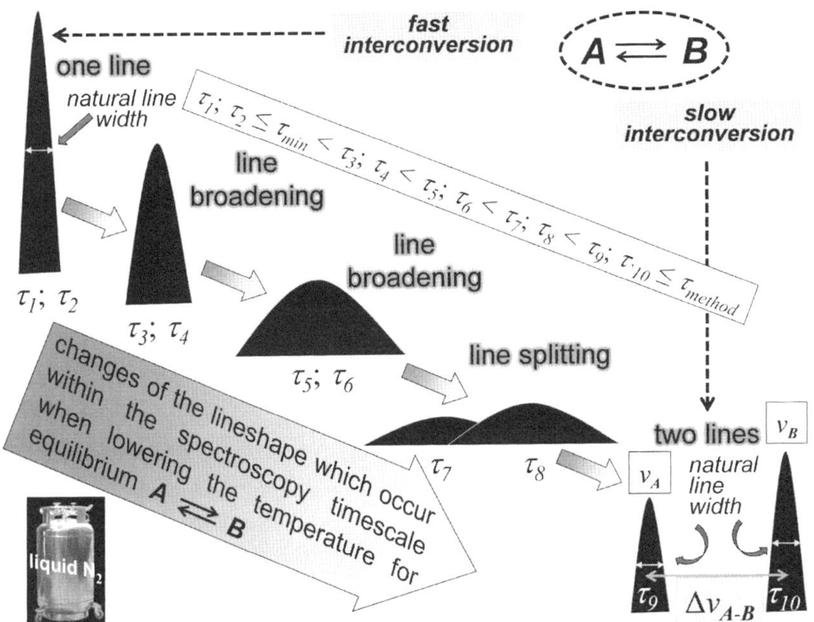

Figure 60. Evolution of the spectral lineshape triggered by varying lifetimes τ_i and τ_j of two interconverting molecular structures *A* and *B* (dotted arrows indicate the temperature decrease). The presence of one sharp line of natural width (not sharpened further when increasing the temperature) shows that the stereoisomer interconversion is fast in the timescale of the spectral method; the line corresponds to a virtual (not related to an energy minimum), time-averaged molecular structure. The presence of two individual spectral lines that arise from splitting the parent line show that the interconversion is slow in the timescale of the spectral method; each line corresponds to an individual molecular structure. Such spectral changes are observed in experiments, which change lifetimes τ_i and τ_j by manipulating the temperature, *e.g.*, in temperature-variable NMR experiments.

period of the quanta emission). Therefore, any line in any spectrum of a structurally static molecular ensemble has some width (so-called natural line width; Fig. 60). The broader the line is, i.e., the energy of the emitted electromagnetic quanta is more uncertain, the shorter the timescale is, i.e., the time period of the molecular emission process is more certain. This energy and the indicated, energy-related time period are different for different physical structural methods. In order to tentatively arrange various instrumental methods according to their timescales, first, one should compare their time constants τ_{method}. This constant τ_{method} is derived from Eq. 7:

$$\Delta\nu_{width} \times \tau_{method} = (2\pi)^{-1} \tag{7}$$

where τ_{method} is the time constant of the spectroscopic method (in seconds), and $\Delta\nu_{width}$ is the natural line width (in Hz) for this physical method.[a]

The method-specific value τ_{method} shows the upper limit of the timescale (time resolution) for the given physical method. Molecular systems with lifetimes, which are larger than the τ_{method} value for the observation method, are static (unchanging) for the observer who is using it. That is, the used analytical method cannot provide information regarding slow molecular dynamic processes if occurring in such systems. For instance, ideally, $\Delta\nu_{width}$ of proton resonance signals is $\sim$0.01 Hz.[b] Then, τ_{method} for ^{1}H NMR is estimated by Eq. 7 as $\sim$ 10 s. What does this value show for molecular structures that are in the thermodynamic equilibrium? It shows that a molecular dynamic process, which occurs slower than a 10 s time period passes,[c] does not alter ^{1}H NMR spectra [i.e., chemical shifts (resonance frequencies) and line shapes], and, therefore, cannot be studied by this spectroscopy. In other words, lifetime τ that is larger than the 10 s value (the lifetime that satisfies condition $\tau > \tau_{method}$) represents the molecular structure as static (non-converting) in the timescale of ^{1}H NMR. In this timescale, e.g. pentahelicene with τ of approximately 63 min (Section 2.5) is an ensemble of rigid molecules.

The natural line width in IR is $\sim$0.1 cm^{-1} (3×10^9 Hz), i.e., it is significantly broader. According to Eq. 7, the time constant τ_{method} of IR is $\sim$10^{-11} s, and, in order not to affect IR spectra of interconverting molecules, their interconversion should take a longer time period than this short one.

---

[a]The resolution in frequency ν should be high for the used instrument to enable it to accurately measure the $\Delta\nu_{width}$ value.

[b]Due to some anisotropy of the magnetic field generated in NMR instruments, the actual value of $\Delta\nu_{width}$ for ^{1}H resonance signals is $\sim$0.05–0.1 Hz for degased non-viscous solutions. Therefore, the actual τ_{method} of ^{1}H NMR is $\sim$1–3 s.

[c]I.e., a weighty majority of molecules in the molecular ensemble do not undergo molecular transformation in this time period.

One may notice that peaks in UV and fluorescent spectra are impressively broad. However, this circumstance should not confuse us. This immense width does not mean an enormously short time constant of these spectroscopies. For instance, in UV spectra, these peaks are a superposition of many partially overlapping individual lines of transitions from vibrational level v_0 with many rotational levels (Fig. 42) of the ground electronic state to different vibrational levels (with many rotational levels at each vibrational level) of the excited electronic state.

It is clear that holding of inequality $\tau < \tau_{method}$ is a necessary condition for enabling to study the molecular flexibility of an organic compound (τ is the lifetime of at least one stereoisomer to be detected, and τ_{method} is the time constant of the chosen physical method). Nevertheless, values τ_{method}, which characterize relative time resolutions of different spectroscopies, are insufficient data for choosing a suitable spectral tool. The only condition $\tau < \tau_{method}$ permits stereoisomers to have any short lifetimes and still be spectrally trappable. We know that physical methods are restricted in their abilities to detect short-lived molecular structures, and, hence, the discussed condition is insufficient for characterizing the time resolution of a spectroscopy. Obviously, one should also figure out what lower limits of time resolution for different physical analytical methods are. This limit τ_{min} for a spectroscopy is the time period that is as long as the shortest lifetime of molecular structures which are still detectable by this spectral method. Molecular structures with $\tau < \tau_{min}$ do not affect related spectra (i.e., line frequencies and lineshapes), similarly to molecular structures which convert to others too slowly in respect to the time constant of the spectral method ($\tau > \tau_{method}$).

According to the definition of τ_{min}, individual (unchanging) molecular structures escape observation, if condition $\tau > \tau_{min}$ is not fulfilled, i.e., $\tau < \tau_{min}$. What does this mean for thermodynamic equilibrium? That is, which molecular structures are observed when exciting rapidly interconverting molecules? Although organic researchers deal with spectral information in analyzing each experiment, the answer probably will surprise many of them. Only the virtual structure that results from spectrally averaging the parent individual structures is "detected"; in contrast to individual molecular structures, this in-time-averaged molecule does not correspond to an energy minimum in the PES. Thus, molecular flexibility of a chemical compound is spectrally detectable if lifetimes τ of its weighty (i.e., sufficiently ponderable in the equilibrium) re-arranging stereoisomers fall within the upper and lower time limits of the timescale of the chosen spectroscopy, i.e., $\tau_{min} < \tau < \tau_{method}$.

What are these lower limits for different spectroscopies, in numbers? The dynamic line broadening is proportional to the square $(\Delta\nu_{A-B})^2$ of the difference $\Delta\nu_{A-B}$ of the ν_A and ν_B frequencies, where ν_A and ν_B are related to individual molecular structures A and B, respectively, in the time-resolved spectrum (Fig. 60).[a] Since the spectral line splitting is the extreme case of the dynamic broadening, one can write for these structures

$$\Delta\nu_{A-B} \times \tau = (2\pi)^{-1} \tag{8}$$

where τ (in seconds) is the lifetime of the shorter-lived stereoisomer, and $\Delta\nu_{A-B}$ (in Hz) is the difference $\nu_A - \nu_B$. Values $\Delta\nu_{A-B}$ are, of course, measurable when "freezing" the interconversion of molecular structures (Fig. 60). Thus, Eq. 8 supplies analytical physical methods with the lower limits of their timescales because τ from this equation is actually equivalent to τ_{min}. That is, the possibility of detecting individual molecular structures A and B at the equilibrium $A \rightleftharpoons B$ is $\Delta\nu_{A-B}$-dependent; the more this value is for a spectral line in the spectrum provided by an analytical physical method, the faster molecular dynamics the method detects.

Table 1 shows the timescales for the spectroscopies which are most frequently used in organic and bioorganic research. These timescales indicate the time limits for individual molecular structures of being detectable by these instrumental methods. Any structural molecular reorganizations (including those for bio- and synthetic polymers) at equilibrium conditions (i.e., stereoisomerizations, chemical transformations and intermolecular associations) as well as short-lived molecules generated at non-equilibrium conditions may be studied by a spectroscopy if the involved individual molecular structures have lifetimes which fall within the time limits of this spectral method. Molecular structures, which have longer lifetimes than the upper limit of the method timescale, are static from the viewpoint of this spectroscopy. Structures, which have shorter lifetimes than the lower limit of the method timescale, are undetectable by this physical method. Comparison of the timescales of these analytical methods

---

[a]These structures are not necessarily stereoisomers. This explanation is related to any pair of interconvering molecular structures, i.e., chemical isomers, molecular associates, and stereoisomers.

Table 1. Analytical physical methods and their timescales as maximal and minimal values of lifetime τ for molecular structures of detectable dynamics.

Physical method	Timescale (s)
NMR (different nuclei)	$10^0 \div 10^{-9}$
ESR	$10^{-4} \div 10^{-8}$
Microwave spectroscopy	$10^{-4} \div 10^{-10}$
IR and Raman spectroscopy	$10^{-11} \div 10^{-13}$
UV spectroscopy	$10^{-14} \div 10^{-15}$
Photoelectron spectroscopy	10^{-18}
X-ray diffraction	10^{-18}
Neutron scattering	10^{-18}
Electron diffraction	10^{-20}

clearly shows why NMR spectroscopy, which is astonishingly informative in determining structures of organic compounds, coincidentally appears the most suitable analytical tool for studying intramolecular dynamics in stereorearrangements (stereoisomerizations) and chemical isomerizations (e.g. π bond migration, skeletal rearrangements) as well as intermolecular dynamics in association-dissociation of molecules and chemical reactions, at equilibrium conditions.

Since values of the difference $\Delta \nu_{A-B}$, in principle, are dissimilar for spectral lines of different compounds and even for different lines in the same spectrum of a compound (e.g. 1 and 10 ppm for resonance signals of chemically non-equivalent nuclei in the ^{13}C NMR spectra of two lowest energy conformers of alkaloid **44**; Section 6.3.1, Fig. 100),[a] the lower limit may somewhat vary for the same physical method. The larger the maximal "splitting" difference $\Delta \nu_{A-B}$ is for spectral lines of a compound,[b] the wider

---

[a] In other words, different spectral lines in a spectrum of a conformationally mobile compound are broadened to a different extent when lowering the temperature.

[b] Among such values for different lines in the spectrum of rapidly interconverting molecular structures (non-observable as individual molecular species). Of course, line splitting appears only in the spectrum registered at the "equilibrium-freezing" conditions (e.g. at low temperatures). Then, we are capable of measuring the line-specific difference $\Delta \nu_{A-B}$ (Fig. 60) that is "hidden" in each unsplit spectral line in the initial, structure-averaging spectrum.

timescale is allotted by the used spectroscopy for examining dynamics of this molecular structure at the equilibrium (e.g. conformer interconversion).

Therefore, the timescale of a spectral method also depends on the analytical instrument (spectrometer) used. The higher the instrument resolution is for frequency v, the more suitable the instrument is for studying molecular dynamics. For instance, NMR spectrometers with higher resonance frequencies expand the NMR timescale (permit to study faster molecular dynamic processes or, equivalently, to detect molecular structures of shorter lifetimes and measure higher rates of molecule interconversions). Diverging resonance signals v_A and v_B in NMR spectra (increasing the value of Δv_{A-B}), such instruments diminish the value of τ_{min}, according to Eq. 8.

Let us illustrate how a resonance signal (for simplicity, a singlet) in temperature-variable NMR spectra shows that lifetimes τ_A and τ_B for two abstract interconverting molecular structures A and B increase when lowering the temperature of the sample (Fig. 60). This temperature-induced increase of the lifetime is described by the Eyring equation (Section 3.5.3, Eq. 14; recall that $\tau_A = 1/k$, where k is the kinetic constant for conversion $A \rightarrow B$ in equilibrium $A \rightleftharpoons B$, and τ_A is the lifetime of A).

Suppose that, at the initial, sufficiently high temperature, both τ_A and τ_B (lifetimes of molecular structures A and B, respectively) are less than the value of τ_{min} for the resonance spectroscopy of the monitored magnetic nuclei and are much less than the τ_{method} value ($\tau_{method} \gg \tau_A; \tau_B < \tau_{min}$). The tracked singlet is sharp at these conditions, and higher temperatures do not sharpen it further. It corresponds to a virtual molecular structure that results from spectrally "in-time-merging" A and B which interconvert so rapidly that NMR has no time to register them as individual structures (Fig. 60). Notably, if we did not know that this spectrum, according to the preliminary statement, is of molecular structures at conditions of rapid equilibrium ($\tau < \tau_{min}$), we could not decide whether it is the spectrum of these detection-escaping (rapidly interconverting) structures or the spectrum of an individual molecular structure.

With essentially decreasing the temperature of the spectral experiment, lifetimes τ_A and τ_B become appreciably longer, and the resonance signal is broadened. The width of the registered line is a sum of the natural line width and an *additional* width (the dynamic line broadening), due to an increase of the Heisenberg uncertainty for the energy of nuclear spin reorientation.

Therefore, in chemical practice, if no nuclear spin–electron spin interaction (e.g., caused by free radicals) takes place, an unordinary broadening of a spectral line is an experimental evidence of the occurrence of a dynamic process that is associated with changes of the molecular structure.

With continuing the temperature decrease, the singlet spectral line is split in two lines (two singlets). This means that lifetimes τ_A and τ_B (τ_7 and τ_8 in Fig. 60) are sufficiently large to enable registration of two *individual* short-lived molecular structures A and B. Further decrease of the temperature both sharpens the singlets to signals with natural line widths and changes the resonance frequencies to some constant frequencies ν_A and ν_B (structures with lifetimes τ_9 and τ_{10} in Fig. 60). These frequency values ν_A and ν_B characterize the resonance signal of the same nucleus for individual A and B;[a] the above discussed difference $\Delta\nu_{A-B}$ for the monitored spectral line is derived exactly from these frequency values. Thus, at this temperature, A and B are detected as static, non-interconverting molecular species and, hence, exposed to exploring their molecular structures by NMR of the nuclei of the isotope that provides resonance signals ν_A and ν_B. Starting from some temperature in our thought temperature-variable NMR experiment, further cooling of the sample (in other words, the increase of lifetimes τ_A and τ_B to values larger than τ_{method}) does not affect the monitored spectral line.

Now, we can return to the "dual" situation with interpreting experimental results of different spectroscopies. Let us call spectroscopies with τ_{min}, say, larger than 10^{-5} s slow spectral methods. For organic molecules flexible at room temperature, these afford a non-existent structure of virtual geometry, which results from spectrally averaging the geometries of interconverting conformers according to their statistical weights (see Eqs. 9 and 12 below). For instance, there is only one resonance signal in the ^{1}H NMR spectrum of cyclohexane (structure **18a** in Fig. 28). This indicates the spatial equivalence of all its protons. However, these spectral data do not correspond to a chair shape of the ring of **18a**; they are compatible only with a hypothesis of a planarity of the ring. Thus, a planar cyclohexane is that virtual 3D structure which this slow spectroscopy provides for **18a** at room temperature.

---

[a]When detecting the temperature-induced NMR signal splitting (Fig. 60), one should relate lines ν_A and ν_B to two individual *conformers* A and B, if it is known that no other molecular transformation, i.e., chemical reaction (e.g., skeletal rearrangement, bond reorganization with keeping chemical connectivity) or VDW association of molecules, occurs.

Two ^{1}H resonance signals separated by 27 Hz are registered for **18a** at 198 K. The ring definitely is not flat. Understanding how a spectral line changes with temperature decrease (Fig. 60), we find out that this molecule undergoes a stereorearrangement; in other words, a molecular shape-converting dynamic process occurs in **18a**. The proton signal from the room temperature ^{1}H NMR spectrum is split in two in the low-temperature spectrum, and the detected difference in frequency is the difference Δv_{A-B} related to each proton of cycle **18a**. The flexibility of **18a** explains how the apparent planarity of the ring appears in room temperature NMR spectra. A rectangular six–carbon hexagon with geminal substituents that are symmetrically oriented relative to the ring plane is a "3D average" of two chairs of inverted geometry (Fig. 28) since they formally are enantiomeric structures.

Using Eq. 8 and the experimental value for difference Δv_{A-B} from low–temperature ^{1}H NMR experiments, the reader reveals why the most frequently used spectroscopy does not detect "true," chair-shaped conformers when registering a routine spectrum for this compound at room temperature. As we know, if a molecule changes its 3D shape very rapidly, the detection method does not have time to "photograph" individual conformers. In Section 2.4, it is mentioned that the lifetime τ of a chair-shaped conformer of monocycle **18a** is about 10^{-5}–10^{-4} s at ambient conditions. A slow method, ^{1}H NMR, is characterized by $\tau_{min} \sim 6 \times 10^{-3}$ s for the measured Δv_{A-B} value, as Eq. 8 shows. This means that ^{1}H NMR spectroscopy can resolve two lines separated by this Δv_{A-B} only for molecular species which have $\tau > 6 \times 10^{-3}$ s. As indicated, cyclohexane chairs are not such a case.

In PES terms, this spectral method does not "discern" two separate energy minima for basic cyclic system **18a** at room temperature, and supplies it with a "false" PES that (*1*) has a single minimum and (*2*) delivers an irrelevant molecular geometry associated with this minimum.

And what about fast methods? We understand that they do detect individual molecular structures (e.g. conformers), i.e. they present these structures at a thermodynamic equilibrium as a mixture of unchanging molecular species (e.g. rigid stereoisomers). For instance, a routine room temperature IR spectrum of monodeutero cyclohexane, by showing two bands of similar intensity separated by $10\,\text{cm}^{-1}$, informs that there are two individual stereoisomers of similar thermodynamic stability. Both bands correspond to the same molecular oscillation that occurs in both stereoisomeric structures. Hence, the registered frequency difference is the difference $\Delta\nu_{A-B}$ that figures in Eq. 8; thus, τ_{min} is $\sim 10^{-13}$ s. In examining neat compound **18a** at this temperature, IR registers five bands in the 1350–700 cm^{-1} region. This IR spectrum indicates that monocycle **18a** has a chair-shaped molecular geometry (Fig. 28) in the liquid phase: exactly this number of IR-active fundamental bands in this spectral region is expected for this molecular geometry of symmetry point group D_{3d}.[a] One can conclude that, at room temperature, this spectral method locates the lowest energy stereoisomers of **18a**, but, if questioned regarding their interconversion, only informs that the stereoisomer lifetime at this temperature is longer than 10^{-13} s.

Molecular flexibility, of course, does not depend on the detection method. Combining the NMR and IR structural evidences for cyclic system **18a**, we understand that the chair-shaped stereoisomers interconvert sufficiently rapidly (i.e., they are conformers), but the interconversion rate is too slow to be registered by IR. One can even conclude that the conformer lifetime is shorter than 6×10^{-3} s and longer than 10^{-13} s $(10^{-13}$ s $< \tau < 6 \times 10^{-3}$ s). The dissimilar "portraits" of the molecule appear due to the different time resolution of these instrumental methods.

The example of the illusory (planar) geometry of **18a** deduced from NMR experiments at room temperature and its phantom rigidity delivered by IR measurements at the same temperature illustrates an important methodological principle that may save many synthetic experimentalists from drawing incorrect conclusions. *Typical (i.e., narrow line width) spectra do not indicate intramolecular flexibility or rigidity.* Indeed, spectral detection of only one or several stereoisomers does not provide relevant information about the lifetime(s) τ of the stereoisomer(s). Detection of conformational mobility requires experiments that alter conformer lifetimes or use instrumental methods of distinct timescales. For instance, temperature-variable NMR experiments, where the conformer lifetime is prolonged by decreasing the temperature (this idea is illustrated in Fig. 60), easily establish that the classical conformational model — cycle **18a** — is flexible at room temperature.

---

[a] See Ref. 19g for the theory of predicting the number of IR- and Raman-active molecular vibrations for molecules of different symmetry.

In daily practice, synthetic chemists record NMR spectra at room temperature and do not carry out temperature-variable NMR experiments. As stressed above, "one temperature NMR spectra" are not informative in regard to the flexibility or rigidity of organic molecules. Only if spectral lines are abnormally broad and (*1*) no reversible chemical process occurs; (*2*) no intermolecular aggregation takes place; (*3*) molecules are rapidly re-oriented (i.e., there are no viscosity and other molecule-orienting effects in the examined solution); (*4*) the solution does not contain paramagnetic contaminates, one can conclude that the "one temperature NMR spectrum" indicates the occurrence of an intramolecular 3D rearrangement. In fortunate cases, such an observation may assist to roughly estimate molecular flexibility without varying the experimental temperature. Significant broadening of resonance signals, which still is not signal splitting, in room temperature ^{1}H NMR spectra shows that the kinetic barrier of the stereorearrangement, which causes the broadening, is higher than 13 kcal mol^{-1}, but lower than 17 kcal mol^{-1}.

As pointed out in Section 2.6, most organic compounds under ambient conditions are conformers that are in fast equilibrium. Let us reiterate what their rapidly changing molecular geometries are when viewed through the prism of "sluggish" instrumental methods. The above example of a virtual, NMR-delivered planarity of **18a** suggests that conformational mobility "blurs" molecular geometry in the observation that uses a slow spectral method. This example is representative — *any method of observation supplies an averaged, virtual 3D structure for molecules that change their geometry rapidly relative to the method timescale.* Let us call the molecular geometry of this virtual structure *spectral molecular geometry.*

What exactly is the spectral molecular geometry? That is, what exactly are molecular geometries, which are revealed in spectroscopic structural studies of flexible organic molecules that do not "disintegrate" these molecular systems into conformers? Physical methods operate with quantitative values, e.g., geometry parameters, spectral frequencies, line intensities, and multiple electrical moments, and the above question is worth generalizing. Which values are measured by a detection method for molecular quantities of stereoisomeric structures that are in a rapid conformational equilibrium on the method timescale? Any measurable physical quantity Q or any geometrical parameter r, which is related solely to the physical properties of individual conformers and not to the medium-associated properties (e.g., NMR resonance frequencies, dipole moment, extinction coefficient, but not, e.g., density, diffusion coefficients and thermal conductivity of the bulk), is composed from additive contributions of individual conformers.

For conformational equilibrium, measured value Q_{av} (or r_{av}) results from averaging values Q_i (or r_i, respectively), where i runs over M conformers. Contributions Q_i or r_i into Q_{av} or r_{av}, respectively, are proportional to the statistical weights of the conformers (Eq. 9). That is, the "molecule-from-spectra" is a "linear combination" of individual structures 1, 2, ..., M taken with coefficients $p_1, p_2, \ldots, p_M$.

$$Q_{av} = p_1 Q_1 + p_2 Q_2 + \cdots p_i Q_i + \cdots + p_M Q_M \qquad (9)$$

In practice, Eq. 9 is not important because reconstructing the spectral molecular structure; it is important because it supplies organic researchers with original molecular "images" that form this virtual portrait of the molecule. That is, such spectra afford virtual values of molecular quantities, e.g. parameters of spectral molecular geometry, frequencies of lines in spectra of rapidly interconverting conformers, and time-averaged constants of nuclear spin–spin interactions registered by NMR; Equation 9 "splits" these virtual values into the values that are inherent of individual conformers. Thus, this equation shows a way for extracting information about individual, though rapidly equilibrating conformers from spectral data.

Measured values of chemical shifts, constants of spin-spin interaction and the rates of cross-relaxation and spin-lattice relaxation for systems, which are flexible in the NMR timescale, result from averaging over the values of the related physical quantities for individual conformers; the averaged values are given by Eq. 9. Neuhaus-Williamson monograph[72] explains that NOE enhancements recorded for such non-static systems are not the averages in the sense of Eq. 9. They correspond to the values that are mathematically derived from the averaged values of the relaxation rates and not from the NOE enhancements for individual conformers.

An example of using Eq. 9 is provided by the case which is most often considered in conformational analysis. For many organic systems, two distinct, rapidly interconverting conformers are highly favored at a certain temperature (e.g., room temperature). Neglecting other, very minor conformers, one can consider a two–position conformational equilibrium. Then, calculations of the conformer ratio are very simple if both quantitative properties Q_1 and Q_2 (e.g., the line frequencies) are known for individual conformers *1* and *2*. Measuring the averaged value Q_{av}, i.e., recording a routine spectrum at this temperature, we come to a set of two linear equations (Eqs. 10 and 11).

The conformer fractions p_1 and p_2 can be trivially extracted therefrom.

$$Q_{av} = p_1 Q_1 + p_2 Q_2 \tag{10}$$

$$p_1 + p_2 = 1 \tag{11}$$

With these values p_1 and p_2 in hand, one can discover molecular 3D structures of both individual conformers by applying Eqs. 10 and 11 to another molecular quantity Q which may be associated with a parameter of molecular geometry (e.g. vicinal constants $^3 J_{H,H}$ of proton spin–spin coupling interactions that are numerically associated with the related torsion angles via the commonly used Karplus equation or constants $^n J_{C,H}$ of spin carbon–proton coupling associated with various torsion angles via less known semi-quantitative and quantitative relationships).

In the general case $M > 2$, populations $p_1, p_2, \ldots, p_M$ of conformers $1, 2, \ldots, M$ are obscured from our view. Equation 9 provides a linear relationship between the measured value Q_{av} and M unknowns Q_i together with M unknowns p_i (totally, $2M$ unknowns). As a rule, we do not have $2M$ independent linear equations of type Eq. 9 or another one, which relate $2M$ unknowns to each other.[a] In simple words, there are more unknowns than consistent linear equations with these unknowns. It is impossible to solve these equations analytically and extract conformer fractions $p_1, p_2, \ldots p_i, \ldots, p_M$ from the averaged value Q_{av} of molecular quantity Q given by experiments. For instance, we need one additional independent linear equation that includes these unknowns, but does not introduce new ones, when experimentally determining populations p_1, p_2, p_3 in the case of three interconverting conformers ($M = 3$). If a fast spectroscopy (e.g. IR) experiment may measure the ratio p_i/p_j (i.e., provide the value of equilibrium constant $K_{eq} = p_i/p_j$) for one pair of conformers,[b] the problem of determination of absolute values p_1, p_2, and p_3 is solved: we have the third, desired equation.

The reader may recall that determination of relative populations of conformers is actually equivalent to determination of their relative thermodynamic stability. From basic

---

[a] Unconditionally, there is one additional (to Eq. 9) independent equation, $p_1 + p_2 + \cdots p_i + \cdots + p_M = 1$. Exactly this equation (as Eq. 11) makes the case of two conformers ($M = 2$) soluble in finding out their populations p_1 and p_2 (see above).
[b] At the temperature of the experiment which has supplied the value Q_{av} of molecular quantity Q.

thermodynamics, we know that the content p_i of the i-th component in a mixture at equilibrium conditions is

$$p_i = \exp(-G_i^0/RT)/\sum_i \exp(-G_i^0/RT) \tag{12}$$

where G_i^0 is its Gibbs energy at temperature T (K), i runs over all M equilibrating components and R is the universal gas constant. It follows from Eq. (13) that the ratio of two stereoisomers in the mixture is determined by the difference of their Gibbs energies as well as the temperature (Eq. 12).

$$p_i/p_j = \exp(G_j^0 - G_i^0)/RT = \exp(-\Delta G_{i-j}^0/RT) = K_{eq} \tag{13}$$

where K_{eq} is the equilibrium constant for transformation $i \rightleftarrows j$. Clearly, values p_i/p_j are easily calculated for different temperatures for any given value of ΔG_{i-j}^0. Thus, the Gibbs energy difference ΔG_{i-j}^0 is a simple descriptor of relative stability of stereoisomers (of course, including conformers).

It is notable that experimentally determined relative stability of stereoisomers (as values of ΔG_{j-i}^0) differs from that supplied by the minimal energy points in the PES (values of ΔE_i from Eq. 6, Section 3.1). In reproducing experimental values ΔG_{i-j}^0 by non-empirical calculations, ΔE_i values are often corrected to include both vibrational energy and *entropy* (Section 5.3.4).

Clearly, the most convenient case is when there is a high predominance of one conformer, i.e., p_1, is near unity (so-called anancomeric equilibrium). If the molecules of the major and minor conformers have no drastic discrepancy in their response to the outer disturbing factor, e.g., photons, the molecular physical properties of such a system practically are the properties of the totally predominated conformer, as Eq. 9 shows. Therefore, in this case, the spectra of the entire system are practically identical to the spectra of the major conformer.

In NMR spectroscopy, a minor conformer with $p_1 < 0.005$, i.e., with the content less than 0.5%, does not affect either resonance frequencies or line shapes of the resonance signals to an appreciable extent if the rate of its conformational transformation slows down or accelerates (the so-nicknamed hidden partner case).[a] Let us suppose that both the major and minor conformers of such a system interconvert rapidly relative to the NMR timescale.

---

[a] However, the chemical or biochemical behavior of the system is not necessarily associated with the predominated conformer. Conformational analysis of apparently "one conformer systems" is therefore not purposeless.

The spectral, NMR-supplied molecular geometry of this compound carries valuable information. This virtual 3D structure is practically identical to the real structure of the major conformer in spite of the instability of the latter within the detection timescale. If these stereoisomers were interconverted very slowly (i.e., were rigid), the recorded NMR spectrum would be the same!

The identity of the NMR-supplied "portraits" of molecularly flexible and rigid compounds with one highly prevailing stereoisomer is a source of a typical misinterpretation of NMR spectra. Many synthetic chemists consider the spectral detection of the single structure of expected geometry as a clear evidence of the rigidity of the molecule and report these "results."[a] A textbook level example of *tert*-butylcyclohexane **18b** (Fig. 31) explains that sole experimental studies of such compounds are insufficient in making binary decision *rigid* or *flexible*. Under ambient conditions, its equatorial conformer predominates so totally that any NMR experiment, using NMR spectrometers of the 600–900 MHz generation equipped with their "standard" computers, would detect only the equatorial conformer. The ring of this monosubstituted cyclohexane is, in fact, flexible to the same extent as the ring of the parent cyclohexane **18a** (Fig. 28), i.e., at room temperature, monocycle **18b** exists in solution or in the gas phase as a mixture of rapidly interconverting conformers. Merely, the steady state concentration of the axial conformer is so subtle that this participant in the conformational equilibrium does not affect the NMR or other spectra of **18b** at all. *By any spectroscopy, conformational mobility is practically undetectable for compounds whose stereoisomers are in an anancomeric equilibrium with the minor component content less than 0.1%.*

Notice that the detection of a very minor component is practically impossible even in the case of rigid (non-interconverting) stereoisomers. The problem is not the NMR sensitivity that, according to a widespread delusion, is low. An experimental spectrum, which is provided by a digitally operated instrument, is not an abstract continuous line. In fact, it is a finite set of points which appear in the spectrum as a line due to a high 1D density of the line-forming points. Computers, which control 300–900 MHz NMR spectrometers used in chemical and biochemical research, cannot provide a *supertight* point registration of resonance frequencies (i.e., cannot record a near-continuous line of free induction decay). As a

---

[a] Such misinterpretations are still scattered throughout research periodicals where NMR professionals are occasional reviewers.

consequence, fine "details" of the physically detected signal, which carry information about trace amounts of molecular species in the sample, disappear, and the eventually registered (Fourier-transformed) spectrum is deprived of the signals of these minor components. This means that a compound at a low concentration cannot be detected even if scanning the sample any number of times. ▌ In other words, a high sensitivity of modern NMR instruments is not utilized because of limitations of their computer architecture. Under these unfortunate circumstances, trace stereoisomers remain unidentified in experimental organic research.

At this point, it is appropriate to stress the significant advantage of theoretical modeling. Theoretical calculations are capable of analyzing any molecular system, and only technical difficulties, e.g., limitations of computer resources, may prohibit solving structural problems by means of computations; however, such difficulties rarely take place in modeling organic molecules of ordinary size. *Non-empirical calculations are a universal instrument.* They can model any molecular structure or molecular property while possibilities of experimental determinations are always limited. For instance, the hidden partner case (a significant difference in Gibbs energy for a pair of equilibrium participants) is not problematic for computational conformational analysis, in contrast to NMR. Both molecular mechanics (Chapter 4) and non-empirical (Chapter 5) calculations locate non-flattened energy minima as well as saddle points in the PES with success that does not depend on their relative energy.

Synthetic specialists traditionally have not trusted results of theoretical calculations or merely ignored them. This should no longer be the case. Due to the progress of theoretical methods in chemistry, techniques in computational mathematics, and, most important, computation/data storage facilities (i.e., computers), *the single condition for success in contemporary computational studies of organic molecules is the choice of the relevant calculation methodology.* The choice is easy when exploring conformational space (see Chapters 4 and 5): stereoisomers are the simplest target for molecular modeling while they are not that for spectral tools of organic chemists, e.g. for NMR, as we learn in this Section. Having, as a rule, command on neither computational chemistry nor intricate NMR methodologies, synthetic researchers, nevertheless, often resort to conformational explanations. What their practice in indicating conformations is and what its limitations are is the topic of the next Section.

3.4 Constructing Conformers Mechanistically: Conformers of Basic Systems as 3D Templates

The conformational concept focuses the attention of synthetic chemists mainly on the question of which conformers of an organic compound are chemically more reactive. Usually, they analyze the reactivity issue by considering with "ready-made" conformers. However, how do we come to know of structures of conformers for different compounds?

A simple idea of 3D templates is exploited in synthetic groups. The number of organic compounds from different chemical classes, for which the conformations and the rates of conformer interconversion have been established, is huge. According to this idea, one can single out some basic systems (Figs. 61 and 95), which have been analyzed in detail, and transfer their conformational features to structurally similar fragments of larger molecules.

In this mechanistic representation, the latter are "LEGO molecules," whose geometries are assembled from polyatomic fragments that adopt some canonical conformations. Put simply, a molecular geometry is an assembly of some different, but "standard" 3D building blocks (canonical conformations).

What exactly are these canonical conformations? Several open chain and cyclic hydrocarbon/olefin systems that are flexible under ambient conditions represent basic conformational systems (Figs. 15, 61 and 97): the conformations of these molecular systems are used by organic experimentalists as 3D templates for quickly constructing 3D molecular frameworks of organic systems. Indeed, geometries, when adopted by structurally similar molecular fragments of different compounds, very often vary insignificantly in their geometrical parameters. If significantly, these variations still correspond to other common molecular geometries that also are inherent to conformers of simple parent compounds from the above indicated small set. Thus, the idea that *conformations of many compounds may be described in terms of canonical 3D geometries* has appeared very fruitful.

Let us call canonical geometries of conformers of parent basic systems (Figs. 15, 61 and 97) *conformons*, by analogy with the term *synthons* of synthetic chemistry. In this mechanistic design, all possible combinations of relevant conformons provide 3D geometries of all possible conformers

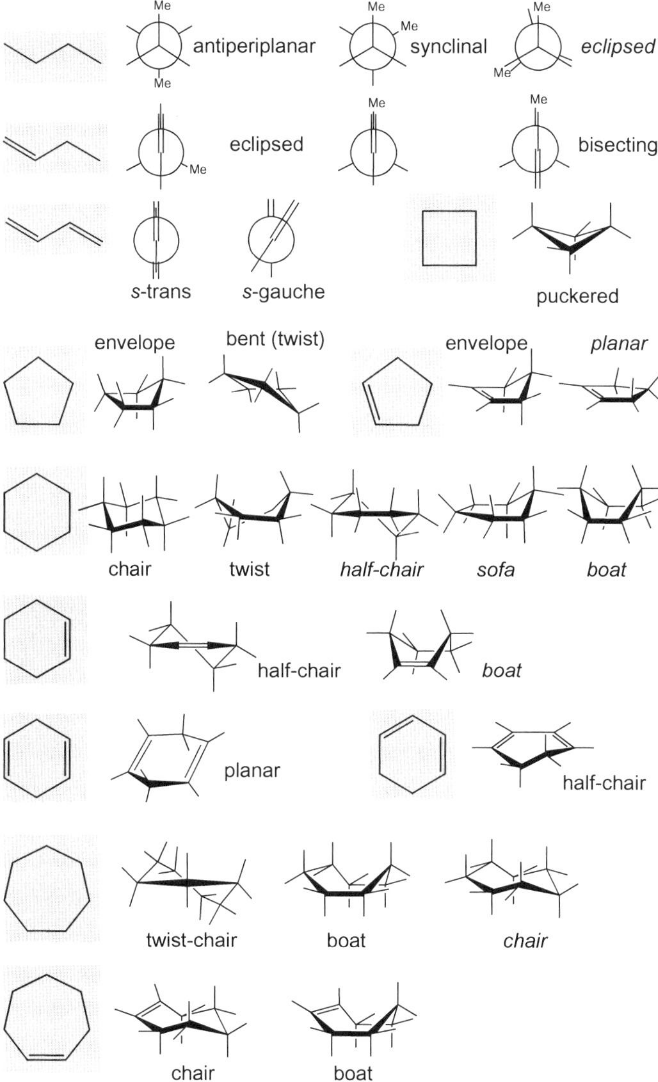

Figure 61. Chemical systems (indicated by gray background) that are most often considered as parent in conformational analysis: their conformations (shown) are used as molecular 3D templates in modeling conformations of organic molecules. Trivial names of the 3D geometries of conformational transition states are in italics. The conformation that is shown in the nearest proximity to the corresponding Lewis structure points out the favored conformer of the parent compound (excluding the cyclopentane case).

of the integral system of interest. For instance, the ring shapes of the *cis* and *trans* stereoisomers of azabicycle **20** (Fig. 41) result from the formal fusion of two azetidine rings of identical puckered geometry (Fig. 61) when combining these conformons in two different possible fashions. Since there are neither other conformons of the four-membered saturated ring nor additional geometrically distinct fusion of the two available identical conformons, one can anticipate that the bicyclic ring of each stereoisomer of **20** adopts a single geometry.

It seems that, when using such molecular 3D templates,[a] no *studies* are required in order to have verisimilar molecular geometries of conformers of non-exotic organic compounds. Moreover, this attractive pen-and-pencil modeling, which, in essence, uses only structural analogies, is ubiquitous in considering conformations in synthetic organic research. Is this practice justified? It is indicated in the beginning of this Section that "the idea that *conformations of many compounds may be described in terms of canonical 3D geometries* has appeared very fruitful." One should precisely understand this practical verdict of the conformational concept. Canonical conformations mirror a tendency of certain (even many) organic compounds to adopting certain molecular geometries. This tendency is not absolute: concerning molecular geometry, other organic compounds have other tendencies, and there is no structural criterion that could differentiate between the former and the latter molecular structures. Thus, the practice of considering conformons instead of studying a compound is not solid when used in *research*. The researcher never knows *a priori* whether molecular fragments of a new molecular system adopt exclusively the conformations of the related parent molecular structures. As any structural analogies, conformon-based predictions of 3D geometry may be incorrect even for simple molecules. Let us call some examples into play acknowledging how unreliable this primitive modeling is.

---

[a]Template-based modeling of 3D structures of organic molecules may be easily performed by involving special computer programs (see Ref. 38b for a general review on principles of this computer-assisted 3D design). The main "philosophy" of these programs consists of assembly of molecular structures from some basic 3D molecular fragments (templates) and, sometimes, examination of resulted "LEGO molecules" by means of empirical structural criteria. In organic *research*, this methodology may be considered only as an auxiliary, pre-starting tool (see below).

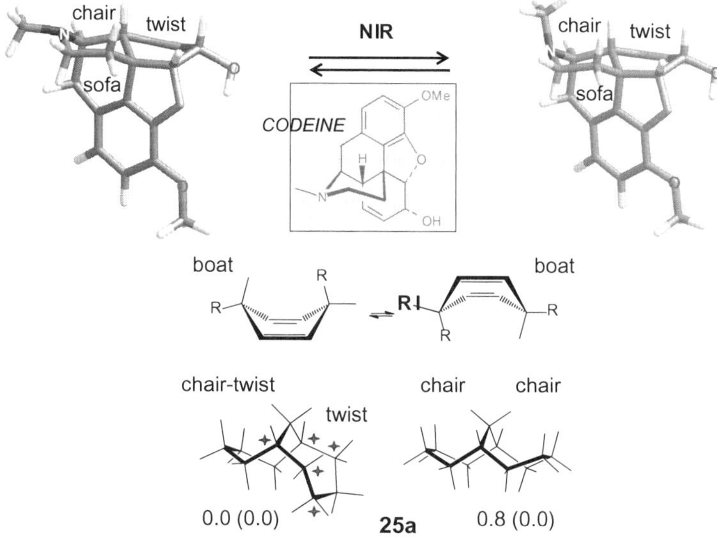

Figure 62. Stereoisomers (conformers at room and some lower temperatures) of codeine, substituted 1,4-cyclohexadienes, and bicyclo[4.4.1]undecane (**25a**). Indicated trivial names are related to the geometries of individual rings. Optimized geometries [MP2/6-31+G(d,p), kcal mol^{-1}] are given for the conformers of codeine and bicycle **25a**. Stars indicate five coplanar ring carbons in the twist-shaped ring of one of the conformers of this bicycle. Numbers show values of *ab initio*-calculated relative molecular energies ΔE_0 (kcal mol^{-1}) and relative steric energies ΔE_{ster} [MM3(96), kcal mol^{-1}, in parentheses].

Both the cyclohexene rings in the molecule of codeine (Fig. 62) are conformationally atypical. One of these rings adopts a slightly distorted twist geometry of cyclohexane while another one is sofa-shaped. Unexpectedly, these unsaturated fragments of morphines "borrow" canonical geometries of the six-membered *saturated* ring. The fact that covalent connectivity "locks" these six-membered rings in atypical conformations has been revealed for the morphine backbone only *post factum*, as a result of experiments as well as true (read: relevant theoretical) modeling. If we employed the conformon-based approach, we would draw incorrect conclusions regarding conformations of these rings: a single conformon of the thermodynamically stable molecule of cyclohexene is a half-chair geometry (Fig. 61).

Let us bring into focus a system of higher rotational freedom around ring bonds. As theoretical calculations demonstrate,[9] one of the cycloheptane rings in the lowest energy conformer of bicyclo[4.4.1]undecane (**25a**, Fig. 62) has a twisted geometry that is neither twist-boat nor boat (conformons of cycloheptane itself; Fig. 61). That is, this twist geometry is not inherent to molecular geometry of cycloheptane for any stationary point of the PES of this parent monocycle.[38c] It appears that molecular 3D templates from the canonical set are unsuitable for modeling this stereoisomer at all. Furthermore, both the cycloheptane rings of another conformer of bicycle **25a** have the chair geometry (Fig. 62).[9] This conformation corresponds to the geometry of a conformational *transition state* of cycloheptane (Fig. 61). Thus, both the conformons of the parent compound (i.e., cycloheptane) do not predict the geometry also of this conformer of **25a**. These examples make clear that a medium size ring may adopt other, non-canonical, conformations, if its geometrical freedom is somewhat restricted by embedding the ring into a polycyclic backbone.[a]

The next example demonstrates that incorrect conformon-based predictions of molecular geometry are associated not only with polycyclic systems. While the single conformon of 1,4-cyclohexadiene is a planar geometry (Fig. 61), various substituted 1,4-cyclohexadienes are boat-shaped (Fig. 62).[38d,e] This substituent-induced alteration of the ring conformation should alert numerous users of the 3D template concept to unexpected failures, which may be faced if neglecting experimental or theoretical conformational *studies*.

At first sight, it seems that the principal way to amend the easy design tool of molecular 3D templates is to extend conformon sets. *E.g.*, for 1,4-cyclohexadienes and bicyclo[4.4.1]undecanes as well as their heteroanalogs, this way means that the mentioned boat (the cyclohexadiene case), twist and chair (the case of bicycle **25a**) conformations should be included into the related set of conformons. However, methodological corrections of this kind have no prospects. Many compounds behave "normally" — structural fragments of their stable stereoisomers adopt only the geometries of the conformers of the related parent systems. In the 3D template modeling, incorporation of atypical geometries into conformon sets

---

[a]For difficulties in prediction of rotamer structures in acyclic compounds, see Section 1.5.

would result in generating "fake" conformers (thermodynamically unstable molecular 3D structures assembled from both canonical and new conformons) together with "true" ones. Of course, this mechanistic tool provides no way to distinguish between them.

The five-membered saturated ring merits special attention. The envelope and twist geometries (Fig. 61) are sole 3D structures, which are ascribed to conformers of cyclopentane type compounds in hundreds research articles, *not* modeling these concrete molecular systems theoretically. This practice of conformational analysis is on shaky ground. *One cannot straight away indicate which geometry, envelope, twist or an intermediate one, is adopted by thermodynamically stable conformers of the five-membered saturated ring a compound unexplored in this regard.* Let us see why this marvelous comment is valid.

Geometrical transformation of this ring, which does not include its inversion trough a planar intermediate structure, is described as pseudorotation.[a] Considered as a process, pseudorotation geometrically is a synchronous change of endocyclic torsion angles associated with some alteration of endocyclic bond angles, which takes out each ring atom, one after the other, of the plane of three adjacent ring atoms. The classical graphical illustration of pseudorotation shows this transformation of the ring geometry as interconversion of ten envelope-shaped rings and ten twist-shaped rings (Fig. 63).

Unfortunately, this ubiquitously distributed graphical sketch (sometimes called pseudorotation wheel) that presents 20 alternate envelope and twist geometries[b] is a source of a widespread misunderstanding. First, these envelope and twist geometries are almost always understood as sole conformational alternatives for the five-membered saturated ring; that is inaccurate (see below). Second, many organic researchers take this *abstract* description for a *conformational scheme* (Section 2.5) and associate these 20 structures with 10 conformers and 10 related transition states, or, more regrettably, with 20 interconverting conformers. In fact, there is no basis for this association. Although unveiling 20 certain conformations, this formal scheme only shows how the shape of the five-membered saturated ring is altered in a molecule of closed electron shell in inverting this non-planar ring. It does not indicate the ring conformations inherent to either chemical structure. Thus, when applied to a concrete compound, this notional scheme merely is incorrect. In contrast to a conformational

---

[a]This inaccurate term [pseudorotation in five-membered saturated rings as well as pseudorotation in six-membered saturated rings (Figs. 37 and 53)] should not be mistaken for formally the same inaccurate term pseudorotation that, however, has another content. The latter pseudorotation is related to spatial rearrangement of ligands in elementary molecular fragments of trigonal bipyramidal geometry (Sections 1.7 and 1.8).

[b]Characterized by certain φ values. Pseudorotational phase angle φ is one of two geometrical parameters in the algebraic Cremer–Pople description of exact 3D geometries of cyclic structures.[38f]

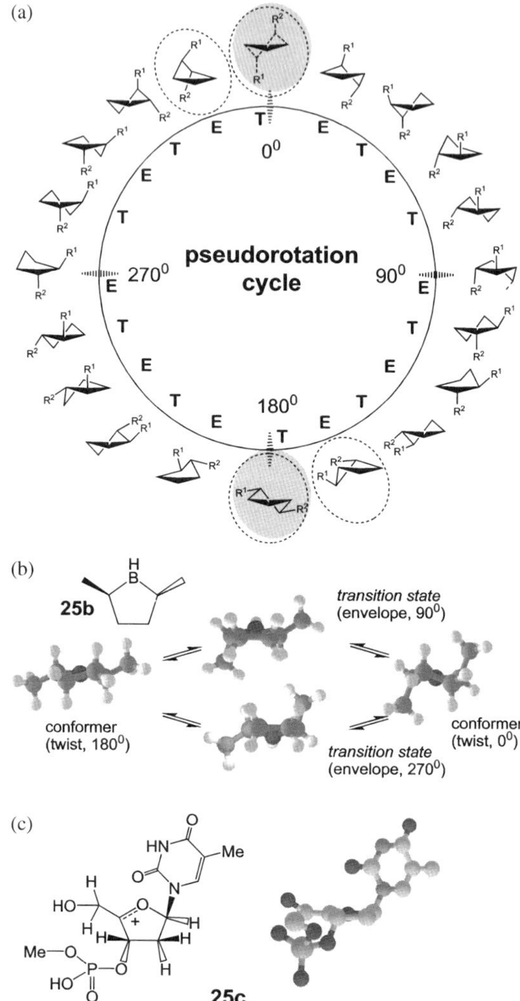

Figure 63. (a) The formal scheme of pseudorotation in an abstract saturated five-membered ring with 20 alternate envelope (E) and twist (T) conformations (for clarity, *trans*-1,2-substituents are indicated). Neighboring conformations *envelope* and *twist* are separated by the difference of 18° in values of geometrical parameter ϕ. Dotted shadowed ovals indicate conformers of *trans*-1,2-dibromo- and *trans*-1,2-dichlorocyclopentane as well as borolane **25b**; other dotted ovals mark conformers of N-Me pyrrolidine (herein, the N-Me substituent is R^1). (b) Conformational scheme[9] for borolane **25b** with computationally derived molecular structures. (c) Calculated[9] molecular geometry of 4′-dehydro nucleotide cation **25c** (hydrogens are not shown).

scheme (e.g., that for borolane **25b**; Fig. 63), it exposes either both relevant and irrelevant conformations, or just irrelevant ones (see below).

Let us leave this formal delineation of pseudorotation and return to real organic molecules, starting from cyclopentane. Two geometries correspond to distinct energy minima in its PES, they indeed are the envelope and the twist conformons shown in Fig. 63 schematically. However, neighboring minima *envelope* and *twist* are separated by very low barriers (the estimates vary from 0.2 to 0.6 kcal mol^{-1}) and, therefore, cannot be related to individual stereoisomers (Chapter 2). For the cyclopentane PES, one has to consider all minima for the *envelope* and *twist* geometries as one broad minimum. Then, the shape of this molecule is characterized by a *single* molecular geometry. However, we do not know, which PES point corresponds to the probability maximum in this broad well of potential energy, or, equivalently, we do not know, which exact geometry is most probable for cyclopentane molecules. One can only say that it is in between the envelope and twist geometries. The next conclusion is methodological: the parent cycle, cyclopentane, does not supply us with conformons.

Conformations of many other compounds that bear a five-membered saturated ring, however, grant a possibility of considering the indicated envelope and twist geometries as conformons. For many substituted cyclopentanes and their heteroanalogs, energy minima in their PESs are separated by essentially higher barriers, i.e., conformers do exist for these compounds. For instance, two (and only two) distinct twist-shaped conformers ($\varphi = 0°$ and $180°$, Fig. 63) interconvert in the case of *trans*-1, 2-dibromocyclopentane,[38g] *trans*-1,2-dichlorocyclopentane,[38g] and *trans*-1,3-dimethylborolane **25b** (Masamune reagent for stereoselective hydroboration). Theoretical calculations easily provide quantitative estimates for conformational dynamics in small molecular structure **25b**: the calculated value of free energy is 5.7 kcal mol^{-1} for the kinetic barrier of RI [the free energy difference for the transition state of RI and the minimal energy conformer (the conformer with diequatorially oriented Me's)].[a] Four other,

---

[a] The latter molecular structure has been examined by *ab initio* calculations at the MP2/6-31++G(d,p) level (Sections 5.2.3 and 5.2.4 explain these abbreviations) with using the Gaussian09 program.[9] Only two conformers of twist ring geometry and different energy and two first-order transition states of envelope ring geometry and equal energy have been located (Fig. 63). By estimating free energy (Section 5.3.4), the 1,3-diequatorial conformer is more stable than the 1,3-diaxial one by 1.9 kcal mol^{-1}.

envelope-shaped conformers participate in pseudorotation of the ring of N-Me pyrrolidine (**25d**).[38h] If exclusively considering pseudorotation in **25d**, there are only two conformers, where $\varphi = 172°$ characterizes the first conformer, while the second one is characterized by $\varphi = 342°$ (Fig. 63). The kinetic barrier of interconversion of these diastereomeric conformers is 3.2 kcal mol^{-1}, according to molecular mechanics calculations.[38h] However, an additional molecular motion — a concerted ring inversion-N-inversion-Me rotation (RINIR) — doubles the number of both conformers and conformational transition states for cycle **25d**; that is, two identical pairs of diastereomeric conformers appear (Fig. 63 spots conformers from one pair). The barrier of 5.3 kcal mol^{-1} characterizes this RINIR. Thus, all four energy minima in the PES of **25d** are sufficiently deep to keep the oscillating molecular structures as individual conformers. We see that conformers of five-membered saturated cycles may have both an envelope and a twist molecular geometry.

What is important in understanding conformational dynamics in such cycles, the number of conformations which the five-membered ring adopts in real molecules (considering both conformers and transition states) appears essentially less than 20 (the number of conformations supplied by the traditional graphical explanation of pseudorotation in this ring; Fig. 63). Consequently, the number of conformers or that of conformational transition states appears less than ten. In PES terms, the energy landscape for these molecular structures is not too complicated: the number of energy minima and the equal number of the related saddle points are small. For instance, a rhomb of a two–position conformational equilibrium (two conformers of distinct molecular geometry and two transition states of equal molecular geometry;[9] Fig. 63) is what remains from the illustrative icosagon of pseudorotation (ten conformers and ten related transition states; Fig. 63), when approaching stereodynamics of cycle **25b** non-formally.

Nevertheless, the envelope and twist conformations are not universal conformons for this ring. Both conformers (i.e., thermodynamically stable molecular structures) and conformational transition states (i.e., thermodynamically unstable molecular structures) may correspond to any position (any value of phase angle φ) in the pseudorotation cycle, depending on the compound. In other words, both conformers and conformational transition states for cyclopentane derivatives or heteroanalogs of closed electron shell may have an envelope, twist, or geometrically intermediate geometry. For instance, there are four conformers of the same intermediate geometry ($\varphi = 45°$, $135°$, $195°$ and $315°$, i.e., non-envelope and non-twist) in the conformational equilibrium of 1,3-dioxolane.[38i]

We conclude that (*1*) there is no ten–position conformational equilibrium for derivatives and heteroanalogs of cyclopentane; such an equilibrium for these compounds includes much less conformers; and (*2*) molecular geometry of stable conformers and conformational transition states of such compounds is certain in the sense that we know what is the spectrum of possible conformations for a molecule, but do not know which of these conformations it chooses to adopt. An obvious methodological conclusion is "to forget" the illusive ten–positional conformational equilibrium, which, as mentioned, is a misinterpretation of the twenty–conformation graphical illustration of pseudorotation in the five-membered saturated ring. Conformational analysis of systems containing this ring appears much easier to perform than it may be thought if being based on the formal scheme of pseudorotation reproduced in Fig. 63. Nevertheless, although the number of conformers involved in the actual equilibrium is small, it varies from compound to compound, and no geometry of the five-membered saturated ring may be considered as a favored conformon. The latter circumstances lead to the conclusion that there is no simple tool capable of indicating conformations of a conformationally unexplored organic system, whose skeleton incorporates this ring.

How can synthetic researchers solidly approach conformers of new cyclopentane derivatives and heteroanalogs? Experimental conformational analysis is difficult for systems, if there is no preliminary information regarding the number (M) of detectable conformers. Besides, NMR studies of conformationally non-degenerate systems with $M > 2$ are outside the competence of specialists in organic synthesis.[a] Theoretical modeling is a convenient as well as reliable alternative! Relevant theoretical calculations are capable of locating minima

---

[a]Therefore, an assumption of two–position conformational equilibrium is frequently used in identifying conformers of saturated heterocyclic five-membered ring systems (e.g., nucleoside or nucleotide ring conformers) by NMR. This simplification is, nevertheless, technically complicated in practical use. For reaching a reasonable accuracy for the molecular geometry of the ring in such conformers, this assumption is applied to several constasts of spin-spin interactions $^3J_{H,H}$ of vicinal ring protons in the form of a multiparameter Karplus type equation (the Altona equation). Fortunately, these laborious calculations have been computerized.[38j,k] The later program[38k] is more convenient because it uses the ubiquitously known Matlab package.

in PESs for non-polymeric organic compounds relatively easily, i.e., to reveal the number of stable stereoisomers and determine their structures without serious obstacles (see next Chapters). Concerning conformational analysis of five-membered ring systems, it is obvious that the "organic chemist-friendly" solution consists of employing molecular mechanics calculations (Chapter 4) followed by non-empirical theoretical computations (Chapter 5). Molecular mechanics calculations locate energy minima that correspond to real conformers or to molecular 3D structures that are geometrically similar to conformers. By refining the obtained geometries, reliable non-empirical calculations provide reliable 3D structures and qualitatively estimate relative energies of the conformers, in a quite tolerable time period.

To a larger extent, predictions of *conformational preferences* (relative thermodynamic stability of conformers) which are made on the basis of the known conformational behavior of similar or basic systems, are problematic. As Eq. 13 shows, small changes in the difference of Gibbs energies for conformers lead to a significant shift in their relative content; the conformer ratio may be even inverted. It is absolutely unclear *a priori* whether differences in molecular structure in a pair of structural analogs do not lead to such a small, but essential change in the relative stability (in other terms, free energy difference) of their structurally corresponding conformers. This uncertainty deprives the conformon approach of the ability to reliably predict predomination of either conformer for an unexplored compound. When transferring conformational preferences from compound to compound, we presuppose the exclusively "happy end" scenario with relative Gibbs energy for their conformers!

Synthetic world-wide abundance of this practice of the 70's has led to incorrect conformational assignments for many compounds. It is worth presenting a demonstrative example that shows vulnerability of this primitive tool of structural analogies. A high predominance of chair-shaped conformers of cyclohexanes is almost a postulate in organic chemistry. Twist conformers (Fig. 61) are favored rather rarely, and this inversion of conformational equilibrium requires the presence of 1,3- or 1,4-positioned very bulky *trans* or *cis* substituents of the ring, respectively.[43] Let us predict the favored conformer for cyclohexane **26a** (Fig. 64). In the cyclohexane series, we have several models of established conformational preferences, monomethyl-, different dimethyl- (including *gem*-dimethyl-) as well as other polymethylcyclohexanes (favored chair geometry) and some di-*tert*-butyl cyclohexanes (favored twist geometry). There are no bulky

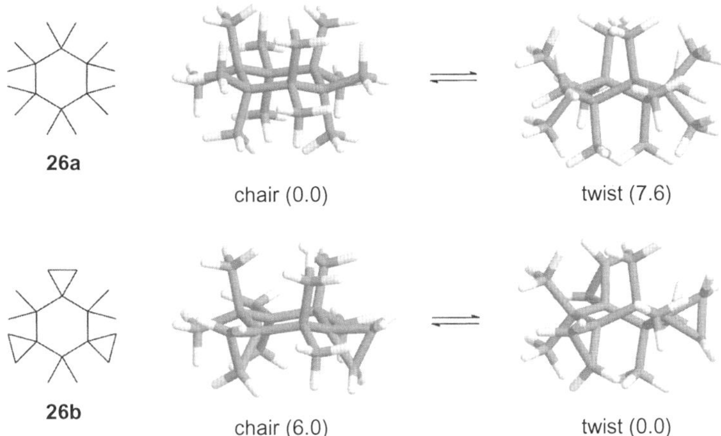

Figure 64. Conformers of cyclohexanes **26a,b**. Optimized geometries are shown and supplied with values of relative steric energies ΔE_{ster} [in parentheses; kcal mol^{-1}, by MM3(96) (Chapter 4)].

substituents in **26a**; hence, the model of di-*tert*-butyl cyclohexanes[43] is irrelevant and, according to the analogy to "normal" cyclohexanes, the preferred conformer should have chair geometry. This estimate appears correct: as MM3-based calculations (Chapter 4) show,[9] there is an immense predominance of the expected conformer (Fig. 64). No contemporary NMR or IR instrument would detect the corresponding minor twist conformer [for self-training, use Eq. 13 to evaluate the conformer ratio for **26a** at room temperature].

Now, let us try to predict the favored conformer of carbocycle **26b** by considering conformational preferences in related cyclohexanes. The closest structural analog for which the favored conformer has been identified is cyclohexane **26a**; as we know, this conformer is chair-shaped. An additional model with known conformational preferences is spiro[2.5]octane. The predominant conformer of this cyclohexane that bears one cyclopropane ring has the molecular geometry of chair. Di-*tert*-butyl cyclohexanes (a case of twist form predominance in six-membered saturated cycles), of course, are irrelevant structural analogs. Hence, our guess is that the favored conformer of **26b** adopts the chair geometry. As theoretical calculations[9] demonstrate, this conclusion is absolutely incorrect. The twist conformer is absolutely predominant (Fig. 64), and the minor conformer (here: of chair geometry) would not be detected by NMR- or IR spectroscopy. Our analogy-based estimate "succeeds" in missing the $7.6 + 6.0 = 13.6 \text{ kcal mol}^{-1}$ inversion in conformational preferences! We should conclude that this simple prediction tool (conformons) is unreliable both in respect to molecular geometry of conformers and their relative stability.

Remarkably, a search for the literature on analogs of **26a,b** and their conformational preferences in the CAS e-databases and the followed analysis of this information took ∼1 h. For comparison, the execution of molecular mechanics calculations for these compounds took less than 1 s for each stereoisomeric structure. Preparation of the initial structure for calculations ("on the screen" drawing by means of a program for 3D molecular graphics) added ∼5 min for each stereoisomer.

In addition, the discussed empirical tool of conformons is based on postulated analogies between 3D shapes of closed electron shell molecules. It is obvious that this generalization cannot be merely extended to predicting 3D geometries of reactive intermediates, i.e., organic cations and anions, radicals, cation-radicals, anion-radicals, carbenes, nitrenes, silylenes, etc. The well-studied five-membered saturated ring presents an interesting relevant example. It seems at first sight that the geometries, which this ring may adopt, are covered by non-planar conformations arose in pseudorotation (Fig. 63). According to the "conformational analysis by structural analogies," the furanose ring, which localizes a carbocation, should be thought only as either twist-, or envelope-shaped. However, rejecting this seemingly robust prognosis, non-empirical calculations demonstrate that $4'$-dehydrothymidine-$3'$-(O-methyl)phosphate cation (**25c**; Fig. 63) has an almost planar deoxyribose ring.[9] A previously unknown geometry for the saturated five-membered ring appears in this oxolane, where delocalization of the cation essentially changes the lengths of both C–O ring bonds and actually produces a "new," structurally atypical, furanose ring with its own, "new," conformation. Clearly, this distortion of the bonding in the ring cannot be predicted without involving molecular modeling, which is based on a strict quantitative chemical theory (Chapter 5), and, thus, is capable of taking into account "invisible" electronic factors, which predetermine the geometry of any molecular system.

The reader probably realizes that a new situation has emerged in conformational analysis. For organic systems of reasonable size (approximately of 20–100 atoms), it is much more convenient to perform appropriate theoretical molecular modeling of conformers than to analyze the literature or to undertake experimental studies. Reliable identification of low energy conformers by means of theoretical calculations is close to becoming an ordinary procedure in the contemporary organic laboratory. The computational procedure is not more complicated than registration of, e.g., a $^1H \times$ ^{13}C HMQC spectrum. Calculation results (molecular geometries and relative energies of modeled stereoisomeric structures) are clear straight away, while spectral data always require interpretation (sometimes very sophisticated) to provide structural information.

3.5 Conformational Dynamics

3.5.1 *The problem of identification of the detected intramolecular motion*

Conformational analysis deals with *conformers* (in adiabatic approximation, with structures that correspond to energy minima of the PES; Sections 2.4 and 2.5). Let us refine the meaning of conformational dynamics. This term describes transformations of conformers, i.e. what alterations of molecular structure are when converting a thermodynamically stable molecular geometry to another stable one. Within the Born–Oppenheimer model, this *what* means the related transition state, i.e. a certain structure of intermediate geometry that corresponds to an inflection point of the PES. The *what* includes the kinetics of conformational transformations since the transition state is characterized by a certain potential energy (the energy value that corresponds to this inflection point in the PES). The geometry of the transition state says which molecular fragment changes its geometry, or, in dynamics terms, which intramolecular motion (rotation, inversion, or a concerted one) takes place. The energy of the transition state says what the rate of the corresponding transformation is at a concrete absolute temperature T (Section 3.5.3, Eq. 14).

Studies of conformational dynamics in ordinary sized organic molecules (containing $\sim$10–200 atoms) focus on identification of the intramolecular motion that occurs upon transformation of individual conformers and on determination of the rate of this stereoreorganization. Probing of giant molecules, i.e., biopolymers, usually deals with the dynamics of changes in the conformational motif (a certain shape of the main or a side chain of the polymer); examples of such changes are protein folding/unfolding, changes of the secondary structure of nucleic acids, and spatial rearrangements of oligosaccharide chains. These dynamic processes include many sequential concerted motions (rotations) of small fragments of the polymer chain; these motions, due to their multiplicity and complexity, are approached as one intramolecular motion. If a small segment of the chain of a biopolymer is conformationally independent from other segments, one does succeed to analyze the dynamics of the 3D reorganization of this biopolymer fragment by describing its individual conformations and related conformational transformations. In such a description, this fragment of the polymer structure is actually considered as a molecule of "normal" size, and the dynamic process can be understood in the accurate frame of concrete intramolecular motions. In studying a concrete biopolymer, it is, however, difficult to ascertain that biologically relevant conformational alterations in the selected segment of the chain are not geared with those in adjacent or distal segments.

One should notice that studies of intramolecular dynamics, either experimental or theoretical, are outside the competence of synthetic organic chemists. Nevertheless, we can glance how conformational dynamics is studied, at least for organic molecules of regular size.

In organic chemistry, temperature-variable NMR (often, but not quite successfully, called dynamic NMR, or DNMR) is the major experimental tool for distinguishing conformers and rigid stereoisomers, measuring rates of conformational transformations and identifying conformers (i.e., establishing their 3D molecular structures).[a] The simplest NMR approach to examining conformational dynamics (establishing 3D molecular structures of conformers and measuring the rates of their interconversions) in a new system is as follows. NMR spectra of such an individual, molecularly non-self-aggregating compound are recorded in some range of temperatures (sometimes a broad one), in a non-viscous solution. Detection of significant line broadening followed by line splitting (Fig. 60), which are induced by decreasing the temperature of the NMR experiment, indicates that a stereoisomerization occurs. Line splitting may lead to (*1*) appearance of two sets of resonance signals which (the sets), as a rule, have a different intensity for the signals of the same (identically positioned in the chemical structure) nuclei; or (*2*) one set of resonance signals with a larger number of them relative to the number of resonance signals in the spectrum recorded at the initial temperature. In case (*1*), each set of resonance signals that result from line splitting is related to an individual diastereomer ["frozen" conformer; e.g. the *anti* or the *syn* conformer of azabicycle **13a** (Fig. 22)]. In case (*2*), this single set of NMR signals is related to a pair of "frozen" enantiomeric or identical conformers with magnetic nuclei from diastereotopic molecular fragments which interchange their diastereotopicity-related positions in interconversion

---

[a]As indicated, NMR leads in probing conformational dynamics, because the lifetimes of conformers of most organic compounds fall within the NMR timescale (Section 3.3, Table 1). In kinetics terms, temperature-variable NMR is capable of measuring the rates of molecular dynamic processes with kinetic barriers lying in the range of 4.5–5.0 to 30–31 kcal mol^{-1}, when using conventional NMR spectrometers.

of the conformers from this pair [e.g. N-invertomers of azabicycle **5a** (Fig. 2)[a]]. Exactly these stereoisomeric structures (i.e., detected at low temperatures within the experimental temperature range) are conformers which are represented in the equilibrium so weightily that NMR can detect them.[b] Of course, they are conformers also at higher, "pre-splitting" temperatures; however, NMR cannot detect them as individual stereoisomeric structures (Section 3.3).

The rate of the monitored conformational transformation is calculated by applying so-called lineshape analysis of NMR spectra with broadened resonance signals. This long-used methodology does not require professional expertise in NMR. Theoretical description of the NMR lineshape is not simple when involving into consideration the impact of molecular dynamics on this shape. Nevertheless, the related theoretical equations have been implemented in various computer programs of different complexity and applicability (e.g. in Bruker's TopSpin NMR), and the user may perform lineshape analysis with only possessing some considerable skills in this filed.

Similarly, if an increase of the experiment temperature causes line broadening followed by line coalescence, the conclusion is that, at room temperature, we deal with (*1*) diastereomeric conformers or, probably, rigid stereoisomers (rigid at this temperature) if there are two sets of resonance signals in the room temperature NMR spectrum; or (*2*) the above stereochemically characterized identical or enantiomeric conformers, or, probably, such rigid stereoisomers if only one set of resonance signals (in ^{1}H NMR spectra, of higher multiplicity than "post-collapse" signals) is in the spectrum. In case (*1*), each set of NMR signals in the room temperature spectrum characterizes an individual conformer, or, probably, a rigid stereoisomer. In case (*2*), a single set of split resonance signals in the spectrum characterizes a pair of identical or enantiomeric conformers with magnetic nuclei in dynamically interchangeable diastereotopic fragments of the molecule [e.g. bowl invertomers of benzylcorannulene[c]], or probably, a pair of rigid stereoisomers [e.g. bowl invertomers of corannulene **4d** (Fig. 68)]. If NMR signals (or, at least, some of them) do not merge pairwise when increasing the temperature, one can claim that the examined diastereoisomers or enantiomers are rigid at least at initial temperatures.[d]

---

[a]Geminal *methylene* fragments of bicyclic amine **5a** are diastereotopic, and its N-invertomers are identical (one can say that they formally are enantiomers). In interconverting these conformers, NIR exchanges spatial positions of these groups relative to the rest of the molecular backbone.

[b]The "hidden partner" equilibrium cannot be studied by using NMR spectrometers whose computers do not record a near-contionous line of free induction decay (Section 3.3). Notably that serial NMR instruments are equipped with exactly such computers. Also, if conformational enantiomerization occurs for molecules which do not have diastereotopic fragments (e.g. bowl-to-bowl inversion in corannulene **4a**; Fig. 68), both resonance frequencies and signal lineshapes are not changed in NMR spectra provided by temperature-variable NMR experiments. In solution, NMR cannot detect such conformational mobility when using achiral media.

[c]The methylene protons are diastereotopic in thermodynamically stable conformations of this compound.

[d]Recall that rigid stereoisomers are defined here as molecular 3D structures with lifetime $\tau > 24$ h (Section 2.5). Stereoisomers with τ that lies in the range 10 s — 24 h are rigid in NMR representation

If conformers of known molecular geometry interconvert via a single possible stereorearrangement, the detected intramolecular motion may be quickly identified by comparing molecular geometries of the conformers, and studies of conformational dynamics are reduced to determination of the rate of this motion. For instance, interconversion of N invertomers of amine **13a** (Section 1.8, Fig. 22) may only occur via NIR. However, sometimes distinct intramolecular motions, with distinct rates, interconvert the same conformers. For instance, the same stereoisomerization in imines (Fig. 32) occurs either as in-plane N-inversion or C=N rotation. Another example is the classical RI (ring inversion) in cyclohexane **18a** (Fig. 28). There are two paths of RI in this compound (Fig. 53), and they both come about as concerted restricted rotation around C–C bonds associated with some changes of endocyclic bond angles. These concerted rotations are dissimilar. The six endocyclic torsion angles that characterize the geometry of the ring of **18a** are different for any pair of transient ring geometries that are related to the "parallel" paths of RI. This includes the alternative transition states of RI: they are geometrically dissimilar (half-chair and sofa, Fig. 53). Thus, RI in cyclohexane occurs via different rotational motions, although they both lead to the same conformer. A similar example is supplied by piperidine **14a** (Fig. 28). For this heteroanalog of carbocycle **18a**, NIR and RI afford structures **14i** and **14j**, respectively. Formally (i.e., if we enumerate the ring atoms), stereoisomers **14i** and **14j** are enantiomers, but, in fact, they have identical 3D structures — atoms in real molecules cannot be labeled. One should conclude that both NIR and RI lead to physically the same stereoisomeric structure of **14a**. These examples show that stereochemical outcomes may be the same for distinct dynamic intramolecular processes. How could we identify the intramolecular motion that is experimentally detected in such a system?

Unfortunately, identification of the detected dynamic process is not a trivial problem. As explained (Section 3.4), NMR spectroscopy "supplies" either the structures of individual stereoisomers or the virtual (spectral) structure resulted by averaging the geometries of these individual stereoisomers (Eq. 9). In both cases, molecular structures of only thermodynamically stable isomers "form" NMR spectra. Therefore, if

---

(i.e., in the NMR timescale; Table 1); nevertheless, they are conformers in the sense of the above definition.

alternative motions provide the same stereochemical outcome for a molecular system, NMR, detecting an intramolecular motion, does not identify it. Similarly, NMR spectra do not indicate whether the monitored conformer interconversion occurs via isolated (individual) intramolecular motions or it is a result of a concerted motion. Neither the detection of an intramolecular 3D rearrangement, nor the measurement of its rate, nor the determination of the structures of interconverting conformers identifies the intramolecular motion involved. For instance, if NMR is the only source of information, it is impossible to conclude whether the signal broadening/splitting in low temperature NMR spectra of the above mentioned piperidine **14a** indicates slowing of NIR, or RI, or both.

In principle, this problem is soluble by involving ultrafast time-resolved physical methods (femtosecond timescale spectroscopies). A transition state in the corresponding saddle *point* of the PES (more accurately, the averaged molecular structure over the saddle point; Fig. 58), as a structure of infinitely short lifetime, cannot be experimentally traced. Experimental methods have some *finite* time limits for detecting any molecular structure (Section 3.3, Table 1). This principal limitation leaves a single option for "observing" transformation of molecular geometry along the MEP. Molecules should be spectrally monitored during their continuous "movement" (i.e., structural change) in some *region* of the PES. In context of conformational dynamics, the most interesting PES region is near the saddle point; therefore, the experiment consists of tracking the in-time evolution of the transition state (a process that has a finite, though extremely short, time period, Section 2.2).[36c,d] One should, nevertheless, stress again that such ultrafast time-resolved spectral experiments are absolutely nonroutine so far; simply put, they are separate, professional studies in chemical physics. At this point, one may recall that theoretical modeling is a universal and feasible alternative that may supply organic and bioorganic researchers with conformational information without seriously disturbing experimental routine.

3.5.2 *A simple solution of the identification problem: calculated conformational barriers*

One can succeed to reliably identify the observed dynamic process relatively readily by combining elementary theoretical calculations

(Chapters 4 and 5) and experimental kinetic data. Let us consider the simplest case of two energy minima (i.e., two stereoisomers) separated by one ridge of energy maxima in the PES (Section 3.2, Fig. 59). Suppose that interconversion of these stereoisomers is not slow (i.e., they are conformers), and its kinetic parameters may be measured by NMR. Suppose further that, using DNMR, we determined the kinetic barrier of this interconversion and, using theoretical methods, calculated the energies for the transition states of all alternative intramolecular motions (i.e., calculated alternative kinetic barriers) that interconvert these conformers.

In the simplest approximation, calculated barrier (more accurately, half-barrier) $\Delta E^{\#}_{el}$ is the difference between the calculated potential energy of the transition state and the calculated potential energy of the stable structure that has the lowest energy among other stable structures related to this transition state.

The values of the experimental ($\Delta G^{\#}_{exp}$) and calculated barrier $\Delta E^{\#}_{el}$, in principal, are unequal, since potential energy E_{el} and Gibbs energy are distinct quantities. In practice, these values may accidentally appear near equal because of calculation (for $\Delta E^{\#}_{el}$) and experimental (for $\Delta G^{\#}_{exp}$) errors. This does not often happen; usually, the $\Delta E^{\#}_{el}$ value does not reproduce the measured barrier well. Therefore, one should prefer not to approximate free energy difference $\Delta G^{\#}$ by $\Delta E^{\#}_{el}$ and model this difference $\Delta G^{\#}$ itself (Section 5.3.4) if computer resources permit.

It is not difficult to apprehend that the lowest *calculated* barrier corresponds to the saddle point in the MEP, i.e., to the saddle point of the *lowest* energy. This lowest energy saddle point in the PES indicates the geometry of the transition state of the *preferred* intramolecular motion, the corresponding energy minima are associated with geometries of the related conformers. Thus, by identifying the favored (in energy) transition state, we identify the favored molecular motion (the favored change in molecular geometry). Even visual comparison of the transition state geometry with geometries of the initial and final related conformers permits us to "trace" the motion. The reader can examine the sequence *initial conformer–transition state–final conformer* of three "snapshots" (images of modeled molecular geometries) for amine **12** in Fig. 21. These molecule images clearly show that inversion of the N configuration occurs in parallel with the rotation of the N-Me group. That is, this intramolecular motion is NIR.

This principle *lesser increase of potential energy — more favorable change of molecular geometry* suggests a general computational methodology for identifying the kinetically preferred conformational mobility (i.e., the favored intramolecular motion). *One can reveal which intramolecular motion (the change of molecular geometry in a specific way) that interconverts two certain conformers is favored among alternative ones, by calculating the barriers of intramolecular motions in these alternative pathways* (e.g., barriers of C=N rotation and N-inversion in imines; Fig. 32).

And what about the relationship between experimental and calculated barriers? For any molecular system, NMR experiments detect the preferred (the lowest energy) intramolecular dynamic process for the observed conformational transformation; in PES terms, NMR registers that intramolecular motion which corresponds to the MEP. Thus, in the simple case of two neighboring PES minima separated by a ridge with saddle points of different energy (Fig. 59), the lowest *calculated* barrier ($\Delta E^{\#}_{el}$ or $\Delta G^{\#}_{calc}$) is that barrier which corresponds to the *experimental* (NMR-measured) value $\Delta G^{\#}_{exp}$. Since this lowest calculated value of $\Delta E^{\#}_{el}$ or $\Delta G^{\#}_{calc}$ is closer to the value of $\Delta G^{\#}_{exp}$ than the values of other (higher) *calculated* barriers, the identification of the detected intramolecular motion is elementary by comparing the experimental and calculated barrier values. If, for a two-conformer system, there is a reasonable closeness (say, less than 1 kcal mol^{-1}) between the experimental barrier and the lowest calculated barrier while other calculated barriers are appreciably higher, one can be sure that the detected intramolecular motion is assigned correctly.[a]

Theoretical calculations herein may be characterized as the simplest PES modeling [location of stationary points (Section 3.1) in the PES]. It is remarkable that one can identify the fastest motion in a two-conformer molecular system whose conformers are interconverted via alternative intramolecular motions of distinct rate by only modeling the PES, without employing experimental methods (the lowest among the calculated barriers points out the favored intramolecular motion). However, when combing

---

[a]If there is no closeness between the experimental value and any calculated value or the value of a non-lowest calculated barrier is closer to the experimental value than the value of the lowest calculated barrier, *the performed modeling is incorrect*. Here, the term *incorrect* does not mean a technical error. It means that an inappropriate model (e.g., a PES with one or several saddle points missed or a PES for vacuum instead of a PES for solution) has been involved or an insufficiently accurate calculation method has been exploited.

PES modeling and DNMR (in other words, verifying the theoretical calculations performed), we identify the motion *reliably*.

One can extend this combined methodology DNMR — PES modeling to identification of the favored intramolecular motion(s) in compounds that are characterized by multi-minimum, multi-saddle point potential energy landscapes. Suppose that a dynamic process was detected for an organic compound by temperature-variable NMR, and the kinetic barrier of this fast stereoisomerization (conformational transformation) was measured in this experiment. What is the intramolecular stereorearrangement detected? Let us understand the formal answer to this seemingly naive question which challenges experimentalists who meet flexibility of non-trivial molecular systems in their own research. The question may be split into three parts, i.e., (*1*) among many interconverting conformers, what are conformers which are detected experimentally? (*2*) what is the MEP for this system? and (*3*) among all saddle points in the PES, what is the saddle point which corresponds to the experimentally detected barrier?

(*1*) The answer to this question is not obvious. As we know, only low energy (i.e., sufficiently weightily represented in the equilibrium[a]) conformers manifest themselves in NMR spectra. NMR registers such conformers as individual stereoisomeric structures when freezing their interconversion or detects a virtual stereoisomeric structure that is a "linear combination" of these conformers (Eq. 9) when unfreezing the interconversion. Conformers detected in DNMR experiments are the lowest energy conformers if they interconvert more slowly than other low energy conformers [e.g., the chair-shaped (detected) *vs.* twist-shaped (not detected) conformers of cyclohexane; Fig. 53]. If conformers of minimal energy interconvert more rapidly than other low energy, but still populationally weighty conformers, the latter are individual conformers registered by DNMR first, while the former are registered as a virtual, in-time-averaged structure (a "linear combination" of these conformers). Further slowing of the interconversion rate exposes the minimal energy conformers as individual stereoisomeric structures. In PES terms, NMR detects conformers which belong to deep energy minima, irrespective of how many energy ridges separate these minima in the PES and how many energy minima are there (if the number

---

[a] Say, each not less than 3%.

of the minima is not large[a]), according to whether the ridges are high or low in the saddle points.

(2) The answers to questions (2) and (3) are illustrated in Fig. 65. As often occurs for a multi-minimum, multi-saddle point PES, the molecule, step-by-step, adopts several conformations in going from a minimal energy conformer to its counterpart in the NMR-detected conformational equilibrium (i.e., to another minimal energy conformer); most of these conformations are adopted by stable structures (conformers), and others are of transition states that separate these conformers. Consider a pair of conformers. In general, a sequence of conformational transformations, which includes other (intermediate) conformers and, of course, the corresponding conformational transition states, provides interconversion of the conformers from the pair; let us call such a sequence conformational pathway or conformational itinerary. There are frequently several alternative conformational pathways in the PES that provide interconversion of the same two conformers (including those of lowest energy; Fig. 65): there, some or all conformers as well as transition states in a sequence differ from conformers and transition states, respectively, in another sequence. Clearly, there is a saddle point in each pathway that has higher energy than other saddle points in this conformational itinerary. Such highest energy saddle points that belong to alternative non-overlapping pathways, in general, have different energies[b] since conformational transition states that correspond to these saddle points have different 3D geometry. The pathway, whose highest energy transition state is lower in energy than the transition state of highest energy from any other alternative pathway, obviously is the pathway in which the molecule minimally increases its potential energy in interconverting its two certain conformers. Saying this, we recognize in this favored conformational itinerary the MEP (Section 3.2); here, it connects a pair of energy minima separated by several energy ridges in the PES. Thus, the MEP for a pair of conformers in a multi-minimum, multi-saddle point PES is the pathway, whose highest energy saddle point has a lower energy than the highest energy saddle point of any other alternative pathway.

---

[a]Otherwise, the entropy contribution should be taken into account. Equation 12 (Section 3.3) reflects the impact of the number of conformers (energy minima) on the occupation of the i-th minimum.
[b]Only if there is a partial overlap of alternative itineraries, the same saddle point may appear of highest energy for them.

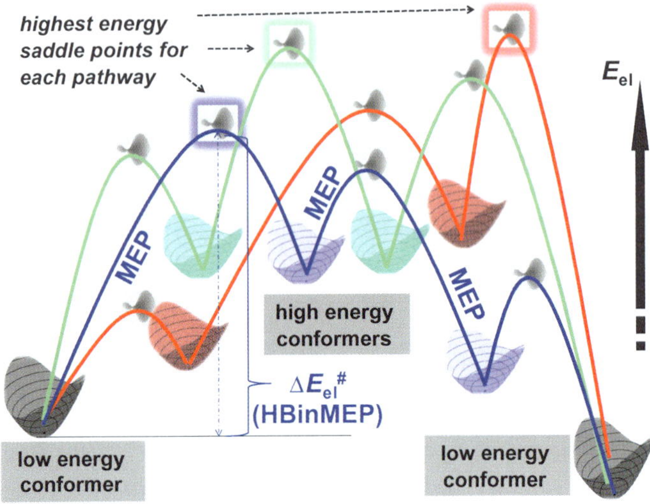

Figure 65.   Three alternative itineraries (each shown in a different color) between the lowest energy minima (shown in gray) in an abstract PES; blue indicates the MEP. The blue rectangle denotes the transition state which relative energy (kinetic barrier HBinMEP) is measured in kinetic NMR experiments.

(*3*) In summary, there are many conformational alternatives for the molecule to change its shape in interconverting the lowest energy conformers [see (*1*) and (*2*) above], and there is one barrier measured by DNMR in some temperature interval (see Section 3.5.1). The favored pathway for the molecule (more accurately, for the molecular ensemble) in interconverting these conformers is the MEP that connects them in the PES. Metaphorically speaking, molecules travel in this way and may be spotted only there.[a] The answer to question (*3*) is therefore obvious. The NMR-measured barrier corresponds to the highest energy saddle point in the MEP, because the slowest rate determines the overall rate of the process. Thus, *NMR measures the highest barrier (HB) in the lowest energy pathway*. Molecules "find" the energetically preferable path that connects the deepest minima in the PES, while NMR "sees" the highest point in this itinerary. Briefly stated, the NMR-measured barrier corresponds to the HB in the MEP (HBinMEP).

---

[a] Statistically.

The answers to questions (*1*)–(*3*) grant a general understanding what is the kinetic barrier measured in DNMR experiments for conformational equilibrium, but do not advise on how to reveal which intramolecular motion is associated with the barrier measured for a concrete compound. Nevertheless, they make it clear that a rational methodology of identifying the detected fast intramolecular motion consists of theoretical modeling of the PES (i.e., location of conformers as well as transition states and establishment of all pairs *conformer$_i$ – related transition state$_{i-j}$* and *conformer$_j$ – related transition state$_{i-j}$*; Chapters 4 and 5). Location of energy minima identifies all conformers of the system of interest low energy conformers including the minimal energy ones. Location of conformational transition states exposes all intramolecular motions including concerted ones that occur in the studied molecular system. Establishment of all pairs *conformer–related transition state* is equivalent to displaying all the itineraries in the PES (itineraries that, through inflection points in the PES, connect an energy minimum with any other energy minimum). The first result of this PES modeling is that one can indicate conformers whose interconversion is detected by NMR for the studied compound [see answer (*1*) above]. The next, desired result is that both the MEP and the HBinMEP-corresponding transition state [see answers (*1*) and (2)] are recognized straight away *in the MEP modeled* for this molecular structure[a] if *M* is less than 15–20 (*M* is the number of conformers). If *M* is larger, location of this indicative transition state *in the modeled MEP* may require use of computerized graph analysis.

Quantitative conformational schemes (actually, graphs; see, e.g., Fig. 53, Section 2.8 or Fig. 66, this section), where stereoisomeric structures are supplied with ΔE_{el} (or, better, ΔG_{calc}) values or more or less equivalent values of ΔE_{ster} (Section 4.2.3), are a convenient graphical representation of all possible itineraries for translocating the molecule from an energy minimum to another minimum in the PES (i.e., for interconverting conformers). Using these graphs, the final, post-calculation step in assigning the experimental barrier to a certain intramolecular motion is "amateurishly" simple. In order to locate the MEP, alternative itineraries between the lowest energy minima in the PES are screened in such a graph by merely comparing relative energy of the transition states of highest energy. In the MEP, such a transition state corresponds to the desired HBinMEP. Being compared with

---

[a]Clearly, location of this transition state in a reliably modeled PES is equivalent to identifying the intramolecular motion detected by DNMR.

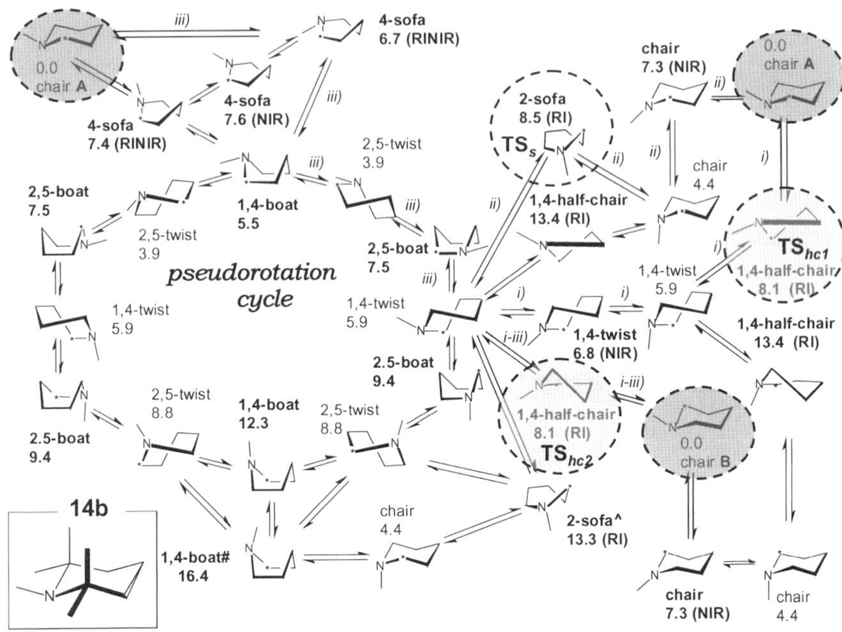

Figure 66. A fragment of the conformational scheme for piperidine **14b** (obtained by MM3, Ref. 37) that includes low energy pathways *i)*, *ii)* and *iii)* of interconversion *chair conformer* **A** — *chair conformer* **B** detected by DNMR. For clarity, α-Me groups are not shown. Numbers show relative steric energies ΔE_{ster} (relative to the lowest energy conformers **A** and **B**, kcal mol^{-1}). **A** and **B** are indicated by dark grey ellipses. Circles in light gray specify transition states $\mathbf{TS}_{hc1}$ and $\mathbf{TS}_{hc2}$ of alternative low energy pathways *i)* and *iii)*, respectively, while the white circle points out the transition state $\mathbf{TS}_s$ of an additional pathway of low energy (pathway *ii)*). Index # depicts the second-order transition state, asterisk is a formal label.

the values of calculated barriers, the value of the experimental barrier provides an accurate (numerical) verification of the assignment.

To illustrate this identification methodology, an example for a relatively small organic molecule is presented (Fig. 66). The nitrogen fragment and the ring of piperidine **14b** are flexible at ambient conditions and, therefore, only one resonance signal of C-Me groups is present in ^{1}H as well as ^{13}C NMR spectra. At low temperatures, the signal is split into two signals of equal intensity. This splitting is a direct indication that a fast intramolecular dynamic process takes place in this compound. DNMR measurements afford a 7.8 ± 0.2 kcal mol^{-1} value for the kinetic barrier of this process at 236 K [i.e., a $(3.2 \pm 0.7) \times 10^5$ s^{-1} value for the corresponding rate].[37] Obviously, there are several concerted intramolecular motions in this amino compound; they are RI, NIR, RINI, ring inversion–isolate rotation of a substituent (RIR), and ring inversion–N inversion–C–N rotation (RINIR). Any of them may be assumed as the rate-determining for the detected fast conformational dynamics in cyclic amine **14b**. Which of these motions, in fact, occurs with this rate?

The chair-shaped conformers **A** and **B** of identical molecular geometry are the only populationally weighty conformers of this compound. This means that, among all conformational transformations in **14b**, NMR tracks the interconversion of **A** and **B**. The conformational scheme[37] supplied by molecular mechanics calculations (Chapter 4) shows that there are three low energy pathways that connect these lowest energy conformers [(*i*), (*ii*), and (*iii*); Fig. 66]. By chance, two pathways (*i*) and (*iii*) have a common highest energy transition state (transition state **TS**$_{hc1}$ of half-chair geometry). Coincidentally, another transition state (**TS**$_{hc2}$) in itinerary (*i*) has the same calculated energy as transition state **TS**$_{hc1}$. If the probability of non-passing kinetic barriers is neglected, both pathways (*i*) and (*iii*) are equal from the viewpoint of the kinetics of the *conformer* **A**–*conformer* **B** interconversion. Pathway (*ii*) has the highest energy point which corresponds to the transition state of 2-sofa geometry (**TS**$_s$). This transition state is slightly higher in energy (by 0.4 kcal mol^{-1}) than transition states **TS**$_{hc1}$ and **TS**$_{hc2}$. Hence, two pathways (*i*) and (*iii*) are two energetically equal MEPs. The equal values of ΔE_{ster} (steric energy relative to the lowest energy conformer **A** or **B**; for steric energy, see Section 4.2.3) for both half-chair-shaped transition states **TS**$_{hc1}$ and **TS**$_{hc2}$ are two HBinMEPs of the same height.

At first glance, pathway (*ii*) may not be considered further as an MEP. However, the accuracy of energy calculations by MM is usually not better than 0.5 kcal mol^{-1}. One cannot exclude the possibility that sofa-shaped transition state **TS**$_s$ is, in fact, of the same energy as transition state **TS**$_{hc}$. For caution, itinerary (*ii*) is also worth considering as the MEP for conformational interconversion *chair–chair* in **14b**. The PES modeled for this cycle is a remarkable case: we have three MEPs, (*i*), (*ii*) and (*iii*), instead of the expected one.[a]

These three transition states $\mathbf{TS}_{hc1}$, $\mathbf{TS}_{hc2}$ and $\mathbf{TS}_s$ are of RI that, as mentioned, is a concerted restricted rotation around endocyclic bonds accompanied by some alteration of endocyclic bond angles. Thus, solely by means of theoretical modeling, we have identified the NMR-detected motion in azacycle **14b** — it is RI. Since the HBinMEP value is known from experiments,[37] we are able to prove our assignment. The two calculated barriers and the experimental barrier are very close in their values; they are 8.1, 8.5, and 7.8 kcal mol^{-1}, respectively.

The reader may see how accurate theoretical modeling is: the experimental data firmly support the "calculations-made" assignment of the kinetic barrier for this conformationally non-primitive system. Indeed, the pathway-associated numerical closeness of the calculated and the experimental values is a weighty evidence. There are twelve different pathways for the interconversion *chair conformer* **A**–*chair conformer* **B** (Fig. 66). The point is that the calculated value of the only HBinMEP (and not of another barrier) is close to the experimental value. Let us estimate the probability for this event to incidentally occur by considering two independent events: (*1*) the values of the calculated HBinMEP and the experimental barrier [it has probability $P_{(1)}$] are close and (*2*) the calculated highest energy barrier, which value is maximally close to the experimental value, lies in the MEP [and not in another conformational pathway that connects **A** and **B**; this event has probability $P_{(2)}$]. By probability theory, if two events are independent, then the probability $P_{(1)+(2)}$ of their both occurring is the product of the probabilities of each event's occurring. Clearly, $P_{(2)} = 1/12$. One cannot estimate probability P_1 since the initial assumption in considering probabilities here is that we do not know how successful the performed theoretical modeling is. Let us suppose (intentionally deteriorating the situation) that this probability is not small, say, $P_{(1)} = 3/4$. Then, the estimated probability $P_{(1)+(2)}$ is 0.0625. That is, the probability that the modeled

---

[a]The situation with this PES actually is a didactic example. A similarity of the rates of concurrent intramolecular motions is not a rare case for organic compounds which are characterized by a multi-minimum, multi-saddle point PES. It is not surprising that some highest energy transition states from alternative conformational pathways may appear near in energy if there are many saddle points in the PES.

HBinMEP is *incidentally* numerically close to the measured barrier (that is, appears the genuine HBinMEP by *accident*) is only ~6% even in a bad case scenario.

Unfortunately, in chemical practice, the closeness of the experimental barrier to *one* calculated barrier is often assumed to be sufficient evidence when identifying an intramolecular motion. Now, after the discussion of the reliable methodology of barrier assignment, we can easily understand why the "one calculated barrier argument" is not solid. Its key hypothesis — the closeness is not accidental and other motions have higher barriers — is in no way verified by this approach. In fact, values of several calculated barriers may be close to the experimental value, because kinetic barriers of several alternative intramolecular motions often differ only by 1.5–2.5 kcal mol^{-1} for the same molecular system. Misleadingly, the calculated kinetic barrier for any motion from this set may be accepted as suitable, if considered alone *vs.* the experimental value.

3.5.3 *The Eyring equation*

A noteworthy aspect of our discussion of conformational dynamics is the relationship between the quantities *kinetic barrier* and *rate*. Indeed, these meanings often are equivalently used when describing kinetics of intra- or intermolecular transformations at constant temperature. This equivalence means that researchers imply a single-valued relationship between *kinetic barrier* and *rate*. This relationship is described by the well-known Eyring equation (Eq. 14) that is the central algebraic expression in the commonly used concept of activated complex (transition state), and one can find it in almost every stereochemistry textbook or monograph on conformational theory.[a] It connects the measurable kinetic rate (as the rate constant of the transformation *conformer→transition state*) and the value of the Gibbs energy of activation ($\Delta G^{\#}$ or *barrier*) of this transformation, as follows:

$$k = \chi \times k_B T / h \times \exp(-\Delta G^{\#}/RT) \qquad (14)$$

where $\Delta G^{\#} = G_{transition\ state} - G_{conformer}$, kcal mol^{-1}; χ is the transmission coefficient (usually assumed to be equal to unity); T is the absolute temperature; and constants k_B (Boltzmann constant) h (Planck constant), and R (the universal gas constant) adopt their usual values.[b]

---

[a]Notice that this concept is meaningful only within the limits of the adiabatic approximation of molecular structure (Section 3.1). Otherwise, one should consider states of different probability and not a continuous PES.

Why does the Eyring equation have exceptional significance for studying conformational dynamics? In order to understand the potential of this expression, we probably need to inspect it in some detail. The rate of a change of concentration in respect to the time variable (dC/dt) is a measurable characteristic. In principal, researchers can follow changes of the concentration in time for any molecular structure. In contrast, Gibbs energy (free energy) has no "material carrier" (a measurable physical quantity) that, if monitored, can indicate changes of free energy straight away. The general manner in which we think about free energy is that it is a fundamental algebraic function that quantitatively describes the state of a system (its tendency to be changed and the direction of the changes), depending on physical macroparameters of the environment (e.g., temperature, pressure, composition, and the presence of an external field) and its own features (e.g., potential energy of interactions between elements of the system, and their distribution in space and energy). In short, free energy is always implicitly determined by many external and internal parameters. How do experiments provide this "invisible" energy, or, more accurately, its change ΔG? ΔG for two states of a molecular ensemble (in PES terms, the ensemble states that correspond to two points in the PES; Section 5.3.4) is calculated using the quantitative relationship of this difference with measurable quantities that depend on experimental parameters, if, of course, the dependence is known.

The Eyring equation is an example of exactly such a relationship.[a] This equation affords kinetic barriers $\Delta G^{\#}$ for inter- and intramolecular transformations deriving them from experiment data (i.e., the rate k), and makes it possible to obtain the difference $\Delta G^{\#}$ (a numerical value) with a level of accuracy that depends only on the accuracy of the measurements of the contributing parameters. Note that the rate k is a measurable "macro quantity." Thus, the significance of this prominent algebraic expression is that it provides an important *molecular* characteristic, $\Delta G^{\#}$ (according to the activated complex concept, the free energy of the transition state relative

---

[b]$k_B = 3.2995 \times 10^{-24}$ cal K^{-1}; $h = 1.5836 \times 10^{-34}$ cal s^{-1}; $R = 1.9872$ cal K^{-1} mol^{-1}.

[a]Another example is the classical Van't Hoff isobar $\Delta G° = -RT \ln K$, where K is the equilibrium constant at conditions of unchanging pressure. Herein, two states of the molecular ensemble are related to two energy minima in the PES. The Eyring equation describes the situation with the free energy difference, where a saddle point and one related energy minimum are involved.

to the free energy of the related thermodynamically stable molecular structure) operating with the "macro characteristic" k. The Eyring equation is, therefore, very useful in conformational considerations. For instance, Eq. 2, which specifies the border line that separates between the conformational and rigid states of molecular structures at different temperatures (Section 2.6, Fig. 47), is a particular case of the Eyring equation written by taking the certain value of $k = 8.4 \times 10^4$ s as a numerical parameter.

3.6 Flexible or Rigid? Assumptions in the Synthetic Laboratory

Synthetic chemists do not unilaterally conduct studies of conformational dynamics. However, even if supported by NMR specialists in detecting intramolecular motions and measuring their rates, they cannot firmly identify the registered intramolecular motion, because theoretical modeling (e.g., see Section 3.5) is still not an operating tool of organic experimentalists. How are conclusions regarding conformational mobility of either organic compound drawn in synthetic and bioorganic research groups? The answer is discouraging: a "simple look" of structural analogies is usually the only tool used. Molecular structure is viewed as an assembly of structural fragments that have certain flexibilities. In this projection, a system does not require additional analysis unless it has an exotic chemical or 3D structure.

How does this "tool" work? A molecular system is formally split into recognizable fragments that are associated with the fragment-specific mobility. Such structural fragments are common elementary molecular units of different chemical structure (e.g., C-, N-, P-, Si-, and S-centered units), as well as differently substituted structural fragments of the ethane type [i.e., $R_1(R_2)(R_3)C-C(R_4)(R_5)R_6$; $R_1(R_2)(R_3)C-X(R)_n$, where X = heteroatom]. The number of these typical fragments is relatively small, and kinetic constants of internal rotation and geminal ligand mutation (and, thus, kinetic barriers $\Delta G^{\#}$) at various temperatures are known for many of them. Since kinetic barriers have only a weak dependence on temperature, these flexible and rigid "conformons" are associated with the corresponding barriers that are supposed to have constant values over a wide range of temperatures. In practice, these numerical values (barriers) have been actually replaced with two opposite qualitative characteristics of flexibility of these "conformons," rigid or flexible. Molecules are "two-colored" in this mechanistic representation, i.e., they are assembled from molecular building blocks assigned to two stereodynamic labels — either flexible or rigid.

For instance, it is always assumed that rotation of Me is free until very low temperatures are reached. What does this ubiquitously implied assumption mean? This generalization for *rotation of Me* actually postulates that Me rotamers *in any compound* have a very short lifetime. In other words, any Me group in different compounds is formally replaced with a generalized Me group that rapidly rotates around its C_3 axis. Extending the category *flexible* from the C–Me fragment to the C–C unit of simple substituted ethanes, organic chemists reasonably consider branched and unbranched alkyl groups at ambient conditions as flexible chains. Experimental data regarding low barriers of C inversion in model carbanions supply organic chemists with approximation that all trialkyl units $R(R^1)C^-$ (R^2) are configurationally labile above ~ 180 K. Similarly, experiments show that amide fragments in dipeptide sequences *amino acid–Pro* have essentially increased lifetimes relative to other amide fragments of the polypeptide chain (Section 2.5). This observation permits to bioorganic chemists to supply proline-containing domains with a "label" of long lifetime and to suggest that these dipeptidic fragments are rate-determining in folding and unfolding chemically intact protein chains. In light of this generalization, these biopolymers are linear chains with amide fragments of only two rotational flexibilities. Amide fragments of the *amino acid–Pro* sequences are "rigid" to some extent, while other C–N fragments are flexible.

Flexibility is actually self-exposed in this primitive modeling, because structural fragments are designated *a priori* as either flexible or rigid frameworks, at the level of customary Lewis structures, without experiments or theoretical calculations. Such a simple look at diphosphine **21** (Fig. 45) correctly indicates that, at room temperature, the pholpholane rings are flexible (undergo pseudorotation; Section 3.4), and there is rotational freedom (Section 1.5) for ring substituents and for the central C–C bond. In a similar manner, one can reasonably conclude that all elementary molecular units in this compound, C- as well as P-centered units, are configurationally stable. Without any studies, we have supplied all fragments of **21** with the "labels" *flexible* and *rigid*, implying room temperature.

As one can expect, this "simple look" *certain chemical fragment–certain (characteristic) rate* is not a valid tool for qualitative estimates of molecular flexibility when dealing with even diminutively unusual molecular structures. It is limited much more than the VSERP ruling in static stereochemistry (Section 1.1). The reason for these limitations is absolutely obvious. A molecule is not an additive assembly of structural fragments since all of its electrons interact with each other. Due to this cooperativity of electron-electron interactions, properties of many different organic molecules cannot be modeled when assembling these structures from given structural fragments with characteristic properties. Let us consider some examples of compounds, whose Lewis structures "have nothing to report" about the molecular flexibility.

The amino fragment in many trialkylamines undergoes a fast NIR at temperatures that are below 273 K (0°C) for several tens of degrees. The N unit in, e.g., tertiary amine **12** (N-methyl-N,N-diisopropylamine, Fig. 8), is flexible at the temperature of bath methanol–liquid nitrogen (175 K, or −98°C) and even at lower temperatures. Within the limits of the mechanistic association *certain chemical fragment–certain rate*, the N fragment in alkylamines is thought to have configurationally flexibility even at low temperatures. This association *alkylamine nitrogen–fast NIR* dictates that the N fragment of another N-methyl-N,N-diisopropylamine, bicyclic amine **5a** (Fig. 2), should be flexible. However, this bicyclic amine is a "perfidious" system. The so-called bicyclic effect (the Lehn effect) is attributed to the skeleton of 7-azabicyclo[2.2.1]bicycloheptanes; as Lehn assumed, this effect consists of exceptionality of this skeleton in slowing down NIR for amines. Indeed, the N fragment in bicycle **5a** is rigid[a] at the temperature of the methanol — liquid nitrogen bath.[44] The label *flexible* cannot be "copied" from the "conformationally general" amino fragment to the amino unit of bicyclic amine **5a**.

Maybe, azanorbornanes are unordinary amines and, therefore, are an exception to the general "rule" *amine nitrogen — fast NIR*? However, in stereodynamic aspect, 7-azabicyclo[2.2.1]bicycloheptanes are not exceptional trialkylamines; e.g., 7-azanorbornane **5d** (Fig. 2) has a low barrier of NIR.[4a] As shown for tertiary alkylamines, the NIR rate is determined by the 3D geometries of the N and the adjacent C elementary molecular

---

[a] In the sense of rigidity defined in Section 2.5 and illustrated in Fig. 47.

fragments;[27] if these molecular fragments have some geometrical distortions,[a] the NIR barrier of the alkylamine is high. In this light, the bicyclic effect merely does not exist. It "arises" from the traditional association *alkylamine nitrogen — fast NIR* that is true only for a certain range of geometries of the N and C tetrahedrons of the amino fragment.

Continuing with "perfidious" amines, it is worth mentioning morphine compounds. Among substituted piperidines, codeine (Fig. 62) looks to be an ordinary N-Me piperidine. There are no anomalies in the 3D geometry of its piperidine ring, and one should expect a "standard" 7–9 kcal mol^{-1} barrier of NIR for this amine. However, its NIR barrier is so low that conventional temperature-variable NMR cannot measure it.[45] This failure of the customary association *certain chemical fragment–certain rate* demonstrates that quick, Lewis structure-based analogies may mislead in predicting the kinetics of stereoisomerizations. On this occasion, we remark that quantum mechanical *calculations*[27] predict this unordinary fast NIR in morphines and disclose the crucial role of orbital energies. This example of morphine compounds reminds us that "invisible" factors of quantum mechanical origin escape attention of the mechanistic understanding of flexibility of molecular frameworks.

Another example comprises compounds with elementary molecular units of distorted geometry (Section 1.1). These fragments cannot be supplied with characteristic rates of geometrically normal elementary fragments; simply put, the "copy-paste tool" of structural analogies does not work for estimating their molecular flexibility at all. Do monocyclic aromatic systems possess rotational freedom for endocyclic C–C units? Keeping in mind the "reference standard," i.e. a planar molecule of benzene, the obvious answer is "no." In fact, exactly a concerted restricted rotation around endocyclic bonds provides interconversion of $M-$ and $P-$stereoisomers of helicenes (the non-terminal rings of these polycyclic aromatic systems are non-planar; see, e.g., Fig. 45, Section 2.5). *Ortho*-disubstituted benzene **27** (Fig. 67) well illustrates this motion that occurs in many polycyclic aromatic compounds with tight, ring-deforming steric contacts of fragments. The ring of **27** is non-planar, and the *trans* geometry of the *tert*-Bu substituents mimics the geometry of the fusion of non-planar rings in helicenes. A concerted restricted rotation around the endocyclic bonds coupled with

---

[a]These molecular geometries even are not strikingly distorted ones.

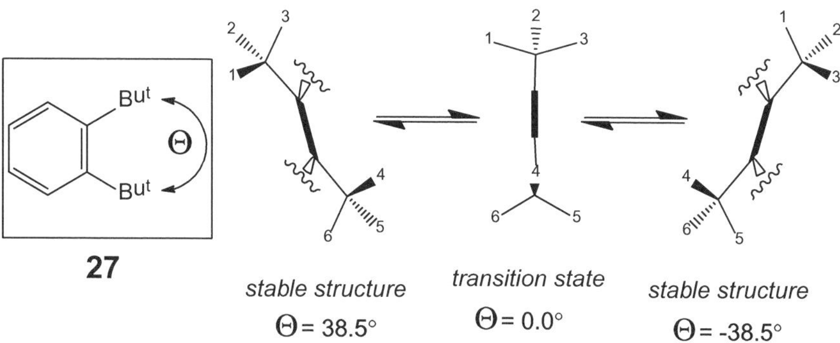

Figure 67. Concerted *ring inversion–tert-Bu rotation* in benzene **27**. Angle Θ is the torsion angle between the *ortho*-positioned *tert*-Bu groups.

rotation around the exocyclic C–C bonds inverts the ring; simultaneously, both the *tert*-Bu's invert their orientations relative to the aromatic ring.[46a]

Also, without preliminary information, what could one conclude about configurational stability of *pyramidal* carbons in aromatic systems, e.g., in buckybowls (Fig. 68)? The "simple look" of structural analogies cannot be applied to these compounds at all, because the canonical geometry for aromatic carbons is planar. Let us suppose that we know what the real situation with the central carbons of corannulenes is — these elementary molecular fragments are pyramidalized. This information does not disclose whether these aromatic carbons of atypical geometry are configurationally stable; clearly, additional data are required. *Experiments* have shown that the bowls of the parent corannulene **4a** as well as mono-/disubstituted corannulenes rapidly invert at 0°C (273 K), i.e., non-planar carbons of these compounds undergo a fast pyramidal inversion.[47a]

Additional examples of configurational mobility of pyramidal aromatic carbons are inversional conformational transformations in antracene **19a** (Figs. 35 and 68) and acepentalene **4e** (Fig. 68).[36f,47b] The established configurational instability of C pyramids in several pattern compounds of different skeletons suggests that non-planar aromatic carbons are configurationally unstable at the ice bath temperature. In the "binary" association *flexible or rigid*, this suggestion is equivalent to ranking this structural fragment as *flexible*.

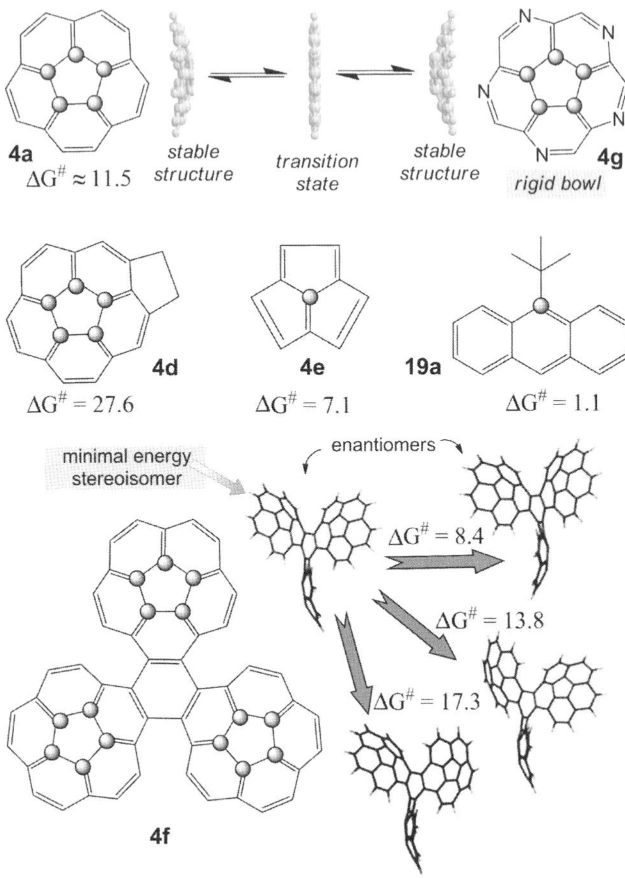

Figure 68. Examples of aromatic systems with pyramidalized aromatic carbons (indicated by grey balls). Bowl-to-bowl inversion is schematically shown for compound **4a**. For trimeric corannulene **4f,** inversion of each bowl in the lowest energy conformer occurs with a distinct kinetic barrier $\Delta G^{\#}$ (calculated;[47c] kcal mol^{-1}) and leads to a distinct conformer. The presented molecular geometries of conformers are calculated geometries.[47c]

Let us use this conformational cliché *non-planar aromatic carbons are configurationally flexible* in further demonstrating the unreliability of the quick, "copy-paste" look at molecular flexibility. The above association of pyramidal aromatic carbons with kinetic instability attributes flexibility to any buckybowl compound including, e.g., corannulenes **4d** and **4g**. However, "ignoring" the prediction, compound **4d** shows a high barrier of the bowl-to-bowl inversion[47a] (Fig. 68). The structure is rigid, e.g., at the ice bath temperature (273 K; see Fig. 47 for the *rigidity–flexibility* demarcation line). Pentaazacorannulene **4g** is rigid at room temperature, as *ab initio* calculations unveil. This heteroanalog of flexible parent molecular system **4a** could not be under suspicion in being rigid, as the "copy-paste flexibility" rules on pyramidalized aromatic carbons; of course, we become sceptical about this rule.

Trimeric corannulene **4f** presents even a more demonstrative example of the weakness of the "flexibility transfer" from a conformon to the molecular geometry of a structural analog. As theoretical modeling shows,[47c] this compound has significantly different barriers in three pathways of inversional stereoisomerization of the lowest energy conformer that lead to three different stereoisomers (Fig. 68). If we had considered the parent corannulene **4a** as a conformon for buckybowls which is labeled *flexible*, we would anticipate an evidence of equal or near-equal flexibility for corannulene rings in this conformer of **4f**.

If one could weigh structural factors which predetermine conformational rigidity of buckybowls, we should be able to predict whether a new corannulene overrides or follows the mentioned prescription for conformational mobility (*non-planar aromatic carbons are configurationally flexible*). Therefore, let us discuss structural factors that affect the barrier of bowl-to-bowl inversion.

As indicated, by attaching the ethylene bridge to the corannulene rim in structure **4d**, the inversion barrier is significantly increased. This shows that the strain[a] essentially contributes to the barrier. The change of aromaticity[b] going from the stable aromatic structure of large curvature to the planar aromatic transition state (Fig. 68) is an additional barrier-determining factor.[46d] Possessing different mutual orientations of the corannulene fragments, the alternative transition states of bowl-to-bowl inversion, which converts the lowest energy conformer of **4f** into three other conformers[c] (Fig. 68), differ in the backbone strain and, probably, aromaticity. Most frequently, neither Lewis structures, nor information about structural deformations of some fragments, nor accurate molecular geometries throw light both on the strain in the whole molecular backbone and the ring aromaticity.[d] ■ That is, both of

---

[a]As known, *strain* is a convenient energy-related, quantitative molecular characteristic, despite the absence of a strong physical basis in defining *strain* for molecules.

[b]Similarly to strain, aromaticity is a non-physical, but convenient molecular characteristic; it describes a certain chemical bonding, with some stress on its energetics.

[c]Saying conformers, we imply the conformational equilibrium for **4f** at room temperature.

[d]For instance, could the reader reasonably decide which stereoisomer is more strained and which one is more aromatic among the computationally located[47c] stable stereoisomers of **4f** (Fig. 68)?

the flexibility-determining structural factors (the strain and the aromaticity extent) cannot be reliably compared for two aromatic bowl-shaped structures when analyzing them in a primitive, non-computational manner. We would be unable, therefore, to predict whether the bowl inversion in corannulene **4f** or **4g** is fast or slow by comparing changes of their molecular geometries and aromaticity, which occur when converting the stable stereoisomer to the related transition state, with the same changes for the conformon supplied by the flexible molecule **4a**.

Thus, we cannot draw *a priori* conclusions regarding the conformational flexibility for either corannulene of unfamiliar chemical structure. Also a more general statement *nonplanar aromatic carbons are configurationally flexible* should not rule researchers who encounter molecules with such pyramidalized carbons.

The above demonstrated difficulties in assigning molecular flexibility reflect the common situation with the quick, analogy-based estimation *flexible* or *rigid*. The concluding remark is as follows: "easy-to-make" qualitative predictions of flexibility are reliable only for molecular fragments that are structurally plain analogs of parent compounds which conformational behavior has been studied in depth.[a]

In addition, a whole class of intramolecular dynamic processes is missed by the "simple look." They are concerted motions. Correlation of intramolecular motions in a molecule means that they have the common transition state; in simple words, they occur simultaneously. The overall, resulting motion is termed concerted (also, coupled or correlated); dynamic gearing is an equivalent term that means coupling of two or more intramolecular motions. Motion gearing deprives conformationally typical structural fragments (i.e., those which geometries correspond to conformons; Section 3.4) of "standard" rates associated with individual conformational transformations of these fragments. One can say that a concerted motion, by involving two or more these conformon-related molecular fragments (sometimes separated by several bonds), formally combines them into one larger fragment. Obviously, no rates, which may be attributed to this new, "no conformon-associated" molecular fragment, are in the bank of "standard" rates that characterize transformations of conformons.

---

[a] Here, parent compounds may be compounds of any structural complexity that are trivially composed of fragments of established flexibility or rigidity.

Correlated rotation around two or more chemical bonds[a] and NIR in alkylamines (Fig. 21, Section 1.8) are the most studied concerted intramolecular motions. For instance, the ubiquitously known RI in different cyclic systems actually occurs via a concerted, restricted (in rotational angles), rotation around endocyclic bonds, which is coupled with some changes of endocyclic and exocyclic bond angles.[b] This complex, one transition state motion is the only possible motion in RI. Isolate intramolecular motions (rotations around either ring bond) would disrupt these bonds, if were occurred. Only cooperativity in geometrical changes provides structural integrity of the flipping ring for cyclic systems.

Concerted dynamic processes occur in organic molecules much more frequently than it is usually thought. However, no "pocket" approach (e.g., Newman projections, mechanical molecular models, 3D images of molecules of changeable geometry on the screen, structural analogies) could descry them. Let us assume that we have a very precise mechanical elastic model of a molecule. Could it predict dynamic gearing? Upon stereochemical transformations, geometrical changes are developed as prescribed by a "mysterious" wave function, in no way obeying any purely theoretical or empirical mathematical law of mechanics including elasticity representations. Mechanical thinking cannot locate the MEP due to an elementary reason — molecules are not mechanical systems. ▉

The geometrical outcome of a concerted motion is easily recognizable: the resulting molecular geometry cannot be provided by any individual (isolate) intramolecular motion, but is formally provided by the involved individual motions occurred in a sequence. Thus, motion gearing cannot be established from considering molecular geometries of the interconverting conformers. The only criterion of predominance of dynamic gearing over occurrence

---

[a] Triptycenes are ubiquitously known in this aspect.[46b,c] This dynamic gearing (often called gear effect) is amazingly "picturesque" in these molecules, where two large geminal substituents are geared similarly to two meshed three–teeth mechanical gears.

[b] Sometimes changes of bond angles are significant in the inverting ring. Stereoisomerization *trans* → *cis* in strained cycloalkenes, e.g., in cycloheptene **4b** (Fig. 2), is an example of RI that changes the configurations of ring atoms (here: of the two olefinic carbons). This fast concerted rotation, which includes a formal 90° rotation around the C=C bond (in fact, there is no π bonding in the transition state of this rotation; Ref. 46e), is accompanied by depyramidalization of the olefinic carbons.

of sequential isolated intramolecular motions is location of the transition state with a 3D structure that (*1*) is a result of geometrical changes caused by two or more (but not one) individual motions (see, e.g. the transition state of NIR for amine **12** in Fig. 21); and (*2*) is the highest energy transition state in the MEP (Section 3.5.2, Fig. 65). We again come to the conclusion that modeling of the PES is necessary when studying flexibility of organic molecules.

Can structural analogies assist at least in revealing whether a concerted intramolecular motion is predominant for a system of interest? A clear answer is provided by the *o*-xylene and hexamethylbenzene example. For the "copy–paste" tool of structural analogies, the gearing of rotation of Me groups is equally possible in these benzenes. However, the estimate appears incorrect even for these non-puzzling patterns. Theoretical modeling readily shows that Me rotation is correlated in the hexamethyl compound, while the rotating Me's are not geared in the dimethyl analog.[46a,b]

Concerted intramolecular motions are, so far, relatively rare objects of conformational studies. Moreover, an ill notion is scattered in organic circles — many researchers believe that concerted intramolecular motions have higher barriers than isolate motions and, therefore, can be ignored. In fact, the situation is just the opposite. In many cases, a concerted intramolecular motion favors over sequential individual motions, because their coupling often diminishes destabilizing interactions in the transition state (e.g., steric hindrances).

One can learn from the discussion in this Section that association of molecular fragments with "standard" rates of related conformational transformations may be seriously misleading. Recognizable molecular 3D fragments (conformons) may appear "non-standard" in the stereo-dynamic aspect, if they are structural components of not-quite-trivial systems. In final instance, a customary estimation of the mobility of molecular frameworks by splitting Lewis structures into fragments of known flexibility is unreliable. We have ascertained again that just a "simple look" has a doubtful value in conformational considerations. When dealing even with minimally complicated molecular structures, flexibility conformons may be called for drawing only preliminary conformational hypotheses.

The reader has probably realized that a positive trend in the conformational field is to give no credit to primitive or obsolete tools in analyzing molecular shape.[a] It should be absolutely clear to organic and bioorganic experimentalists that *computational molecular modeling, due to its reliability and procedural simplicity, ultimately replaces different improvised tools (not based on a solid quantitative theory) in estimating static molecular 3D geometry and molecular flexibility of organic systems uncommon to the slightest degree*. The next three Chapters discuss how non-computational chemical researchers can perform valid computational explorations of molecular shape.

---

[a] When calculation methods and computational facilities had not been well-developed, organic chemists had a serious justification for using such primitive tools of conformational predictions. Today's organic experimentalists do not have it for continuing this often speculative practice.

4

A "Fast" Computational Method: Molecular Mechanics

4.1 Computational Methods in General

By providing relevant understanding of molecular systems of interest and often prompting further research strategy, molecular modeling becomes an essential component of deep, systematic chemical and biomolecular studies. Regrettably, the vast majority of synthetic organic and bioorganic chemists still have a dim idea about computational chemistry (quantitative molecular modeling) and, obstinately ignore it, drastically narrowing down their possibilities in conducting effective contemporary research.[a] As a rule, they do not know what the essence of method of molecular modeling is and, therefore, even have no measure of how to accept calculation results obtained by other researchers, as trouble-free estimates or as speculative data that cannot prove or disprove the initial hypothesis. This lack of awareness and incapability of performing elementary theoretical calculations (calculations of molecular potential energy and, thus, modeling of molecular geometry) are incompatible with the ability to solve problems in frontier organic and bioorganic chemistry. Let us start from a very general description of methods of computational chemistry in correcting this gap in chemical knowledge of the ambitious reader.

Theoretical calculations of four types are used in molecular modeling — molecular mechanics (accepted abbreviation: MM), molecular dynamics

---

[a]Amusingly, many researchers are capable of relying on primitive molecular models. For instance, conformational conclusions are not rarely drawn by manipulating with Newman projections (as explained in Section 1.5, this primitive "theoretical" modeling is not a research tool).

(accepted abbreviation: MD), quantum mechanical (accepted abbreviation: QM), semi-empirical calculations, and non-empirical QM calculations; they are all characterized below in this Section.

Strictly speaking, the term non-empirical calculations comprises of two "subtypes": so-called *ab initio* (Sections 5.2.2 and 5.2.3) and DFT (Sections 5.2.5 and 5.2.6) calculations. However, the term *ab initio* calculations is often used as a synonym of non-empirical calculations; in this not-quite-accurate terminology, DFT methods fall under the category of *ab initio* methods.

Molecular mechanics. In very superficial consideration, MM models of molecules are based on principles of classical physics. Within the limits of this simplified view, MM builds a molecule as a system of integrated abstract spheres of different radius and dissimilar elasticity; the integrity is provided by many distinct sphere-sphere bonds. Each sphere is strongly, but non-rigidly, bonded to at least one other sphere, forming an elastic framework that includes all spheres. Elasticity of the spheres themselves does not permit to non-bonded spheres to significantly deform each other if the framework geometry brings them into contact. Transparently, these different MM spheres mimic different atoms with relevant VDW radii, sphere-sphere bonds represent chemical bonds, and sphere-deforming contacts of non-bonded atoms model VDW interactions. Similarly to a mechanical system of elastic balls connected by short springs (Fig. 69), this abstract non-rigid 3D framework — the MM-modeled molecule — is even capable of virtually oscillating with frequencies which may be associated with vibrational frequencies in the corresponding real molecule.

In mechanical model *balls and springs*, distinct values of mechanical potential energy (deformation energy) characterize different distances between the balls. If (*1*) the springs are excessively neither stretched, nor compressed, nor screwed; (*2*) the angles between any two springs connected to the same ball are not distorted; and (*3*) neither non-connected balls dent each other, the deformation energy is minimal. If the balls deviate from their optimal positions, the deformation energy increases. Using the force terminology, we would say that the elasticity force causes the springs and balls to proportionally resist changing the geometry of the framework or any of its fragments. Similarly, any 3D geometry of the modeled molecule

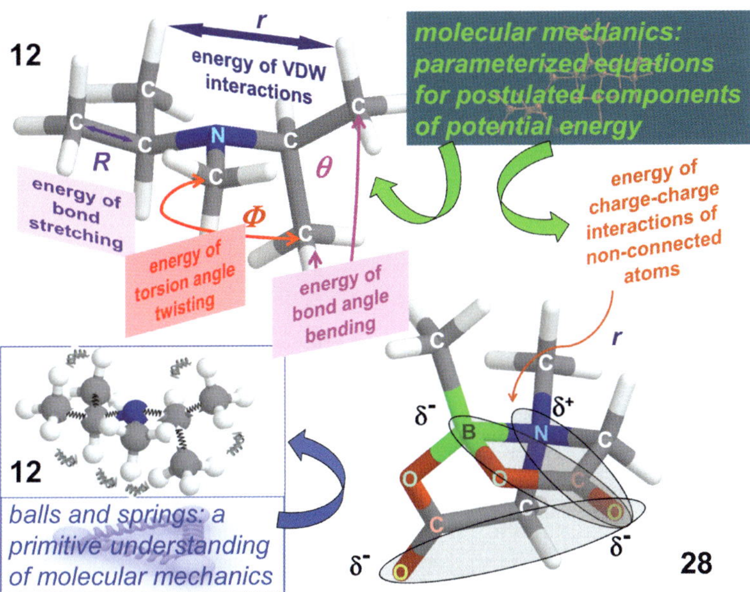

Figure 69. MM representation of molecules exemplified by MM-modeled molecular structures of amine **12** and mixed anhydride **28**. The MM-modeled molecules are composed of atoms of different formal types (atoms identically colored in the modeled molecular framework are atoms of the same chemical element, while identically colored atom symbols indicate the same type of MM atoms). Four atom-atom dispositions (connected, geminal, vicinal, and, any non-connected atoms including geminal and vicinal ones) are usually postulated in MM to provide five atom-atom interactions (indicated by tilted text). The related energies are components of total MM potential energy (steric energy). Arrows indicate four corresponding parameters of molecular geometry (bond length R, bond angle Θ, torsion angle Φ, interatomic distance r for non-bonded atoms) that are independent variables in five corresponding parametric equations *distance (angle)–energy* (e.g., Eqs. 15, 17–19, and 24; Section 4.2.2).

is associated in MM with a certain energy (so-called steric energy). If we deviate the spheres from their spatial positions in this geometry, we increase the steric energy of the system. One can formally speak about forces that require a certain (optimal) 3D geometry of the MM molecule; this molecular geometry corresponds to a minimum steric energy. Thus, the molecular model in MM externally is similar to mechanical model *balls and springs* both in the structure and the energy-deformation relationship.[a]

What is important in conformational context is that, by calculating steric energy for stereoisomers, we establish the intriguing relationship *molecular geometry–potential energy*. Simply put, by using MM, one can easily estimate relative stability of stereoisomeric structures including conformational transition states and, thus, build quantitative conformational schemes (Section 2.8).

MM-modeled molecules look trivial — they have atoms, bonds, and different recognizable (e.g., VDW and electrostatic) interactions between non-bonded atoms.[b] However, this easily interpretable result of MM modeling does not explain why such virtual molecules are true molecular models in many aspects, i.e., why they have 3D geometries and some quantitative characteristics (e.g., dipole moments, vibrational frequencies, and relative stability of stereoisomers) that sufficiently and accurately reproduce those of real molecules. What, in essence, is the methodology of MM that transforms the primitive idea of an elastic molecular framework to a true molecular modeling? Algebraic equations of type *deformation–energy* with empirically or quantum mechanically derived coefficients (Section 4.2.2) are the core of MM. These theoretical equations with empirical elements (called parameters; they are the mentioned coefficients) make this method of molecular modeling solid. These simple mathematical expressions establish the desired quantitative relationship *molecular geometry–steric energy*. In other words, they relate a certain value of steric energy to a certain geometry of a molecule. Because the parameters are adjusted empirically, MM is classified as an empirical method of molecular modeling.

---

[a]This inadequate, but simple, idea of a similar (here: elastic) energetic response of mechanical system *balls and springs* and molecules when non-drastically deforming them is reflected in the name of this quantitative molecular modeling-molecular mechanics.

[b]This structure reflects the traditional, simplified, view on chemical structure of organic molecules: a chemical bond connects two atoms, and such pairs form the molecular skeleton by chemically connecting their terminal atoms. In fact, real molecules do not have bonds. The latter are an abstraction (in contrast to atoms) that permits to interpret chemical structure of thermodynamically stable, electron-held atom conglomerates of certain atomic composition and certain relative disposition of the atoms — molecules — as a 3D framework of certainly bonded atoms. Remarkably, chemical structure of QM-modeled molecules is absolutely realistic: as composing elements, they have atoms (i.e., nuclei and electrons as a whole), but do not have interatomic interactions (specified as, e.g., chemical bonds, electrostatic and VDW interactions).

We remember that molecular systems are absolutely not mechanical. Nevertheless, MM frequently reproduces properties of real molecules well. The success of this "mechanical" modeling arises from the constructively developed core of MM (MM equations) and empirical parameterization of MM equations which is undertaken when developing finished, working tools of MM (force fields; Section 4.2.5). The parameterization is intended to compensate the unphysical nature of the elastic model of molecular structure.

Is MM a universal method of molecular modeling? Relationship *molecular geometry–steric energy* is, in general, valid for compounds from a considerably large set of chemical classes of organic compounds since empirical parameters in MM equations are adjusted so that these equations accurately model many selected molecular structures (reference structures) which have either of these chemical functionalities. This empiricism restricts MM. One cannot develop empirical parameters for MM equations to suit them to *any* chemical structure. Hence, this molecular modeling cannot be applied to molecular systems which chemical structures are not encompassed by the current empirical parameterization.

As indicated, MM provides relationship *molecular geometry–steric energy*; using MM, one can therefore obtain some close analog of the PES (Section 4.2.3). Hence, MM may be an excellent tool in building quantitative conformational schemes (Section 2.8) for organic compounds of "proper" chemical structure (i.e., the structure that falls under category *covered by parameterization*[a]). A brilliant feature of this quantitative molecular modeling is that computations in MM are very fast. The elementary computational procedure in MM — calculations of the steric energy for a given 3D geometry of a molecule — requires only 10^{-2}–10^{0} s. The procedure (termed geometry optimization or energy minimization; Section 4.2.4) of location of one energy minimum starting from an arbitrary molecular geometry (i.e., modeling of a molecular structure, which corresponds to a local minimum of steric energy)[b] takes from 0.5 s to 1–2 min, depending on the

---

[a] Section 4.2.5 provides general information regarding chemical structures encompassed by empirical parameterization in MM.

[b] It includes a sequence of elementary computational procedures (iterations). Usually, this sequence consists of no more than two or three dozens of iterations when optimizing the geometry of a "normal size" molecule.

size of the molecular system, its initial 3D geometry, the performance of the computer, and, to a lesser extent, the force field chosen. For instance, geometry optimization for a 50–80 atom structure takes less than 3–30 s by applying even a fine force field and using an ordinary desktop PC. It is amazingly simple to perform a successful computational experiment by using MM.

Molecular dynamics. MD models intramolecular dynamics. It is a combination of MM and classical descriptions of particle (herein: MM spheres) movement in the potential field of energy; this movement depends on the particle velocity. MD simulates motion of bonded MM spheres (atoms) in time, i.e., time-dependent reorganization of the MM-modeled framework. Why are simulated trajectories of atoms feasible? MM provides quantitative relationship *interatomic distance–energy* for any pair of spheres (i.e., atoms) of the system, and Newton's second law connects force F on particle Z and its velocity dL/dt for translation L.[a] From classical physics, the force on Z is known in each point of potential field E (expressed as $F = -\mathrm{grad}\, E$). This allows one to obtain a differential equation for the trajectory of Z that does not include force F. In other words, for a particle, this equation supplies particle coordinates and energy for each time moment (i.e., the particle trajectory and the energy in each point of the trajectory). In principle, the result (the coordinates and energy of particle Z at time t) is an exact solution because, for any particle, classical physics enables simultaneous determination of its location in space and its energy with any accuracy selected in advance. MD, by solving such equation for each moving MM sphere (i.e., atom) for time $t_0, t_1, t_2, \ldots t_n$ [$t_{i+1} = t_i + \Delta t$, where Δt is a very short time period (frequently, 1 or 2 fs) and n is the chosen number of time periods Δt (for instance, several thousands)], models practically continuous in-time-evolution of the whole molecular skeleton during the preselected period of time $n\,\Delta t$. One could ask how the atoms of the static modeled molecular framework get directions of their motions as well as velocities when subjected to computational MD experiments. The principal answer is that the directions of atom motions come from skeletal oscillations of the MM-modeled molecule. Atom velocities in MD are

---

[a]The force F is proportional to the first derivative of the particle velocity with respect to time, where the proportionality coefficient is mass m of the particle. In mathematical notation, $F = m \times d^2 L/d^2 t$.

"regulated" by varying the temperature in such virtual (computational) MD experiments.

At this point, we stop further description of MD methodologies. Strategies of MD in conformational studies are not trivial; understanding of assumptions and their consequences in different MD calculations requires more extensive theoretical background than synthetic chemists have. Incorrectly designed computational MD experiments mislead. In sum, MD is scarcely a tool for purely synthetic experimentalists, and we omit consideration of this method.

Quantum mechanical calculations. QM-based (widely used synonyms: *ab initio*, non-empirical) methods of theoretical calculations[a] are, in principle, the most reliable and most versatile tool for analysis of molecular systems. Indeed, the latter are micro-objects, and QM is microworld-oriented. We know that properties of microsystems, e.g. molecules, are described by a "mysterious" wave function (Section 3.1). *Ab intio* methods are based on mimicking this cryptic function by means of other, reasonably chosen functions. These functions, in contrast to the wave function, have certain mathematical forms and are designed so that they meet two general requirements: (*1*) to be sufficiently adequate in reproducing molecular quantities (e.g., molecular energy and molecular geometry); and (*2*) to be convenient in numerical calculations. Of course, the work on mimicking the wave function has been done by theoretical physicists as well as chemical theoreticians, and not by QM users. Theoreticians have succeeded in creating different approaches that deal with the polyelectron wave function and are capable, with different accuracy, to model molecules very feasibly. Also, theoretical calculations that are based on QM-derived *density functional theory* (DFT) succeed in modeling organic molecules with great accuracy (Section 5.2.5); this accuracy is often comparable with that of advanced *ab initio* methods.[b]

There are different methods of calculations in QM (Sections 5.2.2, 5.2.3, and 5.2.6); what are they? Non-empirical calculation methods of QM

---

[a]The general term *QM method(s)* means different level approximations (Sections 5.2.2 and 5.2.3) of a strict physical theory (the QM theory).

[b]Since DFT is based on QM, DFT methods belong to QM methods. On the other hand, unavoidable empiricism in constructing DFT functionals (Section 5.2.5) prevents us from calling DFT methods *ab initio* (non-empirical) methods. Therefore, the separate term *DFT method(s)* is used here.

and those of DFT are merely different level approximations of QM. All calculation methods supplied by QM are able to afford values of measurable physical quantities for molecules or molecular associates (e.g. electric multipole moments, vibrational frequencies, NMR chemical shifts, constants of nuclear spin-spin interaction, and parameters of molecular geometry) as well as their "unmeasurable," model-specific quantitative characteristics (e.g. electronic energy E_{el} discussed in Section 3.1, atomic charges, and orbital energies). In principle, *any* molecular property may be modeled by QM. However, the accuracy of the result (the obtained molecular model) is predetermined by the used QM calculation method; low level QM approximations more often supply very inaccurate and even irrelevant results.

Possibilities of *ab initio* modeling of molecular systems are limited mostly by computer performance: QM calculations are "heavy" consumers of computer resources and demand very fast processors and vast physical memory. However, since the appearance of available high performance computers, these methods tend to lead in routine modeling of non-polymeric organic molecules including reliable conformational explorations of these systems. The next reason for the occurring shift MM → QM in using computational methods in organic chemistry and small molecule biochemistry is that, being a universal and reliable molecular modeling, OM, a "hard nut" to learn, has become a practical research tool with the development of QM calculation programs (e.g., Gaussian, Q-Chem, Jaguar, ADT, and Turbomole); those are simple in use and, technically, do not require professional knowledge in computationally chemistry. After a short, circumstantial training, synthetic chemists also can perform QM calculations if using these commercially available computer programs; however, for planning computational QM experiments and correctly interpreting the modeling results, one should be somewhat versed in the basics of QM computational chemistry. Therefore, considerable attention is devoted here (Chapter 5) to explain the QM matter in a "non-computational chemist-friendly" manner.

Semi-empirical calculations. So-called semi-empirical methods[a] are basically QM calculations with some essential approximations that simplify computations of the trial (wave function-mimicking) orbital function. For

---

[a]In order to make these methods at least recognizable to non-specialists, trivial names of the most known semi-empirical approximations are presented without explaining the approximation essence in each case. Those are CNDO, INDO, MINDO/3, MNDO, AM1, and PM3. The last two methods are most popular in theoretical studies that use semi-empirical calculations.

instance, all core (non-valent) electrons are neglected. The most significant simplification consists of replacing numerical integration of QM with estimates taken from experimental parameters that are related to molecular energetics. The philosophy of semi-empirical calculation methods is simple: fewer computations with minimal sacrificing of their accuracy.

The reason for this obvious decrease of calculation accuracy is guessable. *Ab initio* calculations require solid computer resources to run with reasonable efficacy in time. Even if a computer [e.g. a high performance computer (HPC)] provides a smooth occurrence of the computational process (e.g., interruptions associated with the size of the HPC memory[a] do not happen), a considerably long time nevertheless is necessary to complete such computations. Semi-empirical calculations are an attempt to weaken the dependence on computer performance with attempting to keep the universalism and reliability of QM modeling as maximal as possible. Of course, the cost of this compromise is an unpredictable decrease of reliability of results. For instance, *tert*-butyl group is a textbook example of the ring substituent that, due its bulkiness, always *highly* prefers equatorial orientation in six-membered saturated cycles that are not locked covalently in a ring geometry with an axially oriented *tert*-butyl substituent. This conformational rule is absolutely correct. However, AM1 calculations show[48] *great* predominance for the axial conformer of N-*tert*-Bu piperidine over the equatorial conformer! It is not surprising that, in parallel to the rapid progress of computer processors, this molecular modeling loses its significance in organic research relieving place to non-empirical and DFT-based computations. Therefore, semi-empirical methods are not considered here. MM (Chapter 4) and QM calculations (Chapter 5) are our focus.

4.2 Molecular Mechanics

4.2.1 *The "skeleton" of MM*

Let us start from the easier tool, MM. The underlying concept of practical MM may be formulated in three tenets. (*1*) Chemically different atoms in real molecules are represented by model atoms of distinct types, where the

---

[a]Numerical integration is an attribute of computations in QM. If the modeled molecule is not small, there are many integrals to be simultaneously computed. The memory of the exploited HPC may appear insufficient for operating with them.

atom types are universal for different compounds; atoms that are considered being bonded in real molecules, are considered as bonded (connected) in MM-modeled molecules. (*2*) It is postulated that potential energy of a molecule is a sum of a few certain components, where each energy component is associated with an atom-atom interaction in all pairs of atoms that have a certain mutual disposition in the MM molecular framework. (*3*) For any two model atoms that contribute to the same energy component, potential energy of their interaction is described by the same parametric algebraic equation *molecular geometry–energy*, where numerical parameters for atoms of distinct types only are different.[a] The energy contributions are additive; the sum of all the contributions for all energy components represents the total potential energy in the MM sense (termed steric energy; E_{ster}) for the modeled molecule. Thus, molecules in MM are 3D frameworks of connected model atoms, where the connection of two model atoms means chemical bonding between them (Fig. 69). Non-connected model atoms in these frameworks represent chemically non-bonded atoms. Any geometry of such a molecular framework is characterized by the related value of steric energy. Steric energy is a sum of energies of postulated interactions in pairs of connected as well as non-connected atoms.

MM postulates interactions of atoms within atom pairs and describes these interactions quantitatively. According to a widespread general explanation of MM, this method models molecules as mechanical systems, where hard spheres (atoms) are connected by springs (bonds and interactions of non-bonded atoms) of different elasticity; the elasticity of the mechanical model reflect the flexibility of the related fragments of a real molecule (Fig. 69). This understanding is a notorious oversimplification. MM equations, due to empirical or QM-based parameterization performed with referring to molecules (and not to truly mechanical systems), model molecular geometry and potential energy in a "cryptic" manner, i.e., not relying on any mathematical law that describes elastic behavior of either mechanical object.

Concerning MM equations, one can remark that the relationship *molecular geometry–energy* that they establish is exactly what molecular modeling of rapidly interconverting stereoisomers (conformers)[b] is expected to provide. In PES terms, MM calculations supply

---

[a]There is a clash of two different senses of the term *parameter*. Herein, *parameters* mean numerical values that are chosen according to either criterion and then inserted into a mathematical equation as coefficients, exponents of a power, etc. In describing molecular geometry, *parameter* means either interatomic distance, or bond angle, or torsion angle, or another geometrical characteristic of molecular structure.

[b]As we remember, stereoisomers, whose lifetimes fall within a certain time window, are defined as conformers (Sections 2.4 and 2.5).

us with some potential energy (as mentioned, steric energy E_{ster}) for any non-absurd 3D geometry of a molecular structure, i.e., with the PES in the MM representation (see Section 4.2.3 for the meaning of SES).

These structural principles of practical MM do not require much comments. Only one meaning from item (*1*) needs an explanation. It is clear that model atoms are atoms that form an MM-modeled molecule. However, what should we understand under types of model atoms? For instance, are primary, secondary, tertiary, and quaternary carbons MM atoms of the same type? Before answering this question, one has to note that MM is a practical approach to molecular modeling that has a rigorous structure [items (*1*)–(*3*)], but does not define elements of this structure (MM atom types, the number of MM equations and their mathematical forms) uniformly. Saying merely, these elements are differently defined by different MM developers. As a result, there are different calculation methods in MM; they are called force fields (see below).[a] While keeping the principal structure of MM calculations [items (*1*)–(*3*)], force fields differ in expressing the *molecular geometry–steric energy* relationship (i.e., they differ in MM equations) as well as in defining types of model atoms.

Thus, our question should be: what are the types of model atoms defined in either force field? In some force fields, MM atoms of the same type correspond to the atoms of a chemical element that, being incorporated into the same molecule or into different molecules, have the same chemical bonding with the directly connected atoms (the neighbors from the first chemical surrounding). For instance, the methyl, methylene and methine carbons in compounds **12** and **28** (Fig. 69) are of the same type, while the carbonyl carbons in mixed anhydride **28** are related to another formal type. In anhydride **28**, there are two atom types of carbons (the four-coordinated and three-coordinated C's) as well as two types of oxygens (the carbonyl and ester O's). Considered similarly, the oxygens in free radicals **9a** and **9b** (Fig. 3) should belong to the same type of MM oxygens despite the dissimilar hydrocarbon skeletons;[b] this type differs from the MM oxygen

---

[a]At this point, it is sufficient to say that force fields are exactly what researchers use when undertaking MM modeling (Section 4.2.5).
[b]Current force fields do not include parameterization for N-oxyl radicals.

type to whose two oxygens of the corresponding $N–O^-$ hydroxylamine anions belong.

This crude approximation in types of model atoms does not provide a considerable accuracy in MM modeling of organic molecules, and additional subdivision in atom types, e.g. according to the chemical structure of the first and, sometimes, the second covalent surrounding, is an attribute of other, modern force fields. For instance, nitrogens in alkylamines, hydroxylamines and hydrazines are usually assessed there as model atoms of different types. Similarly, ester and ether oxygens or carbonyl oxygens in esters and amides are usually defined in MM as distinct atoms. Even the mentioned tetrahedral CH_3, CH_2 and CH carbons may be assigned to different model atoms. Also, lone electron pairs often are approached in MM as pseudoatoms (also called dummy atoms). Then, distinct types are attributed to lone pairs (i.e., dummy atoms) of different heteroatoms.

A set of certain model atoms together with a set of certain parametric equations [mentioned above in item (3)] is termed *force field*. These equations are not rigorously predetermined by chemical structure of molecules; they are composed by different theoreticians according to their own criteria. Therefore, there are different force fields of distinct quality and applicability. E.g., a force field developer may set up aminic nitrogens in **12** and **28** as model atoms of the same type (amino nitrogens). On the other hand, there are no restrictions to accept only the nitrogen in **12** is as an aminic N. Then, the nitrogen in **28** is of another formal type, say, the aminic N with an additional coordinate bond, and we would have two slightly distinct force fields. One can suppose that, among them, plausibly the force field with more specified atom types would be more successful for complexes with amino ligands. It would treat these molecular structures (calculate their steric energy and molecular geometry) more accurately due to more scrupulous atom specification (that is, more "structure-sensitive" parameterization.

In general, there is no question that significant decrease of MM atom types sharply diminishes the accuracy of MM calculations. On the other hand, a sufficiently large number of atom types, say, from 50 to 150 for main classes of organic compounds composed from elements C, H, O, and N, provides an acceptable accuracy in modeling these molecular structures, as many years of experience of computational chemistry has shown. The further increase of the number of MM atom types scarcely would improve the accuracy of MM calculations undertaken for these core chemical classes. A more weighty factor, the quality of MM equations [mentioned in item (*3*)], sets its own limitations.

Structural element (*3*) is the "heart" of MM calculations. This "heart" is a quantitative dependence *molecular geometry–energy* in the form of simple algebraic equations. There are many such equations in MM.[a] Nevertheless, the situation with them is not too complicated: in general, MM equations fall under five main categories. Let us look briefly at them.

4.2.2 *The MM-sculpting equations*

Any MM equation relates a geometrical parameter (e.g., a distance) for two atoms (both bonded and non-bonded) and postulated potential energy of this atom pair in one-dimensional space of this parameter.[b] In other words, any MM equation is a function $E_x = f(x)$ (called energy potential), where E_x is some potential energy, and independent variable x is a parameter of molecular geometry associated with a certainly composed atom pair. Thus, in calculating MM energy (steric energy) for a molecule in a "frozen" geometry, many atom pairs that are composed both from bonded and non-bonded atoms are the subject for these equations. These equations are algebraically different for differently composed atom pairs; more concretely, each MM equation $E_x = f(x)$ is applied to the relevant pair. Energies obtained for the pairs are further summed, yielding the total potential energy (the steric energy E_{ster} of the molecule in this conformation).

What does the relevant pair of atoms mean in our consideration of MM equations? In dealing with atom pairs in MM, five energy potentials are usually postulated. They are potential energies associated with bond stretching or shortening (energy considered for pairs of chemically bonded atoms), bond angle bending (energy considered for pairs of geminal atoms), torsion angle altering (energy considered for vicinal atoms), VDW interactions (energy considered for pairs of non-bonded atoms), and electrostatic

---

[a]Returning to the meaning *force field*, one may recall that, more than the difference in MM atom types, the difference in MM equations is what specifies a force field.

[b]Probably, one should be reminded that potential energy of a physical object is the energy that only depends of its location in space (i.e., its coordinates x, y and z).

interactions (energy considered for pairs of non-bonded atoms; Fig. 69). Consequently, each postulated energy potential $E_x = f(x)$ uses one of the four geometrical parameters: interatomic distance R (for bonded atoms), bond angle Θ, torsion angle Φ and interatomic atomic distance r (for non-bonded atoms; Fig. 69). Thus, the relevant atom pair for an equation $E_x = f(x)$ is the pair of atoms that this energy potential considers, i.e., two bonded atoms for the bond stretching potential, two geminal atoms for the bond bending potential, two vicinal atoms for the torsion angle potential, and two non-bonded atoms for the VDW potential or the potential of electrostatic interactions.

As mentioned, five energy potentials are traditionally considered in MM. The first energy potential deals with bond stretching, and the second one similarly treats bond angle bending (see below). These equations establish a single-valued relationship between parameter q of molecular geometry (q is the bond length or bond angle) and the energy of its deviation from its optimal value q_{eq}. Energy that corresponds to this value q_{eq} is accepted to be zero. In the ground of these equations, one could find a simple idea that, for any pair of atoms of given MM types, each parameter q has an optimal value q_{eq} that characterizes a hypothetical molecular geometry with absolutely unstrained bonds and bond angles. In organic molecules, many bond lengths and bond angles usually have values which somewhat deviate from MM values q_{eq}, due to steric, electrostatic or other interatomic interactions. Then, an optimized molecular geometry (corresponding to an energy minimum) provided by any force field is associated in MM with an increase of energies of bond stretching and bond angle bending in the considered system relative to a zero energy value in an abstract molecule with unstrained bonds and bond angles. Even if deviations from different "ideal" values q_{eq} are small, they significantly increase the value of total (molecular) steric energy E_{ster}.[a]

There is no hypothetical non-strained geometry in three other categories of MM equations. Energy potentials, which are described by them, do not have zero energy for any value of the corresponding geometrical parameter. As a result, even the most stable geometry is characterized

---

[a]Clearly, the more atoms are in a molecule, the higher the E_{ster} value is. Therefore, it is meaningless to compare steric energies calculated for molecular structures of different chemical composition.

by a non-negligible, often substantial, energy for torsion potential, VDW interactions, and electrostatic charge–charge or dipole–dipole interactions. Moreover, the energy of electrostatic interactions (see below) is different from zero for any charge–charge separation.[a] All these equations "link" potential energy and an absolute value of a parameter of molecular geometry and not the energy and the change (relative value) of this parameter.

After discussing MM energy potentials in general, it is worth glancing at mathematical forms of these five categories of MM equations. Dependence *geometry parameter–energy* in classical physics may be expressed as a continuous algebraic function, due to continuity of a classical potential field. It means that, in principle, it is possible to select a convenient algebraic function and to equip it with numerical parameters that enable an accurate representation of this dependence. This way had been traversed by MM founders and chief developers (Westheimer and Mayer, Hill, Hendirckson, Wiberg, and, later, Schleyer, Allinger, Sheraga, Brooks, Karplus, MacKerell, Jorgensen, Kollman, van Gunsteren, Halgren, Sun). They have recruited a simple idea of the well-known Taylor expansion that any continuous single-valued function $f(x)$ may be represented in point x by an infinite sum $f(x_0) + df/dx \, (x - x_0) + [2! \times d^2 f/dx^2 \, (x - x_0)^2]^{-1} + [3! \times d^3 f/dx^3 \, (x - x_0)^2]^{-1} + \ldots$. Employing the Taylor expansion for unknown function *potential energy–geometry* in the 3D Cartesian space, truncating it after a few terms, and supplying these truncated expressions with empirical numerical parameters, they have obtained the core equations of MM. Due to this empirical refinement, MM functions are capable of providing reasonable values of MM potential energy (steric energy, as mentioned) for molecular systems in spite of truncating Taylor's sum for components (i.e., energy potentials) of this total energy.

Nowadays force field developers either add new energy potentials to the five traditional ones or modify these common functions, hunting for higher calculation accuracy. They also use these energy potentials unchanged when adopting the parameterization to new chemical structures. Of course, MM users do not feel that a thorough, long-term work is beyond each commonly known force field and exploit force fields as finished products.

---

[a]Coulomb's law (see Eq. 24) states that charges interact at any distance. Therefore, the energy of electrostatic interactions is zero only at infinite remoteness of point and, consequently, dissipated charges.

Bond stretching. The relationship between interatomic distance R_{AB} of non-bonded atoms A and B (i.e., bond length; Fig. 69) and energy E_{AB} may be expressed as

$$E_{AB} = 1/2 \times k_{AB}(R_{AB} - R_{\text{eq}})^2 \tag{15}$$

where k_{AB} is the force constant (a parameter; kcal/mol $\times \text{Å}^2$), R_{AB} is the instantaneous length of bond A–B (Å), and R_{eq} is the length of the unstrained bond A–B (a parameter; Å). Here, as well as in other equations below, we can see how types of model atoms are specified in MM. Parameters k_{AB} and R_{eq} in these equations are specific for a pair of atoms A and B that is different from other pairs by types of its atoms. This example of MM equations shows why algebraically the same MM energy potentials are suitable for modeling differently functionalized molecules. In MM, the same equation, e.g. Eq. 15, is applied to pairs of *any* equally positioned[a] atoms in various molecular structures, and empirically adjusted parameters are only different for the chemically distinct pairs.

Energy values supplied by Eq. 15 strongly depend on numerical parameters in this equation. For instance, the energy cost is $\sim$1.5 kcal mol^{-1} for a 0.05 Å deviation of the C–H bond length from its optimal value. It is clear that different lengths of bond A–B may correspond to the same minimal stretching energy if values of this parameter are varied. In other words, force field developers can set up MM spheres (model atoms) to be "softer" or "harder." What "softness" or "hardness" is selected? A single criterion for choosing successful parameter values in Eq. 15 is the correspondence of the calculated molecular geometry to that from high accuracy experiments or from high level *ab initio* calculations. The same selection principle is also used in parameterizing other equations below.

However, Eq. 15 appears problematic when applied to the case of large deviations from R_{eq}. If the interatomic distance is increased infinitely ($R_{AB} \to \infty$), stretching potential E_{AB} becomes infinitely positive. A modified equation, which also includes the cubic and the quartic terms[b] (Eq. 16), removes this limitation. Also, the presence of more parameters adds more flexibility to Eq. 16 and adapts it to apply to different classes of organic compounds. Not incidentally, this equation is used for calculating the bond

---

[a] E.g., pairs of connected atoms in the case of the bond stretching-related equations.
[b] It is derived by inserting these terms of the Taylor expansion for the stretching potential function in Eq. 15 (see Ref. 33b).

stretching energy potential in rigorously parameterized MM3, MM4 and MMFF94 force fields (Section 4.2.5). One can see that the four–parameter (k_{AB}, k_{AB-3}, k_{AB-4}, and R_{eq}) Eq. 16 is not terrifyingly complex:

$$E_{AB} = 1/2[k_{AB}+k_{AB-3}(R_{AB}-R_{eq})+k_{AB-4}(R_{AB}-R_{eq})^2] \times (R_{AB}-R_{eq})^2 \tag{16}$$

Equations 15 and 16 do not exhaust the pool of empirical equations for bond stretching that have been proposed for utilizing them in MM (e.g. the commonly known Morse potential); some of these other equations are components of a few force fields. These functions $E_{AB} = f(R_{AB})$ are not presented here since they are of interest of only computational chemists who develop force fields.[a] For MM users, it is only useful to get to know the principle of how such equations associate molecular geometry and energy than to analyze their details. It is sufficient to select and exploit a *suitable* "standard" force field (Section 4.2.5) for using MM in organic research.

Angle bending. The next energy potential to be considered is related to bond angles. Similarly to the previous case, potential energy E_{Θ} of distortions of bond angle Θ (Fig. 69) is given by the Taylor expansion of the potential energy function where derivatives are replaced with numerical parameters (Eq. 17):

$$E_{\Theta} = 1/2[k_{\Theta}+k_{\Theta-3}(\Theta - \Theta_{eq})+k_{\Theta-4}(\Theta - \Theta_{eq})^2 + \cdots] \times (\Theta - \Theta_{eq})^2 \tag{17}$$

Herein, k_{Θ}, $k_{\Theta-3}$ and $k_{\Theta-4}$ are parameters, Θ_{eq} is the value for the unstrained angle Θ (a parameter), indexes $\Theta-3$, $\Theta-4$, etc. indicate the cubic, quartic, etc. terms, respectively. In different force fields, this expansion is truncated at different terms. For instance, the widely used "general organic" MM3 force field (Section 4.2.5) includes all terms from k_{Θ} to $k_{\Theta-6}$ while the not less known protein-oriented Amber force field (Section 4.2.5) uses only the quadratic term [$1/2\, k_{\Theta} \times (\Theta - \Theta_{eq})^2$].

Since Eq. 17 does not model the angle bending energy for linear atom triads ($\Theta = 180°$) even fairly, sometimes the quadratic term is modified to [$k_{\Theta} \times (\Theta - \Theta_{eq})^2]/2 \sin^2 \Theta_{eq}$.

---

[a]Descriptions of mathematical "skeletons" of force fields may be found in the specialized chemical literature, e.g. in Refs. 49a–e.

Torsion potential. Potential energy that reflects energy of interactions between vicinal atoms is termed torsion potential (E_Φ); mathematically, it is described by Eq. 18. There, the independent variable is torsion angle Φ, which shows reciprocal orientation of two vicinally positioned atoms. For instance, the CNCC angle characterizes mutual geometrical orientation of two methyl groups of amine **12** (Fig. 69). The related MM torsion potential E_Φ shows the energy of interactions of the carbon atoms of these Me groups as a function of Φ.

$E_\Phi(\Phi)$ is a periodic function. Therefore, this function is simpler represented via the periodic Fourier expansion:

$$E_\Phi = 1/2 \sum_j A_j[1 + (-1)^{j+1} \times cos(j\Phi + \Phi_j)] \tag{18}$$

where j is periodicity (e.g. $j = 3$ for Me), A_j is the amplitude (the maximal value which potential E_Φ can adopt), and Φ_j is the torsion angle.

Van der Waals (VDW) interactions. From the QM perspective, VDW interactions (interactions of non-bonded atoms) are a long range electron correlation. At first glance, this complex phenomenon scarcely could be approximated by an empirical function. Fortunately, potential energy of these interactions E_{VDW} between non-bonded atoms A and B is sufficiently accurately represented by a function which, due to its success, even has its own trivial names "Lennard–Jones potential" or "6–12 potential." In mathematical notation, this empirical function is

$$E_{VDW} = a_A/r^{12} - b_B/r^6 \tag{19}$$

where a_A and b_B are parameters related to atoms A and B, respectively, and r is the distances between these atoms. Very often, Eq. 19 is written in another form where parameters a_A and b_B are replaced with parameters σ_{AB} and ε_{AB}:

$$E_{VDW} = 4\varepsilon_{AB}[(\sigma_{AB}/r)^{12} - [(\sigma_{AB}/r)^6] \tag{20}$$

In Eq. 20, parameter σ_{AB} has units of length. If this function $E_{VDW}(r)$ is differentiated in respect to variable r and the derivative $\partial E_{VDW}/\partial r$ is set equal to zero (the condition of the minimum of the "6–12 potential"), one could discern that $r = 2^{1/6} \times \sigma_{AB}$. In other words, parameter σ_{AB} has physical sense. It shows (with proportionality coefficient $2^{1/6}$) the distance between non-bonded atoms A and B when the energy of their interactions is minimal.

Of course, the Lennard–Jones potential is not the single function that is used for describing VDW interactions in MM. Its refined forms (see, e.g. Ref. 33b) are also used. For instance, in the Amber force field (Section 4.2.5), this potential is combined with the term of the energetic contribution of electrostatic interactions (Eq. 21; q_A and q_B are partial charges with which atoms A and B are supplied, i.e., they are parameters; ε is the dielectric constant).

$$E_{\text{non-bonded}} = a_A/r^{12} - b_B/r^6 - q_A q_B/\varepsilon r \tag{21}$$

For polarized (in the molecule) atoms, the Born–Mayer–Huggins potential (Eq. 22) is applicable; it is written as

$$E_{\text{VDW}} = -L \times \exp(-M \times r) - a_A/r^6 - b_B/r^8 \tag{22}$$

where L, M, a_A and b_B are parameters.

Some force fields roughly approximate H-bond H–B (B is a proton acceptor) as an interaction between non-bonded atoms H and B. They include a "softened" Lennard–Jones potential (Eq. 23):

$$E_{\text{H-bond}} = a_H/r^{12} - b_B/r^{10} \tag{23}$$

To improve the quality of this approximation, sometimes the electrostatic term (as it is in Eq. 21) is added; e.g. this modeling of H-bonding is present in the original version of the MM3 force field (Section 4.2.5). However, an explicit drawback of these approximations is the lack of directionality for the modeled H-bond. Strong H-bonding of type A–H–B (one energy minimum) or A–H $B \rightleftharpoons A$ H–B (two energy minima) is characterized not only by its length H–B or A–H but also by the bond angle AHB. Therefore, more accurate MM models, e.g. the MM3(94) version (MM3 released in 1994), use more complicated parametric functions in order to provide directionality of the modeled H-bond. Nevertheless, one can easily recognize "fingerprints" of the Lennard–Jones potential there (see Ref. 49a).

Electrostatic interactions. In MM, similarly to Mulliken atomic charges in QM, partial point charges are assigned to atoms of the molecular framework. The "trick" is, of course, unphysical. Atoms in molecules do not carry certain charges because no electron from the molecular electron cloud is localized at any point of the molecular space. In other words, a continuous charge distribution takes place for molecules, and non-existing atomic charges (point charges) are a primitive model of this distribution. Nonetheless, this "illegal" *point charge–point charge* model of electrostatic interactions of non-bonded atoms in MM, due to parameterization, succeeds to take into account electrostatic energy. In the simplest approximation, the energy of electrostatic interaction $E_{Coulomb}$ between two atoms A and B that have formal point charges q_A and q_B, respectively, is described trivially if applying the Coulomb law to these charges (Eq. 24):

$$E_{Coulomb} = 1/4\pi \times q_A q_B / \varepsilon r \tag{24}$$

where r is the interatomic distance, ε is the dielectric constant (relative permittivity; $\varepsilon = 1$ for vacuum). Transparently, q_A and q_B are parameters here. This equation is used for a rough modeling of solvent effects by MM. From classical physics, we know that Coulomb forces between point charges are decreased by factor ε if the charges are located in the dielectric medium with dielectric constant ε. Some force fields take $\varepsilon \neq 1$ and use in Eq. 24 other, solvent-specific values of ε when modeling molecules in either solvent.

Also, higher order electric moments are taken into account by advanced force fields. Polarized chemical bonds may be approached as dipoles. Then, the energy E_{dipole} of dipole–dipole interactions between these bonds A–B and C–D is

$$E_{dipole} = (\mu_{AB} \times \mu_{CD})[\cos \chi - 3cos(\alpha_{AB} \times \alpha_{CD})]/(\varepsilon \times r^3_{AB/CD}) \tag{25}$$

where μ_{AB} and μ_{CD} are magnitudes of dipole moments vectors, χ is the angle between lines that connect the mid of bond A–B and the mid of bond C–D, α_{AB} and α_{CD} are the angles between the line, which connects the mid of bond A–B and the mid of bond C–D, and the vectors of the corresponding dipole moments. Often dielectric constant ε is included as an atom pair-specific parameter (see Ref. 33b for details).

Cross-terms. In fact, the above five separate categories are insufficient for an accurate molecular modeling. This is not surprising. Postulation of only five components of potential energy unpredictably restricts verisimilitude of 3D molecular structures modeled by MM. In order to diminish this roughness in MM fundamentals, the sixth, seventh, etc. potentials, so-called cross-terms, are added in solidly designed force fields to the mentioned five energy potentials. These additional equations combine separate energy potentials by postulating, e.g. stretch-bend, stretch-torsion, stretch-stretch, and bend-bend coupling. For instance, Eq. 26 introduces stretch-bend energy $E_{AB/\Theta}$ for atom pair A and B:

$$E_{AB/\Theta} = 1/2 \times k_{AB/\Theta}(R_{AB} - R_{eq}) \times (\Theta - \Theta_{eq}) \qquad (26)$$

A few cross-terms are frequently sufficient to supply a force field with an acceptable calculation accuracy. For instance, the MM3(94) version has three cross-terms: stretch-bend, stretch-torsion and bend-bend. Therefore, MM calculations with using this force field are of high quality for, e.g. cyclopropane compounds, because the stretch-torsion coupling potential is very beneficial when modeling strained systems. The previous Allinger's force field, MM2, has only the stretch-bend cross-term, and it is not surprising that MM2 models cyclopropanes unsatisfactory. Widely used CHARMM and MMFF force fields (Section 4.2.5) have only one cross-term. A force field may not include cross-terms at all as, e.g. force field Sybyl. Instead, some diligently developed force fields (e.g. Amber, CHARMM, and MMFF) have another additional term, so-called the out-of-plane angle potential (the inversion potential). This function describes the energy of deviation of model atom D from the plane of a triad of connected atoms A–B–C where atom D is bonded to the central atom B of the triad.

Steric energy. The overall MM energy E_{ster} of the system is a sum of the above explained energy contributions for all pairs of atoms I and J. For instance, Eq. 27 represents such a sum:

$$E_{ster} = \sum_{I-J} E_{stretch} + \sum_{I-J} E_{bend} + \sum_{I-J} E_{torsion} + \sum_{I-J} E_{VDW}$$

$$+ \sum_{I-J} E_{electrost} + \sum_{I-J} E_{cross} \qquad (27)$$

The number of summed terms is unequal in different force fields. For example, E_{ster} in MM3 is a sum of energies (Eq. 27) of nine distinct interactions (including cross-terms) while E_{ster} in MM4 involves additional six ones. Note that molecular geometry supplied by MM characterizes one "geometrically averaged" molecule from a molecular ensemble, while related numerical values of steric energy E_{ster} are for a 1 mol ensemble of molecules. That is, values of E_{ster} are calculated in kcal mol^{-1} of the traditional metric system of energy units (not recommended by IUPAC for using) or kJ mol^{-1} in SI (Systeme Internationale; the International System of Metric Units) units, which is the most thought-out metric system (recommended by IUPAC).

4.2.3 *Steric energy*

In conformational excurses, steric energy is the most relevant energetic quantity supplied by MM: the difference of values of steric energy (ΔE_{ster}) for two stereoisomeric structures quantitatively shows their relative stability. This statement that compares *energies* of stereoisomers seems self-evident; however, for competently interpreting results of MM calculations, we probably need to understand what steric energy actually is. Definitely, potential energy E_{ster} is not potential energy E_{el} from QM (see Section 3.1) since deterministic Eqs. 15–26 have no relation to the truly physical Eq. 5. MM equations are built somewhat arbitrarily in accordance with a sole practical criterion: to provide the correspondence between calculated molecular quantities and experimental ones. As we have seen (Section 4.2.2), even their mathematical form is sufficiently flexible. Also, steric energy is neither enthalpy nor free energy.

Then, what is steric energy in respect to these fundamental thermodynamic functions? Steric energy summands, i.e., potentials given by Eqs. 15–26 or similar ones, contain numerical parameters. In developing a force field, parameter values are accepted when reaching a fit of calculated elements of molecular geometry (e.g. bond lengths) and corresponding *measured* geometrical elements for reference compounds (simple non-functionalized or monofunctional compounds).[a] Measurements (e.g., by electron diffraction) supply us with molecular geometries that are averaged over vibrational and rotational levels of molecules of reference compounds at the temperature and pressure in the experiment. In thermodynamic aspect, this averaging in real molecules is characterized by Gibbs

---

[a] This scrupulous, step-by-step development of many consistent parameters for MM equations is what is called force field parameterization. Note that not all parameters in a force field are usually derived using experimental data obtained at identical physical conditions.

energy at corresponding physical conditions. Thus, in searching proper parameter values, the calculated geometry in MM is an averaged geometry of molecules of a statistical molecular ensemble, and, consequently, steric energy implicitly "contains" some part of free energies of reference compounds. Therefore, to some uncertain extent, steric energy represents free energy when MM calculations are applied to other, non-reference compounds.

On the other hand, many parameters, e.g. those related to the torsion potential, are obtained by means of QM calculations that do not supply the Gibbs energy for an ensemble of molecules. If temperature-dependent distributions of chemically and geometrically identical molecules (i.e., identical stereoisomeric molecular structures) into different intramolecular energy levels (Section 5.3.4) are not taken into account, the calculated energy is the internal potential energy of one modeled molecule. If QM computations calculate these distributions of molecules, the resulting energy is Gibbs energy when considering it only roughly. Such QM modeling is always performed for a single molecule, and, therefore, includes neither the energy of intermolecular interactions even for one distribution (relative disposition) of molecules in space, nor the entropy resulted from many of such distributions. Thus, when supplying MM equations of a force field with parameters, QM calculations do not make the force field to reproduce Gibbs energy of molecules in the statistical ensemble.

It turns out that steric energy is an "inseparable mixture" of different thermodynamic energies in an absolutely uncertain ratio. If so, what should we understand under this artificial diffuse "construction" (steric energy), whose components (MM equations) are independently chosen and modified at the request of different architects of force fields?

In order to answer this question, let us return to the basic assumption of MM that is taken from classical physics. Atom–atom interactions in MM interpretation formally are forces (only electrostatic interactions are real physical forces). In classical physics, forces, which affect physical objects, are associated with potential energy of these interacting objects. The energy in MM equations is therefore potential energy, and the sum of these energies — steric energy — is potential energy that is related to a system of specifically bonded atoms (a molecule). Hence, a telling interpretation of steric energy probably would be the statement that *steric energy is a lax analog of potential energy* E_{el} *(for a molecule) of the Born–Oppenheimer approximation in QM* (Section 3.1), where the analogy is additionally disturbed by the circumstance that steric energy to some extent incorporates free energy.

Relating molecular geometry and potential energy, MM operates in Euclidean space whose points have molecular geometry and steric energy coordinates. One can term an infinite set of values of steric energy E_{ster} (a surface in this space), which characterize different geometries of an integral (intact chemical structure) molecular system, the steric energy surface (SES) of this system. That is, the SES is a potential energy surface (see Fig. 70 for a SES example). In light of the above indicated analogy of potential energies E_{el} and E_{ster}, one can say that the SES is a lax analog of the PES in the E_{el} sense (for description of such a QM-defined PES, see Section 3.1).

One can reasonably assume that the E_{el} PES and a successful[a] SES plausibly have a similar shape for the same molecular structure. This similarity is important exactly in computational conformational studies. It means that, in principle, one can use both MM and QM for modeling molecular geometry and quantitatively estimating relative stability of stereoisomeric structures since, as indicated, both tools are supposed to yield approximately the same model (the surface of potential energy) of the flexible 3D shape for a molecule. Of course, the MM- and QM-modeled energy landscapes should not be identical in details. Note that, in contrast to minima points in the QM-provided PES, such points in the SES that is modeled by an empirically parameterized force field correspond to averaged molecular structures (Section 4.2) detected in experiments. One cannot, therefore, expect that the geometries of a molecule that are supplied by QM and a reliable force field are identical. They should be slightly distinct. However, for high quality QM modeling and MM calculations that use a well-parameterized force

---

[a]A general characteristic *successful* is used here not accidentally. Indeed, a SES is not an attribute of a molecular system. It is a model provided by a force field, and force fields differ in their capabilities of modeling molecular structures accurately (Section 4.2.4). In other words, each force field supplies its own SES for the same molecule. The shapes of these SESs may appear appreciably different for the same molecular system. In this case, some SESs do not describe the relationship *geometry–energy* of the system even satisfactory, i.e., they may be considered as unsuccessful, "false" SESs.

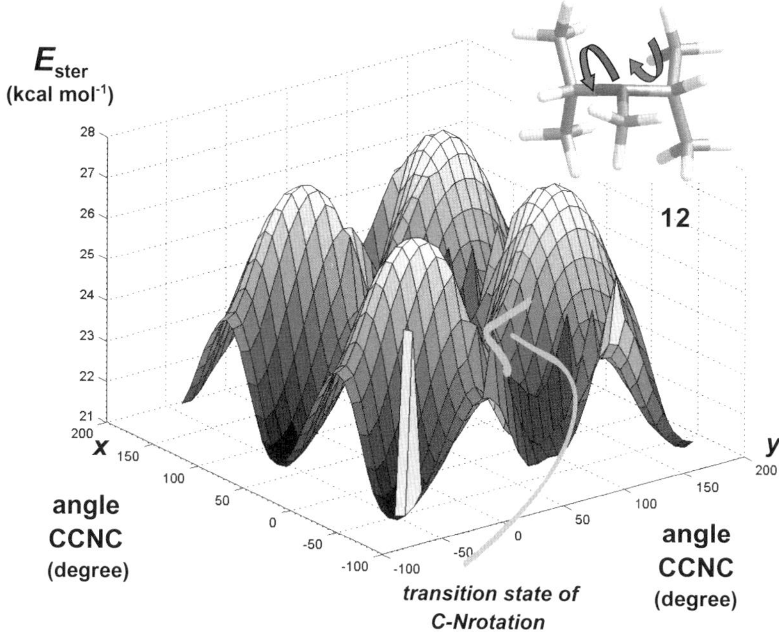

Figure 70. The SES region that corresponds to rotation of the *iso*-Pr groups of amine **12** [modeled by the MM3(96) force field]. Independent variables x and y are two torsion angles CCNC. Location of one of the transition states of NPr^{iso} rotation in the SES (a saddle point; one among four ones in this region) is indicated by the grey arrow. The four striking maxima are related to four molecular geometries where the NMe substituent is eclipsed with two 1,3-positioned CMe groups.

field, the discrepancy between the QM- and MM-derived geometries is often unessential.

If parameters of a force field are obtained using solely QM calculations (as it is in force field MMFF94), does this mean that steric energy is potential energy E_{el}? One should not forget that MM equations are absolutely not QM equations. The QM quantity E_{el} cannot be exactly reproduced using physically non-strict models; it may be only approximated by them. That is, E_{ster} is never E_{el}.

Since minima of steric energy and inflection points in a successful SES of a molecule correspond to the same structures in the strictly defined Born–Oppenheimer PES, these stationary points of the SES correspond to the geometry of conformers (or rigid stereoisomers) and transition states, respectively. For instance, a region of the SES for amine **12** shown in Figs. 70 and 84 describes the *rotation angle–steric energy* relationship for rotation of both *iso*-Pr N-substituents (rotation around two bonds C–N). Minima "are" stable rotamers, and saddle points "are" transition states of rotation (eclipsing of NMe and CMe vicinal groups). If the structures that correspond to these stationary points of the SES are subjected to geometry optimization using QM methods (Chapter 5), the resulting molecular geometries are only insignificantly different from the MM-delivered ones, and the new (i.e., QM-derived) minima again "are" stable rotamers, while the new saddle points "are" transition states of C–N rotation.

And what about energy calculations in MM, i.e., the heights of energy ridge passes (saddle points) and depths of minima in the SES landscape? These height and depths are relative values, i.e., they are differences ΔE_{ster} between stationary points of higher and lower energy in the SES. These differences are most important in MM energy calculations — the difference ΔE_{ster} for a pair of stereoisomers has physical sense of the relative stability of these stereoisomers. For a pair of stable stereoisomers at thermodynamic equilibrium, the corresponding ΔE_{ster} value reflects their relative thermodynamic stability. The more the ΔE_{ster} value is, the more stable the stereoisomer with the lower value of E_{ster} is relative to the second stereoisomer from the pair; this quantitative relationship is described by Eq. 13 (Section 3.3) if replacing $\Delta G°$ with ΔE_{ster} there. If a stable stereoisomer and the transition state of stereoreorganization of this 3D molecular structure are in focus, the related ΔE_{ster} value represents the kinetic barrier (or, equivalently, the kinetic rate at this temperature; Eq. 14, Section 3.5.3) of this stereoisomerization, i.e., kinetic stability of the stereoisomer at the given physical conditions. Therefore, relative steric energies ΔE_{ster} for stereoisomeric structures — quantitative estimates of relative stability of stereoisomers — are the ultimate target of energy calculations in MM.

Differences ΔE_{ster} are trivially derived from values E_{ster} supplied by the same force field. For instance, the MM3(96) force field provides the value of E_{ster} of 98.33 and 105.92 kcal mol^{-1} for the chair and twist conformer of cyclohexane **26a** (Fig. 64, Section 3.4), respectively; the corresponding ΔE_{ster} is 7.6 kcal mol^{-1}. This value of ΔE_{ster} shows (Eq. 13, Section 3.3) that six identical chair conformers drastically prevail over their six identical twist-shaped counterparts in the twelve–position *chair–twist* equilibrium of **26a** at normal conditions (for qualitatively understanding interconversion of chair and twist conformers in cyclohexanes, see Fig. 53, Section 2.8). Or, reading the contour map for the SES region for alkylamine **12** in Fig. 84, we elucidate that the barriers of *iso*-Pr rotation for our pattern are very low: all related differences ΔE_{ster} do not exceed a 3.5 kcal mol^{-1} value. This low height indicates a free rotation of the isopropyl groups even if lowering the temperature very significantly (Fig. 47, Section 2.6).

Beginning users of MM may pay attention to the circumstance that E_{ster} values themselves have no strict physical sense and, in fact, are worthless. Moreover, steric energy is calculated differently by different force fields or even by distinct versions of the same force field (Section 4.2.5). A specificity of a force field is in the first instance constituted by the steric energy-"sculpting" equations (Section 4.2.2); as we remember, exact differences in the set of MM equations, together with differences in defining MM atom types, generate distinct force fields. Values of E_{ster} for a molecular system, which are obtained by means of different force fields, are therefore unequal. Therefore, differences ΔE_{ster} show relative stabilities of the related molecular structures only if the steric energy E_{ster} of each structure is calculated using the same force field.

It is important to know that one cannot compare E_{ster} values of *any* two molecular systems in order to obtain a meaningful, telling ΔE_{ster} value. Values of E_{ster} result from summing the contributions that are related to pairs of model atoms (Sections 4.2.1 and 4.2.2). The more atoms the system contains, the higher the sum E_{ster} is. Hence, the first condition for a valid comparison of steric energies E_{ster} for a pair of molecular structures is that they should have identical chemical composition.

Furthermore, since force field equations reflect relationship *molecular geometry–potential energy*, they do not concern energy of chemical bonds excluding H-bonding.[a] Therefore, the second condition for comparing values of E_{ster} is that, as a rule, only isomers with the same chemical connectivity and the same chemical functionality can be considered. For instance, one can compare steric energy of any of four glucopyranoses (D-α, D-β, L-α, L-β) with steric energy of any of four galactopyranoses but E_{ster} of any of these cyclic hemiacetals cannot be compared with E_{ster} of any of their open chain forms (hydroxy aldehydes). Or, it is meaningless to compare

---

[a]As we remember, it is modeled in MM as some interaction of non-bonded atoms.

steric energies of a ketone and its enol: values E_{ster} calculated for chemical isomers carry no information about their relative stability, in contrast to OM-delivered values E_{el} (Section 3.1). Even for positional chemical isomers, e.g. keto tautomers of 1,2- and 1,3-cyclohexadienone, or geminally and vicinally difunctionalized systems, the relative thermodynamic stability cannot be estimated by calculating their steric energies. Thus, *only stereoisomeric structures may be considered in estimating relative stability by MM. Relative stability of chemical isomers may be estimated by MM only if there is an evidence that the integral chemical bonding (the set of all chemical bonds including C–C, C–H, C=C and aromatic bonds, irrelevantly to the position of the identical bonds in the molecular skeleton) is the same in both isomers.*[a] Examples of such chemical isomers are 2,4-dioxo- and 2,6-oxoadamantane, adamantane and twistane, methylcyclohexane and ethylcyclopentane, but not 1,2-dimethyl- and 1,3-dimethyladamantane, or 1,3- and 1,4-cyclohexadiene.[b]

An additional instruction for MM beginners is that the chemical structure cannot be deduced from considering an MM-modeled molecule. Any MM atom (MM sphere; Section 4.2.1) is associated with a certain chemical type, e.g. aminic nitrogen, iminic nitrogen, ammonium nitrogen, and amide nitrogen. Therefore, a molecular system is recognized by a force field as a given chemical structure, and its steric energy is calculated with using parameters for MM atoms (Section 4.2.2) related to this chemical structure. The latter is not altered in optimizing the molecular geometry of this system since this iterative calculation procedure (geometry optimization; Section 4.2.4), of course, does not change MM atom types in the initially recognized structure.

Relative thermodynamic stabilities of stereoisomers in equilibrium are, of course, inner, "observer-independent" molecular features. However, one cannot expect that different force fields afford equal values of the difference ΔE_{ster} for a pair of stereoisomeric structures. Such values of ΔE_{ster} should not be equal because different force fields "construct" steric energy by similar, but not identical equations for E_{ster}-composing potentials (Section 4.2.2). Thus, force fields have appreciably different accuracies in modeling the SES (in other words, in calculating E_{ster} for the same molecular

---

[a]This condition also means equal chemical functionality of the isomers.

[b]The chemical bonding is in part different in these isomeric dienes. It is commonly known that the bonding between two adjacent carbon atoms in alkanes differs from the bonding between the C-2 and C-3 carbons in 1,3-butadiene.

geometry).[a] Values of ΔE_{ster} therefore reflect the relative thermodynamic stability of stereoisomers only if these structures are modeled with using the same force field; otherwise, they reflect nothing. Moreover, one should necessarily choose a proper (i.e., sufficiently accurate) force field for modeling either molecular structure by MM (Section 4.2.5).

As indicated, a successful SES for a molecular structure (i.e., an MM model of its energy–geometry relationship) is similar to an adequately modeled Born–Oppenheimer PES (Section 3.1) for it, i.e., to a QM model of the molecule. That is, *in principle, QM and MM are near-equivalent in quantitatively describing relationship molecular geometry–molecular potential energy (i.e., in reproducing molecular geometry and relative stability of stereoisomers as well as transition states of their interconversions).*[b] This near-equivalency is important. It means that one can verify results of MM modeling by comparing them with the related results of accurate QM calculations, e.g., comparing MM- and QM-derived geometries, frequencies of molecular vibrations, as well as relative stabilities of stereoisomers.

One should not forget that these differently modeled surfaces (the SES and adiabatic PES) are, nevertheless, not identical. For instance, contrary to the Born–Oppenheimer PES shape, the SES shape is also determined by the temperature which is related to the modeled molecule. Relative stability of the same two stereoisomers are not exactly equal when calculating it by QM and MM. That is, both the difference ΔE_{el} and the ZPE-corrected difference ΔE_0 (Section 5.2.6) are not exactly equal to the difference ΔE_{ster} of MM, vibrational frequencies supplied by MM differ from those calculated by QM methods, etc. In practical aspect, this means that one cannot expect an absolute identity of MM and QM calculation results, when examining the validity of MM estimates via an accurate QM modeling of the same molecule (Section 4.2.6). The only requirement in verifying MM results for relative stability of stereoisomers consists of an indicative similarity between the related MM and QM quantitative estimates. Under *indicative similarity* for a pair of stereoisomers, we may

---

[a] These non-equal values of ΔE_{ster} supplied by different force fields for the same stereoisomers may, nevertheless, appear close when using rigorously parameterized force fields (Section 4.2.5).

[b] For stable stereoisomers, relative thermodynamic stability (relative depths of energy minima); for transition states, relative kinetic stability (relative heights of kinetic barriers).

understand the difference between ΔE_{ster} and ΔE_0 that does not exceed 0.5 kcal mol^{-1}.

4.2.4 *Geometry optimization in MM*

Providing a certain steric energy E_{ster} for any non-absurd molecular geometry, MM equations themselves do not show whether the considered 3D structure is or is not related to an energy minimum or to a saddle point in the SES. That is, by using these core equations, one can calculate E_{ster} for a molecular geometry, but one cannot say where this SES point is relative to other points of this surface: we have only one SES point. In principle, one may calculate E_{ster} for the second, third, fourth, etc. molecular geometries. It would be difficult to assume that, when undertaking conformational analysis by MM, the SES is modeled in the absolutely non-rational, even no-prospect, point-by-point manner. Then, it is valid to ask: if we have a structure of an arbitrary 3D geometry (a typical starting situation in conformational analysis), how does MM find its conformers and conformational transition states? In terms of computational chemistry, how does MM locate minima as well as inflection points in the SES?

In fact, this problem does not originate from MM. It is a classical problem in analytical geometry: how to find the nearest minimum in a surface starting from some point there, if the surface curvature is analyzable at any point of the surface? There are different ways to solve this problem. Several of them are used in MM, and two "chief" algorithms for locating an energy minimum in the SES (in other terms, for optimizing geometry[a]) are very briefly explained below. Note, the related computational procedures are computerized in force field packages, therefore mathematical meanings below should not terrify non-computational chemists.[b]

Optimization procedures. In a simpler algorithm, which is called the steepest descent method, gradient vector $grad[\partial E_{ster}/\partial x_1, \partial E_{ster}/\partial y_1, \partial E_{ster}/\partial z_1, \ldots \partial E_{ster}/\partial x_i, \partial E_{ster}/\partial y_i, \partial E_{ster}/\partial z_i \ldots \partial E_{ster}/\partial x_N, \partial E_{ster}/\partial y_N, \partial E_{ster}/\partial z_N] = \nabla E_{ster}$ is used [$x_1, y_1, z_1, x_2, y_2, z_2 \ldots, x_N, y_N, z_N$ are

---

[a]Geometry optimization and energy minimization are synonyms, both widely used in computational chemistry.

[b]This mathematics only shows the principles of geometry optimization in MM. Corresponding numerical computations are what is required for optimizing the molecular geometry for a concrete molecular structure by MM; they are executed by MM calculation programs.

Cartesian coordinates of atoms $1, 2, \ldots, N$, ∇ (nabla) is the Hamilton operator]. This vector is oriented oppositely to the direction of the maximal downward slope of the SES. Therefore, we actually need the oppositely oriented vector $-\nabla E_{ster}$. Partial derivatives of type $\partial E_{ster}/\partial x_i$ can be easily computed since force field functions (see, e.g. Eqs. 13–26) are differentiable analytically. Thus, we have the desired $-\nabla E_{ster}$. Then, going down by steps along this vector, steric energy E_{ster} is calculated in the terminal point of each step until values of E_{ster} start to increase. That is, we have found the minimal energy point in this line. At this point, components of the vector $-\nabla E_{ster}$ are recalculated, and its new direction is thereby determined. Note that this direction is perpendicular to the previous one. Repeating the step-wise descent along this new direction, we locate the minimum of E_{ster} also in this line. Then the procedure is repeated, starting from determination of the direction of the next descent. In this iterative way, descending in zigzags in the SES toward the E_{ster} minimum, we eventually locate it (Fig. 71). The gradient is unity at this point.

Importantly, energy minimization in MM is a downhill method: the molecular structure, upon this procedure, is "fallen down" into the closest energy minimum in the SES and not into the global minimum. In other

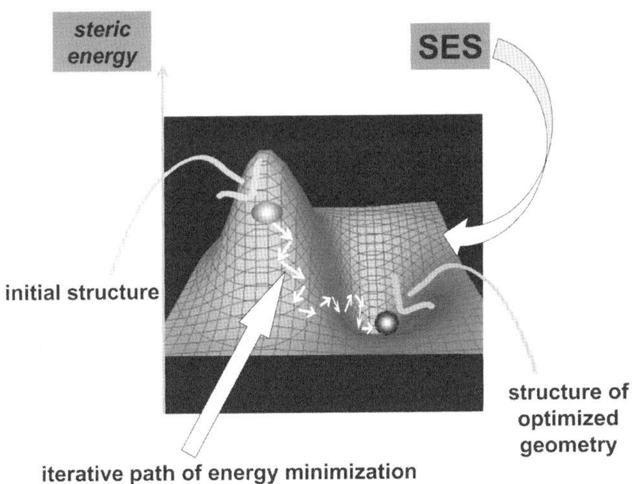

Figure 71. Energy minimization by the steepest descent method leads the molecule to descend in zigzags to the nearest energy minimum in the SES.

words, the modeled molecule does not cross SES barriers when optimizing its geometry. An explicit drawback of this method is that too many iterations are required to reach the E_{ster} minimum if it is flat. If the chosen step is too large, no convergence will be achieved at all — the optimization procedure "steps over" the minimum in any iteration.

A well-known Newton–Raphson algorithm is more effective. Scarcely synthetic and bioorganic experimentalists need a detailed mathematical explanation of it. One could mention that in MM, this method is related to determination of the gradient ∇E_{ster} as well as the curvature of the SES. That is, the Newton–Raphson method effectively "improves" the molecular geometry by selecting the way of maximal descending slope at every point of the SES where the gradient and the curvature are calculated. The curvature is defined by the full Hessian matrix (a matrix of all second derivatives of type $\partial^2 E_{ster}/\partial^2 x_i$, $\partial^2 E_{ster}/\partial^2 y_i$, $\partial^2 E_{ster}/\partial^2 z_i$ for the *i-th* atom; its dimension is $N \times N$; often called the Hessian). Since this matrix determines the surface curvature at any point of the gradient direction, multiplication of the inversed Hessian matrix H_0^{-1} by gradient ∇E_{ster} affords a new vector $H_0^{-1} \times \nabla E_{ster-0}$ that leads to the region of the minimum (index 0 indicates that the matrix as well as the gradient is calculated for initial coordinates of atoms). Performed iteratively, this mathematical operation provides molecular geometry that corresponds to the nearest minimum of steric energy.

In contrast to the steepest descent method, efficacy of this procedure is increased when approaching the minimum. On the other hand, if the geometry of the initial structure and the final one are very dissimilar, the number of Newton–Raphson iterations is sufficiently large when beginning the geometry optimization. Besides, the Newton–Raphson procedure is computationally expensive since second derivatives are calculated as well as the Hessian matrix is inverted in each iteration. Therefore, it is sometimes recommended to start the geometry optimization with using the steepest descent procedure. When the initial geometry has been improved, the optimization process can be continued by means of the Newton–Raphson method. This recommendation is not always worth following. Modern computer processors that are characterized by astronomic numbers of flops[a] may readily "digest" organic molecules (however, not biopolymers) in

---

[a] Flops: floating point operations per second.

the Newton–Raphson way. Besides, the number of iterations normally is less when this procedure and not the steepest descent minimization is employed.

Strictly speaking, the Newton–Raphson method seeks for the *nearest* stationary point that not necessarily should be a minimum. At a stationary point, the gradient is zero. How could we know which stationary point, a minimum, a maximum, or an inflection point, has been located? This is indicated by the Hessian. If the point is a minimum, all elements of the Hessian are positive. All negative eigenvalues of the Hessian mean location of a maximum. The presence of both negative and positive eigenvalues corresponds to a saddle point. If the Hessian has one negative eigenvalue, the located saddle SES point is of the first order (Fig. 56).[a]

Moreover, it is frequently important to calculate all eigenvalues of the Hessian in *each* iteration (to undertake so-called full matrix minimization) in optimizing molecular geometry.[b] In this manner, we can reach a saddle point (and not "pull" the molecule into the closest energy minimum), if the initial point in the SES is near this important inflection point. This means that one can locate *transition states* by means of the Newton–Raphson algorithm. The ability to locate both energy minima and inflection points of the SES is a distinct feature of this method of geometry optimization in MM: it opens a way to the ultimate goal in conformational studies of molecular systems, i.e., to an exhaustive description of conformational space as an arranged set of conformers and conformational transition states (Sections 2.8 and 3.5.2).

Dealing with geometry optimization for molecules of different size and running several computations simultaneously, computational chemists try not "to overstress" computers. Computational time in MM is increased drastically with the increase of the number of atoms in a molecular system. For instance, if this number N is increased only by 1, the number of intrapair interatomic interactions rises N-fold. Therefore, in order to perform MM calculations in a time- and computer resource-effective regime, a simplified Newton–Raphson algorithm known as the block diagonal

---

[a]Of course, professional MM programs compute the eigenvalues and display these numerical results in the output (summarizing) file.

[b]This is practically easy: full matrix minimization is an attributive calculation option of professional MM programs.

Newton–Raphson minimization often is used. The economy is reached by means of a simple mathematical trick. All elements of the Hessian are defined as zero except a block on the diagonal related to one atom. This simplification means that coordinates of only one atom are subjected to a change in each iteration. And indeed, the block diagonal Newton–Raphson minimization is essentially more time-effective than the full matrix one — it is $\sim$10–20 times as faster for 50–100 atom organic molecules. However, this calculation algorithm leads to exclusively locating a minimum of the SES.

Another way to a "quick" optimization of molecular geometry is decreasing the number of pairs of non-bonded atoms for which certain energy contributions are calculated. This decrease is absolutely justified. For large organic molecules, most of interatomic interactions assessed in MM are interactions of non-bonded atoms which are essentially distant one from another. The Lennard–Jones potential (Eqs. 19 and 20) is a very fast descending function. Therefore, energy contributions of VDW interactions into E_{ster} are miserable for the pairs of very distanced atoms and can be neglected. This ignorance of remote VDW interactions in MM is called cut-off. MM calculation programs usually neglect VDW interactions of atoms, which are separated in space by more than a certain specified value (often 8 or 10 Å); many MM calculation packages give to the user the possibility to choose the cut-off distance at his/her own discretion. This time-saving cut-off restriction is practically necessary when optimizing geometry of non-small molecules, e.g., peptides, oligosaccharides, dendrites.

Electrostatic interactions are of long range. Therefore, a plain cut-off is problematic in this case: the real balance of attractive and repulsive Coulomb forces between distal positive and negative point charges would be disturbed in a large MM-modeled molecule, e.g., a polyelectrolyte or a protein. The Fast Multipole Method (FMM),[a] which compensates electrostatics cut-off by taking into account multipole-multipole interactions, solves this problem in part. Electrostatic cut-off is still a problem in MM. In contrast to specialists in structural biochemistry, organic chemists do not need to use it because their objects are smaller molecules.

---

[a]For instance, FMM is implemented into late versions of the continuously developed force field CHARMM.

Convergence. Mathematically, first derivatives $\partial E_{ster}/\partial x_i$, $\partial E_{ster}/\partial y_i$ and $\partial E_{ster}/\partial z_i$ are zero (index i depicts the i-th atom) if the molecule "is" in a minimum in the SES — any change of spatial coordinates of any atom leads to an increase of E_{ster}. These derivatives are near zero in the vicinity of the minimum, i.e., $\partial E_{ster}/\partial x_i \rightarrow 0$, $\partial E_{ster}/\partial y_i \rightarrow 0$ and $\partial E_{ster}/\partial z_i \rightarrow 0$ upon $E_{ster} \rightarrow min$. In minimizing the steric energy E_{ster} of the molecule, any change of steric energy $\Delta(E_{ster})_{k+1}$ for two consecutive iterations k and $k+1$ $[\Delta(E_{ster})_{k+1} = (E_{ster})_{k+1} - (E_{ster})_k]$, as a difference of two numbers, is finite. However, the condition $E_{ster} \rightarrow min$ requires the difference $\Delta(E_{ster})_{k+1}$ to be an infinitely small value. This means that an infinite number of iterations would be required for reaching the bottom of the potential well in the SES. Therefore a numerical threshold for $\partial E_{ster}/\partial r_i$ (r_i is a coordinate of the i-th atom of the molecule) is introduced in MM. It is a rms (root mean square) average of values of derivatives $\partial E_{ster}/\partial r_i$ for all atoms and is called convergence criterion. It actually determines which near-minimum point in the SES is accepted as the minimum point. According to this criterion, calculations are stopped — the minimum is accepted as being reached — when this average does not exceed an *a priori* given small value, e.g., 0.01 kcal mol^{-1}/Å. An additional convergence criterion is the maximal absolute value for numerically computed derivative $\partial E_{ster}/\partial r_i$ for each i-th atom; an acceptable value may be 0.001 kcal mol^{-1}/Å. Of course, any convergence criterion may be numerically specified before running geometry optimization. Strictly speaking, a SES minimum that is located by MM calculations is not unique: results of energy minimization for the same molecular structure (i.e., the resulting molecular geometries and E_{ster} values) would be slightly different if distinct convergence criterions are used.

In some relation, this Section deorients organic experimentalists. They would think that MM users should manage with these theoretical notions every time when performing routine MM calculations. This would be essentially difficult for non-specialists — considerable efforts are required in order to find equations and parameters for a concrete force field in reviews, manuals or books which have been written by computational chemists. Therefore, it is necessary to emphasize here that synthetic chemists do not need conceptual as well as technical details of MM to perform MM calculations. Moreover, these details are hidden from such MM users; for them, force fields are finished computer programs — related equations have been implemented in their computational core. The latter is surrounded by an interface that enables the on-the-screen dialog with the program by means of program-specific commands. In commercial computational packages — and practically only they are useable for organic researchers — there is no need even in this command line[a] dialog. These convenient tools of molecular modeling are equipped with a graphical interface that displays 3D images of molecules and has virtual buttons on the screen — say, *Draw*, *Erase*, *Select*, *Calculate*, etc. Calculation results (e.g., final molecular geometry as a 3D image, values of calculated energy potentials and steric energy) are displayed on the screen, and the program writes them down in numerical form in a simultaneously generated output file (a text file). By means of these facilities, the elementary calculation procedure in MM is reduced to three consequent actions — to draw the structure on the screen, to select the force field from the list, and to push the *Calculate* button. This childish simplicity and impressing performance — one geometry optimization for a 100–atom molecule only takes several seconds on a serial laptop — have a side effect. Due to "overaccessibility" of MM calculations, many research articles contain untrustworthy or even incorrect structural analysis based on MM calculations. There, the problem frequently is not in MM; it arises from an incompetent use of this method of molecular modeling. Next Sections in this Chapter guide the beginner in practical methodology of MM modeling of small molecules and explain difficulties that appear in the way to a successful MM model of either organic molecule.

4.2.5 *To choose the force field*

Now, one could say that the key features of MM are in general clear for us. We know what *force field* means, what steric energy E_{ster} is, what stationary points in the SES are and how they are located in theory. Our next question is, how do MM users, equipped with these theoretical basics of MM, proceed further?

---

[a]Program-specific textual commands given by the user; he/she controls the text correctness by typing the commands on the screen.

As mentioned (Section 4.2.4), MM calculations with using specialized computational programs are technically very simple to execute. However, a skilled MM user is not transformed into a technician. Different force fields are of very distinct quality in modeling molecular parameters since force fields are products of empirical, non-coordinated,[a] various solidity adaption of structural principles of MM modeling (Section 4.2.1) to differently oriented MM tools. No force field can equally successfully model organic molecules that have distinct chemical structures, and different force fields are not equivalent in their capability of modeling chemically the same molecule. For the same compound, some force fields may supply a distorted, untrue molecular geometry, some others may supply sufficiently precise molecular geometries, and some force fields may be unable "to read" (to accept for executing calculations) the chemical structure of this molecule at all. Also, a force field may be successful when providing molecular models for some compounds but it may appear feebleness if applied to compounds from other chemical classes. MM-supported considerations in dozens of publications in organic chemistry journals suffer from unprofessional recruitment of either force field. Obviously, the choice of the suitable force field is crucial in "obtaining" the virtual (i.e., modeled) molecule that has geometry and energy characteristics as close as possible to those of the real molecular system. Thus, before shooting an MM computational procedure, chemists have to take an essential methodological decision: they have to select a force field that will provide an adequate MM modeling of molecular systems of interest.

3D geometry and relative energy of stereoisomeric structures are the main molecular features that MM supplies for conformational considerations. There is no universal prescription for *a priori* selecting the MM tool that would be most accurate in modeling these characteristics of either functionalized organic molecule; a verification or another justification of obtained results of MM calculations is necessary (see at the end of this Section). Nevertheless, let us briefly discuss the main force fields in order to have an orientation that may assist in selecting at least a relevant force field.

---

[a] I.e., performed by research groups that are independent one from the others in developing force fields.

Force fields and their versions. Under a force field, one should understand a set of MM equations of certain distinct form (Section 4.2.2) and related numerical parameters of certain values together with a set of solitarily defined model atoms (Section 4.2.1). As emphasized in the previous section, development of a force field is a prerogative of computational chemists; for experimentalists, force fields are finished products in the form of computer programs. That is, non-computational chemists can only choose a proper MM tool from the pool of existing force fields. This pool is not vast: it contains several main force fields.[a] One may say that they are families, where each family comprises related, very similarly designed force fields (versions; see below). Force fields have their own names, e.g., force fields MM2, MM3, MM4 (by Allinger's group), OPLS (by Jorgensen's group), CHARMM (by Brooks and Karplus groups), Amber (by Kollman's group), Gromos (by van Gunsteren's group), MMFF (Merck Molecular Force Field; by Halgren's group), COMPASS (by Sun's group). After introduction of the these main force fields for common use in the period of the 80's–90's, appearance of a new force field, e.g., GLYCAM06 (by Woods) or ECCEP-05 (by Scheraga), is a rare event.

Thus, commonly known names of force fields — MM3, MM4, OPLS, Amber, CHARMM, Gromos, etc. — actually indicate the *force field family*, i.e., each name specifies a group of force fields of almost identical architecture (called *versions*) and does not specify a single unique force field. Nevertheless, differences between versions are semantically ignored and the term *force field* and not a more accurate term *force field family* is used.

Under *version*, one can understand an extended (in parameterization and/or atom types) or slightly modified (in force field equations) force field relatively to some parent force field. That is, versions do not contain principal changes in the architecture of the parent force field of the family. Versions appear as a result of further step-by-step development of the parent version of a force field of original architecture; they are released with the time progress. Thus, later versions are improved or re-targeted[b] releases of

---

[a]For the list of the most known force fields see e.g., Refs. 33b. These general names, however, can sometimes mislead because their significantly modified versions are often mentioned in the literature using the name of the parent force field. Also general, non-detailed descriptions of computational packages of MM may list the names of implemented force fields without specifying the version of either force field.

[b]Versions that target other chemical classes of organic compounds.

the initial version. Such a set of versions actually is what the term *force field* combines. Note that a version is what the researcher uses in concrete MM modeling.

The newer version is either more accurate in modeling molecular structures encompassed by the previous parameterization, or is much more accurate in modeling molecular structures of a certain, targeted chemical class (e.g., proteins), or is capable of modeling additional functional chemical groups. Therefore, different versions of the parent force field may supply somewhat distinct MM models of the same molecule; in this case, the latest version of the same targeting obviously delivers a better molecular model.

Versions may be hidden under the same common name (family name) or may receive their own names. To some extent, this chaos with names may confuse chemists not involved in MM calculations deeply. For example, CFF91, CFF93 and CFF95 are versions of CFF (Consistent Force Field). Many experimentalists do not know that the unspecified name MM3 may mask any of four successive versions of this force field.

As mentioned, a version, in the form of a computer program, and not the related force field (a family of versions) is what the researcher uses when modeling either molecular structure. Nevertheless, in describing such a computational MM experiment, one traditionally (but inconsistently) says that the force field is used.[a]

Remarkably, although versions are fully functional and are exploited in research as individual MM tools (computational MM programs), they frequently are not ultimate products. Force field developers continue to insert some minor changes (both technical and architectural) in the version used. This tiny improvement finally results in appearance of an equally functional subversion which has a better performance, e.g., due to removing a few program bugs, or has higher accuracy, e.g., due to some minimal changes in parameterization. Unproductively, in MM practice, such a subversion is also called *version*, hampering our orientation in the force field pool.

As indicated, a later version contains next developments (e.g., an extended set of atom types and parameters for them, additional terms in calculations of steric energy) and sometimes technical corrections as well. Let us mention some examples of force field versions. For the Amber force field, the Amber-ff99 version is a continuation of the Amber-ff96 and Amber-ff94

---

[a]Reporting in research articles and reviews about the force field used, chemists often do not indicate its version. Notice, the accurate "address" of the employed MM tool is *force field* +*version*. Non-identical results, e.g. for relative stability of conformers, may be obtained for a system, if it is modeled with using distinct versions of the same force field.

versions. In addition to refining torsional parameters for the peptide back-bone, it extends the set of atom types. Version Amber-ff99SB improves the Amber-ff99 version by further corrections of these torsional parameters. Concerning MM3, the MM3(96) version is a product of development of the MM3(92) version and contains new, thoroughly developed parameters for amines and sulfones as well. MM3(2000) adds explicit parameterization for phosphates, organogermans, fluorinated hydrocarbons and oxocarbe-nium ions $-O-(H)C^+- \leftrightarrow -O^+ =C(H)-$ as well as re-parameterization for tertiary ammonium ions. In this version, aromatic fragments are treated by *ab initio* MP2 calculations (Section 5.2.3).

If cardinal changes are introduced in the structure and number of the force field equations and parameter set, the modified force field actually is a new force field. This sequel gets a new name as, e.g. MM3 *vs.* MM2, CFF *vs.* CVFF, PCFF *vs.* CFF, CHARMM General Force field (CGenFF) *vs.* CHARMM. One can regard such a sequel as the first version of a new force field (a new family). Appearance of both new force fields and new versions of existing force fields reflects evolution of MM methods.

The main force fields in organic and biochemical research currently are MM3, MM4, MMFF (as the MMFF94 version), Amber (ff94 and later members of the Amber family), OPLS-aa, CHARMM (different versions), GROMOS, and COMPASS.[a] Among them, CHARMM, Amber, GROMOS, MM3, and MMFF94 are most widely used.[b] Further development of a majority of "noble" MM families (e.g., Amber, CHARMM, GLYCAM, MM4, GROMOS) is continuing. Permanent MM users follow their progress; nevertheless, episodic MM visitors and beginners certainly can obtain good results by selecting the suitable force field and using even not the latest, but sufficiently advanced version of the chosen force field. Probably, CHARMM (including CGenFF; see below), Amber and GRO-MOS currently are the most dynamic and versatile force field families (see Section 4.3.1 for their website addresses). For instance, CHARMM includes versions that are renewed from time to time and separately targeted to modeling peptides (CHARMM22), lipids (CHARMM36), carbohydrates

---

[a]The well-known MM2 may be regarded as an obsolete force field.

[b]Including less known versions of these force fields e.g., MM3[a] and MM3hc (MM3 versions), CHARm (a CHARMM version).

(CHARMM35), as well as nucleosides, nucleotides and nucleic acids (CHARMM27). Such force fields, which are oriented to a certain biochemical class and should have a superior accuracy when modeling compounds of this class, are the trend in the development of MM for needs of molecular biochemistry.

Computational drug design operates with both small molecules of quite different chemical structures and biopolymers. It requires of MM sufficient universality in reproducing various organic structures, and, probably, this demand stimulates force fields to evolve in the opposite direction, from narrow-oriented force fields to broad-spectrum-minded ones. Borrowing main structural elements of CHARMM, a new "general organic" force field, CGenFF (by MacKerell),[49e] has appeared recently; this CHARMM-based force field is under further development, and the current subversion of CGenFF (version 2b8[a]) replaces the previous ones. In GLYCAM06, the original version predominately targets carbohydrates, while the latest version is designed as a "general purpose biomolecular" force field.

Force field quality. For the user, force fields, in the form of either version, are interactive (program–user dialog-providing) computer programs. In other words, when performing geometry optimization by MM, we use a stand-alone calculation program or a calculation program that is a component of a hierarchically organized cluster of programs (molecular modeling package) and are able to give program-specific commands for executing either computational MM procedure. Suppose, our computer facilities may deal with any MM and QM calculation program or package. In this ideal situation, we would say that, in selecting the force field, the single requirement to selecting the MM tool is its accuracy in the modeling of molecular structures of organic compounds.

Let us see force fields in that light. We can single out three elements that predetermine the quality of force fields: (*1*) the number of types of model atoms; (*2*) the success of *geometry–energy* equations; and (*3*) the quality of force field parameters. When using a force field, the quality of its key equations (energy potentials; Section 4.2.2) as well as parameters in these equations, i.e., the quality of accuracy prerequisites (*2*) and (*3*), is difficult to estimate. Indeed, parameterization of equations of a force

---

[a]See webpage *http://mackerell.umaryland.edu/~kenno/cgenff/download.html.*

field is achieved by matching probe values of the parameters iteratively. The parameter variation is continued until some MM-calculated quantities[a] fit either experimental data or values that are supplied by accurate QM calculations. In other words, force field parameterization is a result of a laborious, time-consuming calibration that takes more effort than a deep MM-based study. Examination of the calibration quality for the force field would mean re-parameterization, and, for an uncertain time period, would freeze out the force field from using it with confidence. Testing of potential energy functions is equivalent to creation of a new force field. Therefore, one can leave consideration of two accuracy factors (*2*) and (*3*) and turn to the remaining factor (*1*), which is much easier to estimate.

There, we speak in a clear language of numbers. The principle of force field comparison is self-evident: the more detailed description of the molecule is in terms of MM atom types and numerical parameters for energy potentials, the better the force field should be. The larger number of MM atom types means an increase of the number of parameters for MM energy potentials (exemplified by Eqs. 15–26), although the parameter number is not strictly predetermined by the number of atom types. An infinitely detailed specification of molecular structures is, of course, impossible, and any force field has its own number of atom types as well as parameters. Comparison of force fields shows that a sufficiently large number of model atom types (more then 50) for a relatively small number of chemical elements (i.e., basic "organic" elements: C, H, O, N, Cl, Br, P and S) as well as a large number of parameters (more than 1000) is present for only a few force fields. Let us introduce an estimate n_{av}, which describes formal quality of a force field, and term it *averaged element specification*. This estimate n_{av} is defined as quotient $n_{av} = $ *number of parameters in MM equations* $\times$ *number of MM atom types/number of chemical elements*, where *number* is related to the indicated architectural element of the force field. This characteristic n_{av} reflects how detailed an "averaged" chemical element is defined in a force field. Clearly, the higher n_{av} is, the more detailed the specification of the force field atoms is. Let us call force fields, which are characterized by $n_{av} > 8000$, fine force fields.

---

[a]Those usually are molecular geometry, dipole moments, conformational barriers, oscillation frequences.

For example, MMFF94 includes about 9000 parameters[33b] for 19 chemical elements and 99 atom types ($n_{av} \approx 46900$), MM2 has 3722 parameters for 30 atom types and 12 elements ($n_{av} \approx 9300$), CGenFF uses 5129 parameters for 139 atom types and 14 chemical elements ($n_{av} \approx 50900$).[a] For comparison, an early Sybyl (Tripos) version has only 37 atom types and 425 parameters for 15 chemical elements ($n_{av} \approx 1050$). The Dreiding force field has only about 200 parameters comprising all "organic" as well as main group elements. The averaged element specification is terribly low ($n_{av} \approx 14$) for UFF,[b] a general purpose force field. It assigns 126 types of model atoms to 105 chemical elements and uses 12 numerical parameters for each atom type. Thus, beginning MM users can reasonably conclude that only fine force fields are solid research tools of MM.

Fine parameterization in MM is laborious. As mentioned above in this Section, development of a force field *for research* is a result of tedious efforts of experts. One can even say that a continuous, several years work of a research group on development of a force field is an outer but remarkable sign for a high quality of the force field. Probably, since the number of different chemical functions in various combinations is much greater than the number of serious MM developers, *existing fine force fields are very limited in modeling molecular structures of relatively rarely visited chemical classes.* So we learn that one should not try to find a suitable force field for modeling such chemical structures (Section 4.2.7); it is worth transferring to QM methods (Chapter 5) straight away.

Specialists in molecular mechanics quickly recognize chemical structures that are not "digestible" for standard, commonly available versions of force fields (versions not equipped with supplemental, separately developed parameterization that targets a new chemical functionality). Other researchers should learn (from the relevant chemical literature; often available in the internet) what chemical classes covered by either force fields are. When starting their first computational MM experiments, inexperienced MM users may be guided by the following non-scientific principle: organic structures of closed valent shell from chemical classes that are

---

[a]For basic versions.
[b]Universal Force Field (developed by Rappé *et al.*).

described in introductory courses of organic chemistry are treatable objects for fine force fields; non-metalorganic structures from chemical classes that are described in advanced courses of organic chemistry are worth trying to model them by MM, while other chemical structures rather cannot be subjected to MM calculations (see Section 4.2.7).

Many professional MM programs do not allow overestimation of their own possibilities. For instance, all original (not implemented into a calculation package from another source) versions MM2 or MM3, force field CGenFF, as well as the fine force fields within the shell of the MacroModel or Yasara packages of MM (Section 4.3.1) deny to minimize steric energy for chemical structures that are not covered by their parameterization. Thus, fine force fields of MM (as computational programs) assist users: they notify about missing parameters and often indicate (by generating the relevant text) the molecular fragment not recognized by the specified force field. To provide at least some structural model, some computer programs of MM modeling are supplied with an option to use for a molecular unit, which is not covered by parameterization, existing parameters for a related structural unit. If the user decides not to supply the problematic fragment with "alien" parameters, the optimization procedure is not continued. If the user agrees to apply these parameters to this untreatable unit, geometry optimization is performed. Of course, the final judgment of the result is problematic. The probability, that molecular geometry modeled in this way is realistic at least in bare outlines, is unclear. The obtained value of steric energy is not worth the researcher's attention at all.

Not all MM programs use exclusively original (or minimally modified) solid force fields. For example, the MMX force field (computational package PCModel) is designed to treat any organic system by using parameters of MM2 together with additional, so-called generalized parameters that are supposed to comprise a maximally broad set of organic structures. Such an unlimited "omnivorousness" is, of course, impossible; therefore, MMX does not afford research-scale significance to calculation results for non-banal systems.[a] To caution MM beginners against a misleading use of "light" force fields, let us look at MMX results for some compounds. Note that "craftiness" of many MM-derived structures is not recognizable until they are studied experimentally or modeled by means of a universal tool, QM (Chapter 5).

The S fragment of sulfurane **17a** (Section 1.8, Fig. 25) has a virtual trigonal bipyramidal geometry.[b] QM calculations firmly confirm this (Fig. 72). Original force fields, e.g., MM3 (any version) and CGenFF (version 2b8[c]), "refuse" to deal with this molecule since they have no parameters for modeling this molecular unit.[d] MMX, that also does not have specific

---

[a]Unfortunately, one could encounter, in research articles, conclusions based on MM calculations with using MMX.

[b]Or, using the stereochemical jargon, a wedge geometry (Fig. 24).

[c]The currently most advanced version of CGenFF that, in turn, is a force field sequel of CHARMM.

[d]Excluding MMFF94. It accepts structure **17a** for energy minimization and supplies the molecular geometry for it that is a similar to the 3D geometry modeled by MMX (Fig. 72). One should note that

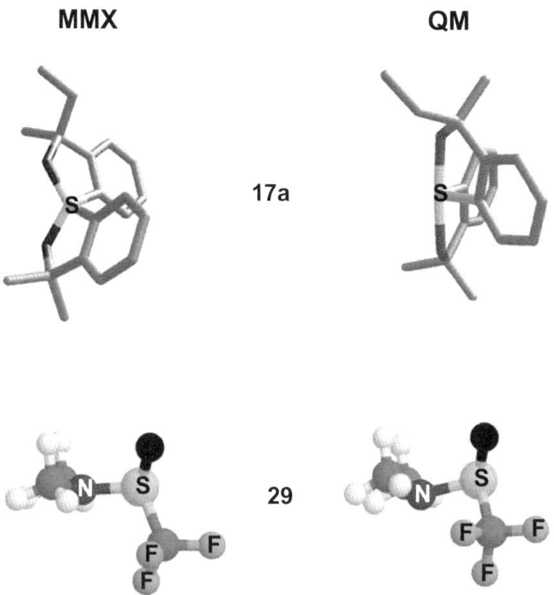

Figure 72. 3D geometry of sulfurane **17a** as well as sufinamide **29** by MMX and QM [MP2/6-31+G(d,p)].[9] For simplicity, H atoms are not shown for **17a**. The S electron lone pair in **17a** is shown schematically.

parameters for sulfuranes, manipulates with some "generalized" parameters, not informing the user about the unavoidable penalty of this "generalization" and not estimating (in contrast to scrupulously developed force fields) how "bad" the used MMX parameters are. Not quite surprisingly, a strikingly corrupted molecular geometry of **17a** is produced by this force field (Fig. 72). It is a "successful" case of MMX-misleading: because the MMX-generated tetrahedral S geometry ignores the lone electron pair of the sulfur atom in this sulfurane, attentive researchers can immediately catch up that the MMX result is dubious.

"Generalization" of the geometry of amide nitrogens as planar units leads to the failure of MMX in the case of sulfinamide $Me_2NS(O)CF_3$ (**29**; Fig. 72). Its N fragment is planar by MMX. In fact, it is pyramidal. This failure of MMX probably would remain undiscovered for inexperienced researchers since the MMX-modeled 3D geometry of the N-fragment of **29** does not seem suspicious.

No current fine force field would "agree" to optimize geometry for molecular structure **28** (Figs. 69 and 73) from the group of MIDA mixed anhydrides (protecting group of the

---

MMFF94, in contrast to other fine force fields, sometimes abuses its parameters, recognizing chemical structures that, in fact, are not covered by its parameterization and then delivering "false" molecular geometries.

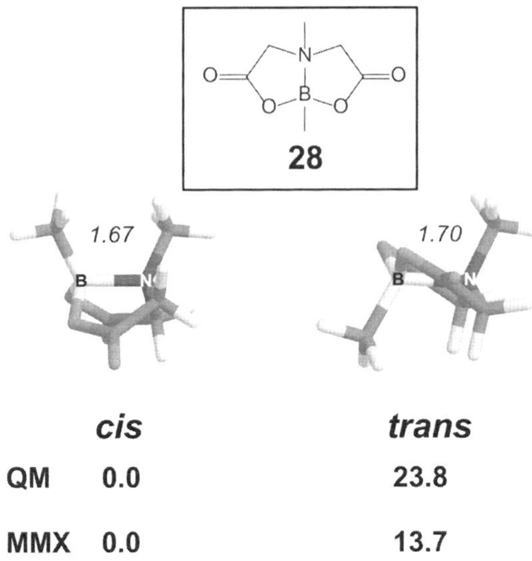

Figure 73. QM-optimized [B3LYP/6/31+G(d,p); Section 5.2.4] geometries of stereoisomers of bicycle **28**. Numbers indicate relative energy (ΔE_{ster} for MMX and ΔE_{el} for QM; kcal mol^{-1}). Numbers in italics show QM-calculated interatomic distances B–N (Å).

boronic acid moiety in the Suzuki-Miyaura coupling[51]). Research-oriented force field do not have parameters for the uncommon fragment, B–N, which, in addition, is a structural component of the strained system **28**. Because of its "generalization," MMX accepts this structure for modeling.

Molecular geometry is very close for QM and MMX models of the *cis* stereoisomer of **28** as well as of its *trans* isomer. However, energy estimates are a more sensitive indicator of molecular modeling quality than calculated parameters of molecular geometry. MMX failure is exposed if comparing relative stereoisomer stabilities supplied by MMX and QM. The *cis* stereoisomer totally predominates over the *trans* isomer, as MMX does indicate.[9] However, QM calculations demonstrate that predomination of the *cis* isomer is greater by 10.1 kcal mol^{-1} than MMX estimates it (Fig. 73). This two–digit kcal mol^{-1} inaccuracy in energy estimates exemplifies what high inaccuracy may be masked by a very similar molecular geometry when calculating steric energy as defined by an uncritically developed force field.

Molecular structures, which have electron-deficient chemical bonding, are problematic for any force field (Section 4.2.7). Also, MMX invariably misses this bonding and delivers a false molecular structure. The idea of broad generality embedded into MMX in an intolerably careless form leads this "light" force field to significantly distort the molecular geometry of, e.g., Li amide **11** (Fig. 74). While most fine force fields "honestly" reject to treat this chemical

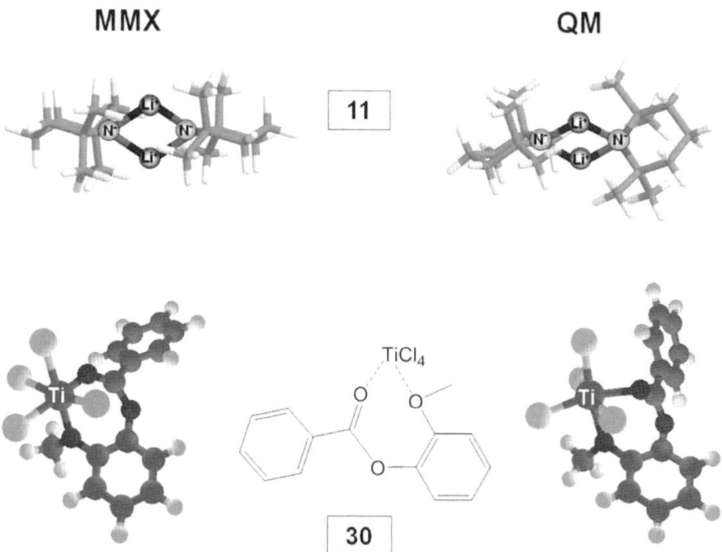

Figure 74.　3D Geometries of lithium amide **11** and bidentate Ti complex **30** by MMX and QM [optimized at the MP2/6-31+G(d,p) level (see Section 5.2.3) for vacuum conditions].

structure, MMX does not hesitate to minimize energy of this molecule and, as a result, models a symmetrical geometry with piperidine rings in the same plane. However, these rings are actually turned one relative to another by almost 90°, as QM-based calculations show. In addition, MMX parameters supply both nitrogens with an incorrect, nearly regular tetrahedral geometry, instead of a highly distorted tetrahedral geometry (Fig. 74).

Simplest post-HF *ab initio* calculations (Chapter 5, Section 5.2.3) are capable of modeling molecules of metal complexes including organic complexes of non-heavy transition metals. MM modeling of the latter is still problematic, and specialized, organometallics-oriented force fields (e.g., LFMM, SIBFA, VALBOND and VALBOND-TRANS) are far from providing reliable molecular geometry for non-trivial metalorganic compounds. In this light, MMX that shows to the user a preparedness to model metal complexes is, mildly speaking, too ambiguous. A simple example of chelate **30** (Fig. 74)[a] illustrates this. When applying this force field to the molecular structure of **30** optimized at the MP2/6-31+G(d,p) level of QM (Section 5.2.3), the overall molecular geometry is distorted by MMX so significantly (see Fig. 74) that resulting local inaccuracies (e.g., no difference in lengths of apical and equatorial Ti–Cl bonds of the MMX-modeled molecule) might escape our attention.

---

[a]Such bidentate Ti complexes probably are intermediate compounds in the TiCl_4-catalyzed addition of nucleophiles to esters bearing an *o*-methoxy- or *o,o*-dimethoxyphenyl fragment.

In defense of MMX, one should say that also other "light" force fields, e.g., UFF, even demonstrating coordination of the $TiCl_4$ molecule by the two oxygen ligands, cannot predict the accurate geometry of the molecule of **30**. However, researchers, in contrast to educators, cannot "excuse" force fields that pretend to be universal. Computational chemists merely would not apply MM to chemical structures that contain "non-parameterized fragments." The readiness of MMX, with its minimal parameter resources, to afford molecular geometry for a broad set of organic structures may be a trap for MM beginners. To their note, *"light" (low n_{av}) force fields are not research tools.* Nonetheless, freely distributed in the Internet (e.g., Argus, ChemSketch[a]) or commercial computational MM programs (e.g., PCModel, StruMM3D) that include such force fields are convenient for generating molecular structures of rough 3D geometry useable as initial one in further, geometry-refining QM calculations.

There are force fields that are of *a priori* significantly decreased accuracy, but, nevertheless are used in research; those are united-atom force fields. They do not have individual model atoms, in contrast to all-atom force fields discussed above. To diminish computation expense, model fragments, e.g., CH_3, CH_2, NH, etc., are used in united-atom force fields instead of model atoms. Such force fields are required to analyze very large molecular systems (e.g., biopolymers, ionic liquid media) computationally. Ordinary size structures, if modeled by MM, are subjected to calculations by means of all-atom force fields.

Could we use the averaged element specification n_{av} for comparing the quality of fine force fields and point out "the best force field"? Unfortunately, this simple estimate n_{av} scarcely could be a precise indicator. Moreover, there is no such indicator at all. Analysis of a vast massif of calculation results collected during the thirty-year history of intensive use of MM in chemical research could not indicate one or two of the most successful fine force fields. Such an analysis would lead to a somewhat discouraging conclusion: fine force field cannot be arranged according to their accuracy in modeling. *The accuracy of any fine force field varies with applying the force field to chemically different molecular structures.* In other words, any fine force field cannot be characterized as more/less accurate than another one without regarding to chemical structure of modeled molecules. One cannot single out "the best force field" from the pool of force fields because such a force field does not exist. Also, there are no two or three "force field champions."

---

[a]This computer graphics program for building molecular organic structures on the screen may optimize their geometry with using an essentially truncated CHARMM version.

One may deduce from this brief consideration that structural specificity (in other terms, applicability) of any fine force field is limited: some molecular structures can be modeled accurately, while others cannot. Structural specificity of a force field cannot be exactly known *a priori*. Indeed, this practically important characteristic of force fields means a set of core chemical structures for which the force field can provide molecular models of acceptable accuracy. Such a set obviously is renewed with progress in experimenting with the force field and sometimes exposes both unexpected applicability of the force field to some molecular structures and its unexpected failure in regard to other ones.

Force field selection. One should not think that the quality as well as applicability of a fine force field are absolutely unclear to the researcher before probing the force field in modeling of molecular structures of interest. There is a rough classification of applicability of fine force fields, and it is helpful for selecting a *group* of such force fields that may model a concrete chemical structure more or less successfully; in the next step of the selection, one can choose either force field from the selected group (see below). Due to an exceptional role of biopolymers (proteins, nucleic acids, polysaccharides) in living organisms, organic compounds may be formally divided into two groups — biopolymers and others.[a] Since biochemical research occupies a very weighty position in contemporary science, it occurs that MM tools are developed "in accordance" with this amorphous classification. That is, modern fine force fields are designed with an orientation to be either "biopolymeric," or "organic."[b] They undergo a purposeful parameterization and terminologically are characterized either as bioorganic (biomolecular), or as general organic force fields, respectively (see, e.g., Ref. 33b, 49c,d and 50a–c,g). Examples of force fields of general organic purpose are MM2, MM3, MM4, MMFF94, CGenFF (CHARMM General Force Field), GAFF (General Amber Force Field), OPLS-aa (an all-atom version of OPLS) and QMPFF3 force fields. Families CHARMM (as a family of distinct versions, where each version targets certain biomolecules), Amber, GROMOS, OPLS, and GLYCAM06[c] are

---

[a]These "others" are low molecular weight organic compounds as well as synthetic organic (i.e., not of biological origin) polymers.

[b]MM applications to inorganic or material chemistry are not considered.

[c]Note that some versions of GLYCAM06 have been designed as a general organic purpose force field.

known as the most widely used bioorganic force fields. Importantly, one cannot suggest "the best force field" in each of these two groups of fine force fields (see below). Also, there is no force field that is tuned equally fine for several classes of organic compounds.

Organic fine force fields have many MM atom types for carbon, hydrogen, oxygen and nitrogen (in contrast, "rarer organic" chemical elements, e.g., phosphorous, sulfur, iodine, are represented essentially weaker[a]). This parameterization is associated with different chemical functionalities of these elements but not with certain chemical classes of compounds. For instance, 29 types for the carbon, which are associated with various molecular C-containing fragments, are present in CVFF. Clearly, organic compounds of different skeletons are the main target of this force field. In contrast, bioorganic force fields are focused on chemical classes — MM atom types and, thus, related parameters are essentially extended in these force fields with regard to either biological compounds, e.g., either proteins, or nucleic acids, or oligosaccharides. For instance, among 20 carbon types in CHARMM22, 12 types are related to amino acids. Even an inexperienced researcher would conclude that this version of CHARMM is sharply peptide/protein-oriented.

The obvious consequence should be that organic force fields are less successful if applied to biopolymers or even to their precursors (amino acids, monosaccharides and nucleotides). On the other hand, force fields, which are designed as modelers of biopolymers, should be in a disadvantage in relation to non-specific organic fine force fields when applied to "foreign" objects — open chain hydrocarbons, carbo- and heterocycles (in particular, to strained systems), keto compounds and their derivatives, etc. Is the MM user in a convenient position to select the force field that should be suitable for modeling of his/her own molecular objects? Unfortunately, rational characteristics of force fields *organic* or *biomolecular (bioorganic)* given by their developers do not guarantee that an organic or biomolecular force field will provide better accuracy if applied to a "purely organic" or "biomolecular" structure, respectively.

---

[a]For instance, CVFF has only one type for P. This force field would equally treat a planar P in cyclic aromatic phospholide anion $C_4H_8P^-$ a pyramidal P in phosphites $P(OR)_3$, a tetrahedral P in phosphates $O{=}P(OR)_3$ and a trigonal bipyramidal P in phosphoranes PX_5.

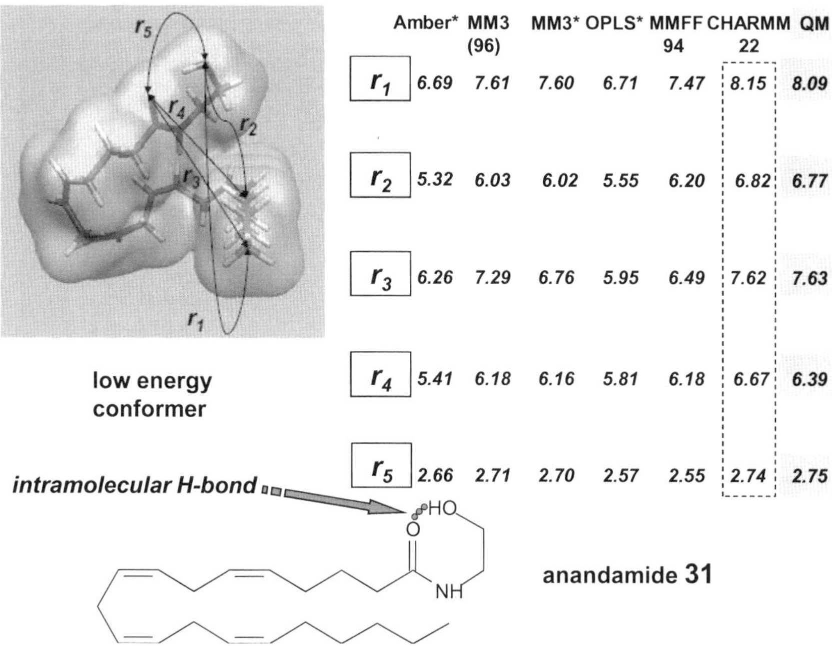

Figure 75. Selected interatomic distances (Å) for a low energy conformer of anandamine **31** for the geometry optimized by using different force fields as well as QM calculations, for vacuum conditions [herein, QM calculations mean energy minimization at the MP2/6-31+G(d,p) level (Section 5.2.3)]. Star-indexed force fields are versions of the parent force fields implemented into the MacroModel6.5 package. Dotted rectangle indicates the best results for geometrical similarity of the MM and QM structures.

Results of MM calculations for anandamide **31** (an endo cannabinoid[a]) and N-caproyl-(Gly)$_3$GABA-O-NHCH$_2$CH$_2$OH (**32**) are a demonstrative illustration. Monoamide **31** is a compound of "ordinary organic" structure (Fig. 75) while **32** is a peptidic analog (Fig. 76).[b]

There are many rotamers for each of these compounds (Section 1.6). Herein, the molecular geometry of some low energy conformer of **31** as well as for some low energy conformer of **32** (Figs. 75 and 76) have been optimized by QM calculations;[9] for simplicity, these

---

[a] Δ^9-THC (Fig. 16), the main psychoactive component of marihuana, is an exo cannabinoid.
[b] GABA: 4-aminobutyric acid; Gly: glycine. Both amino acids are inhibitory neurotransmitters in the brain. There are no suggestions in the literature that compound **32** is a naturally occurring peptide.

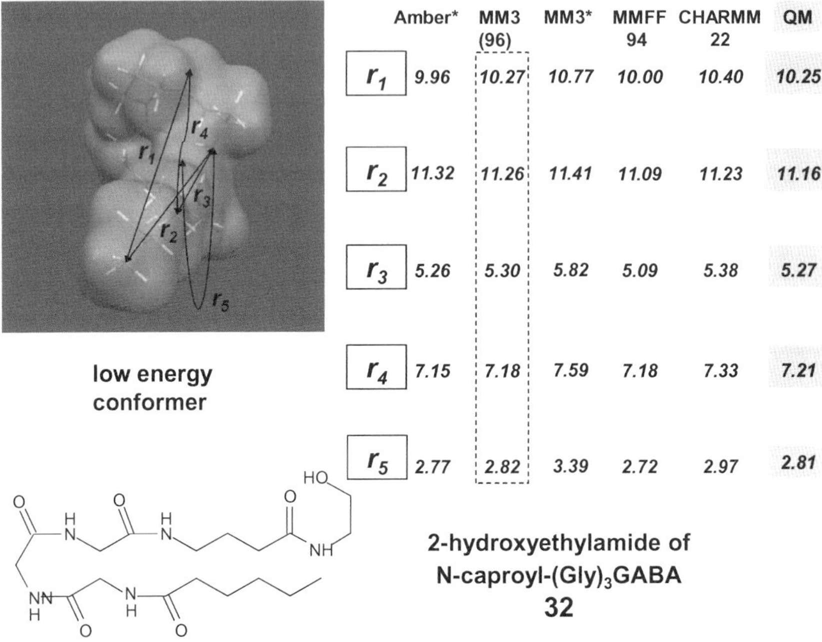

Figure 76. Selected interatomic distances (Å) for a low energy conformer of anandamide analog **32** in geometries optimized by "organic" and "biomolecular" force fields as well as QM calculations at the MP2/6-31+G(d,p) level, for vacuum conditions. Dotted rectangle indicates the best fit results.

molecules have been modeled for vacuum conditions [i.e., not affected by other (e.g. solvent) molecules].

The resulting geometries of the considered rotamers of **31** as well as **32** have been re-optimized by separately employing the Amber[a] (MacroModel's[a] version of Amber), CHARMM22, OPLS[a] (MacroModel's version of OPLS-aa), MM3(96), MM3[a] (Macro-Model's version of MM3), and MMFF94 force fields that are considerably frequently used in MM molecular modeling. Among them, Amber[a] and CHARMM22 are of peptide-targeted parameterization, and OPLS[a] is a force field that has an extended parameterization for nucleoside amino bases. The other force fields, MM3, MM3[a] and MMFF94, are universal organic force fields without specific parameters for peptides, carbohydrates or nucleosides.

Thus, we have the QM, Amber,[a] CHARMM22, OPLS,[a] MM3, MM,3[a] and MMFF94 molecular models of the considered conformer of **31** and those of **32**.[b] We can estimate the

---

[a] MacroModel: a professional program for MM calculations (Section 4.3.1).

[b] Similar modeling with employing the Sybyl force field is notable. In contrast to the others, this force field does not "think" that these conformational motifs (3D geometries of the main chain) of the

quality of these MM calculations *with regard to these structures* by comparing distances between selected (probably, essential for biorecognition) atoms in these MM-optimized geometries and the corresponding distances for the QM-derived geometry. For the "ordinary organic" molecule **31**, "biomolecular" force fields Amber[a] as well as OPLS[a] fail while "organic" MM3, MM3[a] and MMFF94 do not (Fig. 75). Nevertheless, the best fitting is supplied by the "biomolecular peptidic" CHARMM22!

Of course, "peptidic" parameters do not abolish the "general organic" part of parameterization that is present in CHARMM22. One could speculate that an excellent geometry for arachidonyl amide **31** supplied by CHARMM22 is due to a successful "general organic" parameters of this force field. Following this logics, we could not expect that "general organic" force fields would be admirable for the peptidic analog **32**. The undertaken MM calculations are a surprise: all used fine force fields show a very tight closeness of geometries by MM and QM (Fig. 76). Moreover, "general organic" force fields MM3, MM3[a] and MMFF94 provide a better 3D geometry than the "biomolecular" Amber.[a] Although the "peptidic" CHARMM22 affords for peptide **32** a 3D structure that reproduces the QM-derived structure more closely, the best fit is provided by MM3, a "general organic," peptide-not-oriented MM tool.

A convincing example of how organic and biomolecular force fields may not meet researchers' expectations is supplied by MM experience with carbohydrates. One could anticipate that no general organic force field can successfully compete with carbohydrate-oriented versions of biomolecular force fields. Ironically, benchmark studies have shown that no biomolecular force field may systematically surpass MM3[50d] that, as we remember, has been designed as an "organic" force field. Even the carbohydrate-targeted versions of CHARMM and GLYCAM06[50f] have not pushed MM3 aside, and this force field, due to the theoretically well-thought out architecture, continues to occupy the surprisingly high position in the "foreign society" of biomolecular force fields.[50e] It is unknown yet whether the recent corrections in the CHARMM course to carbohydrate modeling[50g] have led this highest rank biomolecular force field to stably precede MM3 in such calculations.

It appears that, if we are based on characteristics *organic force field*; *biomolecular force field*, we may be unable to foresee which fine force fields will model either polyfunctional chemical structure better. By imparting to a force field versatility in MM atom types, which aims to define a certain, goal chemical class as detailed as possible, force field designers do succeed in orienting this concrete MM method to an accurate modeling of chemical structures from this class. However, this approach does not set a

---

considered conformers of **31** and **32** correspond to an energy minimum: in re-optimizing the QM-derived geometry of the molecules, it appreciably re-folds the molecular backbone. These MM exercises with simple molecular structures with using different force fields indicate that Sybyl is rather not suitable for solid research.

rigid frame of applying this fine force field to such only structures. Equally, a fine force field designed with targeting some chemical class does not guarantee that any other force field which does not have an extended chemical class-specific parameterization would always supply the worst MM models of molecules from this class. MM users should understand this general specification of force fields as strong recommendations, but not as strong directives. Nevertheless, in research practice, one certainly can be guided by these general characteristics. Contemporary versions of bioorganic force fields model version-targeted biopolymers of long chain much more accurately than purely organic force fields, while the latter are more precise in modeling organic molecules of chemical functionality which differs from that "hunted" by biomolecular force fields.[a]

Thus, we can easily choose the *group* of fine force fields that are suitable for a concrete chemical structure. The next step is selection of the "best fit" force field, and it is not trivial. The problem is that, among fine force fields of the same structural specificity (i.e., oriented to modeling of molecules from the same chemical classes) one cannot point out a more accurate force field. For instance, one cannot say which force field from the "noble" families of biomolecular force fields, CHARMM, Amber, or GROMACS, models a concrete protein better. In the best case scenario, one can assert that a certain fine force field *usually* supplies more accurate models of molecules of either chemical structure than other force fields of the same general orientation. For instance, one can claim that, statistically, MM3(2000) and MM4 provide better results for neutral cyclic organic molecules of ordinary size than other force fields provide. One cannot be sure that these MM tools are superior for a concrete molecule since there is a considerable probability that another force field, e.g., MMFF94, will turn out more accurate for this chemical structure. Thus, the organic experimentalist who tries to "pick up" the proper fine force field by solely comparing known molecular models delivered by different force fields may view the problem of force field selection from the probability perspective: *in selecting the MM tool by solely assessing force fields (and not experimenting with them), one can*

---

[a]Note that, in the last decade, specialists in computational chemistry merely have not used force fields of "wrong" applicability in modeling biopolymers or small organic molecules.

select it with only fair probability of its superiority with regard to concrete chemical structures.

What does this *fair probability* mean in practical aspect? If an appropriate force field should be selected for molecular modeling of "conventional" organic compounds, i.e., compounds that are not biopolymers, sharply chemical class-oriented force fields, e.g., peptides/protein-targeted Amber-ff03, Yamber2 (a version of Amber-ff99), ECCEP-05, may be immediately excluded from consideration. The probability that such force fields will supply better results than MM3 or other "organic" force fields is low, and, re-phrasing, one can consider these chemical class-specific force fields as unsuitable for modeling of, in particular, carbo and heterocycles as well as polycyclic systems. Similarly, fine force fields of general organic purpose may not be taken into account if a highly polar charged biomolecule in solution, e.g., a peptide in water, is the modeling object. The accuracy of these MM tools appreciably yields to the quality of models that are supplied by peptide-targeted versions (e.g., CHARMM22, PFF02 or Amber-ff03) of "biomolecular" force field families. Sometimes, one may be certain regarding a concrete force field. For instance, CGenFF, a general purpose organic force field, cannot be successfully applied to biomolecules at all. A number of specific parameters of parent CHARMM that are related to peptides, carbohydrates, nucleic acids and carbohydrates have been removed from CGenFF.[49e]

Thus, non-computational chemists are situated uncomfortably when trying to understand the accuracy of fine force fields without experimenting with them. Let us run through a quick summary of what was discussed above. The specificity of each force field is embedded in atom types, mathematical forms of equations and parameter values. These inner variations are invisible for MM users and, more importantly, their effect on the quality of models of various organic molecules is predictable only in terms of qualitative probability (probability within some qualitatively characterized range). Force field labels of kind "for proteins," "for peptides," "for carbohydrates" and "for organic compounds" may assist in rejecting force fields that are *most likely* to be unsuccessful in molecular modeling of concrete organic systems (see Table 2 taken from Ref. 49f). No preliminary analysis may indicate which force field among fine force fields with the same nickname *organic* or *biomolecular* is more accurate for modeling the "organic"

Table 2. Sample mapping from probability ranges to verbal expressions of uncertainty (verbal probabilities were derived from empirical studies).[49f]

Probability range	Adjective	Adverb
0.0	*impossible*	*never*
0.0–0.1	*very unlikely*	*very rarely*
0.1–0.25	*unlikely*	*rarely*
0.25–0.4	*fairly unlikely*	*fairy rarely*
0.5	*as likely as not*	*as often as not*
0.5–06	*more likely than not*	*more often than not*
0.6–0.75	*fairly likely*	*fairly often*
0.75–0.9	*likely*	*commonly*
0.9–1.0	*very likely*	*very commonly*
1.0	*certain*	*always*

or "bioorganic" molecular object of interest, respectively. The information regarding a successful use of a force field for certain chemical structures does not guarantee transferability of its accuracy to their structural analogs to be analyzed. That is, only qualitative expectations of kind *fair*, *likely*, etc. are our prerequisites. With such loose estimates of relative accuracy of force fields in hand, how to proceed with selection of the "best fit" force field?

We definitely require the selected force field to treat given chemical structures more accurately than other force fields treat them. The way out of the situation of intolerable uncertainty is simple: one has *to examine* the accuracies of alternative force fields with respect to concrete chemical compounds to be studied. Simply put, one requires to model molecules of two or three probe structures selected from the series of compounds of interest by means of both MM and QM calculations. The quality of MM calculations with regard to any "ponderous" structure(s) is revealed by comparing results, which are supplied by each examined force field, with results of sufficiently accurate QM calculations [i.e., comparing molecular geometries supplied by MM *vs.* those supplied by QM as well as ΔE_{ster} *vs.* ΔE_0 values for either stereoisomers modeled for the same conditions (usually, the temperature and the medium); Section 4.2.6].

In practice, one requires to have a 4–5 member set of fine force fields that includes as late as possible versions of both organic and biomolecular fields, say, MM3, MMFF94, CHARMM, OPLS-aa and Amber, as well as a computational package that can provide a sufficiently broad spectrum of QM calculations, e.g., GAUSSIAN, Q-Chem, Turbomde, Jaguar or ADF. Such a set permits to non-computational chemists to select a suitable force field and, using it, to obtain trustworthy MM models of organic molecules which chemical structure does satisfies the selected MM calculation tool.

In research routine, examination of the accuracy of force fields[a] by QM calculations is quite possible because nowadays energy minimization by a satisfactory accurate QM method takes a reasonable time (say, 2–5 days of running on an eight processor computer when minimizing energy for a $\sim$30 atom system). Using even a lower performance computer,[b] MM minimization procedures are completed within the timescale of seconds. Thus, values of ΔE_{ster} for two stereoisomeric structures of such molecular system for vacuum or solution conditions are obtained in a few minutes, for several force fields.

This methodological requirement — to examine or to verify the force field quality for concrete chemical structures to be modeled by MM — also says that published MM calculation results for hundreds of functionalized organic compounds are of doubtful validity. Merely, this requirement (often demanding to perform additional calculations) has not been fulfilled in many chemical studies. A typical reasoning in such publications is: "Force field calculations show that the ring has a distorted conformation of ... Therefore ...", or "MM energy calculations show that the major conformer is ... Therefore ...". This "therefore" is unbased. One has to remember that, if applied to chemical structures which analogs of the same chemical functionality and the same fusing of rings have not been modeled in reliable (i.e., verified) MM computational experiments, MM modeling of molecular geometry and MM estimates of relative stability should be considered as *conditionally* accurate. The condition is an experimental or QM-provided proof of molecular models supplied by MM. Such a proof may be supplied by pattern structures that have

---

[a]We assume that several fine force fields are available to the researcher.
[b]One may mention even tablet PCs. iSpartan (Wavefunction, Inc.) is an elementary, freely distributed molecular modeling program for running on iPad tablets. However, its MM module includes only MMFF94. YASARA's MM and MD programs (commercial) can be run on some Android-operated tablets (see website *http://www.yasara.org/android.htm*).

molecular skeletons of both the same chemical functionality and the same fusing of rings. Of course, when undertaking molecular modeling by MM, researchers should follow this methodological requirement of proving their modeling. *Results of non-verified MM calculations obtained for non-elementary chemical structures are an insufficient basis for further conclusions.*

One should stress that, due to reasoned quantitative relationships *geometry–energy* (Section 4.2.1) associated with multiple thoroughly selected and correctable empiric parameters of fine force fields, MM calculations are absolutely not speculations, in contrast to primitive "tools" of 3D structure probing (Sections 1.5 and 3.4) practiced in experimental organic laboratories. Molecular structures supplied by MM and not examined by comparing them with experimental or QM-derived structures have to be viewed as fairly likely candidates (Table 2) to be successful molecular models that should be verified and not be neglected straightaway.

4.2.6 *Quantitative estimates in force field selection*

The accuracy of force fields is not a single-valued meaning. It may be related to distinct quantitative molecular features calculated by MM (geometrical parameters, heat of formation, relative stability, vibrational frequencies), and the accuracies in calculating these distinct molecular parameters may be different. Any of these quantitative characteristics may be used for comparison of MM results with the "etalon"; which of them should be used in verifying MM calculations?

Molecular geometry is the first indicator of the quality of MM calculations. In addition to the accuracy for the geometry itself, it reflects the accuracy of these calculations for all molecular quantities "modelable" by MM. If the 3D molecular geometry supplied by a force field is not accurate for a molecular system, one cannot expect that this MM calculation tool would be satisfactory in modeling other molecular properties, e.g., E_{ster} or vibration frequencies of this system. However, this mirror (the accuracy in molecular geometry) is not perfect. An accurate MM-delivered molecular geometry does not necessarily indicate that other characteristics of the molecule are modeled well.

In starting an MM study of a non-trivial chemical structure or chemical class, we need to know which force field, say, from two force fields **A** and **B**, is more applicable. We use molecular geometry as the first, convenient indicator of better accuracy for the molecular structure(s) to be studied. In other words, in the initial stage, we perform MM calculations using both **A** and **B** and applying them to the same stereoisomer. The 3D geometry of a molecular system provided by different force fields is not identical. A considerable discrepancy in MM modeled molecular geometries shows straightaway which of the force fields, **A** or **B**, is less accurate and should be abandoned when modeling the considered molecular system.[a]

Sometimes significant discrepancy is observed for molecular geometries when applying such force fields to the same 3D structure.[b] This discrepancy may not indicate that the conformer, which is located by force field **A**, is absent in the conformational space modeled by force field **B**. The SES modeled by **A** and the SES modeled by **B** may have the same general shape, but are shifted one relative to the other along a geometry coordinate axis (Fig. 77). Geometry optimization of the initial structure by force field **A** leads the molecule "to fall" into the nearest energy minimum. Because of the mutual shift of the two SESs, the initial (pre-optimized) structure may be nearer another energy minimum in the **B**-modeled SES, and energy minimization with using force field **B** "pushes" the molecule into this minimum.[c]

It may occur that different force fields supply satisfactory, very similar 3D geometries for a molecular structure. In SES terms, the two modeled SESs are "horizontally" shifted to some insignificant extent one relative to another. In this case, molecular geometry is not a successful criterion in selecting a more accurate force field. Accurately MM-modeled 3D geometries of conformers cannot guarantee that the obtained values of steric energy difference ΔE_{ster} estimate the conformational equilibrium well.

What is the molecular characteristics that is a more susceptible indicator of quality (in other terms, accuracy) of MM modeling than molecular geometry? Figure 77 reflects the typical situation with SESs, each of which is modeled by a distinct force field: the related hills in these energy landscapes are of somewhat different height. That is, while providing realistic

---

[a]As explained (Section 4.2.5), this may be cleared by comparing the MM-modeled geometries with the corresponding results of QM modeling.

[b]i.e., force fields **A** and **B**, in optimizing the same initial molecular geometry, "produce" different stereoisomeric structures.

[c]This slippage to another energy minimum caused by replacing the force field with another one is highly probable for organic systems that have conformers of similar 3D geometry, e.g., long chain-containing compounds.

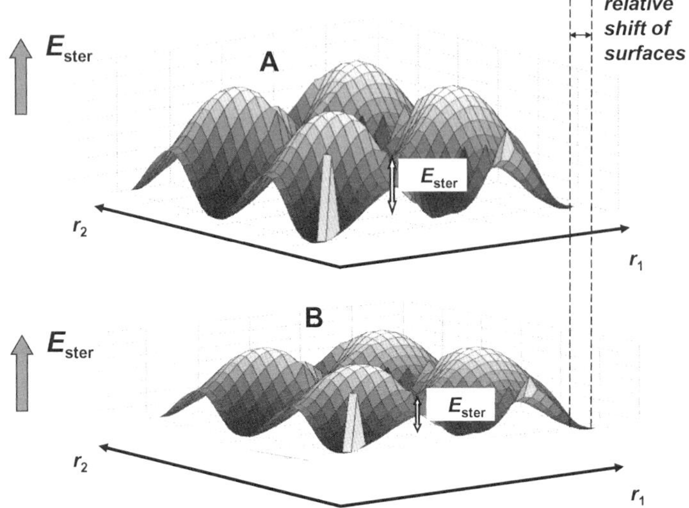

Figure 77. A typical situation with two SESs (indicated as SES_A and SES_B) each of which is modeled by using a different fine force field: the SES_A and SES_B have a similar general shape, are somewhat shifted one relative to the other (r_1 and r_2 are parameters of molecular geometry), and the energy differences ΔE_{ster} in the SES_A and SES_B are not equal for any pair of the same stereoisomeric structures.

molecular geometries for stereoisomers, fine force fields differ in calculating values of steric energy E_{ster} (Section 4.2.3) and, consequently, rather often supply different values of ΔE_{ster} for the same pair of stereoisomeric structures. It is obvious therefore that energy estimates ΔE_{ster} are a sensitive probe for the force field applicability. The more accurate force field better reproduces experimental values for relative stability of stereoisomers or experimental values of conformational barriers.

Let us glance at the example of (+)-sparteine (**33**, Fig. 78).[a] Sparteine rings are flexible even at significantly decreased temperatures.[52a] Among the lowest energy conformers, the chair-boat-chair-chair conformer is highly preferable, as QM calculations (Fig. 78) and NMR experiments show.[36e,52a,b] The molecular geometry of this conformer is practically the same if optimized by QM or MM [e.g., by MM2(91), MM3(96) and MMFF94 force fields; Fig. 78]. This is true also for the all-chair conformer.

---

[a]The enantiomer of **33**, (-)-sparteine, is widely used for stabilization of Li carbanions in stereoselective additions.

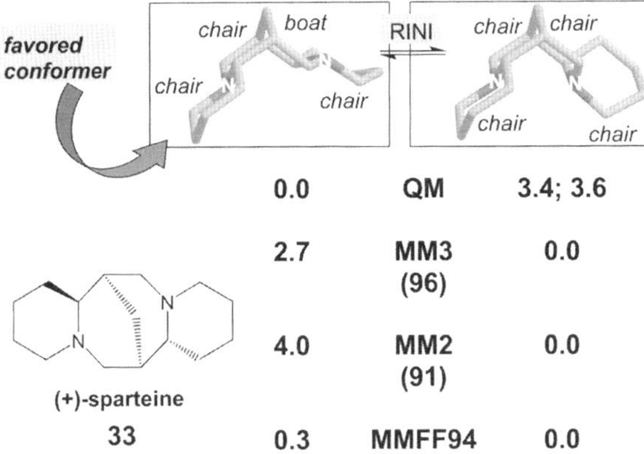

Figure 78. Two most stable conformers of tetracyclic diamine **33**. Their relative stability is represented by values of potential energy ΔE_{el} calculated by QM (kcal mol^{-1}; Refs. 36e,51a). Values of ΔE_{ster} supplied by the considered force fields (kcal mol^{-1}; Ref. 9) mislead: they show a reverse (incorrect) stability order for these conformers.

However, MM calculations[9,52a] indicate a reverse stability order for these conformers. Moreover, the first two force fields demonstrate even a high predomination of the all-chair conformer. Of course, this is an incorrect result. We have to conclude that these force fields, despite a satisfactory representation of molecular geometry, are very inaccurate in modeling the molecule of diamine **33**. This systematic inaccuracy indicates that, in modeling this system by MM, some intramolecular interactions (see below) are taken into account by neither the force field-building equations (Section 4.2.2), or MM atom type sets (Section 4.2.1).The systematic fault of force fields in treating chemical structure **33** suggests to organic researchers that, instead of MM in the form of standard (additionally not parameterized[a]) force fields, QM calculations should be applied to polycyclic systems, which bear lone electron pair-possessing atoms in spatial proximity, instead of a quick modeling by a force field. In principle, computational chemists may modify a fine force field with targeting any certain narrow class of organic chemical structures.

Experimental values of free energy differences for conformers frequently are not available. Then, the assistance in selecting the suitable force field is provided by QM calculations. Let us assess the quantitative comparison of

---

[a]In principle, computational chemists may modify a fine force field with targeting any certain narrow class of organic chemical structures. Fortunately for organic experimentalists, such laborius work is not required: molecules of size of polycycle **33** may be modeled by QM methods within a reasonable time [e.g. in running routine DFT calculations (Section 5.2.6) overnight].

MM and QM estimates of relative stability of stereoisomeric structures. In this force field examination, geometry optimization by a sufficiently accurate QM method (e.g., MP2; Section 5.2.3) with calculating total molecular energy E_0 (Section 5.3.4) should be performed at least for two stereoisomeric structures, for vacuum conditions; similarly, these structures should be modeled with using each examined force field. These computations supply us with two QM-derived molecular geometries and one corresponding ΔE_0 value (Section 5.3.4), together with $2n$ molecular geometries and n values of ΔE_{ster} delivered by n force fields. If (*1*) the QM- and MM-modeled molecular geometries of both stereoisomers are practically the same; and (*2*) the absolute value $\Delta\Delta_{abs}$ of the difference $\Delta\Delta$ between ΔE_{ster} and ΔE_0 ($\Delta\Delta = \Delta E_{ster} - \Delta E_0$ and $\Delta\Delta_{abs} = |\Delta\Delta| = |\Delta E_{ster} - \Delta E_0|$) for two stereoisomers is less than 0.5 kcal mol^{-1}, one can characterize the examined force field as a suitable MM tool for treating this chemical system. If the $\Delta\Delta_{abs}$ value is less than 0.2 kcal mol^{-1}, the force field may be accepted as an excellent MM tool for modeling molecular systems of this chemical class.[a]

Of course, one obtained small value of $\Delta\Delta_{abs}$ may be an unreliable criterion. The more MM-derived structures are compared with the related QM-produced "etalons' (values of ΔE_0), the more reliable the choice of the "best" force field is. One can accept that, if differences $\Delta\Delta$ have the same sign (either positive or negative), three values $\Delta\Delta_{abs} \leq 0.5$ kcal mol^{-1} for three stereoisomers justify the use of the corresponding force field for the series of the related chemical structures. That is, MM modeling with using this force field becomes reliable for structural analogs of the examined molecule which have the same functional groups (including unsaturated units) in the same relative positions in acyclic or "no cage, no strain, no bridgehead heteroatom" polycyclic frameworks. The MM beginner may realize that, by "authorizing" a fine force field, the QM-mediated examination of MM results for one or two chemical structures opens the way for making quickly available reliable molecular geometries and relative stabilities for conformers of each of hundreds of their close structural analogs.

---

[a]Strictly speaking, we do not know whether the QM-supplied molecule is of high or low accuracy. However, the key assumption in examining force field quality by QM is that the concrete QM model is sufficiently accurate when supplied by a QM method of generally proven accuracy.

$\Delta\Delta_{abs}$ is not a strict thermodynamic estimate. It is a difference of two different quantities that result from certain, but absolutely different theoretical approximations (Sections 4.2.1, 5.2.2 and 5.2.5). Therefore, in principle, calculated values ΔE_{ster} and ΔE_{el} or ΔE_{ster} and ΔE_0 should be unequal ($\Delta E_{ster} \neq \Delta E_{el}$ and $\Delta E_{ster} \neq \Delta E_0$). Situation $\Delta\Delta_{abs} = 0.2$ kcal mol^{-1} does not mean that the deviation of the ΔE_{ster} value from the $\Delta G°$ value (the Gibbs energy difference for the real stereoisomers) is 0.2 kcal mol^{-1}: $\Delta G°$ may be either somewhat higher or somewhat lower than 0.2 kcal mol^{-1} or equal to this value. Situation $\Delta E_{ster} = \Delta E_{el}$ or $\Delta E_{ster} = \Delta E_0$, if occurs, does not show an absolute accuracy of the examined force field. Both ΔE_{el} and ΔE_0 values have their own inaccuracies. The situation $\Delta E_{ster} = \Delta E_0$ means that the inaccuracy of the MM model is incidentally equal to that of the QM model.

$\Delta\Delta_{abs}$ value of 0.5 kcal mol^{-1} may be viewed as a threshold for the acceptance of the force field for modeling the conformational equilibrium of this compound. Of course, there is no rigorous basis for choosing this numerical value. This value of $\Delta\Delta_{abs}$ is not deduced from any equation; we merely choose this value. Why could we not set up, e.g., the threshold value of 0.2 kcal mol^{-1}? The point is that MM calculations cannot stably provide this high accuracy $\Delta\Delta_{abs} \leq 0.2$ kcal mol^{-1} in estimating $\Delta G°$ for conformers of any "MM-digestible" chemical structure, while the 0.5 kcal mol^{-1} value of $\Delta\Delta_{abs}$ is achievable.

Moreover, this 0.5 kcal mol^{-1} value of $\Delta\Delta_{abs}$ is still useful in estimating relative stability of stereoisomeric structures. Indeed, molecular structures, which are *considerably* represented in the thermodynamic equilibrium for the room temperature, lie in the free energy range[a] of only 2 kcal mol^{-1}. In other words, for two conformers at ambient conditions, this value of free energy difference $\Delta G°$ means 96% of predomination of one conformer over the other. Therefore, the 1.0 kcal mol^{-1} value of $\Delta\Delta_{abs}$, a half of this energy window, deprives MM estimates of carrying worthy information regarding the quotient of each conformer. Suppose that MM calculations give equal fractions of two conformers at the room temperature ($\Delta E_{ster} = 0$ corresponds to the ratio 50:50), while the $\Delta\Delta_{abs}$ value obtained in examining the used force field is 1.0 kcal mol^{-1}. The uncertainty for the relative amount of the conformers is intolerably large: the maximal estimated population of one of the conformers from this pair

---

[a]i.e., the range of $\Delta G°$ values that characterize their populations (Eq. 13, Section 3.3).

is 84% (calculated for $\Delta G° = 1.0$ kcal mol^{-1}, using Eq. 13), and, hence, the ultimate MM estimate for its population lies in the range 50–84%. It is clear that this spread in the conformer ratio values — from equal stationary concentrations of the conformers to a high predomination of one of the conformers — eliminates the idea of quantitative conformational studies. The considered criterion in force field selection ($\Delta\Delta_{abs} = 1.0$ kcal mol^{-1}) is weak, and the MM-provided conformer ratios, due to their inaccuracies, may be useless in conformational considerations.

In the same situation with the MM-supplied ratio (i.e., 50:50), the 0.5 kcal mol^{-1} value for $\Delta\Delta_{abs}$ gives for the population of the conformer the range 50–68% that is almost twice as smaller as the previous range. Even translated into the qualitative language, this MM estimate — equal or moderately larger fraction of one conformer — is telling. Note that the $\Delta\Delta_{abs}$ value of 0.2 kcal mol^{-1} is not an often event in modeling 30–40 atom functionalized organic molecules by MM. If the preliminary criterion is very ambitious (i.e., $\Delta\Delta_{abs} = 0.2$ kcal mol^{-1}), it is likely that no force field would pass the test of three values $\Delta\Delta_{abs} \leq 0.2$ kcal mol^{-1} where each value $\Delta\Delta_{abs}$ is obtained for a separate pair of conformers. Therefore, the $\Delta\Delta_{abs}$ value of 0.5 kcal mol^{-1} seems a satisfactory threshold in selecting the force field for conformational analysis of either chemical structure.

If several structures are subjected to MM calculations, one can trap and remove the *systematic* inaccuracy of these calculations. The systematic inaccuracy for a series of conformers is uncertain in estimating ΔE_{ster}; it may be from one-tenth to several kcal mol^{-1}. Regardless to its magnitude, it may be eliminated if we have sufficiently large sets of both ΔE_{ster} and ΔE_{el} values for the modeled molecule (i.e., a sufficiently large set of modeled conformers; say, at least five ones). Applying regression analysis to these sets, one can see whether values ΔE_{ster} and ΔE_{el} do correlate (for linear regression analysis, see Section 6.2.2). If there is a good correlation, corrected values ΔE_{ster} are free from the systematic calculation error; in addition, they are the absolute error value-invariant. However, in this way the easiness of MM calculations is lost. To obtain an extended set of ΔE_{el} values, one must minimize energy, using QM methods, for several (at least five, as mentioned) conformers of the system of interest. Then, there is no need in MM calculations for these concrete structures at all.

Notice, the above comparisons are between MM and QM results for molecular systems modeled for vacuum conditions. However, in organic laboratory, one deals mostly with solutions, including determinations of relative thermodynamic and kinetic stability of molecular species at equilibrium, with exploiting DNMR or another experimental method. Then, the use of vacuum conditions in computational modeling imparts an additional inaccuracy to molecular models. Should MM and QM models for solution be employed instead of simplified models for vacuum conditions, when selecting a suitable force field or verifying MM results?

At first glance, comparison of MM- and QM-modeled solvated molecules is a better verification. Nevertheless, scarcely one should often undertake it. Solvation is a complex dynamic process and the solvation shell even in static projection has an intricate, unknown in details structure. From the viewpoint of molecular modeling, system *solute–solvent* is incomparably more complex than a single molecule in vacuum. Surrounding this abstract molecule with molecular environment, we actually create a new model with an increased uncertainty in accuracy. Without deep, systematic studies in each concrete case, it would be unclear which inaccuracy is fatal for our model, that of approximating molecules in solution by molecules in vacuum or that of imperfectly modeling solvation, or both of these inaccuracies, or none of them. For reasonable simplicity, in undertaking conformational analysis of small organic molecules (that are not polymers or oligomers), it is more practical to assume that the vacuum model does represent their conformational behavior at least for non-polar non-liganding solvents.

Solvation models in MM utilize principles of classical physics. Models that take the solvent as a certain macromedium not dividing it into substructures are called continuum solvation (implicit) models. Approaches, which represent the solvent as a set of molecules or other abstract structures, are called explicit solvation models.

The simplest continuum solvation model includes values of relative permittivity ε in Eq. 24 (Section 4.2.2) replacing $\varepsilon = 1$ for vacuum with the ε value for the solvent (e.g. $\varepsilon = 81$ for water). Obviously, this decrease of the electrostatic energy term is a very rough "trick."

The generalized Born solvent accessible model (GBSA/MM) is the most ubiquitous continuum solvation sketch. It is of much higher quality than the above indicated model of the medium; remarkably, it does not significantly extend computation time. The solute molecule is approximated by a solid shape formed by atoms of certain effective radii (Born radii), the solvent is assumed to be a uniform dielectric with dielectric constant ε. This model adds to E_{ster} of the molecule in vacuum an energy term E_{solv} which reflects the energy of interactions of this solute with the surrounding medium. E_{solv} is a sum of three terms E_{cav}, E_{VDW} and E_{pot}. E_{cav} represents the energy of formation of the cavity in the solvent bulk for occupying it by the solute. E_{VDW} approximates the energy of VDW interactions between the solute and solvent molecules. E_{pot} accounts potential energy of electrostatic interactions between these molecules. It is accepted that $E_{pot} = 0$ for an alkane solute. Not giving mathematical expressions for these terms, one should note that they are algebraic functions that include parameters of the solute and the solvent. Among the parameters, there are s (for a sum of E_{cav} and E_{VDW}) of the model atom of type i of the solute, q_i (a formal charge of the i-th atom of the solute), ε (dielectric constant of the solvent) and a_i (the effective Born radius of the i-th atom of the solute; this value is defined differently in different approximations). Parameters $r_{i,j}$ (the distance between the i-th and the j-the atom of the solute) and SA_i (solvent accessible area of the i-th atom of the solute) characterize 3D geometry of solute molecules. Notably, atomic charges q_i may be taken from any reasonable molecular model. Therefore, good QM-based models of charge distribution supply the best values of q_i and in this way significantly improve the quality of the GBSA/MM modeling.

It is transparent that this model ignores specific solvation effects (e.g., H-bonding or stacking interactions of aromatic rings) as well as mutual polarization of solute and solvent molecules. For GBSA/MM, the calculation error in estimating solvation energy may be near 1 kcal mol^{-1} for neutral organic molecules and near 4 kcal mol^{-1} for charge-carrying molecules.[50b,54] However, one may not be too upset because of this significant inaccuracy. In predicting relative stability of conformers, we need *difference* ΔE_{ster}. It is reasonable to assume that the calculation error due to imperfect modeling of solvation is diminished because of subtracting the absolute value E_{ster} for a conformer from that for another conformer.

Nonetheless, it is only an assumption. Usually, we have no information about energy difference in solvation of distinct stereoisomers. Results of MM calculations for conformational equilibrium in solution may be quite useless, if stereoisomers have a dissimilar solvation shell, e.g., because of the presence of intramolecular H-bonding only in a certain stereoisomer. Then the calculation error of solvation energy may be essentially different for this structure and structures with free H-donating and H-accepting moieties exposed to the solvent.

In explicit solvation models, a single molecule is placed in an abstract box (with dimensions not less than $2 \times 2 \times 2$ nm) filled by modeled solvent molecules or their truncated models; the number of such solvent molecules is large (several hundreds or more). There, to reduce the cost of computations, solvent molecules are rigid (not subjected to energy minimization), or, as mentioned, structurally "simplified." Their molecular structure is reduced either to spheres possessing the Lennard-Jones potential as well as dipole moment (e.g., TIP3P and TIP4P models of water molecules) or to MM united-atom models of molecules (for organic solvents). Sometimes polarity of these artificial molecules is augmented with an induced dipole moment; in this case, they are polarizable by their neighbors. In "constructing" these molecular counterparts, a prudent parameterization is performed until this abstract medium well reproduces bulk properties of the corresponding real solvent (density, heat capacity, etc.).

The energy of a whole system *solute–solvent* with an arbitrary arrangement of molecules is calculated using a fine force field. Then this arrangement is changed either randomly or by means of MD simulations, and the energy of the whole system is re-calculated. Multifold recurrence of this procedure gives a large set of energy values that represent the state of the system at different time moments. Averaging them, we get the energy value that actually characterizes this dynamic system *solute–solvent*. For regular organic molecules, this explicit modeling is approximately of the same accuracy as GBSA/MM is.[50b]

Thus, equilibria in low polarity, non-ionizable solvents are a convenient case for estimating conformational equilibrium of molecules in solution by MM: to some extent, the gas phase "imitates" a bulk of such weakly solute-binding solvents. Therefore, MM calculations for vacuum conditions may be used in modeling such conformational equilibria (i.e., estimating relative stabilities of conformers in solution). However, if results of MM calculations are intended to model molecules in a polar solvent or solvent with either specific intermolecular interactions (e.g., H-bond-forming, lone electron pair- or π-donating solvents), a solvation model should be used. Also, solvation energies are significant for large molecules, and separately insignificant local discrepancies in solvation of stereoisomers ultimately may result in a considerable difference in free energies of solvation of the entire molecules. This difference may be sufficiently large to render MM calculations for vacuum irrelevant for estimating relative stability of these 3D structures in solution.

This circumstance hampers selection of the appropriate force field. We do not know whether a poor accuracy, if obtained for fine force fields in modeling a highly polar or ionized molecule in a polar solvent, is because of their inadequacy to the chemical structure of the modeled molecular system or because of the failure of the employed solvation model. An additional examination which would deal with the quality of either solvation model is, of course, possible. However, it would transform routine selection of the suitable force field into a separate laborious, long-term benchmark study. MM loses its attractivity of being a fast and easy research tool. The practical issue from this situation with ordinary size organic molecules may be guessed: when modeling molecules dissolved in such solvents, one may leave MM at the beginning and start with QM calculations using an appropriate solvation model (Section 5.3.4). Then at least one factor of uncertainty, the quality of molecular modeling by means of force fields, is removed.

4.2.7 *When not to use molecular mechanics*

The usual manner in which researchers think about computational conformational analysis is to try MM. However, the reader has figured out that MM, as a pool of finished force fields, is limited in modeling organic molecules by chemical structures that are covered by these force fields. Could one *a priori* identify problematic cases and prevent a useless search of the proper force field and, more importantly, escape "mismodeling" of such molecular systems?

This question is not for computational chemists. They know which structures have been used to define model atoms in different force fields and how successful their equations are. For instance, they know that the most widely used MM tool of pure organic chemists, late versions of MM3, contain a special term as well as special parameters for cyclopropanes but have no similar "key" to aziridines. Therefore, they would not treat by MM3 systems, which bear the aziridine ring, or would originate high quality parameters for such systems, if necessary. Professional erudition of computational chemists and experienced MM users allows them at first sight to discard compounds that are improper subjects for MM modeling by either force field.

However, for beginners, one could specify three general structural features (*1*)–(*4*) of organic compounds that may result or definitely result in unacceptable inaccuracy of MM calculations. The presence of one of these features is sufficient to decline the option of MM calculations for the studied molecular structure and to immediately re-orient the modeling to the QM option. Otherwise, each MM-generated structure should be examined via QM calculations what is equivalent to conformational analysis solely by QM.

(*1*) *Non-covalent bonding.* Not considering electrons in the molecule, MM does not model any molecular characteristics explicitly associated with, e.g., valence electrons. Thus, in modeling molecular systems, MM does not recognize chemical bonding; it models molecules from "chemical templates" (MM atoms: Section 4.2.1) included into either force field.[a] As we know, accurate modeling of molecular properties by a force field is

---

[a] That is, when modeling a molecular system, MM does not analyze what chemical structure it actually has. Only molecules of given chemical structure are subjects for MM calculations; if the used chemical structure is untrue, these calculations do not correct it.

achieved by tuning up its equations and parameters in these equations so that they reproduce these properties for a calibration set of simple (probe) structures of known chemical structure.[a] While covalent bonds (both σ- and π-bonds) between different "organic" atoms are more or less represented in such calibration sets (as MM atom types), non-covalent bonds, e.g., multicentered dative and electron-deficient bonds, ylide bonds,[b] are rather absent there. As a result, this empirical approach has included in MM a few types of chemical bonds. Probably, only canonic aromatic bonds (bonding via delocalized π-electrons in planar rings formed by σ-bonded atoms), ylide type bonds $S^+\!-\!O$ in sulfones and sulfoxides (in obsolete notation, $S=O$) as well as $P^+\!-\!O$ in phosphates, phosphonates and phosphinates (traditionally but incorrectly indicated as double bond $P=O$) are non-covalent bonds that are implicitly embedded in fine force fields. Other, rarer fragments with non-covalent bonds are not covered by original versions of these force fields. Therefore, it is meaningless to apply MM calculations to molecular systems that have fragments with non-covalent bonds, excluding structures from the above mentioned chemical classes.

Many of such MM-indigestible chemical structures are readily recognizable. One could recall that chemical connectivity in, e.g., delocalized carbanions, carbonium ions, carboranes, dimers of boranes, classical ylides, phosphazenes, includes non-covalent bonding. Sometimes, in order to recognize non-covalent bonding, one should overcome inertia of "thinking covalently" and discern that, e.g., widespread notation of phosphazenes $X_3P=NR$ misleads by hiding a ylidic bond between P and N. However, one cannot deliberately identify non-covalent bonding in many molecular structures.[c]

---

[a]This "non-physical" adjustment, *in principle*, allows to MM to model any chemical bonding.

[b]One cannot relate bonds, which are partially represented by a ionic bond $X^+\!-\!Y^-$ between σ-bonded atoms X^+ and Y^-, to covalent bonds.

[c]Any chemical bond actually is an extremely oversimplified implicit *model* of interactions of valent electrons of atoms in the molecule. QM theory explicitly deals both with electrons and nuclei (Section 5.2.2) and provides the physically true molecular model, a 3D nuclear backbone with a certain electron density. Additional QM techniques may "translate" this structure into the unphysical, but absolutely commonly used language of chemical bonds, i.e., may assign the relevant chemical bonding to any molecular fragment. For this purpose, QM calculations oversimplify the QM-modeled "no bond" molecular structure by using either orbital concept, e.g., that of natural bond orbitals.

[4n]Annulenes, e.g., [16]annulene (Fig. 17), are a pretty good example. These non-planar cycles are very simple structures in light of the 4n+2 Hückel criterion of aromaticity (n is the number of π-electrons of ring atoms): they are cyclic covalent structures with alternate double bonds and therefore seem to be routine objects for MM. However, a non-covalency unknown to MM is hidden in these molecules. Moreover, it first appears then disappears: this aromaticity is present in some stereoisomers of these cycles and is absent in others.[53a] This aromaticity is the Möbius 4n aromaticity (resulted from a Möbius strip topology of p-orbitals of ring atoms[53a,b]). The Möbius aromaticity is not revealed by existing force fields because they have not been "trained" for non-planar aromatics of Möbius topology of p-orbitals. Therefore, any force field would not relate aromatic stereoisomers of [4n]annulenes to energy minima, i.e., the current MM is absolutely irrelevant for these compounds. Another example is dialkylphosphite anions. QM calculations show that the relevant chemical structure is an O,P-delocalized anion (Fig. 27). Obviously, no current force field can model molecules of these classical phosphorylating agents.

(2) *Spatial proximity to electron-rich fragments.* Interactions of valence electrons of non-bonded molecular fragments (through-space intramolecular interactions) are of considerable energy if the participating fragments are spatially close. These electronic effects are chemical group-specific; they are known as, e.g., $\pi-\pi$ interactions of aromatic rings, σ-bond–σ-bond interactions, lone electron pair–π interactions, anion–π, cation–π, lone electron pair–lone electron pair, and lone pair–σ-bond interactions. Standard versions of many fine force fields do not take into account these interactions of non-bonded fragments at all.[a] Some force fields calculate energy contributions of either through-space interactions. For instance, CHARMM allows for $\sigma-\pi$ interactions, cation–π interactions as well as stacking $\pi-\pi$ interactions of canonic (not Möbius) aromatic rings. Nevertheless, one cannot rely on MM when modeling molecular structures where non-bonded electron-rich fragments are near in space. Long-range (through-space) interactions of their electrons are rather outside of force

---

[a]*Additional* parameterization succeeds to reproduce energy contributions of some non-covalent interactions by equipying model atoms with additional charges. Then "artificial" dipole-dipole interactions mimic the required energy contribution.

field parameterization or, at least, related parameters of a force field are not of high quality. Examples below illustrate to which extent MM models may be disturbed for molecules that have such spatially proximal fragments.

"General organic" force fields flunk when applied to tetracycle **33** (Fig. 78, Section 4.2.6). MM incorrectly estimates the relative stability for conformers of this diamine. The reason is simple: force fields do not "see" a strong repulsive interaction between the N and N' lone electron pairs in the all-chair conformer. The nitrogen pyramids are proximal and oriented essentially toward each other in this 3D structure. Such a rare orientation obviously is absent in structures from the original calibration sets of these force fields.

Important compounds, carbohydrates, somehow are not suspected by many MM users as "insidious" objects. However, these biomolecules are not that structurally trivial for MM because two of their oxygen atoms, the ring O and anomeric O, are geminally positioned. Interactions of their proximal lone electron pairs are poorly treated by widely used versions of fine force fields, and, thus, MM may fail in modeling these molecules.

For instance, the geometry provided by MM3(96) for the minimal energy conformer of 1-O-methylated β-L-rhamnose (**34**; Fig. 79) deviates from the QM-supplied geometry. Both Amber-ff94 and MMFF94 afford the geometry that is similar to that by MM3. The bond C-1–1-O (bond between the anomeric carbon and anomeric oxygen) is bent over the ring plane (specified by atoms C-2, C-3 and C-5) in the MM model while this bond is perpendicular to the ring plane in the QM model. This deviation is because the QM-modeled pyranose ring is more flattened than that produced by MM. In other words, this pyranose ring is a normal chair by "standardly thinking" MM, while it is a distorted chair by "flexibly thinking" QM. Nevertheless, one cannot claim that carbohydrates are exclusively objects of QM modeling. This example only indicates that, for these chemical structures, one should verify molecular geometry and energy characteristics if obtained with using either force field.

(3) Heteroatomic systems with distorted elementary molecular fragments. Allinger's force fields MM2, MM3 and MM4 are the best MM modelers of strained carbocycles. This means that, more or less successfully, their parameterization covers organic systems that include carbon units of perturbed geometry. However, these as well as other fine force fields

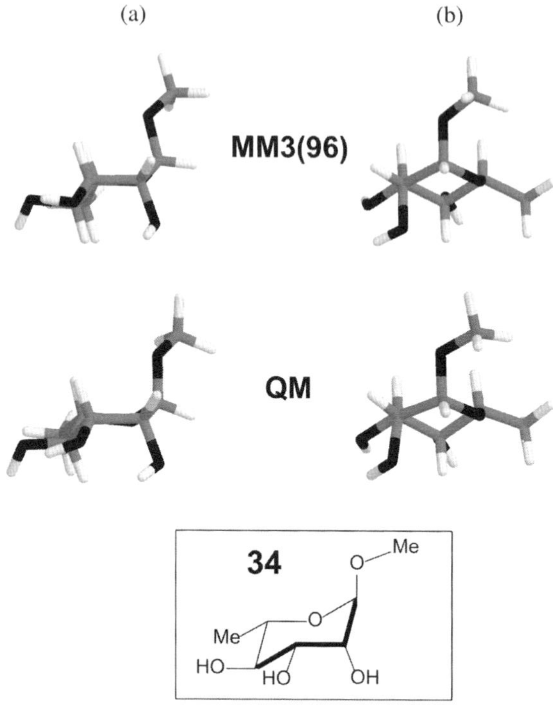

Figure 79. MM- and QM-provided geometries of the lowest energy conformer of rhamnoside **34**: (a) side view (bond C-2–C-3 lies in the ring plane), (b) rear view (bond C-2–C-3 is perpendicular to the chart plane). MM: MM3(96) force field, for vacuum conditions; QM: the B3LYP/6/31+G(d,p) level (see Sections 5.2.4 and 5.2.6), for vacuum conditions.

may not be suitable for heteroatom-containing systems if the heteroatom-centered fragments have distorted geometry. For instance, MM3 demonstrates an excellent prediction of relative stability of the *trans* conformer and the cis_{Nax} conformer of azabicycle **35**, as the QM-based verification shows [$\Delta\Delta_{abs}$ (Section 4.2.6) $= 0.1$ kcal mol^{-1}; Fig. 80].[36e] Also, the barrier of their interconversion is estimated by MM sufficiently accurately (7.0 kcal mol^{-1} by MM *vs.* 7.7 kcal mol^{-1} by QM). Does this mean that this force field is capable of modeling this system?

Results for the closest analog, the cis_{Neq} conformer of **35**, reject this assumption. A 4.4 kcal mol^{-1} inaccuracy of MM suddenly appears. There is no obvious perturbation of the geometry of the cis_{Neq} conformer

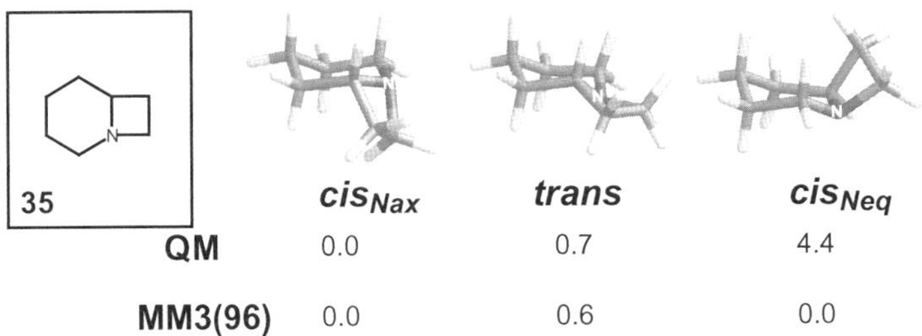

Figure 80. Selected stereoisomers of azabicycle **35** (they are conformers at room temperature). The shown geometries are supplied by *ab initio* calculations [at the B3LYP/6-31+G(d,p) level; for vacuum]. Numbers indicate relative energy [ΔE_{ster} for MM3(96) and ΔE_{el} for QM; kcal mol^{-1}].

vs. two other structures, the *trans* conformer and the *cis*$_{Nax}$ conformer. Nevertheless, one should not be surprised. The considered warning sign (*3*) cautions us: since the azetidine ring is a structure with a significantly distorted N unit (the related endocyclic CNC angle is near 90°), one could expect that MM may afford very inaccurate energy estimates for bicycle **35**.

If this organic fine force field is applied to more strained azetidines, the results are worse. Azabicycle **20** has two fused four-membered rings (Fig. 41, Section 2.3). MM3(96) and OPLSaa(2005) predict that the *trans* stereoisomer of **20** is less stable than the *cis* isomer by 25.1 (Ref. 36e) and 24.6 kcal mol^{-1}, respectively. An accurate estimation of their relative stability in vacuum, which is provided by QM calculations, affords much larger value of 32.8 kcal mol^{-1}. The inaccuracy $\Delta\Delta_{abs}$ (Section 4.2.6) of these advanced versions of MM3 and OPLSaa is 7.7 and 8.2 kcal mol^{-1}, respectively; that is, these MM estimates do not have a meaningful accuracy at all. CHARM family's CGenFF estimates its own parameters for this amine as extremely poor;[a] in other words, this force field is inapplicable to model this molecule. Only MMFF94 delivers a still inaccurate, but more telling estimate of 29.9 kcal mol^{-1} ($\Delta\Delta_{abs} = 2.9$ kcal mol^{-1}) for relative stability of the considered stereoisomers of azabicycle **20**. These examples show that, as a rule, MM alone (as a pool of standard versions of force

---

[a]One can estimate the quality of CGenFF parameters for a chemical structure by uploading the related *.mol2* file to website https://cgenff.paramchem.org.

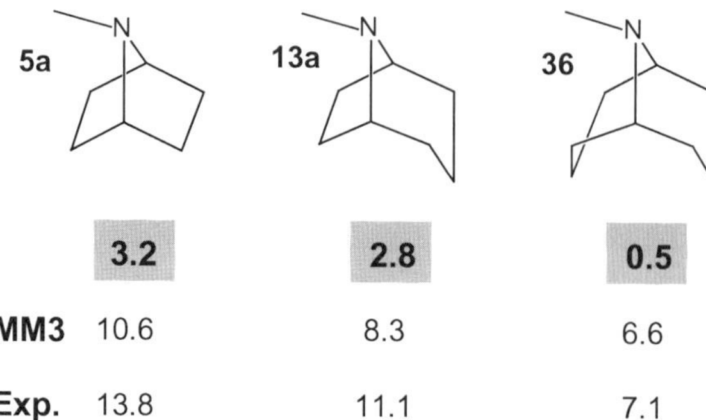

Figure 81. NIR barriers for m-azabicyclo[m.n.1]alkanes **5a**, **13a** and **36** [$\Delta E^{\#}_{ster}$ for MM3(96) and $\Delta G^{\#}$ for DNMR; kcal mol^{-1}]. Numbers in gray rectangles show the differences between the experimental and calculated values.

field families[a]) cannot be used for conformational analysis of systems that contain a fused small heterocyclic ring.

The next example is more puzzling. Azabicycles **5a**, **13a** and **36** (Fig. 81) do not contain either aziridine or azetidine fragments, and their geometries seem similar. Could we use MM for quantitatively predicting their molecular flexibility? At first sight, it seems that there is no contraindication in undertaking this easy computational experiment. MM3(96) or MM3(2000) probably are the best tools for it because these versions of MM3, including an explicit parameterization for amines, usually reproduce NIR barriers for alkylamines at least satisfactory.

In fact, the 3D geometries of the N-centered elementary molecular fragment are different for these amines. The endocyclic CNC angle decreases from a normal value of 112° (by QM) in almost unstrained analog **36** to an unordinary value of 97° (by QM) in azanorbornane **5a**. MM3(96) provides similar values. Both MM and QM inform us about a distorted geometry of the heteroatom-containing (here: nitrogen-containing) unit in bicycles **5a** and **13a**. Nevertheless, the geometrical anomaly, which has been computationally disclosed for the N fragment of these two azabicycles, indicates that

---

[a] Such versions do not provide satisfactory energy estimates even for monocyclic heterocycles of small ring size, e.g., for substituted aziridines and azetidines, since they do not contain parameters for targeting small size heterocycles. Of course, additional parametrization could also comprise these systems. As mentioned, MM is a correctable tool; however, the command of corrections belongs to computational chemists.

one can expect an MM failure in estimating the kinetic barrier of pyramidal N inversion for them and no failure in the case of analog **36**. Indeed, in providing NIR barrier values for these bicyclic amines, MM3(96) shows selective accuracy. For the "normal geometry amine" **36**, the calculated barrier reproduces the experimental value well while calculated barriers are very inaccurate for amines **5a** and **13a** (Fig. 81).[4,27] This inaccuracy is increased in parallel to the increase of the deviation of the endocyclic CNC angle from the "normal" value.[a] The distorted geometry of the N-centered elementary molecular fragment in these azabicycles does give red light to estimation of their NIR barriers by means of MM.

Similar conclusions can be drawn regarding other heterosystems with distorted geometry of heteroatom-centered elementary molecular fragments. One should view structural distortions in such molecules as a serious warning: escape MM calculations for these systems if possible! Use QM calculations instead!

Warnings (*1*)–(*4*) do not include an obvious case of structures that have fragments, for which standard versions of fine force fields do not contain parameters. Parameterization in standard versions (not extended to additional chemical classes) of fine force fields has been more or less thoroughly performed for some set of main functional groups. Unfortunately, functional groups that are "equipped" with parameters of either force field are not listed in chemical literature. In order to exactly know structural specificity (applicability to either chemical structure) of the force field of interest, one should find its specification (i.e., information about types of MM atoms included into the force field). This information is often available in the Internet.[b]

The term *main chemical functions* is a key word for quickly recognizing the structural specificity of the force field. This diffuse term is used here in connotation of the educational tradition in organic chemistry: main chemical functions are those considered in related introductory textbooks, e.g., carbonyl moiety $R^1R^2C(=O)$ and not nitrone group

---

[a] MM, as a non-quantum mechanical approach, equally takes into account geometrical distortions of molecular fragments of the same chemical structure. Therefore, from the MM viewpoint, CNC angle values and NIR values should correlate for alkylamines. However, there is no quantitative correlation between these values including those for **5a**, **13a** and **36**.[4,27] This indicates that the alkylamino fragments in different molecules cannot be approximated by an "universal" alkylamino fragment. Indeed, Lewis structures — a formal chemical structure of molecules in MM — are too rough models of chemical bonding. They, in an oversimplified form, most frequently single out only certain atoms as bonded and ignore some contribution of electrons of other atoms into this bonding.

[b] See, e.g., webpages http://dasher.wustl.edu/tinker/distribution/params and http://www.pharmacy. manchester.ac.uk/bryce/amber benevolently created and maintained by researchers of these universities; also, webpage *http://aten.googlecode.com/svn-history/r1062/trunk/data/ff/oplsaa.ff*.

$R^1R^2C{=}N(O)R^3$, carboxyamide group $RC({=}O)NR^1R^2$ and not iminophosphorane (phosphazene) moiety $R^1R^2P({=}NR)R^3$, phosphate group $(RO)_3PO$ and not phosphite group $(RO)_3P$. Significance of certain chemical classes in organic synthesis (e.g., phosphonium salts $[(R_2N)_3PX]^+{\cdot}PF_6^-$ in peptide synthesis or stabilized Li anions $[R_2CX_{EWG}]\,Li^+$ in C–C coupling reactions) has no relationship with the meaning *main chemical functions* considered herein. Thus, "warranted" patterns for MM calculations usually are monofunctionalized molecular structures of closed valence shell that belong to either main chemical class of organic compounds. A chemical structure belonging to neither main functional class of organic compounds fairly often indicates that it cannot be treated by standard versions of organic or biomolecular force fields.

(4) Stereoisomeric structures separated by very low kinetic barriers. MM is too rough for modeling considerable changes of molecular geometry associated with tiny changes of molecular energy. Such energy changes are of the magnitude of the energy of the first vibration level; however, force fields do not "see" this and may generate a "false" SES region with energy minima or inflection points that, in fact, do not correspond to equilibrium geometries (in other words, to individual conformers) or individual conformational transition states, respectively. For instance, both MM2 and MM3 cannot indicate that the "transition state candidates" for RI in cyclohexene — boat *vs.* two enantiomeric twist forms — possess almost the same molecular energy and are not separated by appreciable kinetic barriers, as QM calculations show.[a]

Let us summarize what chemical structures of closed electron shell usually can and cannot be modeled by using standard (additionally not parameterized) force fields. Structures described here in subsections *(1)*–*(4)* are untreatable by today's MM. In a very general way, other structures may be classified as *(1)* main monofunctionalized systems and polyfunctionalized systems with distally positioned main functional groups; *(2)* polyfunctionalized systems with main functional groups proximally positioned in the molecular skeleton; and *(3)* systems with other, subsidiary organic functions or with scarce (non-trivial) molecular skeletons. As mentioned above, structural cluster *(1)* is more or less successfully treatable by MM. For

---

[a]This means that the molecular geometry of this transition state does not have exact single–valued numerical parameters (Section 2.4). The latter lie in some interval with boundary values that characterize the related enantiomeric twist conformations.

instance, molecular modeling of cyclic and acyclic monoalcohols or cyclo-hexanes bearing 1,4-positioned basic organic moieties is not problematic for MM. Nevertheless, the requirement for examination of force fields (Section 4.2.5) prescribes that MM results for unexplored systems from this group should be examined by comparing them with QM-derived estimates for 1–3 patterns of such structures. For cluster (*2*), the use of MM is justified for compounds that belong to groups of structures of the same function-ality for which a success of MM has been demonstrated. Usually, they are compounds that have a pair of vicinal functional groups (e.g., vicinal diols, dihalogenides and diamines, α-hydroxy ketones) or belong to certain biologically important organic classes, e.g., peptides and proteins. Appro-priate fine force fields are sufficiently accurate in modeling them. One can consider other polyfunctional structures with proximally positioned func-tional groups, e.g., polyketones, carbohydrates, as objects for which the accuracy of MM calculations is uncertain *a priori*. If one can succeed in selecting the suitable force field for concrete structures of this type (Sec-tions 4.2.5 and 4.2.6), there are no reasons to abandon MM when studying them.

MM users, who do not intend to augment either force field with addi-tional parameterization, probably have to view cluster (*3*) as structures that should be subjected to QM and not to MM calculations. If chemical func-tionality of a compound does not relate it to either main organic class, rather no original fine force field has parameters that target this functional-ity. Hence, this chemical structure cannot be treated by means of standard versions of current force fields. It is meaningless to apply standard versions of force fields to, e.g., carbanions, carbenium ions, onium compounds, organic radicals, annulenes, helicenes, polyfluorinated compounds, gemi-nally functionalized compounds (e.g., thioacetals, gem-diols), azides, azo compounds, ylides and pseudoylides[a] (e.g., N-oxides $R^1R^2N(O)R^3$), as

---

[a]Excluding phosphate triesters (a non-nucleophilic character of the phosphoryl oxygen masks an essen-tial charge separation $P^+–O^-$ in their molecules). For instance, Amber (all versions including GAFF), MM3 (all versions), OPLS-aa, and CGenFF have parameters that define P and O atoms of phosphate esters as separate types of P and O MM atoms, respectively, recognizing this functional group as a distinct chemical unit. The quality of these parameters is, nonetheless, insufficient for accurately modeling functionalized phosphate esters, e.g., carbohydrate phosphates, and new, accuracy-refining parameters augment latest versions of Amber and CHARMM oriented to modeling of phosphorylated biomolecules.

well as species with a hypervalent atom even if an unscrupulous MM calculation package accepts such structures for geometry optimization. Even if resulted molecular geometries seem realistic, obtained values of ΔE_{ster} reflect the conformational equilibrium from a corrupted perspective. The researcher is challenged to recruit QM. In this case, it is worth to start calculations with optimizing the geometry by MM (in jargon of computational chemistry, to clean geometry) and re-optimize the obtained crude 3D structure using a QM method (Chapter 5).

In MM modeling of large molecules or molecular complexes, e.g. proteins or protein-protein complexes, the problem of result unreliability (i.e., a significant inaccuracy of modeled quantitative molecular characteristics) appears even in modeling their 3D geometry. These systems do not contain exotic chemical fragments. However, non-covalent interactions of non-bonded fragments are an essential factor of geometrical preferences in these highly organized coils. As mentioned above, no force field takes into account all these interactions. For instance, long-range VDW interactions are "cut" at certain interatomic distances (Section 4.2.2) in MM models of molecular nano-size objects. In order to overcome these problems at least in part, many computational chemists develop their own "object-oriented" (additionally and specifically parameterized) versions of biomolecular force fields. Of course, such a fine tuning is out of the competence of organic experimentalists.

4.3 Pre-starting Notices

4.3.1 *MM calculation programs in brief*

First of all, for MM users, force fields are computer programs. These programs (so-called source codes) cannot be run directly since source codes are not executable files. Therefore, at least, they should be compiled. Thus, a user, who is not familiar with programming, cannot use original force field files.

To make MM calculations available to a majority of chemical researchers, software specialists transform source files into files that are suitable for a concrete operating system (e.g., Windows, Linux) and usually "envelope" the force field core by additional programs. As a result, a hierarchical "superprogram" (computational package) is created. By eliminating the necessity in programming skills, molecular modeling packages provide, to non-computational chemists, a comfortable use of MM (or/and QM). The point is that such packages permit the user to exploit the relevant English and on-the-screen molecular graphics and not a "need-to-learn" language of program-specific commands. This stunning feature is supported by a certain, "enveloping" module of the computational package — graphical user interface (GUI)[a] that is a versatile interpreter of such commands. It provides to both two "participants," the computational and graphical modules of the package and the user, to "understand" each other although each of them only "speaks" his own language.[b] Thus, by means of GUI, the researcher can easily operate with on-the-screen 3D structures and the calculation machinery "hidden" in the computational module. GUI enables, e.g., creation of molecular structures on the screen, their structural modifications there, export and import of generated structures (i.e., files in which molecular geometry or other molecular properties are "written down"), visualization of 3D structures by reading such files, calculation management, analysis of obtained structures (displaying of geometrical parameters, VDW, solvent-accessible and other molecular surfaces on the screen, quantitative comparison of the geometry of molecules, arrangement of them according to a selected molecular feature, superposition of structures, etc.). All modern computational chemistry packages have a spectrum of calculation options that is sufficiently broad in valuable packages. Often, several (from 2 to 5) force fields are included into one computational package.

The largest and most frequently updated reservoir of information about MM programs is, of course, World Wide Web resources. There, even an unprofessional information source, Wikipedia, supplies a list of programs (see for *List of software for molecular mechanics modeling*). To locate relevant sites on the Web, MM beginners could use keywords *molecular mechanics program*, or *force field program*, or similar ones and then get a complete insight into currently available MM software. Websites of commercial and open-source packages provide a general description of these programs; solid producers of chemical software offer an option to clarify

---

[a]GUI may be a stand-alone program.

[b]Otherwise, the use of MM (also QM) is technically more complicate. For instance, original programs from the MM2, MM3 and MM4 families are not equipped with a GUI. One should know the command vocabulary of the installed programs in order to exploit them (to perform MM calculations). One should use external molecular graphics programs in order to easily generate or modify initial molecular structures as well as to visualize or easily analyze the resulting ones.

technical nuances by means of an *e*-mail-provided dialog. Inexperienced MM users should not neglect this opportunity. Information about versions of announced force fields, inserted modifications, the presence/absence of MM parameters for specific chemical classes, the option of adding parameters by the user, geometry minimization methods, the implemented models of solvation, analysis of geometrical similarity in large sets of 3D structures, optimal hardware and installation details is *necessary*. Often, only the package producer can supply this valuable information.

Monographs on computational chemistry, supplying information about MM programs (see e.g., Refs. 33b, 50b, 55), restrict it by giving some technical characterization of the listed software and leave potential users the option to estimate by themselves the quality of either package. In addition, the pool of available computational programs is being changed sufficiently rapidly and overviews in chemical literature are not keeping track of changes. However, for MM beginners, the absence of practical recommendations for selecting a suitable MM package is intolerable — a thought-out successful selection is practically impossible for an inexperienced researcher. In order to relieve synthetic chemists of an uncomfortable way of trials and errors, this Section takes the liberty to judge which products might be preferable for them.

Like much calculation software, MM packages are commercial, non-commercial (open-source) or free of charge for academic users. Free trial MM programs can be downloaded from the related websites. There is a vivid relationship between the quality of current MM programs and the charge requirement. Open source MM packages, e.g., Argus 4.3 and Avogadro 20.7.2 equipped with the own GUI, are terribly limited in calculation options and actually may be utilized only in preliminary acquaintance with MM, graphical representation of 3D structures, etc.[a] To put MM calculations on solid ground, organic experimentalists should focus on the use of truly professional programs.

Currently, a single freely distributed research-oriented MM package with GUI is Tinker (by Jay Ponder, Washington University). It and its GUI, similarly to serious commercial MM packages, are under continuous

---

[a] Some modern commercial programs of *molecular graphics*, e.g., organic-chemist-oriented Chem3D Pro or ChemBio3D Ultra 13 and 14, also have a minimal functionality in providing MM calculations.

development. This multi-platform package (version 6.3; Windows, Mac, Linux; download site *http://dasher.wustl.edu/ffe/*),[a] has a representative set of fine force fields [Amber (versions ff94, ff96, ff98, ff99, ff99SB], CHARMM [versions 19, 22, 22/CMAP], MM2(91), MM3(2000), OPLS (versions OPLS-UAA and OPLS-aa), MMFF94 and AMOEBA (versions 2004 and 2009]. Certainly, this force field set fulfills organic experimentalists' needs in obtaining at least satisfactory MM molecular models. In addition, it has an efficient conformational search, different implicit solvation models, normal mode vibrational analysis [Section 4.4.3, subsection (1)] and several minimization algorithms. It is not surprising that Tinker for Linux is widely used. Another solid MM freeware is GROMACS. Due to a perfectly optimized programming code, it is the fastest computer program for MD simulations; however, it may also be exploited in quick MM energy minimizations. This absolutely professional, Linux-resident program, which exploitation may be facilitated by involving a stand-alone GUI,[b] uses periodically renewed versions of bioorganic fine force field GROMOS, one of the most accurate and dynamic MM tools for modeling of peptides and proteins.[c] Indeed, these chemical structures are the main focus of different versions of GROMOS. GROMACS does not include other force fields and, therefore, scarcely suits organic chemists.

Commercial MM programs are a very convenient but sometimes expensive alternative. MacroModel (by Schrödinger), Spartan (by Wavefunction, Inc.) as well as HyperChem (by HyperCube, Inc.) can pretend to be favorites for synthetic chemists. These commercial packages of computational chemistry are upgraded to new editions from time to time and are the most versatile finished products among Windows- and Linux-operated MM programs. Basic features of these packages are described in the manufacturers' websites and vain detailed information is supplied by their manuals that are also available on the Web. Herein, one may indicate major advantages of these computer programs for MM calculations. A convenient set of nine fine force fields [MM2*, MM3*, Amber,[a] Amber94, MMFF94,

---

[a] Note that there is no GUI for Windows-resident Tinker.
[b] See webpage *http://www.gromacs.org/Downloads/Related_Software/GUI*.
[c] The two others are CHARMM and Amber.

MMFF94-s, OPLS-aa, OPLS(2001), and OPLS(2005)] has been implemented into MacroModel. There are four force fields in HyperChem8 and three force fields in Spartan10. However, some of these force fields are not widely known, and their accuracy is, therefore, *a priori* unclear for the user. Conformational search is very effective in MacroModel; structure analysis options (including clustering and comparison of molecular geometries) are versatile in this package. Also, Spartan10 and HyperChem8 include conformational search options and may sort generated conformers according to their relative energy. With MacroModel embedded into a multipurpose chemical calculation package Schrödinger Suite, MM calculations become practically more convenient: Schrödinger Suite incorporates a successful QM program, Jaguar, and therefore, MM-delivered structures can be verified by QM calculations without involving other programs of computational chemistry. Also, packages Spartan10 and HyperChem8 provide QM calculations of sufficiently high levels of theory; of course, this enables verification of MM results within the same calculation package. Importantly, each of these packages has a convenient GUI developed by the manufacturer of the calculation program; e.g., the GUI for Schrödinger's MacroModel, Jaguar, and Suite is Schrödinger's Maestro (free of charge for academic users).

Other commercial MM packages are essentially restricted in modeling organic compounds. For instance, YASARA (by YASARA Biosciences) is sharply oriented to proteins and includes four related bioorganic force fields (Amber, YASARA, YAMBER and NOVA; the modeling accuracy of the latter three is insufficiently proved) and only one, rarely exploited organic force field (GAFF). Some QM packages for chemical research contain modules for MM calculations that are intended to clean (roughly optimize) the geometry of the molecular structure of interest before subjecting it to QM calculations. However, all such modules are equipped with minimal sets of force fields; some of them often are low accuracy force fields. E.g., the list of force fields in the ubiquitously used Gaussian09 (by Gaussian, Inc.) is exhausted by Amber96, Dreiding and UFF, the MM module of Turbomole (by Turbomole GmbH) only provides UFF, and such a module of SCIGRESS (by Fujitsu) includes two organic force fields of similar architecture — MM2 (an obsolete force field) and MM3. Besides, MM modules of QM packages are incapable of performing conformational search.

Discovery Studio 4.1 (by Accelrys) is another professional MM calculation package that may be used by organic chemists. However, in the first instance this gigantic computational package is designed to suit biological and medicinal chemistry and not organic chemistry. Conformational analysis of organic molecules is only one of the options of the multifunctional Discovery Studio. The set of fine force field families is not large (CHARMM and its original late versions targeted to different biomolecules; CFF and MMFF94). On the other hand, conformational search in this package may be performed by means of six different algorithms which practically excludes missing energy minima for small molecules. Several non-primitive MM solvation models are present. Since this package is aimed at dealing with large molecular structures (proteins), quantitative analysis of molecular geometry for sets of obtained stereoisomers is very diverse and also convenient for small molecules.

Another commercially available MM program for Windows, which, in principle, might be suitable for organic research, is BOSS (by Cemcomco). Solvation modeling by MM is an explicit feature of BOSS; QM calculations are also an attribute of this program. However, this program contains versions of the only OPLS force field. In addition, BOSS does not have its own GUI for Windows.

"Light" MM packages may be an unwelcome surprise. We have learned that force fields are their prime core. Unfortunately, there is no guarantee that original force fields and the core of a package are identical. The core is "invisible" for users, and its changes may indiscernibly affect calculation results. Indeed, without a thought-out examination, one can descry an alteration of results only by an accident. Therefore, *incorporation of well-characterized or original versions of force fields is an attribute of a MM package for research.* For instance, users of the MacroModel are notified by its manual which force fields have been modified in this package; the corresponding force field equations are provided. Behaving oppositely, "light" MM packages should not be ultimate computational tools of molecular modeling if used in research.

Otherwise, MM calculations may mislead. For example, the original, "unpackaged" version of MM3(96) affords the geometry with an intramolecular OH$\cdots$N bond for the minimal energy conformer of aminoalcohol **37** (Fig. 82), and experiments confirm the presence of this H-bonding).[56] Transparently, this geometry should not be changed by geometry optimization if the same version of MM3 is employed. MM3(96) from a "light" package PCmodel (version 7.5; by Serena Software, Inc.) changes the molecular geometry and removes the H-bond! Probably, the original version MM3(96) and the version used by PCmodel7.5 are not identical. In using MMFF94, this PCmodel eliminates the *cis/trans* isomerism from azabicycle **20** (Fig. 41, Section 2.3): it shows a planar nitrogen fragment, in contrast to, e.g., the MMFF94 force field from the "solid" MacroModel.

In principle, the absence of exact information about implemented versions of force fields excludes any MM package from the collection of *research* accessories. This information is not always easily verifiable. A many-page user guide accompanies any serious computational package. Sometimes such manuals contain only common notes or references concerning implemented versions of employed force fields; model atom types and parameters for them

                    Conformational Concept for Synthetic Chemist's Use

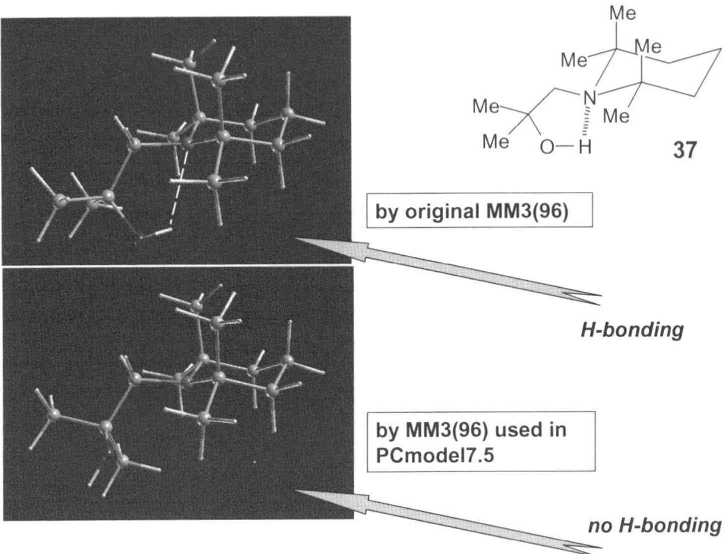

Figure 82.   The correct and the incorrect geometry for the minimal energy conformer of piperidine **37** supplied by the original ("unpackaged") MM3(96) version and the MM3(96) version from the PCmodel-7.5 program, respectively.

are not listed. Therefore, one should stress again the importance of a pre-purchase technical dialog with experienced users or Web-"resident" producers of molecular modeling software.

Usually described in depth by experts in computational chemistry (see e.g., Refs. 49e and 50f), the original version of a force field is a brilliant, but not quite convenient "gift" to users. As noted, such versions are supplied in the form of non-executable source codes, and transformation of such codes (i.e., certain files) in functional programs requires basic skills in handling operational systems. Nevertheless, due to benefits of the use of chief force fields (origin guarantee, availability of the latest versions, no or low cost), it is useful to know their web locations:[a]

- CHARMM: *http://www.charmm.org*
- AMBER: *http://ambermd.org*

---

[a]Regrettably, QCPE (Quantum Chemistry Program Exchange; website http://ftp.ccl.net/cca/info/ QCPE.shtml), where original Allinger's MM3 and MM4 had been available, does not provide these force fields anymore.

- MMFF94: *http://www.ccl.net/cca/data/MMFF94*
- GROMOS: *http://www.igc.ethz.ch/GROMOS/getgromos*

4.3.2 *Important features of a good MM calculation package*

The focus of this Chapter is practical conformational analysis by MM. Professional computational packages present to the user many calculation and structure handling options; which of them are most important in conformational studies? Let us discuss MM packages from this perspective.

Among calculation-related options, an indispensable one is conformational search. Indeed, the first question that is asked by the researcher in conformational studies is: what are stereoisomers for the given compound? What transition states are in interconversions of these stereoisomers is the next question. Concerning a MM calculation package, these issues are transformed into a simple yes/no query: could it generate stable stereoisomers as well as transition states starting from an arbitrary 3D structure? Conformational search is exactly a procedure of generation of these structures. In other words, it is a partial modeling of the SES (Section 4.2.3): it is modeled as a set of "points of interest" — a set of both minima and saddle points (see e.g., conformers and conformational transition states in Fig. 53, Section 2.8) — and not as a continuous surface.

Unfortunately, there is no principal solution for SES modeling, and various methods of conformational search used in computational chemistry locate critical points in the SES with efficacy that tends to 100%. The common principle of exploring conformational space is trivial: a set of arbitrary 3D structures is generated by means of a certain algorithm and then these structures of different geometry are subjected to optimization of the geometry by means of a force field. Thus, each stereoisomer is located in MM by performing two sequential operations, generation of molecular structure of a crude, even arbitrary geometry and the elementary MM procedure, geometry optimization applied to the generated structure. The larger the set of generated intermediate structures is, the higher the efficacy of the undertaken conformational search is.

Methods of conformational search differ one from the other in the fashion of how they generate intermediate (non-optimized) structures. Simple algorithms include predetermined changes of molecular geometry, e.g., stepwise changes of torsional angles. More sophisticated algorithms are based on stochastic changes of geometric parameters, e.g., random changes of torsional angles or atom coordinates. Of course, the condition of integrity of chemical connectivity and "freezing" of initial configurations of elementary molecular fragments is embedded in algorithms of stochastic generation of molecular 3D structures.[a] Different conformer-searching algorithms are not identical in their efficacy. For instance, location of all conformers of piperidine **14b** (Fig. 52) requires generation of $\sim$600 structures by means of the stochastic engine of the original program MM3(96). To reveal these conformers, at least 5000 structures have to be generated if stochastic search from the MacroModel6.5 package is used.

None of conformational search methods guarantees that all stable stereoisomeric structures as well as transition states of their interconversions are located. Nonetheless, one can claim that effective algorithms of conformational search do find all these stereoisomers for organic molecules of ordinary size (say, up to 50–60 atoms) if the system of interest does not contain a long chain (say, $>$ C$_5$) acyclic fragment or several short (say, C$_3$ or C$_4$) chains. Generation of 10^4–10^6 stereoisomeric structures of arbitrary geometries followed by energy minimization is usually sufficient to explore conformational space of any such system completely. For cyclic organic molecules with a 5-membered to 10-membered ring, generation of 5×10^2–10^5 3D structures and optimization their molecular geometries may be the standard protocol that provides location of all stable stereoisomers.

Acyclic molecules have an enormous number of stable rotamers (because of low barriers of rotation, these rotamers are conformers at ambient conditions; Section 2.6). Therefore conformational space of even relatively small systems as, e.g., of eicosanoid **31** (Fig. 75), could not be covered by any conformational search. It means that one could never be sure that the located minimal energy stereoisomeric structure corresponds to the global minimum of E_{ster} for a molecular system with billions and more conformers.

---

[a] Some MM packages suggest to the user to select whether the initial configuration of either molecular fragment is or is not kept in generating molecular structures of arbitrary geometry. The "non-freezing" option is convenient when conformational search is applied to molecular structures with a non-rigid molecular fragment that does not produce identical 3D structures ("formal" enantiomers) by inverting the configuration, e.g., to cyclic amino structures.

In this connection, it is useful to recall the Lewinthal paradox for protein folding. It states that a protein chain cannot fold itself at all. According to its reasoning, an unfolded protein molecule does not "know" how to reach its unique folded form: all conformational pathways are equal to go at the initial moment $t = 0$ of the chain dynamics. This is true also for any "next" protein conformer since, as far as it appears, it is in the situation $t = 0$. Hence, the protein, as a statistical ensemble of molecules that are dissipated in different conformational pathways, has to traverse many of them to adopt the native 3D form of the chain. Transparently, this is impossible because even the number of possible rotamers for a long chain is unimaginably high (Section 1.6). The chain folding would take a longer time period than the time of any known physical process in the universe, even if the kinetics of each transformation step is of the ps time scale. In other words, each protein molecule would "travel, travel, and travel" in the labyrinth of uncountable conformational pathways before reaching the global energy minimum. As known, this does not occur: protein folding is the process of timescale ranging from milliseconds to minutes. The Levinthal paradox actually indicates that protein folding is not a random selection between *equal* possibilities: probably, essentially lower barriers for certain pathways of concerted process *intramolecular H-bond formation/dissociation — internal rotation — solvation shell reorganization* are the directing "signs" of the fastest route to the global minimum in the surface of free energy for molecules of a biopolymer.

This "veto" on equal probability helps to understand why stochastic conformational search cannot be used in *de novo* modeling of biopolymer molecules. Indeed, in practice, the entire conformational space is not modeled for multi-conformer molecules, and the minimal energy conformer (i.e., corresponding to the global energy minimum) is not searched by means of stochastic methods. Generation of molecular geometries by means of a randomity-based algorithm (stochastic conformational search) for open chain molecules may be compared with a hypothetical protein chain that does not "know" about short pathways to conformers and is forced to "travel" in conformational space by randomly choosing itineraries with equal probability for any choice. As we learned, this strategy has no prospect. Then, what is the methodology for locating the lowest energy conformers of biopolymers? A total majority of the conformers play no role in biological functioning of these molecular giants. This circumstance is a conceptual basis of reduction of an unrestrained conformational space to an essentially smaller one when modeling these molecular objects. Experimental data (X-ray diffraction and NMR) determine very restricted regions of the conformational space of a biopolymer molecule/complex (i.e. a conformer or several conformers that are most stable at the experiment conditions). Then only these "hot" regions are computationally explored (usually by means of MD) in order to identify biologically significant conformers of the biopolymer or understand its action mode.

Another essential feature of a good MM package is a multifunctional GUI. At first sight, possibilities of a computational package in operating with given molecular structures seem some convenient, but not necessarily required augments. In fact, they impart to calculation packages the comfort of being used without directly manipulating files resulted from computations, i.e., they deliver easiness and quickness in handling molecular structures virtually (i.e., on the screen). As explained, GUI is the component of a calculation package that provides this handling; for the user, the GUI is the control panel of the computational package. Conformational search may yield thousands of stereoisomeric structures for a molecular system. It would be very inconvenient to extract the required information (e.g., steric energies for these structures) from such a large information pool by reading the related many-page text files (output files) or manipulating with the command language of the used operating system to analyze these files and create a new, short text file where the analysis results are written down. By pushing virtual buttons of this "English-speaking" panel,[a] the researcher can, in a few seconds, arrange the modeled stereoisomers according to the chosen criterion,[b] or group them according to molecular 3D similarity and then see/analyze all modeled structures in the desired order, on the screen. This example shows that, without a multifunctional GUI which, in the first instance, includes molecular graphics options, any computer program of molecular modeling is unsuitable for smooth exploitation in organic laboratories.

A trivial option of any calculation program is to save files that contain information for a selected molecular structure so that this or another program can "read" the molecular geometry (e.g., visualize or take for further optimization). Thus, the initial, optimized, or modified geometry do not disappear from the researcher and may be called back for further manipulations, if necessary. These files are text files that include structural information as a set of Cartesian or internal coordinates[c] of all atoms. Also, information about localized charge (e.g., for ionized functions) may be present in such files. These files may be saved in different formats (offered

---

[a]The language of virtual buttons in the GUIs of such packages is "human" English (e.g., replace atom, optimize geometry, remove geometry constraints) and not the commands and syntaxis of the used operating system (e.g., grep *text* *filename1* > *filename2* in Linux or UNIX).

[b]From the GUI menu. A trivial option from such a menu is arrangement and visualization of located conformers according to the increase of steric energy.

[c]A set of interatomic distances, bond and torsion angles for a molecule; these numerical data are written down in a certain form called Z-matrix. With using GUI-equipped packages of computational chemistry, non-specialists may successfully perform molecular modeling without understanding the formal structure of files of molecular geometry.

by the GUI menu) that are (*1*) force filed- or MM package-specific; or (*2*) "common." The most widespread "common" format is the Protein Data Base (pdb; Section 4.4.1) format;[a] names of related files are conventionally supplied with file extension .pdb. Another one is XYZ (the corresponding file names are supplied with extension .xyz). The pdb format is recognized by any program for molecular modeling or molecular imaging, and it is convenient to save geometry-encoding files as pdb files. An example of a force field-specific format is the file format of the MM3 force field (Section 4.4.1).

As many other formats, (as, e.g., the format of structure-writing CHARMM files) CHARMM card (names of such CHARMM files have extension .crd), the pdb format allows writing and reading information regarding atomic charges.[b] Nevertheless, such files most often do not include these formal data (e.g., when they are related to molecular structures delivered by X-ray diffraction analysis). The point is that this information is rarely useful in MM energy minimizations: in calculating electrostatic energy potential (Eq. 24, Section 4.2.2), a force field by default uses atomic charges of MM atoms defined in this force field.

Computational conformational analysis is frequently associated with dealing with large massifs of stereoisomeric structures. It is therefore important that the MM package exploited in the organic laboratory is capable of doing this. Professional MM packages provide different options for quickly inspecting thousands of modeled molecular structures. Those associated with drug design, e.g. Discovery Studio, focus on revealing relationship *molecular structure–biological activity*, and most of their options for analyzing molecular structure are irrelevant for chemical research. Chemistry-oriented MM programs are less versatile. The advanced among them include clustering of stereoisomeric structures according to the chosen geometry parameter(s), arrangement of the structures in ascending order of E_{ster}, and estimation of molecular similarity [e.g., by using root mean square (rms) values for differences in parameters of molecular geometry]. Nevertheless, this set of options is sufficient for organic and bioorganic chemists.

---

[a]The extension of such file names is pdb, e.g., file *structure1.pdb*.

[b]Partial charges formally assigned to atoms in the molecule. As mentioned in Section 4.2.2, the concept of atomic charges is formal: electrons in a real molecule do not belong to either atom and, hence, any atom cannot be characterized by a certain electric charge. Therefore, for the same molecular structure, the atomic charge for an atom is numerically not identical when supplied by different theoretical assumptions.

Let us summarize which calculation and other features a hypothetical perfect MM calculation package should have if intended for conducting conformational studies of organic compounds. They are:

- The presence of a representative set of late versions from several fine force field families,[a]
- Availability of the list of MM atoms and parameters for each implemented force field,
- On-the-screen-presentation of calculated numerical data [values of E_{ster} in each iteration of energy minimization, final steric energy and final energy of pairwise interactions for each MM atom pair (i.e., all energy potentials calculated for each relevant atom pair; Section 4.2.2); frequencies of molecular vibrations, dipole moment],
- High quality visualization of molecular 3D structures (including exact 3D imaging of modeled structures as well as their oscillations) with on-the-screen structure-building/modifying options,
- A versatile GUI that initiates execution of any package-related command and selection of any package-related option, controls running calculations, and manages molecular graphics,
- A GUI-supported option of changing/adding force field parameters,
- Conformational search options,
- The presence of several minimization algorithms including an algorithm that enables location of conformational transition states,
- A possibility to vary the convergence criterion (Section 4.2.4) for energy minimization,
- A calculation option that establishes relationship between a transition state and the corresponding stable stereoisomers (Section 4.4.3),
- Options of structural constraints for molecular structure undergoing conformational search as well as geometry optimization,
- An option to save structural information in different file formats,
- The presence of several solvation models (Section 4.2.6),
- Options of storing and analyzing large sets of generated structures,
- Generation of different surfaces and plots, e.g., SES associated with selected geometrical parameters (see e.g., Fig. 70), plot of rotation

---

[a] A set that incorporates as late as possible versions of MM3 or MM4, MMFF94, Amber and CHARMM would be admirable.

(see e.g., Fig. 83), equipotential energy contours (see, e.g. Fig. 84), solvent-accessible surface, VDW surface, electrostatic potential surface,

- Reading, saving, exporting and importing molecular structure information as files in many different formats that are readable by other professional packages of computational chemistry (e.g., as input or output files of Gaussian, Jaguar and Q-Chem; Section 5.1),
- Graphics rendering options.

The reader probably will be disappointed — in the market of programs of computational chemistry, there is no MM package that possesses all these features. The number of sophisticated MM packages is not quite large, and basically dissimilar fine force fields are sufficiently represented in only few of them (Section 4.3.1). Some calculation and structure analysis options are present in almost all MM packages, some options are distributed between different ones. Moreover, an important option for establishing the relationship between a conformational transition state and related conformers, which is crucial in building conformational schemes (Section 2.8), is actually hypothetical: such an option is not provided by any current MM program.[a] MM beginners have no other choice than to follow a compromise: the more essential options (successful set of force fields, conformational search, GUI, computerized analysis of multitudes of structures) are collected in a package, the better it is for conformational studies.

An additional restriction of technical character appears in selecting an optimal MM package. When purchasing a package of computational chemistry, the researcher has to proceed from the limiting prerequisite — technical limits of available hardware and software facilities. They predetermine those MM (also QM) programs which may be used in either research group. Indeed, in order to be finished in a reasonably short time period, computations, in accord with their complexity, require certain computer

---

[a] In establishig this relationship by MM, computational chemists use auxiliary "home made" programs that can generate molecular structures that both have lower energy than the MM-located transition state and lie in the MEP. The principle for generating these intermediate (i.e., thermodynamically unstable) structures is simple. MM calculations, providing the transition state, also provide frequencies of its virtual oscillations. The oscillation, which disturbs the transition state so that geometrical changes correspond to the changes along the MEP (Section 3.2) in both descent directions, produces intermediate ("below the saddle point") stereoisomeric structures. Geometry optimization for these structures leads them to "fall" into the minima that corrrespond to the desired stable stereoisomers (Section 4.5).

performance as well as either computer architecture. Computer-operating software may be incompatible with either computational program.

Which of these two factors (hardware and software) should be considered first when selecting a suitable MM package? Hardware is not technically problematic for MM calculations undertaken for "normal" size (non-polymeric) organic systems. Numerical computations in solving MM equations (Section 4.2.2) are not cumbrous. Therefore, processor performance, disk space and memory size are not crucial for a typical step-by-step procedure of MM modeling of such an organic molecule (creation of the 3D structure on the screen → conformational search with geometry optimization → analysis of modeled stereoisomeric structures) — this sequence is completed in the timescale of a launch. Although the GUI appreciably occupies RAM, even an ordinary laptop is suitable, if the computer architecture allows to leave for calculations free 2 Gb of the random access memory (RAM), ~50 Gb on the disk and more than 50% of processor performance.

In contrast to MM calculations, QM computations require much more RAM and, more importantly, much higher processor performance. Therefore, QM calculations usually exploit computer clusters with parallel processing. Of course, computer clusters are preferable also for MM computations (they are *highly* preferable for QM calculations; maybe, even necessary), but, in chemical departments, these "high calibers" are usually purchased by computational chemists. Nevertheless, organic experimentalists may carry out routine QM modeling of small organic molecules with using PC. In order to provide considerably fast occurrence of such computations on a PC, it is better to have a 64-bit desktop computer with 4–8 processors and at least 64 Gb of RAM.

Software is a more serious technical problem when undertaking molecular modeling. The software of the available computer system may not satisfy requirements of the desired calculation package. The major limitation is the platform (operating system; OS) which the MM calculation package requires.[a] There are four basic operating systems, Linux, UNIX, Mac (Macintosh) and MS Windows.[b] If there is no

---

[a] As known, any computer program only runs in a certain operating system.
[b] The listed platform names characterize general types (families) of operating systems; such a family is composed of several versions. Exactly a version of an operating system is what is installed in the computer and manages it. For instance, a modern IBM-compatible PC runs either in the Windows 7, or Windows 8 environment; those are different versions of the MS Windows operating system. Linux

compatibility of the computer-operating platform and that required by the program, the latter cannot even be installed in this computer. A non-computational chemist has to decide what he/she should match, the computational package to the existing OS, or the OS to the selected package.[a]

Any professional package of computational chemistry is compatible with 1–3 of these platforms. That is, such packages are released as platform-specific versions. E.g., Schrödinger Suite (release 2014) and Spartan (release Spartan14) have versions for late versions of Linux, Windows and Mac. UNIX and Linux provide stable running of continuous calculations during a long term period (months) at processor overloading conditions. High performance computers (HPC) employed in chemical research use only these platforms. Probably, one should view Linux as the leading platform for professional applications in the present time and in the near future; one may expect that more and more chemical calculation software will be Linux-supported. However, either Windows- or Mac-equipped PCs are exploited in organic research groups. These platforms may not meet requirements of some research-oriented packages of computational chemistry. Besides, synthetic bioorganic and organic specialists, as a rule, are "Windows-addicted" and are not versed in both Linux and UNIX. These circumstances, to some extent, restrict the number of optional MM calculation packages for organic and bioorganic laboratories. For instance, a versatile drug design-targeted Discovery Studio 4.0 (by Accelrys; the related GUI is Discovery Studio Visualizer) is a multiplatform MM package that, nevertheless, does not have a version for the Mac environment.

Computers have either 32-bit or 64-bit virtual memory addresses; a computational package may use either 32-bit or 64-bit computers. Thus, before purchasing an MM computational package, one should check that the selected version of the package is compatible both with the operating system and the 32- or 64-bit architecture of the computer system intended to "host" the software.

---

versions are, e.g., Red Hat Linux Enterprise 5 and SUSE Linux Enterprise 10. Note that some simplest, tablet-oriented MM programs are designed for using late versions of Android.

[a]Computational chemists prefer Linux; correspondingly, they use Linux-operated software.

4.4 Getting Started with MM Calculations

4.4.1 *Structure preparation*

Technically, MM calculations are impressively easy to perform. For a beginner, it is sufficient to optimize geometry for a couple of molecular structures by using a GUI-equipped MM program to feel free with this calculation procedure. However, geometry optimization for one structure is rarely an ultimate task in conformational studies. To understand conformational behavior of a molecular system, it is necessary to locate all stable stereoisomers as well as transition states of their interconversions (Sections 2.8 and 3.5.2). For solving this problem by MM, one should undertake four sequential technical procedures: (*1*) prepare (on the screen) the initial structure; (*2*) perform conformational analysis (locate stable 3D forms); (*3*) locate transition states of conformer interconversions; and (*4*) establish the relationship between stable stereoisomers and corresponding transition states. Let us look through them.

The first (and the simplest) step in molecular modeling is the input of the structure to be analyzed. There are two ways to prepare the structure for computations — to create a new file that contains information for molecular geometry or to export such a finished file for a similar molecule and modify it. If opened by a text editor program, the content of such files probably would "terrify" synthetic chemists — conglomeration of numbers says nothing to them about molecules. However, the common structure of such files is not puzzling. Let us understand the general structure of the MM3 input file for one of the stereoisomers of our pattern, tertiary amine **12** (Fig. 8).

The first two lines in this text file are command lines, i.e., they contain a description of the molecule (in a free text form; positions 1–60 in the first line) and directives to the MM3 program what it should do with the structure specified very roughly in these lines. For instance, positions 1–60 in the first line are allocated for any text, number 25 (positions 65 and 66 in the first line) displays the total number of atoms in the molecule, the zero in the 72nd position of the first line indicates for the MM3 program operation *to perform geometry optimization*, number 99999 in the end of the first line means the maximal time period (min) of processor occupancy that is allotted for this operation, number 0.00002 in the second line is the convergence criterion (kcal mol^{-1}/Å). Of course, these instructions to the program may be changed or extended by the user.[a]

```
  N,N-diisopropyl-N-methylamine, conformer 1              0    25 1       0 0      99999
  0     4   0.00002                 17   0     0    0     0    0     0    0          1         0
  1     2            3    4    5     0    0     0    0     0    0     0    0     0    0         0
  2     7            0    0    0     0    0     0    0     0    0     0    0     0    0         0
  3     8            0    0    0     0    0     0    0     0    0     0    0     0    0         0
  4     6            0    0    0     0    0     0    0     0    0     0    0     0    0         0
  1     9            1   10    1    11    2    12    4    13    5    14    5    15    5        16
  6    17            6   18    6    19    7    20    7    21    7    22    8    23    8        24
  8    25
 -0.05150   -1.28450   -1.60800    1
  0.30250   -1.20450   -0.11700    1
 -0.34050   -0.06150    0.56700    8
  0.18450    1.27150    0.18700    1
  1.71250    1.33950    0.35200    1
 -0.43050    2.36850    1.07500    1
 -0.03850   -2.52450    0.59400    1
 -1.80550   -0.11250    0.51400    1
  0.49050   -2.12250   -2.10200    5
 -1.13550   -1.46250   -1.77800    5
  0.22750   -0.35950   -2.15800    5
  1.40450   -1.09250   -0.02800    5
 -0.07450    1.48750   -0.87300    5
  2.08650    2.38050    0.23400    5
  2.03650    0.99050    1.35800    5
  2.25350    0.73750   -0.41000    5
  0.05650    3.35350    0.89900    5
 -1.51250    2.52650    0.87300    5
 -0.31550    2.13650    2.15800    5
  0.12050   -2.45050    1.69300    5
 -1.08750   -2.84950    0.42500    5
  0.60850   -3.35350    0.22900    5
 -2.25350    0.57750    1.25500    5
 -2.18750    0.15950   -0.49100    5
 -2.18850   -1.11850    0.77100    5
```

The next seven lines describe chemical connectivity in molecule of **12** wherein numbers are associated with atoms; their enumeration is arbitrary (generated by the MM3 program or another computer program, which can save structural files in the MM3 format) and not of either chemical nomenclature. The first three columns below the connectivity data are x, y, z coordinates (Cartesian coordinates) of atoms relatively to a voluntarily chosen (by the molecular modeling program) coordinate origin. The fourth vertical column shows numbers that are actually symbols of MM atom types in

---

[a]The purpose of each symbol in the command and other lines is explained in the MM3 user manual.

MM3. Note that the notation in such files is line- and position-specific: any symbol is recognized by the MM3 program according to its location in a certain 80-position line.

Now, let us cast a glance at the general structure of the *pdb* file that "encodes" molecular structure of this stereoisomer of **12**. As one can see, it does not differ essentially from the structure of the file written in the MM3 format. After the first descriptive line, one can easily recognize atom numbers given by a program (the third column) as well as Cartesian coordinates of the atoms (next three columns) relatively to an arbitrary defined origin. Last 25 lines are connectivity records. Of course, there are some differences. Files in the pdb format show structural information explicitly, e.g., the last column in the pdf file indicates atomic charges.[a] In the MM3 file, this information is "hidden" in MM3 atom types. The main discrepancy is that the pdf file is exclusively about molecular geometry while the MM3 file also contains operational instructions for the MM3 program. The command feature is typical for files generated by computational packages. A minimal training could supply the beginners with skills to write/modify such text files. By means of such files, MM users "introduce" molecular structures to the calculation program. Modern computational packages can read, interconvert and save input/output files of many different formats.

```
COMPND   Title: N,N-diisopropyl-N-methylamine, conformer 1
HETATM   1    C01 UNK   0    8.232    5.731    -0.846    1.00    0.00    0
HETATM   2    C02 UNK   0    8.586    5.811    0.645     1.00    0.00    0
HETATM   3    N03 UNK   0    7.943    6.954    1.329     1.00    0.00    0
HETATM   4    C04 UNK   0    8.468    8.287    0.949     1.00    0.00    0
HETATM   5    C05 UNK   0    9.996    8.355    1.114     1.00    0.00    0
HETATM   6    C06 UNK   0    7.853    9.384    1.837     1.00    0.00    0
HETATM   7    C07 UNK   0    8.245    4.491    1.356     1.00    0.00    0
HETATM   8    C12 UNK   0    6.478    6.903    1.276     1.00    0.00    0
HETATM   9    H08 UNK   0    8.774    4.893    -1.340    1.00    0.00    0
HETATM   10   H09 UNK   0    7.148    5.553    -1.016    1.00    0.00    0
HETATM   11   H10 UNK   0    8.511    6.656    -1.396    1.00    0.00    0
HETATM   12   H11 UNK   0    9.688    5.923    0.734     1.00    0.00    0
HETATM   13   H13 UNK   0    8.209    8.503    -0.111    1.00    0.00    0
HETATM   14   H14 UNK   0    10.370   9.396    0.996     1.00    0.00    0
HETATM   15   H15 UNK   0    10.320   8.006    2.120     1.00    0.00    0
HETATM   16   H16 UNK   0    10.537   7.753    0.352     1.00    0.00    0
```

---

[a]Atomic charges are not indicated in the shown pdb file; there, zero means *not indicated* (and not zero charge).

```
HETATM   17  H17 UNK   0   8.340   10.369   1.661   1.00   0.00   0
HETATM   18  H18 UNK   0   6.771    9.542   1.635   1.00   0.00   0
HETATM   19  H19 UNK   0   7.968    9.152   2.920   1.00   0.00   0
HETATM   20  H20 UNK   0   8.404    4.565   2.455   1.00   0.00   0
HETATM   21  H21 UNK   0   7.196    4.166   1.187   1.00   0.00   0
HETATM   22  H22 UNK   0   8.892    3.662   0.991   1.00   0.00   0
HETATM   23  H23 UNK   0   6.030    7.593   2.017   1.00   0.00   0
HETATM   24  H24 UNK   0   6.096    7.175   0.271   1.00   0.00   0
HETATM   25  H25 UNK   0   6.095    5.897   1.533   1.00   0.00   0
CONECT    1    2    9   10   11
CONECT    2    1    3    7   12
CONECT    3    2    4    8
CONECT    4    3    5    6   13
CONECT    5    4   14   15   16
CONECT    6    4   17   18   19
CONECT    7    2   20   21   22
CONECT    8    3   23   24   25
CONECT    9    1
CONECT   10    1
CONECT   11    1
CONECT   12    2
CONECT   13    4
CONECT   14    5
CONECT   15    5
CONECT   16    5
CONECT   17    6
CONECT   18    6
CONECT   19    6
CONECT   20    7
CONECT   21    7
CONECT   22    7
CONECT   23    8
CONECT   24    8
CONECT   25    8
END
```

Fortunately, nowadays there is no need to spend time for a laborious work of manually preparing input files. GUIs of MM and QM packages allow creation or modification of virtual 3D molecular structures on the screen. Drawing of a 3D molecule on the screen creates the corresponding structural file in the computer RAM. Option *Save* in the GUI means saving this file (in any chosen format from the menu of the format set of the package) in the computer hard disk, option *Open* means reading saved text files as molecular graphics (not as a text). Versatile additional options of GUIs, as e.g., *Add/Hide hydrogens*, *Replace atom*, *Add/Delete bond*, *Invert configuration*, etc., facilitate "preparation" of a required initial structure. Due to such a comfort, drawing of a 50–100 atom molecule (and, hence, creation of the molecular structure file) takes no more than several minutes.

Its 3D geometry is far from that of any realistic molecular structure, and one should optimize it.

4.4.2 *Conformational analysis in a trice*

Geometry optimization (energy minimization) is initiated elementarily with using the GUI. It is sufficient to indicate the force field, minimization method (e.g., the Newton-Raphson minimization; Section 4.2.4), modeled conditions (vacuum or a solvent model) and cut-off (if necessary; Section 4.2.4). Then command *Minimize* or a similar one in the GUI starts MM calculations, iterations in energy minimization rapidly change the molecular geometry (its evolution usually is displayed on the screen), and a stereoisomeric structure of optimized geometry is the result of this procedure.

MM programs save information about optimization process and optimized structures by generating new file(s). Typically, molecular geometry (i.e., atom coordinates) and different additional geometrical parameters (e.g., distances between atoms of any atom pair of the system) are written down in such a file (output file). This text file also contains energy characteristics of the optimized structure (E_{ster}, and components of E_{ster}) as well as some other calculated molecular properties (e.g., atomic charges, dipole moment, vibrational frequencies). Thus, output files are program-specific "soft" documents that present all calculation results as a text. Such text files, which have been created by a professional MM calculation program, are readable by MS Word. However, it is more important that they are recognizable for other solid calculation packages. This recognition means that a good calculation program may operate with an imported structure of optimized geometry (may display it on the screen, modify, re-optimize the structure using another force field, etc.). In particular, output files can be utilized for generation of new input files. Obeying GUI commands, the MM program extracts the required structural information for the optimized molecular structure from the output file and saves this "3D structure" in the produced input file of any selected conventional format, e.g., in that of the input file for a QM calculation package.

An optimized structure, of course, is not the only possible stereoisomer of the modeled molecule. Geometry minimization forces the system to "fall" into the nearest minimum in the SES. This energy minimum is scarcely global (the lowest in energy in the SES); rather it is a local one (a minimum that is not global). We apprehend that a single elementary procedure of MM calculations — geometry optimization applied to a molecular structure of either shape — does not provide even a token understanding of conformational equilibrium for this molecular system. In contrast to a single experimental observation using a fast experimental method (Section 3.3), a single geometry optimization does not show whether the modeled stable 3D structure is the conformer that appreciably contributes to the real conformational equilibrium. It is quite possible that it is a high energy structure, i.e., a very minor conformer, for modeled conditions. Information about other stereoisomers is still absent. Hence, it is necessary to "find" (i.e. to model) them. In other words, one should locate all other minima in the SES, or, in methodological terms, perform conformational search.

How is it technically performed? It is mentioned (Section 4.3.2) that modern MM packages include an option of conformational search. Conformational searches, in which an algorithm of random generation of arbitrary 3D geometries is embedded, are an effective methodology of computational conformational analysis for organic molecules of ordinary size. Certain mathematical algorithms randomly alter Cartesian or internal coordinates of atoms of the molecular backbone while keeping chemical connectivity and (optionally) original configuration of elementary molecular fragments unchanged. As a result, different distorted molecular geometries are produced; after being generated, each of them is automatically subjected to geometry optimization (Section 4.2.4). Different "fates" await the generated intermediate (non-optimized) stereoisomeric structures in this basic MM procedure. Most structures of perturbated geometry are transformed to either stable stereoisomer. Many of the resulting structures are the same, i.e., duplication of stereoisomers is a usual "side product" in this energy minimization. Some "ugly" molecular frameworks do not undergo energy minimization. Their 3D geometry is distorted so significantly that minimization algorithms cannot restore normal values of geometrical parameters — any geometry change upon geometry optimization leads to further increase of E_{ster}. Thus, upon stochastic conformational search, the "structure waste" is huge; therefore, the number of intermediate structures to be generated should be sufficiently large to produce an entire set of stereoisomers of distinct geometry. Nevertheless, in spite of generating computer time-consuming "waste," random methods are effective in conformational search. The number of generated geometries finally does not appear gigantic to locate all or, at least, almost all energy minima in SESs for small organic (not polymeric) molecular systems. The ratio *generated structures:located energy minima* is, say, in the range of 30–200.

Indeed, for 20–200 atom monocyclic or polycyclic organic molecules, stochastic conformational search is excellent: practically, it does not miss stable stereoisomeric structures (energy minima in the SES) for a molecule if a sufficiently large set (10^3–10^5) of different probe structures is generated. Because each automatic procedure *stereoisomer generation + geometry optimization* usually takes less than 1 s, exploration of conformational space for such systems is completed within a half of hour or, in an unfortunate case, in a few hours. This performance of MM impresses every organic chemist who, for the first time, has encountered location of conformers by MM. Instead of speculative considerations of molecular geometry and conformer stability, computerized conformational search immediately[a] and "by itself" (i.e., not requiring the researcher's accompaniment) provides all possible conformers of accurate geometry and at least their fair relative energies[b] as well.

As explained, the use of stochastic conformational search is meaningless when applied to molecular systems with an astronomic number of stereoisomers (i.e., conformers at ambient conditions). Moreover, no other method (MD, simulated annealing[c]) could provide exhaustive exploration of conformational space of such molecules. Linear chains are a vivid example of molecular structures characterized by such a conformational diversity (Section 1.6). The size of such molecular systems even should not be large — exploiting modern supercomputers, no conformational search could guarantee location of the global energy minimum, e.g., for a linear octapeptide.

During running conformational search, professional MM packages collect stereoisomers of optimized distinct geometry (create an output file that contains structural and energy information for each stereoisomer). Using structure analysis options (controlled by the GUI) of the used calculation package, the pool of these "filed" stereoisomers may be readily arranged, e.g., they may be sequenced according to the increase of E_{ster} (decrease of stability) or grouped according to the chosen geometrical criterion(s). Of

---

[a] Say, in a third of the business day by means of MacroModel (Section 4.3.1) when generating 5×10^4 initial structures for a 25–30 atom molecule and optimizing their geometry.

[b] Of course, under the condition that the employed force field suits the chemical structure studied.

[c] Not explained herein. Sophisticated methods of location of energy minima are usually not required even for macrocyclic molecules; much larger systems, biopolymers, are their target. Nonetheless, one could indicate a remarkable feature of these methods. They provide crossing of energy barriers between energy minima in the SES, in contrast to energy minimization in MM (Section 4.2.4).

course, any resulting stereoisomer may be separately "called" to the screen and analyzed.

With reliable results of computational conformational analysis (3D geometries and relative stabilities of conformers) in hand, one can easily interpret experimental spectra of conformationally flexible compounds.[a] Suppose that a force field well models (locates with supplying accurate geometry and energy values; Section 4.2.6) conformers of a flexible molecular system, which NMR spectra for solution have been recorded at room temperature. Results of this successful molecular modeling permit to easily identify conformers that contribute to the spectra. MM quantitatively resolves the "blurred" experimentally observed conformational equilibrium into its components (i.e., conformers): we merely know them from the MM modeling. Values of ΔE_{ster} for these conformers reflect their relative content (Section 3.2, Eq. 6). As commonly known, analytical methods do not have unlimited sensitivity and therefore cannot detect substances in low concentrations. One may realize that *located conformers* with percentage of more than 1% are exactly those that give the recordable, though averaging-masked, NMR response for the studied compound.

This example also shows how results of computational conformational analysis for a flexible molecule can disclose straight away whether a related room temperature NMR spectrum for solution is (*1*) an in-time-averaged spectrum of the highly predominated conformer and its very minor, "NMR-invisible" conformers (hidden partners; Section 3.3); or (*2*) an average of spectra of near stability conformers. Such a differentiation may be problematic by solely using NMR (Section 6.1.3). In contrast, values of ΔE_{ster} for located conformers immediately indicate which situation, (*1*) or (*2*), the NMR spectrum reflects.

Unfortunately, MM frequently does not supply a very high accuracy in estimating relative thermodynamic stability of conformers of organic molecules in solution. Moreover, ΔE_{ster} is not a precise "model" of ΔG° (Section 4.2.3). Organic experimentalists often forget this and relate to ΔE_{ster} values as to true *quantitative* estimates. Note that these values should be approached rather as more or less reliable *qualitative* estimates. For instance, applying a force field with proven accuracy of 0.5 kcal mol^{-1} for some chemical structures to such a molecular system (Section 4.2.6), we almost certainly establish which of the two conformers with $\Delta E_{ster} > 0.5$ kcal mol^{-1} is major.

Thus, MM-based computational conformational analysis is nowadays a fast, effective and convenient procedure. However, we must not forget some duality of MM in accuracy (Section 4.2.6). These calculations are more or less successful in modeling molecular geometry but may be very rough in estimating relative stability of stereoisomeric structures. It is worth viewing MM-located stereoisomers as structures of identified geometry, but of

---

[a]Recall, most organic compounds have conformational mobility at room temperature. E.g., sterically unhindered C–C, C–O, C–B, S–S, etc. rotamers are conformers even at much lower temperatures.

not quite clear relative stability. The latter has to be refined by trustworthy *ab initio* calculations (Chapter 5) or proved via comparison with experimental values of $\Delta G°$. That is, conformational analysis by MM supplies organic researchers rather with good starting structures for further, refining calculations than with a set of stereoisomers firmly arranged according to their relative stability.

4.4.3 *Location of transition states in MM*

If the selected force field is adequate for a molecule, location of only energy minima for this molecular system does reflect the conformational equilibrium. However, the modeled equilibrium is seen by us from the static perspective: we know which stable stereoisomers do exist and what their relative stability is. We still do not know how rapidly/slowly the stereoisomers interconvert. Therefore, we cannot indicate whether the located stereoisomers are conformers at given thermodynamic and chemical conditions (temperature, pressure, medium) or which of these stereoisomers are conformers at these conditions. To identify conformers, i.e., to estimate τ for each located stable stereoisomer (Section 2.5; in kinetic terms, to estimate interconversion barriers), one should also locate inflection points (Section 3.1) in the SES. In surfaces of potential energy for molecules, they almost always are saddle points (Fig. 56).

Routine NMR spectra do not indicate that stereoisomers interconvert (i.e. stereoisomers are conformers at given conditions) if these intramolecular stereoreorganizations occur rapidly; formally, such spectra only present one 3D structure (Section 3.3). One may mistake the virtual structure, which actually masks a fast equilibrium of two or more conformers, for a rigid stereoisomer. Misinterpretation of NMR data that consists of judging the non-existing spectral structure as a real, rigid 3D molecular structure is a typical methodological mistake of many synthetic chemists. When modeling both stereoisomers (Section 4.2.2) and corresponding transition states (this Section), MM modeling assists in simply escaping it.

Saddle points in the PES (Section 3.1) or, correspondingly, saddle points in the SES correspond to transition states[a] of intramolecular

---

[a]i.e., structures of maximal energy in the MEP (Section 3.2, Fig. 58). There is a settled terminological inaccuracy here. Literally, transition states should be undestood as states. Nevertheless, in the traditional usage of this term, transition states are termodynamically unstable molecular structures and not states.

3D reorganizations (in stereochemical terms, transition states of stereoisomerizations). We remember that relationship *molecular geometry–molecular energy* is not revealed by any primitive model as, e.g., by the principle of minimal geometrical distortions for traveling from an energy minimum to another one [Sections 1.7 and 1.8, subsection (*C*)]. Therefore, for any two located related stereoisomers (i.e., separated by one barrier in the MEP; Fig. 58), we *a priori* know neither the geometry of the transition state for their interconversion nor its energy. Location of transition states requires sophisticated methodologies.

In computational practice, the situation with identification of transition states of stereoisomerizations is nevertheless not so embarrassing. Inadequacy of the principle of minimal geometrical distortions does not mean that, if we follow the minimal distortion path for a stereorearrangement, we will never find the transition state there or in some proximity to this path. Frequently, e.g., for trivial rotations around a single bond, the desired saddle point is in exactly the region of this path in changing molecular geometry. Then, using elementary geometrical considerations, it is easy to reasonably assume what the geometry of the transition state approximately is [see subsection (*A*) below]. In other words, one can suggest a structure that is related to a small region in the PES (or SES) where the transition state lies, i.e., to merely indicate the molecular geometry that probably corresponds to the saddle point region (Section 2.4, Fig. 43). Success in guessing near-transition state structure promises success in locating the transition state itself. As explained, some energy minimization algorithms are capable of locating this stationary point in the SES (i.e., the transition state) if the initial (pre-optimized) structure is geometrically close to the transition state (Section 4.2.4).

(*A*) *Location of transition states of assumed geometry.* Let us consider in more detail modeling of rotation around a single bond. In MM, such intramolecular rotation is approximated by formal rotation around the corresponding σ bond in the MM molecular framework. Formal rotation of a molecular fragment provides motion of its nuclei along the minimal distortion pathway, where only one geometry parameter (one torsional angle) is changed in providing a circular trajectory for all nuclei of the fragment excluding those lying in the rotation axis; bond lengths and bond angles keep their initial values. MM, representing molecules as frameworks of pairwise connected atoms (i.e., as mechanistic models), is capable of modeling such formal rotation of molecular fragments and improves this rigid rotor model by imparting reasonable freedom to bond lengths and angles.

In this well-known procedure, the rotational path (the range of the torsion angle to be changed) is divided into several equal short segments (rotation steps), according to the researcher's choice. Steric energy is calculated after each rotational step; thereby geometry optimization includes a freedom for changing all geometrical parameters excluding the fixed torsion angle (the rotation angle). In other words, simulated rotation is performed gradually, rotation step by rotation step. MM packages reduce these operations to a primitive routine: using the GUI, the bond of interest is selected in the displayed 3D structure, and the initial and final torsion angles as well as the rotation step are elementarily specified by typing their values on the screen.[a] In this way, a set of initial structures is prepared. Geometry minimization with one geometrical constraint (the related value of the rotation angle) is performed for each initial structure.

This sequence of energy minimizations with one structural constraint (one "frozen" torsion angle in each rotation step) establishes a quantitative dependence *steric energy–rotational angle* (Fig. 83). Such a rotational plot is a cross-section of the SES (Fig. 55) that lies in the MEP plane and crosses a saddle point in the SES. It is obvious that rotational procedure locates some region of the SES (along coordinate *rotation angle*) that includes the value of this angle in the transition state of rotation.

The maximum of E_{ster} in the obtained rotational plot indicates the point that approximately corresponds to the transition state of rotation. Consequently, the related value of ΔE_{ster} (the larger ΔE_{ster} value among two ones when the initial and final stable rotamers are of distinct energy) is the calculated rotational barrier. It corresponds to the barrier that determines the rate of rotation around the considered bond in the real molecular system.

Some programs (e.g., original versions of MM3) enable to model simultaneous rotation around two and more bonds. Results *two bond rotation–steric energy* often present in the form of a two-dimensional map of equipotential energy contours. Such a map is actually a superposition of cross-sections of the SES that are parallel to the plane of the

---

[a]Without a GUI, the researcher should be familiar with the structure of the program-related input file. This file has to be modified for specifying details of the rotational procedure to be executed.

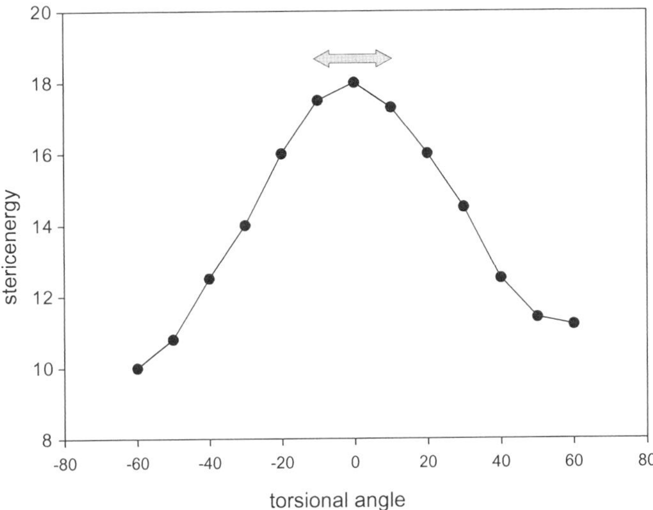

Figure 83. Crude modeling of intramolecular rotation (the magnitude of rotation is 120°, rotation step is 10°). Energy minima are related to a pair of stable rotamers, the transition state of their interconversion lies in the region of maximal energy. The corresponding range of torsion angles (indicated by gray arrow) may be explored with a lesser rotation step in order to locate the transition state more accurately.

used geometrical variables (the xy plane; Fig. 84). Such contour plots are shown for rotation around two bonds in amine **12** modeled by the two bond rotation option of MM3(96) coupled with constrained energy minimization at each rotation step (Fig. 84).

The term *approximately* in regard to locating the rotation transition state is intentionally used in the previous paragraph. This approximate result is not related to the accuracy of MM calculations: it appears due to the stepwise fashion in modeling the rotation path. The modeled trajectory in the SES is fractured while the correct MEP is, of course, a continuous line in the PES. Therefore, the lesser the rotational step taken, the more accurate the value for the rotational barrier. Usually, a sufficiently large rotation step (10–15°) is chosen in the initial stage of calculations. Since the transition state lies in the region of the three highest point peak (Fig. 83), the refining rotational procedure with a much lesser step (e.g., 0.1–1°) is undertaken within this only region. Then the rotational barrier is estimated with a fair accuracy.

CONTOUR PLOT

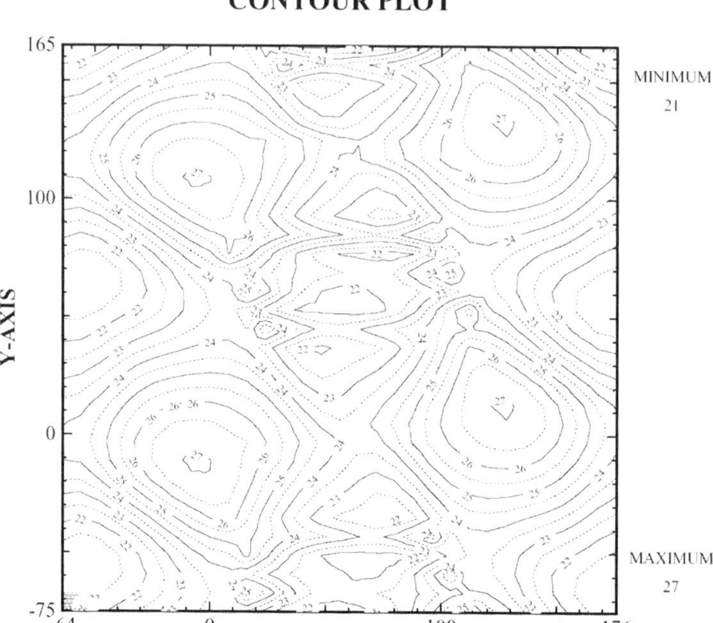

Figure 84.   The SES region from Fig. 70 [a 240° rotation of both N-*iso*-Pr groups in amine **12**] shown in a 2D representation of equipotential energy contours [modeling by MM3(96), rotation step of 8°]. Neighboring contour lines are separated by the difference of 0.5 kcal mol^{-1}.

Note, any internal rotation cannot be modeled by MM. Cotemporary force fields cannot correctly represent the rotation for bonds that do not have cylindrical (C_∞) symmetry of bond-forming orbitals. This chemical bonding is changed upon internal rotation in such diatomic fragments of real molecules. Bonds formed by two π electrons (as e.g., olefins, annulenes) or ylide type bonds (as e.g., in phosphazenes $R_3P{=}NR$ or protonated phosphinoxides $R_3P{=}OH^+$) are clear examples. In C=C bonds, the π bond is formed by overlapping of two single-occupied *coplanar* p-orbitals. No π bonding is present in ylides. The "second" bond in the phosphazene fragment (in addition to an ordinary σ bond P–N) results from overlapping of an unoccupied non-bonding p-orbital of P and a double-occupied bonding p-orbital of N; these orbitals are *coplanar*. In both cases, rotation around a "double" bond disrupts coplanarity of p orbitals of these adjacent atoms — these p-orbitals are orthogonal in the transition state. The integrity of the molecular structure is kept only due to the remaining σ bond C–C or P–N, respectively. Parameterization of current force fields does not comprise such structures. Successful barrier values for such systems might appear by only occasion.

Thus, the *step-by-step rotation–constrained geometry optimization* "trick" in MM (also in QM) provides only approximate values of rotational or rotation-associated (see below) barriers. Could one eliminate principal inaccuracy in approaching the rotation transition state stepwise? The Newton-Raphson full matrix energy minimization is capable of locating saddle points (Section 4.2.4). We could recall that, if MM energy minimization is executed for a structure that is related to a SES point which is near a saddle point (i.e., for a structure that geometry is similar to the geometry of a transition state), the result is location of this transition state. Regarding the full matrix Newton-Raphson method, as Cramer's monograph[33b] noticed: "This can be a pleasant feature, if one is looking for the transition state in question, or an annoying one, if one is not." This is exactly the starting situation we have when the highest energy point is located by means of the step-by-step rotation (Fig. 83). This point in a fractured rotation plot is near the maximal energy point in the hypothetical smooth rotation plot; clearly, the latter point corresponds to the rotation transition state. Therefore, applying the full matrix minimization option[a] to optimizing geometry of such a high energy structure, we remove the single structural constraint (the "frozen" rotational angle) and accurately locate the desired transition state.

Possibilities of this simple rotational procedure in MM are broader than they are usually regarded. If there is no isolated rotation (ISR) in a real system, and rotation is coupled with another intramolecular motion, e.g., with rotation around another bond (gear effect), or with a stereorearrangement of an elementary molecular fragment, this procedure leads to location of the transition state of this concerted intramolecular motion. Then, the top in the generated rotational plot does not correspond to the rotational transition state that, in essence, does not exist in this case. The maximum point of the plot lies in the region of close proximity to the transition state of a rotation-associated *concerted* 3D rearrangement.[b] Coupling of intramolecular motions for either molecular structure may not be even *a priori* assumed by a researcher. However, the gradual rotation around bonds in mechanistic "MM molecules" (Fig. 69) accompanied by energy minimization at each rotation step (with "freezing" the rotational angle in each

---

[a]This option, of course, is present in professional MM packages.
[b]I.e., rotation and another motion occur synchronously and have one common transition state.

minimization) ascertains its occurrence. The final step in locating the transition state for rotation coupled with another intramolecular motion (i.e., the transition state of a concerted intramolecular motion) is the full matrix energy minimization for the structure that has the maximal energy among other structures "from the rotational plot." The ability to descry gearing of internal rotation and another intramolecular motion (e.g., rotation around another bond, configurational inversion, ring reversal) is a genuine discovering feature of this MM procedure *stepwise rotation + constrained geometry optimization* followed by full matrix energy minimization for the highest energy structure from the obtained set of optimized structures with the "frozen" rotation angle.

Indeed, for bonded non-trivial fragments, e.g., two neighboring elementary units of trigonal bipyramidal geometry $A_a(L_i)_5$ and $A_b(L_j)_5$, it is impossible to foresee whether rotation around the bond that connects their central atoms A_a and A_b occurs separately or is geared with stereomutation of geminal substituents L_i or L_j. Only theoretical modeling may solve this problem for a concrete pair of A_a–A_b-bonded trigonal bipyramids. The MM rotational procedure turns out to be a simple effective tool for identifying rotation-associated concerted motions under condition that the used force field can model spatial reorganization of the involved elementary molecular fragment.[a]

For example, rotating the N-substituent in amine **38** (Fig. 85) and minimizing the energy by means of MM3(96), we locate the transition state of a concerted motion, NIR [Section 1.8, subsection (1)].[25b] Obviously, isolated rotation (ISR) of the N substituent does not take place in the N-neopentyl compound due to intolerable steric contacts of the N substituent and the bicyclic backbone upon C–N rotation.

Note that the considered MM procedure *stepwise rotation + geometry optimization* (with final full matrix energy minimization) does not locate the transition state of a rotation-involving concerted intramolecular motion if both ISR and this motion occur there. In this case, this MM procedure only locates the transition state of ISR, and, thus, does not reveal the other stereoisomerization that occurs via the motion gearing. For instance, if the same MM3-assisted rotational procedure is applied to structural analogs **5d** and **38**, the outcome is different. As indicated, this MM procedure applied to amine **38** leads to the

---

[a]Unfortunately, unlike modeling of rotation, a successful MM modeling is not warranted for stereomutations in many chemical polyhedrons (see below).

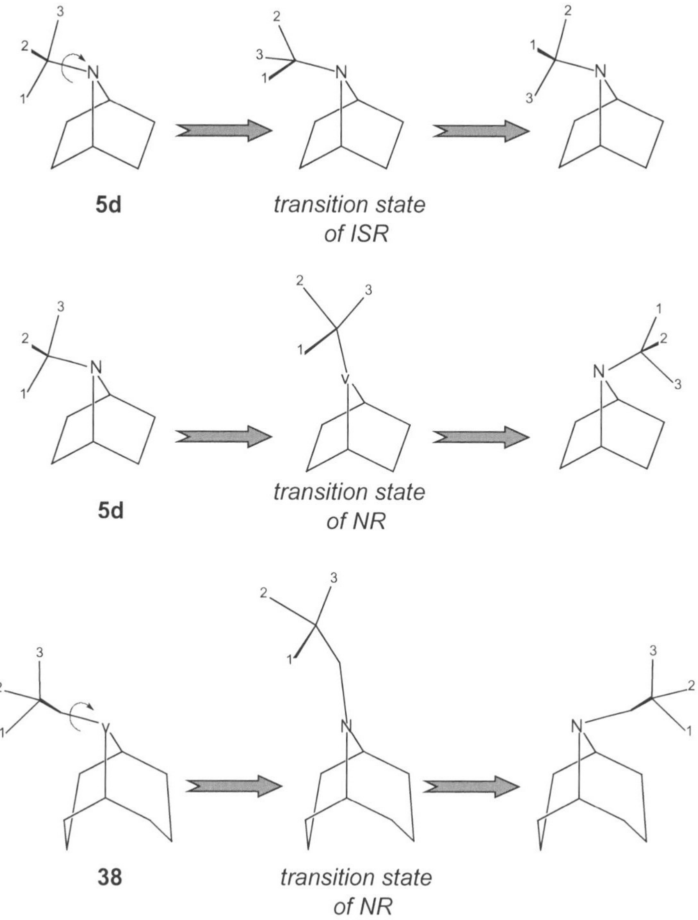

Figure 85. Different outcomes of the rotation — energy minimization procedure [performed using the original version of MM3(96)] for the N-substituent in azabicycles **5d** and **38** (numbers depict Me groups). In amine **5d**, both ISR and NIR may occur while only NIR takes place for amine **38**.

NIR transition state.[a] For the N-*tert*-Bu compound **5d**, this rotational procedure for the N

---

[a]Geometries of transition states of ISR and NIR are dissimilar or, equivalently, the corresponding saddle points have different locations in the SES. The ISR and NIR barrier heights may be close; therefore, by comparing the measured barrier (e.g., by NMR) and *one* calculated barrier (such truncated comparison is a notorious methodological mistake), e.g., the barrier for ISR, one may incorrectly assign the experimentally detected conformational transformation (here: to ISR). Assignment of the

substituent supplies the transition state of its rotation.[4] Location of only rotamers of **5d** via MM-computed rotation does not mean that NIR does not take place in this compound; NIR is an attribute of stereodynamics of most alkylamines, and this N-*tert*-Bu amine is not an exception. As MM calculations show, this concerted conformational transformation occurs in both azabicycles **5d** and **38** (Fig. 85).[4,25b] Merely, location of the NIR transition state for amine **5d** requires another MM procedure than the rotational one (see below).

Thus, rotational mobility of purely σ-bonded elementary molecular fragments "covered" by a force field parameterization may be modeled by MM. And what about internal mobility of these fragments? In this regard, MM is explicitly unpretentious. Two factors reduce MM to an almost impractical instrument for studies of positional exchange of geminal substituents. They are (*1*) non-applicability (insufficient parameterization) of force fields to many chemical polyhedrons; and (*2*) limitations of equations for bond angle bending (see below) in so far developed force fields.

For instance, isolated nitrogen inversion in alkylamines[37] as well as N-inversion-associated concerted motions [e.g. NIR, RINI; Section 1.8, subsection (*A*) and Section 2.2, subsection (2)] in these compounds[4,25b,56] are modeled by MM more or less accurately.[a] E.g., the experimental NIR barrier for amine **14a** is well-reproduced by MM3(96) (Fig. 86).[25b] If a similar intramolecular dynamic process — pyramidal S-inversion — is modeled by fine force fields for thiophene oxide **38** (Fig. 86), difficulties of parameter deficiency appear quickly. MM3(96) as well as CHARMM do not have all parameters needed to optimize the geometry even for the single energy minimum of this compound. If the MM3 program compensates this lack of parameters by resorting to surrogate values (MM3 parameters developed for some other fragments), the calculated barrier is inaccurate — 11.4 kcal mol^{-1} (experimental barrier of S-inversion in such sulfoxides[57] is *ca.* 14.8 kcal mol^{-1} for solution). At first glance, it seems that another prominent force field, MMFF94, may be suitable. Not requiring additional parameters, it does yield an S pseudotetrahedron for the stable stereoisomer of this pseudoylide. However, the resulting value for the S inversion barrier is unacceptable: it is 32.9 kcal mol^{-1}. Obviously, this force field does not take into account an additional dissipation of the p electron density in a planar transition state of **39** (in orbital terms, a maximal overlap of coplanar p orbitals of the sulfur and cyclic carbons). In other words, MMFF94 actually has no proper parameterization for this elementary system.

---

experimental barrier to a certain intramolecular motion (Section 3.2) requires modeling of alternative transition states (here: those of ISR and NIR).

[a]Excluding alkylamines with essentially distorted CNC angles. Parameterization of current force fields is inadequate for modeling NIR for these molecular structures.

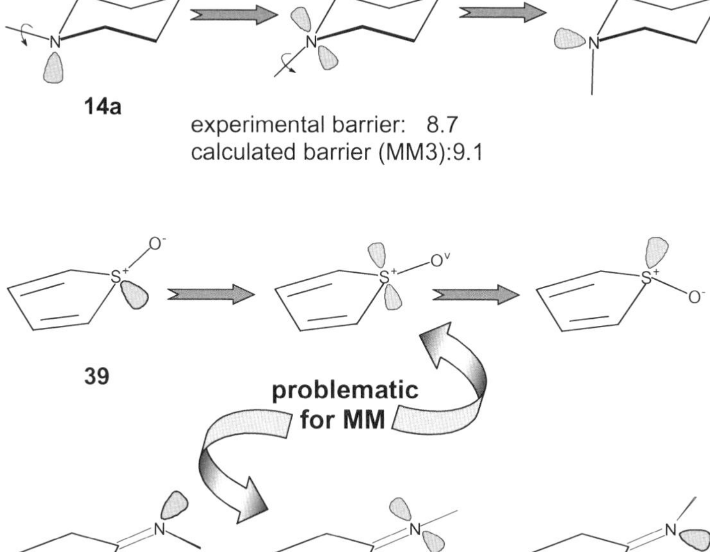

Figure 86. Examples of successful and unavailing models of transition states of 3D rearrangements of elementary molecular fragments by MM: NIR in amine **14a**, pyramidal S inversion in **39** and in-plane N-inversion in **40**. Barrier values are in kcal mol^{-1}.

The second factor is that truncated forms of the Taylor expansion for the potential energy of valence angle bending, i.e., by Eq. 17 (Section 4.2.2) with a few first expansion terms, are essentially inaccurate when deviations of angles from their "normal" values are significant (in contrast to the torsion angle-describing Eq. 18). In addition, if a transition state has a linear geometry (the bond angle under change is 180° in the transition state), the angle bending energy is incorrectly calculated by Eq. 17 at any truncation of the Taylor expansion.

Modeling of in-plane N-inversion in azomethine **40** (Fig. 86) by means of MM3(96) illustrates this. This force field does not have parameterization problems when modeling stable stereoisomers of this compound and affords the anticipated virtual trigonal geometry for the N fragment. Let us take a glance at the transition state of this inversion. QM calculations easily locate the transition state; it has a linear geometry C=N–C. However, this geometry does not correspond to a stationary point in the SES built by MM3. Full matrix Newton-Raphson minimization of energy does not locate any transition state when the QM-derived transition state of linear geometry C=N–C is used as a starting geometry for this calculation procedure in MM3. The force field "forces" this structure to fall into the energy minimum that corresponds to the located stable stereoisomer of **40**. It is transparent that MM3 generates a false SES in the region of a linear geometry of the nitrogen fragment; in other words, it does not "feel" that a saddle point should be there. Another "general organic" force field, MMFF94, does locates a saddle point in this SES region and delivers a better geometry for azomethine **40** (bond angle CNC in this modeled structure is 171.6°). However, it supplies an absolutely unacceptable value for this N inversion (33.5 kcal mol^{-1}). It is quite understandable that also this force field generates for molecular structure **40** a corrupted SES region.

One should not think that the principle of minimal structural distortions, very often predicting the general geometry of transition states for different polytopal rearrangements (Section 1.7, Fig. 18) correctly, may assist in modeling stereodynamics of chemical polyhedrons by MM. At first sight, it seems that, if it is known (e.g., from QM modeling) that the transition state geometry provided by the minimal distortion principle indeed is the geometry of the transition state, one may use this geometry with confidence when modeling conformational transition states by MM. The technique of this modeling is simple: GUIs of solid calculation packages as well as versatile programs for molecular graphics enable rendering of such a verisimilar geometry of the transition state on the screen.[a] Energy minimization with using the full matrix Newton-Raphson method is the next (and ultimate) procedure that should locate the required transition state. Unfortunately, because of the limitations indicated above, such a successful modeling is rather an exception than a rule.

Technically, geometrically simplest pre-optimized (crude) "transition states" may be built on the screen by means of a simple trick. It consists of using, as a template, an elementary molecular unit which *stable* geometry corresponds to the geometry of the modeled elementary molecular unit in the assumed *transition state*. For example, in order to have a linear geometry for fragment C=N–CH$_3$ in azomethine **40** (Fig. 84), one may use allene unit H$_2$C=C=CH$_2$ as a template of linear geometry for a three-atom elementary molecular

---

[a]With using GUI options for molecular graphics (e.g., using structural templates, adding/deleting/ replacing atoms and bonds, numerically specifying bond lengths, bond and torsional angles), geometry is readily manipulated for a virtual molecule on the screen.

fragment. By means of GUI-supported options for deleting/adding/replacing atoms and changing bond multiplicity,[a] the $C{=}N{-}CH_3$ unit in the on-the-screen molecule of **40** is transformed into allene unit $C{=}C{=}CH_2$.[b] Then energy minimization for the modified molecule is undertaken by means of any force field that accepts this chemical structure and keeps the linearity of the allene fragment in optimizing the molecular geometry.[b] The building of the required linear *geometry* is accomplished. The required *chemical* structure $C{=}N{-}CH_3$ is reconstituted by using the mentioned options of molecular building. The resulting 3D molecular structure of azomethine **40** is the desired rough geometry of the transition state of in-plane N inversion in this molecule. Another example is pyramidal inversion. Fragment BR_3 is a convenient template for building a crude (pre-optimized) geometry of the trigonal conformational transition state on the screen.

Such template-built molecular structures, however, do not have a sufficiently good geometry for further locating the transition state by using them as initial structures for the full matrix Newton-Raphson geometry optimization that is supposed to provide the transition state itself. If used, it may rather occur that the molecule "falls" into the nearest energy minimum instead of reaching the saddle point thought of being "right here" in the SES. It is therefore better to improve their geometry and, after improving it, to use this Newton–Raphson procedure for locating the saddle point. That is, the template-built geometry should be brought nearer a true geometry of the transition state (i.e., one should further "draw" the molecule near the saddle point in the SES). Amelioration of the template-built geometry of rotational transition states (see above) is an illustration of such a technique; it may be called geometry annealing. As explained, for annealing 3D geometry of transition states of internal rotation (Fig. 83), the rotation angle(s) is(are) frozen while other geometrical parameters are free to be changed upon energy minimization. In annealing the template-built geometry of the transition state of a geminal stereorearrangement, frozen geometrical parameters are bond angles L_iAL_j of the modified elementary molecular unit $A(L_i)_n$, while the rest of the molecule is free for geometrically changing in energy minimization. That is, the initial, geometry-annealing energy minimization is executed with keeping these geometrical constraints and, of course, using the proper fine force field. The resulting molecular geometry is an excellent initial geometry for the final, full matrix Newton-Raphson energy minimization.

Hence, stereorearrangements of elementary molecular units may be satisfactory modeled by available force fields if (*1*) parameterization comprises the considered unit and also (*2*) changes of bond angles going from the stable stereoisomer to the related transition state are not significant as, e.g., they are upon pyramidal N-inversion in alkylamines. Indeed, in modeling

---

[a]GUI-operated molecular builders insert elementary molecular fragments as VSEPR-dictated polygons or polyhedrons. Also, by choosing the atom of either valency and bond multiplicity on the screen, we make the force fields "know" what the MM atom type of the chemical element is.

[b]E.g., the auxiliary MMX (Section 4.3.1) may be used.

NIR or RINI for these compounds [Section 1.8, subsection (A) and Section 2.2, subsection (2)], one can successfully use the structure with a planar trigonal arrangement of three N substituents (Fig. 20) as initial one for the MM procedure of full matrix energy minimization (as explained, the consummating step in locating transition states by MM).[4,25b,36g]

For instance, the transition state of NIR in azabicycle **5d** (Fig. 87) is easily modeled in this way. All three transition states of NIR are geometrically identical for **5d**. One can reasonably assume that the N fragment is planar, and the Me's of the N-*tert*-Bu substituent are in neither eclipsed, nor staggered conformation relative to the two other N substituents.[a] After building this molecular geometry on the screen, minimizing it with geometrical constraints that keep the N fragment planar, full matrix Newton-Raphson energy minimization affords the desired transition state (Fig. 87).

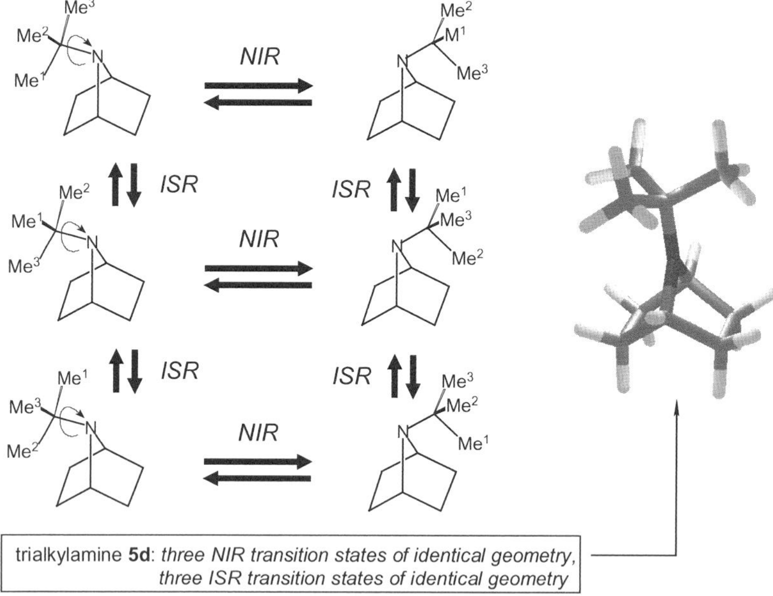

Figure 87. Conformational scheme for amine in amine **5d** and the modeled NIR transition state [by MM3(96)]. Rotational transition states are not shown (there, ISR means C–N rotation).

---

[a]A general conformation of a NIR transition state in alkylamines is shown in Fig. 21, top.

One may expect that, e.g., pyramidal inversion in alkylphosphines could be modeled using MM (the change of each CPC angle is less than $10°$) while edge inversion could not (changes of these bond angles are more than $20°$). Nevertheless, such conclusions are not a methodological advice because the feasible range of bond angle bending from Eq. 17 is unknown in advance, for many chemical elementary fragments. We see again that the physically imperfect basis of MM (Section 4.1) appreciably restricts this simple approach to molecular modeling. *In research, it is always preferable to use QM calculations (at least, at the final stage), when modeling stereorearrangements of elementary molecular fragments in organic molecules of "normal" size.* The role of MM in modeling of this stereodynamics is, nonetheless, not negligible. "Near-transition-state" structures that are produced by MM energy minimizations with freezing either parameters of molecular geometry (see above) may appear successful *initial* molecular structures for further, completing QM modeling of transition states (Section 5.2.2) of any stereoisomerizations. Simply put, such a MM molecular structure that has been obtained by using an appropriate fine force field, is worth subjecting to QM calculations with indicating transition state search according to the apropos option in the used computational QM package (e.g., in either mentioned in Section 5.2.2). If the researcher's guess for the "near-transition-state" structure is correct and the force field supplies this molecular structure, QM rather will locate the desired transition state.

(B) Location of numerous transition states without preliminary considerations of molecular geometry. The above described MM strategy for modeling of stereoisomerization transition states (rotational or others) are convenient if both (*1*) the molecular geometry of such a transition state is in general clear and (2) only one or a few transition states are intended to be located. E.g., rotation of the *tert*-Bu substituent in any conformer of cyclohexane **18b** (Fig. 31) or NIR in azanorbornane **5d** (Fig. 87) are suitable cases for following this strategy.

However, even simple organic molecules may have an *a priori* uncertain conformational freedom. For instance, one cannot guess what the transitions state(s) geometrically is(are) for RI in cyclohexene.[a] Similarly, with

---

[a]Current force fields cannot model RI (ring inversion) for this elementary molecule at all [Section 4.2.7, subsection (*4*)].

only molecular 3D templates in hand (Sections 3.4 and 6.1.3), we have no idea regarding molecular geometries of transition states through which conformers of medium- and large-sized ring molecules, e.g. conformers of cyclooctane (Fig. 95), interconvert.

As we learned, sometimes one can draw verisimilar conclusions regarding 3D geometry of conformational transition states; however, the number of such transient structures may be large for a molecular system. E.g., our conformational sample, amine **12**, has six stereochemically non-equivalent transition states — two different enantiomeric pairs as well as two other transition states — for NIR when considering conformational dynamics for this non-intricate compound (Fig. 88).[a] It would be laborious to build rough structures of dissimilar geometry on the screen in order to further locate each transition state of amine **12** by means of the "magic" full-matrix Newton–Raphson algorithm (Section 4.2.4) of energy minimization. With the increasing size of a molecule, it becomes practically impossible to follow this strategy. Moreover, when using only obvious (i.e., assumed) geometries, one may miss conformational transition states that existence is not indicated by a primitive geometrical insight staggering/eclipsing, planar/ tetrahedral, etc.

Therefore, for "dozens–conformer" organic molecules or molecular systems with simply unpredictable geometrical changes, a successful strategy is a stochastic conformational search combined with the full-matrix Newton-Raphson energy minimization. Numerous stereoisomeric structures of random geometry are generated, and, after being generated, the geometry of such a structure is optimized with using this procedure. Technically, it is not difficult to employ this automatized strategy since the option *stochastic conformational search–full-matrix Newton-Raphson energy minimization* is a component of some multifunctional MM calculation packages; also any version of the original MM3 force field (in user terms, the original MM3 program) is equipped with this calculation algorithm. If a sufficiently large number is set for structures to be generated in stochastic conformational search (as suggested above, 10^4–10^5 structures

---

[a] Rotaion of only one Pr^i group of **12** (Fig. 88) is included into consideration. Conformational dynamics in this compound of course involves rotational freedom for both Pr^i groups. Then the number of stereochemically non-equivalent NIR transition states is 18 (six pairs of enantiomers and six other distinct structures).

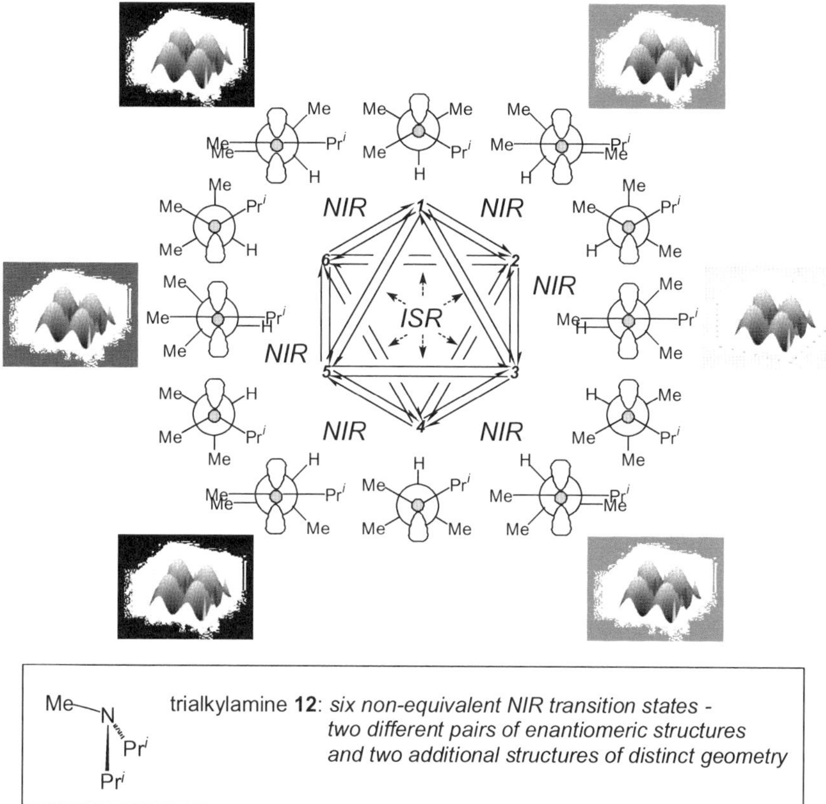

Figure 88. Simplified conformational scheme for amine **12** (in the middle; numbers in the hexagon apexes indicate stable stereoisomers). Newman projections (grey circle depicts the nitrogen atom) of staggered substituent orientation show stable stereoisomers 1–6 (they are conformers at room temperature). Structures with an eclipsed pair of vicinal substituents show six NIR transition states (for clarity, marked with symbolic pictures of saddle point regions; different colors indicate dissimilar geometry of transition states). Enantiomeric pairs of conformers are *1* and *4*, *2* and *3* as well as *5* and *6*. Among NIR transition states, there are two different pairs of enantiomeric structures. Transition states in transformations 1 ⇌ 2 and 3 ⇌ 4 are enantiomeric. Transition states in transformations 4 ⇌ 5 and 1 ⇌ 6 form another enantiomeric pair. Remaining transition states in transformations 2 ⇌ 3 and 5 ⇌ 6 have distinct geometry. Six rotational transition states are not considered [here, ISR (isolate rotation) means C–N rotation].

for a medium-sized ring), one can reasonably assume that the undertaken conformational search leads to location of all conformational transition states.

This number may be reduced to 10^3 when using the original ("non-packaged") versions of the MM3 program, due to a high efficacy of the implemented algorithm of the stochastic conformational search. However, these original programs are not supported by a GUI. As a result, the user deals with MM3-generated output as well as coordinate files of tremendous size that contain structural and energy information for hundreds generated structures. Required data, e.g., Cartesian coordinates for structures of interest, are extracted from the coordinate file by means of simple auxiliary computer programs that are written by "occasional" programmers or users themselves. This makes the use of the versatile original MM3 program unsuitable for chemists who are not experienced in programming or, at least, in manipulations with files using the command language of the operating system. Therefore, as emphasized earlier, GUI-equipped professional MM packages, e.g., MacroModel with its GUI (Maestro), are absolutely preferable for non-computational chemists. Their versatile GUIs, while not requiring special professional skills, provide comparative analysis in vast massifs of modeled geometries.

(C) Recognition of transition states. The optimized structure corresponds to the transition state, if the Hessian has one (the first order transition state) or more (a higher order transition state) negative eugenvalue(s) calculated for this point in SES (Section 4.2.4, subsection *Optimization procedures*; also, Fig. 56). How does the user of an MM program know that the located molecular structure is the sought-for transition state and what is its order?

A "visual" primary indication is molecular geometry. If the initial geometry is kept in common after the full matrix Newton-Raphson procedure, i.e., it is not transformed in the geometry of a stable stereoisomer (e.g., it is of a trigonal and not pyramidal geometry for the N unit in amines), one may conclude that the transition state *probably* has been modeled successfully. However, the researcher should prove that the above quantitative criterion [negative eigenvalue(s)] is satisfied — maybe, the located 3D structure is a high energy *stable* stereoisomer? With no doubt, an inspection is required. It is provided by a separate calculation procedure, the normal mode analysis (NMA; it is called also normal mode vibration analysis) that usually is an attribute of professional MM programs (e.g., Spartan, original MM3, MacroModel). Technically, the inspection is extremely simple. If NMA reveals that there is one or more negative (also called imaginary) frequencies (MM programs display NMA results as a text), we deal with a transition state.

From mechanistic perspective, there are $3N-6$ degrees of freedom for non-linear molecules (N is the number of all atoms of the system; three translational and three rotational modes of the molecule as a whole are excluded). Normal mode vibration is a set of oscillations of *all* nuclei in the molecule so that (*1*) each nucleolus oscillates in a certain trajectory synchronously with other nuclei that oscillate in their certain directions with certain different amplitudes; (*2*) this set of certain oscillations is independent from another set of coherent motions of nuclei (i.e., these oscillations do not cause other oscillations and are not caused by them); and (*3*) these coherent oscillations of atoms do not change molecular energy. In other words, normal mode vibration is an oscillation that disturbs the molecular framework with a certain frequency; the atoms deviate from their equilibrium positions with different, atom-specific amplitudes. An entire set of normal mode vibrations represent all oscillations of the molecule, i.e., all $3N-6$ ($3N-5$ for linear molecules) coherent low amplitude motions of nuclei-in-molecule. In classical description, each normal mode with frequency ν_i means that the molecule behaves as a harmonic oscillator (see below) with frequency ν_i. The in-time-molecule, in this light, is an entire assembly of such independent oscillators of various ν_i. Metaphorically, these frequencies are vibrational "fingerprints" of molecular systems (known in IR spectroscopy as fundamental frequencies).

In a reasonable approximation, normal mode vibrations obey the harmonic law. At each point of the SES, low amplitude displacements lead to changes of energy that are described by the Taylor expansion for potential energy (Section 4.2.2) truncated after the quadratic term. This approximation means some restrictions in NMA applicability. NMA may be applied to small molecules equilibrated at not high temperatures. In such cases, one may consider potential energy wells as near-parabolic in their bottom region and their high energy vibrational levels as populated insignificantly (Section 2.4, Fig. 42).

As we know, the Hessian itself shows the type of stationary points (Section 4.2.4). Why is another mathematical "indicator," NMA, used for this purpose in MM? NMA calculates frequencies ν_i of normal mode vibrations. This procedure is very useful — it provides vibrational "fingerprints" and, in addition, indicates saddle points (transition states). Scarcely synthetic chemists are in need to go deep into mathematical transformations (presented in many monographs on computational chemistry) that show the relation between the Hessian and a set of frequencies ν_n of normal mode vibrations. It is sufficient to indicate that this relation is based on conversion of the Hessian matrix E'' defined in Cartesian coordinates into an equivalent matrix of force constants F. Then frequency ν_i of the $i-$th normal mode is expressed as $\nu_i = \lambda_i^{1/2}/2\pi$, where λ_i is the $i-$th eigenvalue of matrix F. Totally, $3N-6$ values of ν_i are extracted by NMA from the Hessian matrix E'' for an $N-$atom 3D molecular structure. In other words, the Hessian "conceals" frequencies of normal mode vibrations that may be exhibited in the explicit form by means of the NMA formalism. Regarding transition states, this relationship means that one negative frequency ν_i (imaginary frequency) corresponds to the first order transition state, two negative frequencies ν_i and ν_j correspond to the second order transition state, etc. (Section 3.1, Fig. 56). So, the type of a transition state is displayed by the number of negative normal mode vibration frequencies. This recognition of the transition state type is formally equal to the type recognition via signs of the Hessian eigenvalues (Section 4.2.4). The eigenvector that corresponds to such an imaginary frequency indicates descents from the saddle points in the SES [Section 4.5, subsection (D)].

First-order transition states are most important in modeling of conformational dynamics. They are of lower energy than related transition states of higher orders and, thus, the saddle points, which correspond to first-order transition states, lie exactly in the MEP (Section 3.2, Fig. 58). In other words, transition states of the first order are kinetically favored among all alternative ones. Recognition of first-order transition states is, therefore, crucial for quantitatively estimating the kinetics of stereoreorganizations in an organic molecule. Suppose that NMA shows that a transition state and not a stable stereoisomer has been located. The next (and the final) glance in our modeling of transition states is elucidation that the located transition state is of the first order (Section 3.1, Fig. 56). The examination is technically elementary. As indicated above, the presence of one only negative frequency demonstrates that the located transition state has the first order.

Nevertheless, "hunting" exclusively first-order transition states could be misleading. Sometimes, after locating a second-order transition state, researchers do not consider it anymore and focus their attention on the first-order transition states. However, one should not immediately neglect such "ill-looking" transition states. The point is that they may appear to be higher in energy only for an insignificant value, say, of 1 kcal mol^{-1}, and no barrier separates the first- and the second-order transition states (Section 3.1, Fig. 57). This means that the transformation includes also this alternative pathway: a considerable portion of real molecules of the molecular assembly interconvert through the geometry of the second-order transition state. Clearly, both first- and second-order transition states should be taken account in considerations of conformational dynamics for such a system.

(*D*) *Establishment of the relationship between stable stereoisomers and corresponding transition states.* Stochastic conformational searches that include full matrix Newton-Raphson energy minimization (see above) supply the researcher with a set of stereoisomeric structures of two kinds: stable stereoisomers (i.e., conformers at certain physical conditions; Sections 2.5 and 2.6) and transition states of their interconversion. However, after completing the conformational search procedure, it is still unknown which transition states and which stable stereoisomers from this "mixture" are immediately related,[a] i.e., which stereoisomers are directly interconverted through concrete transition states. There is an obvious need to reveal the "genetic" relationship between each transition state and related stable stereoisomers. Indeed, unearthing of all such relationships for an organic system together with determination of relative stabilities for all 3D forms including transition states means a complete quantitative understanding of its conformational behavior.[b]

Thus, the question is how to spot these relationships. Let us recall NMA for transition states [this section, subsection (*C*)]. Again, we escape here mathematical notations and only notice that an imaginary (negative) frequency corresponds to a remarkable vibration mode of the system. That is, one motion of atoms along their oscillation trajectories, among atom motion in other oscillations, is interesting to us. In SES terms, this peculiar motion consists of sliding down from the saddle point, along the MEP, for some small displacement $\pm\Delta r$ (+ and — signs mean opposite directions, and r is a generalized parameter of molecular geometry; Fig. 89). Departing the saddle point *along the MEP*, the system, during each oscillation period, "moves" at one moment toward a related stable structure in the SES and at another toward the opposite related stable structure. The turning points of this low amplitude periodic motion (transient structures that correspond to displacements $\pm\Delta r$; Fig. 89) are determined by values $\pm\Delta r$ as well as the eigenvector associated with this imaginary frequency mode. Since this eigenvector is calculated in NMA, the geometries of these transient structures are in principle available.

Note again, these turning points *are located in the MEP and lie below the saddle point*. As we know, algorithms of geometry optimization in MM force the molecular 3D system, which "is" not in very near proximity to a saddle point, "to fall" into the nearest energy minimum in the SES (Section 4.2.4). Therefore, utilizing the two obtained transient structures as starting geometries for two separate geometry optimizations, we

---

[a] Also the "genetic" relationship between concrete first-order and second-order transition states remains unknown at this stage.

[b] See Section 2.8 for quantitative conformational schemes (e.g., shown in Fig. 53). They actually are a graphic representation of a "network" of these relationships.

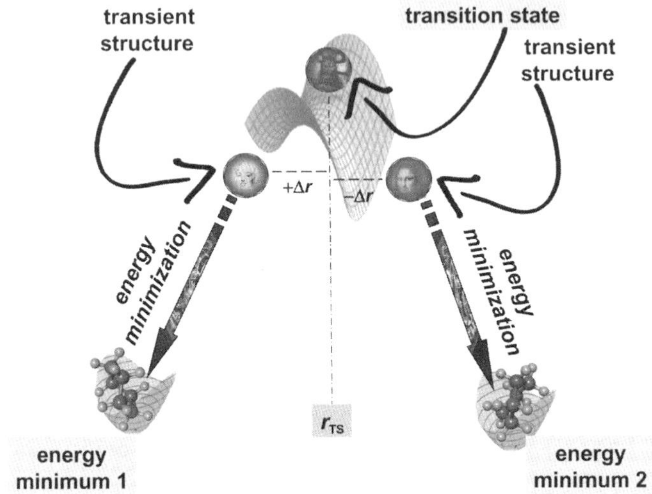

Figure 89. A schematic illustration of the technique for establishing the "genetic" relationship between a transition state and related conformers. NMA provides transient structures that lie in the MEP below the transition state with generalized coordinate value r_{TS}. Energy minimization (arrows) applied to these molecular structures leads to the related conformers.

reach both related stable stereoisomers (in other words, conformers if physical conditions allow the stereoisomer flexibility; Section 2.5).[a] Methodologically, it is better to use another than the full matrix Newton-Raphson minimization procedure, e.g., the block diagonal Newton-Raphson minimization (Section 4.2.4) that guarantees "falling" of the molecule into the nearest SES minima (Fig. 71). Otherwise, the system may return to the initial transition state when optimizing the geometry by calculating the full Hessian matrix (in particular, if deviation $\pm \Delta r$ is very small). Fortunately, research-targeted MM programs include both full matrix and block diagonal Newton-Raphson minimization options.

On the other hand, it is still technically difficult to inexperienced MM users to reveal all relationships *conformer–transition state–conformer* for multi-conformer systems. The problem is that cotemporary packages of

---

[a] A strict algorithm to downhill from a saddle point along the MEP is the Euler steepest descent (ESD) method. Unfortunately, it has not been implemented into most used MM programs.

computational chemistry, even calculating frequencies for transition states, have no computerized option to extract and to save structural information for the $\pm \Delta r$ transient structures (Fig. 87). MM users actually have no input files to proceed with final energy minimization. In practice, chemical researchers have to manage by themselves, using supplementary home-made computer programs that can "extract" these necessary transient structures from NMA-generated files. If the synthetic chemist succeeds in overcoming these technical difficulties, he/she at last can ride high: the tool that often is absolutely necessary in valuable conformational studies is gained. Now, such a researcher, in an easy way, can fully understand the conformational behavior of any "normally sized" organic system that is in the sphere of validity of available fine force fields.

Summarizing this Chapter, let us point out practical aspects of MM modeling that are most important for non-computational chemists. MM is a method of very fast calculations. They quickly provide molecular geometry as well as relative stability of stereoisomeric structures (in terms of steric energy) for organic molecules of normal size (not macromolecules). Conformational analysis of such molecular systems is a simple procedure of a lunch timescale. Modeling of transition states of trivial intramolecular stereoreorganizations via MM is still not difficult although it requires more time.

Researchers can easily perform MM calculations by means of convenient, organic chemist-friendly computational package(s). Definitely, MM is indispensible in quickly modeling geometries of relatively large ($\sim$50–500 atom) molecular organic systems of main chemical functionalities. Such a modeling may be easily performed in the frame of any organic research group.

However, synthetic chemists should be aware that force fields are suitable only for reproducing molecular structures of certain chemical functionalities; their number is not too large. For many other chemical structures, MM calculations either are of rather questionable accuracy, or afford absolutely unacceptable molecular models. For instance, *all* metalorganic compounds fall outside the applicability of currently leading, high quality force fields, while much lesser known metalorganics-oriented force fields can deal with a very limited number of chemical classes of organometallic complexes, with an accuracy not proven broadly.

There is no universal force field that would provide the best or even acceptable accuracy for different chemical structures: as a rule, fine force fields are developed with sharply targeting either chemical class(es). Nevertheless, "chemoselectivity" (i.e., applicability) of either biomolecular force field does not guarantee that another, general organic force field would show a worse accuracy if applied to a "biomolecular structure" (and *vice versa*). Therefore, in order to provide a maximally possible accuracy to MM modeling of a concrete system, the force field should be empirically selected from a relatively small pool of well developed (fine) force fields. One can claim that a set comprising late versions of MM3, MMFF94, Amber and CHARMM force fields would satisfy requirements of an experimental organic research group in successfully performing MM modeling.

MM results for non-trivial organic systems need to be verified. The MM modeling results (e.g., molecular geometry or, better, relative thermodynamic stability of conformers) should be compared with experimental data or relevant results supplied by trustworthy *ab initio* calculations.

Since there are no clear structural criteria for applicability of MM to either chemical system, most molecular structures modeled by MM are in the shadow of some uncertainty. In this connection, it is reasonable to view *values* of ΔE_{ster} for stereoisomers (i.e., their relative stability by MM) of polyfunctionalized, charged, or appreciably distorted molecular organic structures as rather *qualitative* estimates.

5

A "Universal" Computational Method: Quantum Mechanics

5.1 A Growing Trend: *Ab Initio* Calculations

The reader has probably gathered from the previous Chapter that although MM provides faster results, it is, nonetheless, a capricious tool. It requires a balanced tuning (choice of force field, verification of calculation results) for accurate modeling of every new molecular structure. Moreover, the structure of interest may be rejected by the available force fields because they may appear insufficiently parameterized for treating it. If a structure is not accepted, the only alternative for the MM user is to develop additional parameters that will allow the force field to model the "indigestible" chemical system. Development of parameters for a force field is actually a profound and scrupulous work, and only computational chemists can perform it efficiently. These efforts do not guarantee a success of MM calculations if applied to next systems; it has been stressed more than once that there is no transferability *structure A* $\rightarrow$ *structure B* for a force field accuracy when modeling different chemical structures.

In contrast, as mentioned in Section 4.1, QM is based on a strict comprehensive physical theory. Therefore, QM calculations (more often, but not quite accurately, called *ab initio* or non-empirical calculations) have no principal methodological limitations in modeling molecular systems. Owing to a firm physical foundation, modeling by QM is "all-embracing"; it can provide any measurable molecular quantity (e.g., geometrical parameters, heat capacity, electric multiple moments, energies of excited states, oscillation frequencies, spin–spin coupling constants, and chemical shieldings) as well as abstract, model-related properties (e.g., orbital energies and atomic charges). However, the proportionally high cost for the immense possibilities of *ab initio* calculations seriously restricts the use of QM methods. While MM calculations consist of computing only the values of simple algebraic functions for the given values of independent variables (Section 4.2.2) and, in principle, may be performed using an electronic calculator, QM calculations deal with the numerical integration of hundreds and thousands of integrals. This creates a very heavy demand on hardware, in particular, on processor performance as well as memory size.[a] An obvious consequence is that, in practice, molecular systems which can be subjected to QM modeling are significantly restricted in size. This restriction is so strong that geometry optimization by means of *ab initio* calculations cannot be addressed even to one rotamer of a small biopolymer, e.g., tRNA (tRNAs are 75–95-residue polynucleotides) or ribonuclease (a 124-residue protein).

Fortunately, rapid progress in processor performance has opened the door to ubiquitous use of QM applications in organic chemistry because its objects are relatively small molecules. The high computational resource requirements for QM-based computations are satisfied to a large extent by modern high-performance computers (HPCs), at least for the minimal worthwhile modeling (geometry and relative energy of stereoisomers) of 30–100-atom organic molecules. These multiprocessor computer clusters are capable of running several separate jobs (independent calculations that may have various tasks and use the same or different programs) simultaneously.[b] Nevertheless, QM calculations are still time-consuming. For a rough estimation, geometry optimization running on four parallel processors of an HPC takes from one day (for a $\sim$20-atom system) to 20 days (for a $\sim$50-atom system), assuming that a routine DFT (density functional theory) method (Section 5.2.6) is applied to a system consisting of elements

---

[a]Requirements for data storage (hard disks) are significantly lower.

[b]Clearly, the performance of a computer depends on its number of processors. This dependence is not linear; in theory, it is expressed by Amdahl's law: $Perf = 1/[(1 - f) + f/n]$, where $Perf$ is the maximal speedup (performance) of a computer with n parallely operating processors, and f is the fraction of computations that may be parallelized ($0 \leq f \leq 1$). According to this law, the lower f is, the less the n-dependent increase of the speedup is.

from the first three rows.[a] Significantly longer times are required for a PC of similar four-processor architecture, since processors exploited in PCs have lower performance than those of HPCs.

On the other hand, non-empirical calculations are essentially universal. They are relevant to any molecular structure, neutral or charge-carrying, covalently or otherwise bonded, with a closed or open electron shell, and in the ground or excited state. Not surprisingly, QM-based methods have become the major tool in modeling "normal-sized" organic molecules, increasingly supplanting MM and semi-empirical calculations.

Ab initio calculations clearly suggest their use in organic laboratories due to their indispensable significance and chemical universalism. GUI-equipped *ab initio* packages provide technical ease of calculation-associated procedures as well as analysis of their results. Synthetic chemists usually deal with systems that contain ∼30–100 atoms or even fewer. Although the time that is required for modeling such systems *ab initio* is larger than the time required for conducting a synthetic experiment, it cannot be considered as a principal obstacle to conformational excurses in organic research. The time-consuming process of running calculations does not disturb experimental work; it runs "by itself" around the clock in parallel with the laboratory routine. The researcher only reviews calculation results (e.g., molecular geometry and energy); this stage is technical and certainly does not require a considerable time (compare this with the interpretation of NMR spectra of non-trivial compounds!).

The use of theoretical modeling by organic experimentalists offers an additional benefit. Computational chemistry is green chemistry. Computers are not hard consumers of electrical energy. Moreover, they neither require chemicals nor produce environment-polluting compounds during exploitation. Therefore, utilitarian models obtained using theoretical calculations save energy as well as materials that would otherwise be expended in chemical experiments which eventually should lead to the same understanding.

There are no financial reasons not to strengthen an organic research group with its own ability to carry out QM calculations. The cost of a high-performance multiprocessor computer system targeted to modeling non-polymeric organic molecules may equal the cost of an analytical

---

[a] In QM calculations, the overall number of occupied orbitals (and not the number of atoms) is important for the time span of computations. Nevertheless, by counting atoms (of course, this is simple) and not orbitals, one can roughly estimate the relative expenditure of the time that will be consumed by *ab initio* calculations applied to molecular systems of different sizes. The computation time in non-empirical molecular modeling increases proportionally to between the third and fourth powers of the number of atoms.

HPLC system, a little-used decoration of many non-industrial synthetic laboratories. Furthermore, many universities, chemical departments, and research centers have UNIX- or Linux-based, personnel-supported HPCs of either multiprocessor architecture that are intended to satisfy computational needs of the research conducted there. As a rule, researchers have a free secured access to HPCs of their institution. Two or more solid professional QM calculation packages, e.g., Gaussian, Q-Chem, Turbomole, Jaguar, and ADT,[a] purchased under a multi-user license agreement often are residents of these computer clusters. In these circumstances, only the traditional focusing of a majority of the synthesis-oriented organic/bioorganic chemists exclusively on experiments and primitive mechanistic models as well as the lack of information lead them to ignoring the practically simple and truly fruitful molecular modeling.

Let us try to go beyond this obsolete tradition by explaining the essence of QM methods in a very light yet informative presentation, and supplying the beginner with practical recommendations of how to model reliable 3D structures of organic molecules. This task should not be of frustrating complexity for organic experimentalists who operate only with laboratory glassware, chemicals, NMR spectra, and Lewis structures. Quoting relevant key points from the course in basic quantum chemistry (given to all chemistry students) in a step-by-step manner, commenting them if necessary, and connecting these points in a consistent account without the "glue" of mathematical transformations, we will qualitatively explain how QM calculations supply molecular geometries that correspond to stationary points of the PES.

5.2 Understanding QM Computations: A Brief Overview for Beginners

5.2.1 *Atomic units of measurement*

Before we go on to discuss QM methods of molecular modeling, an opportune technical note should be provided. The basic equations of QM adopt a convenient notation if specific units of physical quantities, so-called atomic units (a.u.), are used. In addition, some fundamental physical constants

---

[a]"Light" computational packages with basic options for QM calculations are Spartan and HyperChem (mentioned in Section 4.3.1).

Table 3.   Relationships between basic atomic units (a.u.) and SI units ($\hbar = h/2\pi$; h is the Planck constant in SI units).

Physical quantity (symbol or definition in SI units)	a.u. (name)	SI unit
Mass (m_e)	1	9.109×10^{-31} kg
Charge (e)	1	1.602×10^{-19} C
Length ($\hbar^2/m_e e^2$)	1 (bohr)	5.2917×10^{-11} m
Energy ($m_e e^4/\hbar^2$)	1 (hartree)	4.360×10^{-18} J
Electric dipole moment ($\hbar^2/m_e e^3$)	1	8.478×10^{-30} C $\times$ m

are numerically simple if expressed in a.u. For instance, the value of the Planck constant in a.u. is 2π, and the energy of the $1s$ orbital of the H atom is -0.5 a.u.

Also, it is useful to remember the conversion factor between *hartree* and *kcal*: 1 hartree $= 627.51$ kcal mol^{-1}. Although *calorie* (1 cal $= 4.184$ J) does not belong to conventional SI units, this energy unit is still widely used in chemistry.

Chemical literature and computational programs often supply the results of QM calculations for, e.g., energy, charge, and dipole moment, in a.u. Therefore, it is necessary to know at least the frequently encountered a.u. (Table 3; see Ref. 33b for a.u. for more physical quantities). Note that, in output files, these programs often give intramolecular and intermolecular distances in traditional units, Ångströms (Å; 1 Å $= 0.1$ nm); the Ångström unit does not originate from any systematic nomenclature.

5.2.2 *First division of routine QM methods: The Hartree–Fock approximation*

Synthetic specialists scarcely need to know the main mathematical formalism of QM theory of molecules and different approximations of this theory (QM methods of molecular modeling), e.g., to remember the secular equation or be versed in mathematical forms of different basis set functions. What organic experimentalists do need in their research practice, is to understand the essence and limits of applicability of widely used QM methods. Following this vision, this Section presents the important concepts related to QM calculations without even a minimal mathematical derivation of these concepts. The latter could be found in the literature that explains the basics of computational chemistry in more detail; the monographs by Bachrach[53a] as well as Cramer[33b] are probably among the best mathematically light accounts available.

Conformational studies ascertain the relationship *molecular geometry–potential energy*. Thus, in routine QM-based conformational studies, one needs to model the PES (Section 3.1) for the ground state,[a] i.e., to locate the minima as well as the saddle points of this PES. As we see, the general methodology of exploring conformational behavior should be the same in QM and MM. The difference is that, in MM, these stationary points are of the SES (Section 4.2.3), whereas in routine QM modeling of molecules, we deal with the PES. However, what is the basis of our confidence that QM calculations that are yet obscure for us can provide the PES?

The principal answer may consist of two words: the Schrödinger equation. The electronic Schrödinger equation (Section 3.1, Eq. 5) determines that there is a one-to-one quantitative relationship between molecular geometry and potential energy for molecules in the electronic state p (e.g., in the ground state). Nevertheless, we cannot be satisfied by this principal solution; this fundamental equation is not a practical instrument of molecular modeling. Regrettably to non-computational chemists, there is a lengthy path from the Schrödinger equation to the worthwhile QM calculations of geometry and energy of concrete molecules. Below, we follow this path by leaping from an equation to another, subsequent equation. Tenacious synthetic chemists will be rewarded. Because QM calculations are technically very simple owing to the user-friendliness of modern calculation packages (when installed), these researchers will possess useful skills to obtain a reliable 3D geometry and the corresponding, considerably accurate value

---

[a]The term *ground state* is related to the electronic state for which $2k$ or $2k + 1$ electrons occupy respectively k or $k + 1$ quantized two-vacancy energy levels in a sequential fashion, starting from the lowest energy level and not skipping occupancy vacancies. Non-photochemical organic reactions usually involve molecules in the ground state. The PES for the ground state is the minimal-energy PES, and that for an excited state is a higher-energy PES; these surfaces may significantly differ in shape. Clearly, conformational analysis may be also performed for a molecular system in an uncoupled (energy gap-separated) excited electronic state. If the PESs are similar in energy, the Born–Oppenheimer approximation of the frozen nuclear backbone losses its validity.

of molecular energy for any stereoisomeric structure of any organic system of regular size.

Let us look again at Eq. 5. Why does it provide the desired relationship *molecular geometry–potential energy*, i.e., the PES for each electronic state p? Indeed, using a.u. (Table 3), the Hamiltonian H_{el} in extended mathematical notation is as follows:

$$H_{el} = -\sum_i \hbar^2/2m_e \times \nabla_i^2 - \sum_{i<j} e^2/2r_{ij} - \sum_i \sum_k e^2 Z_k/2r_{ik} \qquad (28)$$

where i and j run over electrons, k runs over nuclei, ∇^2 is the Laplacian operator, $\hbar$ is defined as shown in Table 3, m_e is one electron mass (1 a.u.), e is the charge of one electron (1 a.u), Z_k is the atomic number of the k-th atom, r_{ij} is the distance (bohrs) between the i-th and j-th electrons, and r_{ik} is the distance (bohrs) between the i-th electron and k-th nucleus. The first term in this expression deals with the kinetic energy of electrons, the second is related to the potential energy of electron–electron interactions, and the third describes the potential energy of electron–nucleus interactions. Thus, the electronic energy E_{el} is nucleus coordinate-variant. Also, the energy of internuclear repulsion V_{nucl} (Section 3.1) is determined by internuclear distances, i.e., by atomic coordinates r_{nucl}. The potential energy E_0 in the Born–Oppenheimer approximation (i.e., a point of the PES) is the sum of V_{nucl} and E_{el} for fixed electron coordinates r_{el} and fixed coordinates of nuclei r_{nucl}. Recall that, by definition, wave function Ψ is a continuous single-valued function (Section 1.1). This indicates that there is a single-valued, continuous function of its eigenvalues (energies) of atomic coordinates r_{nucl}, i.e., the PES. Thus, the first step from *basic QM concepts* to the *PES* has been taken; we have discerned that the PES is quantitatively predetermined by wave function Ψ_{el}.

If the analytical mathematical form of Ψ_{el} for a molecular system is known, no principal problems would appear in modeling the PES. By solving Eq. 5, we could find energy values for any molecular geometry. However, as mentioned, the exact solution of this equation is possible only for the hydrogen atom (one electron–one proton system). In these solutions, eigenfunctions are the familiar $1s$, $2s$, $2p$, $3s$, $3p$, $3d$, etc., atomic orbitals (certain algebraic functions) and eigenvalues are the associated energies (usually termed orbital energies). For example, the energy of the $1s$ orbital (an eigenfunction of certain algebraic form) of the H atom is -0.5 hartree (associated eigenvalue). Occupation of this orbital by one electron is the ground state for this system. The PES for the hydrogen atom is actually one point (energy value of -0.5 hartree) because the "molecular skeleton" of this system (one nucleus) has a single "3D geometry." Furthermore, the PES of this system in an excited state is a point of higher energy (the energy of the corresponding occupied orbital).

The wave functions for larger molecular systems (even for molecules H_2^+ or H_2) are unknown; how does QM handle them? Obviously, one must somehow design this mysterious function. At this point, the QM theory of molecules takes the next approximation (as we remember from Section 3.1, the first is the adiabatic consideration: coordinates of nuclei in molecular frameworks are "frozen"). The simplest as well as most famous approximation is the so-called molecular orbital model (also called the independent electron approximation). Note that it is the basis of the majority of calculation methods in chemistry. In the next step, let us understand the first principles of this model.

In general, wave function Ψ_{el} is a function of the coordinates of *all* N electrons of a molecule; in other words, it is a polyelectron function $\Psi_{el} = f(r_1, r_2 \ldots r_N)$, where the lower index runs over electrons. Simplifying the situation with a "cryptic" Ψ_{el}, the molecular orbital model supposes that *(1)* each electron may be supplied with its own wave function χ_i that is independent of the coordinates of other electrons, and *(2)* a combination of such functions χ_i for all the electrons in the system represents the *polyelectron* wave function Ψ_{el}. Such a one-electron function is called a molecular orbital (MO), or a molecular spin-orbital, stressing the presence of the spin variable for each i-th electron. Thus, the MO for the i-th electron (in mathematical notation, χ_i) is the product of the coordinate and spin components,

$$\chi_i = \psi_i(x_i, y_i, z_i,) \times \begin{cases} \alpha(i) \\ \beta(i) \end{cases} \tag{29}$$

where ψ_i is a wave function that depends only on Cartesian coordinates x_i, y_i, z_i of the electron (coordinate component), and α and β are spin wave functions corresponding to electron spin numbers ½ and $-$½ (spin components). Because functions χ_i are independent, the molecular electronic energy E_{el} is the sum of the energies of "separate" electrons. In physical sense, this approximation indicates that each electron is independent of the others and moves in the averaged electric field of all the electrons and nuclei. Note that, in order to escape colossal difficulties (consideration of the polyelectron function Ψ_{el} that includes electron–electron interactions) in modeling molecules, we deteriorate to some extent the quality of the modeling, constructing wave function Ψ_{el} in this manner.[a]

The first appearance of the term *molecular orbital* in the present consideration is noticeable. Synthetic chemists are familiar with the term *molecular orbital*; however, many of them miss its exact meaning. Here, we remind synthetic professionals that the MO is some *function* that predetermines the relationship *location–potential energy* for one certain electron in a hypothetic molecular system where electrons do not interact with each other. For synthetic specialists, who understand MOs as electron spreads in the molecular space, we note that any MO is not a function of the 3D location of the relevant electron. Consequently, a graphical representation of such an MO does not indicate the 3D domain in which the electron has a high probability of being located. As for the coordinates of the electron, one could think that the square of the value of each MO at any point in intramolecular space (and not the MO itself) shows the probability of the location of the electron at this point. However, since the MO model is an essential simplification, one cannot consider the square of the MO at point r_i in coordinate space even as an approximate probability of finding the i-th electron at this point.

Evidently, the above-mentioned combination of spin-orbitals χ_i is not arbitrary. How should these independent functions χ_i be arranged to approximate[b] the "true" polyelectron wave function Ψ_{el}? It turns out that, when satisfying the antisymmetry of Ψ_{el} (Section 1.1), this polyelectron function for the N electron system may be expressed via the spin-orbitals χ_i of separate electrons as a determinant (known as the Slater determinant; Eq. 30):

$$\Psi_{el} = 1/(N!)^{1/2} \times \det |\chi_1(r_1)\chi_2(r_2)\chi_3(r_3)\ldots\chi_N(r_N)| \qquad (30)$$

---

[a] Strictly speaking, the independent electron approximation is incorrect. Electron motions are correlated, and, therefore, there are no individual energy contributions of electrons into the electron energy E_{el} of the molecule (Section 5.2.3). The wave function theory (not considered here) is a stricter approach that approximates the *polyelectron* wave function Ψ_{el} via analytical (explicit form) functions[53c] and, thus, takes into account electron correlation.

[b] By accepting the independence of spin-orbitals χ_i from each other, we neglect electron–electron interactions (Section 5.2.3). Therefore, the wave function constructed from the one-electron wave functions χ_i is an approximation of the "true" unknown wave function Ψ_{el}.

where r_i is a set of spatial coordinates x_i, y_i, z_i together with spin "coordinate" ½ or $-$½. For the first time, we have got some (rough approximate) representation of electronic wave function Ψ_{el} (Eq. 30). However, at this stage, this theoretical equation does not supply us with any practical prescription for energy calculations, since the wave functions χ_i are as obscure as Ψ_{el}.

Keeping aside this "blurred portrait" of the wave function (the Slater determinant) for now, let us examine how we can bypass this elusive function in order to get as close as possible to the eigenvalues in the Schrödinger equation, quantized values of the electron energy E_{el}. The essence of this approach is that Ψ_{el} itself is not reconstructed. It is mimicked by *another*, sufficiently arbitrary function Φ (called the probe or guess function); only general features of Φ are predetermined, and are the same as those of the original wave function: Φ is continuous, schlicht (single-valued), finite, differentiable, orthonormal, and antisymmetric. These features of the probe function allow us to mathematically represent it as an infinite linear combination:

$$\Phi = \sum_i^{\infty} a_i \Psi_i \tag{31}$$

where Ψ_i are eigenfunctions of Ψ, and coefficients a_i are real numbers such that $\sum_i^{\infty} a_i^2 = 1$. Because we deal only with electronic wave function Ψ_{el}, the index *el* is not used hereafter, thus implying that Ψ represents Ψ_{el} and E represents E_{el}.

From Eq. 5, energy E_i (the energy of the i-th electronic state) is expressed as

$$E_i = \int \Psi_i H \Psi_i dr \tag{32}$$

Using Eq. 31, one can demonstrate that a similar integral for the *probe function* Φ shows the relationship between Φ and the eigenvalues E_i of the *original wave function* Ψ (Eq. 33):

$$\int \Phi_i H \Phi_i dr = \sum_i a_i^2 E_i \tag{33}$$

Using the variational principle of mathematics, we can write the functional $\bar{E}$ for the energy supplied by probe function Φ:

$$\bar{E} = \int \Phi H \Phi \, dr \Big/ \int \Phi^2 \, dr \tag{34}$$

For a given Φ, the functional $\bar{E}$ (also termed the Rayleigh quotient) is actually a "function" $f(\Phi)$.[a] In other words, the structural elements of Φ (e.g., parameters, if Φ is parametric) determine $\bar{E}$ (i.e., its numerical values). Minimizing this "function" means finding those values of independent variables (e.g., parameters for function Φ) that provide the minimal value (the functional minimum) for energy E_i. Considering Eq. 33, one can find out that minimization of the functional $\bar{E}$ leads to an energy value $\bar{E}_0$ that is at least as high as the energy E_0 for the ground state predetermined by the original wave function Ψ. That is, the Rayleigh quotient is the upper bound to the exact ground state energy E_0 (i.e., $\bar{E}_0 \geq E_0$ for any approximate wave function). The closer the estimate $\bar{E}_0$ is to the ground state energy E_0, the better is this estimate.

This is a crucial point in understanding the significance of mimicking Ψ via Φ. It appears that one can estimate the energy of a molecular system using a probe function (which we can construct from some selected algebraic functions) instead of the wave function (which cannot be constructed from such explicit functions). Clearly, the accuracy in this estimation (closeness of the calculated value of $\bar{E}_0$ to the "genuine" molecular characteristic E_0) depends on the quality of the chosen Φ. Therefore, selecting an appropriate Φ is crucial for developing a valid calculation method in QM. The best probe function would be that which provides the smallest value for the difference $\bar{E}_0 - E_0$. With this understanding, we arrive near the final point on our path from the basic QM concepts to the PES. We need to apply these energy-related conclusions for Φ to the above-explained MO approximation of Ψ, i.e., to the wave function in the Slater form.

Let us see what this energy-conditioned substitution of Ψ with Φ affords. Returning to the Slater determinant (Eq. 30) and using Eq. 34, one can show

---

[a] In simple terms, *function* $f(x)$ represents a relationship between a set of numbers (say, independent variable x) and another set of numbers (dependent variable y, i.e., function values). *Functional* $F[f(x)]$ denotes a relationship between a set of functions $f(x)$ (independent "variables") and a set of numbers F (functional values).

that the energy functional constructed using Φ for unknown spin-orbitals χ_i (i.e., mimicking them via eigenfunctions φ_i of the probe function Φ) has a simple form as follows (Eq. 35):

$$\bar{E} = \int \Phi H \Phi dr / \int \Phi^2 dr = 2 \sum_i^N H_i + \sum_i^N \sum_j^N (2 J_{ij} - K_{ij}) \qquad (35)$$

where H_i are diagonal matrix elements of the Slater determinant [$H_i = \int \varphi_i(1)^* \times \left(-\tfrac{1}{2}\Delta - \sum_k Z_k/r_{i\_k}\right) \times \varphi_i(1)dr$], J_{ij} are matrix elements of the form $J_{ij} = \iint (|\varphi_i(1)|^2 |\varphi_j(2)|^2 / r_{1\_2}) dr_1 dr_2$ (Coulomb integral), K_{ij} are matrix elements of the form $K_{ij} = \iint (\varphi_i(1)^* \varphi_i(2)(\varphi_j(2)^* \varphi_j(1) / r_{1\_2}) dr_1 dr_2$ (exchange integral), and φ_i and φ_j are eigenfunctions of probe function Φ. Similarly to Eq. 30, Eq. 35 is "silent"; therein, none of functions φ_i is known to us yet. Nevertheless, in contrast to the electron wave functions χ_i from Eq. 30, functions φ_i may be constructed by minimizing the energy functional; we must choose a set of functions φ_i (independent "variables" of the energy functional) that provide a "good" minimum of this functional in Eq. 35. Here, "good" means that the calculated estimate $\bar{E}_0$ is close to the energy E_0 of the ground state of the real molecule. Of course, there are different methods for constructing Φ-related functions φ_i. In an effective and therefore well-known design of φ_i, linear combination of atomic orbitals–molecular orbital (LCAO–MO), each probe MO φ_i, of the N electron system is constructed as a linear combination of m *given* independent functions μ_p (usually indicated in the chemical literature as χ_p):

$$\varphi_i = \sum_1^m c_{ip} \times \mu_p \qquad (36)$$

where i runs over $1, 2, \ldots, N$, and index p that specifies the type of the function μ_p runs over $1, 2, \ldots, m$. Algebraic functions μ_p are called basis functions (or atomic orbitals since these functions are atom-centered), and a set of m basis functions is called a basis set (Section 5.2.4).[a] LCAO–MO is the basis for a majority of routine *ab initio* calculations.

---

[a]A strict approach to the wave function requires an infinite set of summands in Eq. 36. In practice, the m number of functions μ_p that form a basis set is finite and even small. Indeed, starting from a

Thus, in our consideration of the basics of calculation methods, we have finally reached analyzable, correctable functions of explicit algebraic form (basic functions μ_p). However, possessing only these concrete functions (see below), one cannot start the calculation "machinery" of Eq. 35. One also needs to know coefficients c_{ip} that, together with functions μ_p, define "theoretical" functions φ_i in LCAO–MO (Eq. 36) and, thus, enable computing numerical values of these LCAO–MO φ_i. The problem of minimizing the energy functional is reduced to determining optimal coefficients c_{ip}.

This is not a trivial problem. For probe function Φ, coefficients c_{ip} form an $N \times m$ matrix where the i-th column consists of coefficients $c_{i1}, c_{i2}, \ldots c_{im}$. Even the principal mathematical solution for determination of these coefficients c_{ip} (the Hartree–Fock–Roothaan method) is somewhat complicated for synthetic chemists; hence, readers who wish to know more on the basics of computational chemistry may refer to the introductory monographs of Refs. 33b, 53a, or 58a. Here, we only state that these coefficients are adjusted by an iterational procedure [the self-consistent field (SCF) procedure], and, instead of the Hamiltonian, the Fock operator is used to construct the so-called secular equation.[33b,53a,58a] In the first stage, a certain guess matrix of coefficients c_{ip} is taken. This defines the initial (guess) functions φ_i, and they enable computation of the integrals of the Fock matrix. Then, the eigenvalues $\bar{E}_i$ become available including the first, very rough numerical value of $\bar{E}_0$. Using these values, "better" (providing a lower value for $\bar{E}_0$) coefficients c_{ip} are calculated, and the iteration procedure is repeated until an *a priori* specified accuracy criterion (e.g., a very small difference between two $\bar{E}_0$ values obtained from two sequential iterations) is satisfied. The anticipated result of these computations is that we succeed to relate the relevant energy value (i.e., the ultimate $\bar{E}_0$ value supplied by SCF iterations) to a certain geometry of the nuclear framework. This method of *ab initio* calculations is called *restricted Hartree–Fock SCF* (usually abbreviated HF-SCF or RHF; the short notation HF is also widely known).

Note that, using this procedural sequence of *ab initio* calculations (called single-point calculations since the coordinates of the nuclei are not varied), we locate only a point of the PES. This point is scarcely stationary because the SCF procedure does not optimize the geometry and, thus, does

---

certain m, the accuracy of calculations (i.e., for MO energies) does not increase further. This indicates that the so-called HF limit (Section 5.2.3) is reached.

not afford the structure that corresponds to an energy minimum or saddle point. The same calculations of energy performed for another molecular geometry would supply another point of the PES, and so on. The PES is not modeled in such a way.[a] In order to locate the required stationary points (minima and saddle points), the energy gradient- and Hessian-based methods (Section 4.2.4) of geometry optimization (energy minimization),[b] i.e., additional calculations that crucially diminish the number of PES points to be located in searching a stationary point, are used in QM computations. In geometry optimization, the SCF procedure is executed, nevertheless, for many geometries (generated one after the other) of the molecule. Therefore, it usually is the most time-consuming computational process in routine[c] (supplying only geometry and energy) non-empirical modeling of organic molecules.

Thus, we have understood how HF-SCF calculations yield 3D structures that are either stable stereoisomers or transition states. Notice that *ab initio* calculations are much simpler in putting into practice than in learning their basics. Should one mention again that the GUIs of "organic" QM computational packages (e.g., Gaussian, Turbomole, Jaguar, Q-Chem, and ADF) provide easy technical execution of *ab initio* calculations to such a level that skills obtained in middle school would be sufficient for using these fine modeling instruments when installed? In the on-the-screen fashion, *(1)* the structure of interest is assembled from virtual elementary 3D molecular fragments delivered by the GUI (or read from the structure-related file; Section 4.4.1); *(2)* the calculation directives [e.g., geometry optimization or single-point calculations, frequency calculation option, the calculation method, say, HF (for more useful ones, see Sections 5.2.3 and 5.2.6), the basis set (Section 5.2.4), and the solvent model (Section 5.3.4)] are specified; and *(3)* calculations are triggered by the *Calculate* (or similar) option. Calculations results are saved as a multi-page text in the output file; the latter sometimes is readable by a text editor program (e.g., MS Word can open Gaussian output files). However, the GUI of the used calculation package is the best reading tool because it provides interactive analysis of the output file. In addition to the file text reading option, such a GUI is usually capable of displaying 3D images of calculated molecular frameworks and

---

[a]Single-point calculations may be useful in roughly locating transition states of rotation. Several rotamers that correspond to the MEP are specified, and their energy is calculated without optimizing their geometry (Section 4.4.3).

[b]Distinct methods of energy minimization are included into professional QM calculation packages in order to provide the possibility to choose an effective method when optimizing the geometry of a thermodynamically stable structure or a successive method when locating a transition state.[58b] Computational algorithms of energy minimization methods of QM packages are significantly different from those implemented into MM programs.

[c]Only supplying molecular geometry and energy by means of non-high QM theory calculations [e.g. MP2 (Section 5.2.3) or DFT (Section 5.2.6)].

orbitals, selectively demonstrating *ab initio*-derived oscillations of the nuclear framework, rendering parameters of the obtained molecular geometry, modifying 3D structures on the screen, and saving structural information as input files for other calculation programs.

5.2.3 *Toward accuracy: Post-HF (electron correlation) methods*

The quality of HF calculations in energy/geometry calculations is of interest to us. Let us assume that, for organic systems, the Born–Oppenheimer approximation holds, and that the relativistic effects are negligible. Note that this is generally true for systems consisting of moderate-weight chemical elements (say, the first four rows of the periodic table). Are RHF calculations in principle accurate or they contain hidden flaws?

Recall that RHF is based on the physical model that supplies electrons with significant independence. Although this model considers correlations for electrons with parallel spins forbidding them from occupying the same region of space simultaneously, it also approaches electrons with antiparallel spins as independent particles whose movement is unrestricted. This is physically incorrect since, in addition to the Pauli spin correlation, a dynamic (Coulomb) correlation exists between any two electrons. We may state that each moving electron is located at every moment in the center of a spatial region (called a Coulomb hole) where the probability of the presence of another electron there is extremely low. Omission of the Coulomb correlation is the chief explicit inaccuracy of RHF calculations. It leads to an underestimation of bond lengths and overestimation of energy.[a]

These inaccuracies are eliminated partly in the unrestricted HF (UHF) method, which defines MOs for electrons with α and β spin functions as distinct MOs. However, UHF may afford absolutely unacceptable energy values owing to unavoidable, unphysical mixing electronic states of different multiplicity (so-called spin contamination). Furthermore, molecular systems have other correlation effects referred as non-dynamic correlation (long-range correlation). Both UHF and RHF cannot treat non-dynamic correlation. Therefore, UHF-based calculations are becoming increasingly rare.[b]

---

[a]In general, the calculated bond lengths are sufficiently accurate ($\sim 1.5\%$ of the relative error) when HF methods with extended basis sets (Section 5.2.4) are applied to neutral organic molecules whose elementary molecular fragments have undistorted geometry.

[b]Note that separation of α and β spin-orbital functions is necessary when modeling open-shell molecules (e.g., radicals or cation radicals); therefore, UHF and not HF is the correct choice for modeling them.

We have learned that HF does not consider electron correlation. In other words, the accuracy of this method is significantly limited in energy calculations. The most accurate energy value (the HF accuracy limit) is the energy E_{HF}, which could be obtained if we use an infinite set of proper basis functions.[a] The difference between this limit and the true energy E is exactly the correlation energy E_{corr}. In mathematical notation, $E_{corr} = E - E_{HF}$. The correlation energies for organic molecules are not negligibly small. If we strive for a high calculation accuracy, E_{corr} must be taken into account. *Ab initio* methods that also calculate E_{corr} are called electron correlation methods or, less formally, post-HF methods. Unfortunately, none of them can compute the entire correlation energy.

Several computational approaches have been developed in QM for computing E_{corr} as accurately as possible. The best-known are MP2, MP3, and MP4 (perturbation theory-derived methods); CIS and CISD (configuration interaction, CI); CCD, CCSD, and CCSD(T) (coupled-cluster theory), and QCISD and QCISD(T) (quadratic CI including singles and doubles). These methods are based on different expansions of the constructed probe function; for a first-hand view of them, the reader is again directed to Refs. 33b, 53a, and 58a. Nevertheless, in order to select a suitable post-HF method, non-computational chemists need to appreciate the quality of different post-HF methods as well as their computational cost. It is obvious that some post-HF methods have different accuracies in modeling molecular systems, while others provide results of comparable quality. Moreover, the relative accuracy of "comparable" methods may vary when calculations are applied to different chemical structures. We can speak about general relative accuracy in energy calculations. The order of this accuracy for post-HF methods, according to Cramer's monograph,[33b] is as follows:

$$HF < MP2 < MP3 \sim CCD < CISD < MP4SDQ \sim QCISD \sim CCSD$$
$$< MP4 < QCISD(T) \sim CCSD(T) \sim BD(T).$$

When discarding UHF calculations owing to insufficient accuracy reasons, unrestricted methods of refined theory, e.g., UMP2 (a post-HF method; this section) or UB3LYP (a DFT method; Section 5.4.2), are used in calculating molecular structures of such systems. Some caution is required in applying unrestricted DFT methods to biradicals and triplet carbenes; in order to provide satisfactory models, the overlap between the open-shell orbitals should be insignificant.

[a]Of course, a finite number of basis functions forms a basis set (Section 5.2.4).

Clearly, increasing accuracy comes at higher price. If a correlation-treating method is applied, the computation time is drastically lengthened compared with that of HF calculations using the same basis set. For instance, the use of MP2 instead of HF requires $\sim$10 times as much computation time. Roughly, a factor of 10 applies for each step in the accuracy order indicated above. No user has unlimited computer resources; concerning organic experimentalists, one can say that the "post-MP3" calculations usually are beyond their technical possibilities in modeling molecules larger than naphthalene. Fortunately, the "first improvement," from HF to MP2, is almost always sufficient for modeling of PES's of organic molecules. Synthetic chemists who start experimenting with theoretical calculations may merely accept that their first choice *QM method in routinely modeling $\sim$20–50-atom systems is MP2* (see Section 5.2.4 for the suitable basis sets).

Methods beyond MP2 are not applied to building routine molecular models (i.e., to calculating geometry and potential energy of molecules). In organic research, computational chemistry recruits highly accurate electron correlation methods for modeling *reaction* transition states as well as excited states, in thermochemistry, in analyzing electronic structure of molecules, and in obtaining high-quality standards in benchmark (calibration) calculations.

Modern computer clusters with parallel processing are capable of providing technically troubleless energy minimization for larger systems (say, $\sim$50–80-atom molecules) at the MP2 level. However, in practice, several researchers usually share processors of these fast computing machines, and a single user actually cannot utilize the performance of the cluster as a whole. Which QM methods may be used in the organic laboratory for calculating molecular geometries and relative energies of stereoisomers of such non-small systems? Theoretical chemistry has offered organic researchers an attractive alternative to HF-related methods — QM calculations based on the formalism of DFT (density functional theory). The major advantage of routine DFT methods over the HF method is their higher accuracy. In contrast to HF, the calculation accuracy has no limit in DFT; any systematic inaccuracy may be significantly decreased by developing the next generation of DFT methods (Section 5.2.6). In addition, DFT calculations are time-efficient. For instance, the accuracy of B3LYP or B3PW91 (widely used DFT methods; Section 5.2.6) in the geometry and energy calculations is close to that of MP2, and sometimes (rarely) higher. However, B3LYP-

or B3PW91-mediated calculations require much less computational time. Combining these beneficial features (high quality modeling and significant time efficiency), DFT calculations currently prevail in routine modeling of organic molecules. What DFT methods are is discussed in Sections 5.2.5 and 5.2.6.

5.2.4 *Basis sets*

In the beginning, let us understand the command language of computational chemistry. In chemical publications that include QM calculations, acronyms such as RHF/6-31+G(d,p), MP2/cc-pVTZ, and B3PW91/6-311+G(d), which are related to the computational methodology used, will certainly appear. What lies behind these accepted acronyms is absolutely obscure for most organic experimentalists. The attentive reader may discern that symbols before the slash separator should mean the method of QM used (an HF approximation, an electron correlation method, or a DFT functional). The rest [e.g., the above-mentioned combinations 6-31+G(d), cc-pVTZ, and 6-311++G(d,p)] indicates the employed *basis set* of functions that we have encountered for the first time in Section 5.2.2. For instance, the acronym MP2/6-31++G(2d,p) indicates that the MP2 method (a post-HF method; Section 5.2.3) is used in combination with the 6-31++G(2d,p) basis set.

The double slash separator // indicates the use of two sequential calculation procedures. The first is geometry optimization (energy minimization), and the other is single-point energy calculation performed using a higher theory method. In this notation, e.g., the acronym MP2/6-311++G(d,p) //RHF/6-31+G(d) indicates that geometry optimization is executed employing the RHF method combined with the basis set 6-31+G(d), and, in the second stage, single-point energy calculations are performed for the resulting geometry using the MP2 method and the 6-311++G(d,p) basis set.

Now, we can focus on basis sets. Both HF (Sections 5.2.2) and DFT methods (Sections 5.2.5 and 5.2.6) use functions μ_p of exact algebraic form to construct probe MOs or formally similar Kohn–Sham (KS) orbitals (Section 5.2.5), respectively, as expressed in Eq. 36; they are called basis set functions. These functions are primary elements in the QM calculation machinery. They are not derived from any other functions, and theoreticians select them almost arbitrarily, guided by two very general principles. For a function to be selected, it should reflect the electron's distribution (the probability of its location) in intramolecular space as well as possible, and it should be convenient for numerical computations.[a]

Fortunately, theoreticians have a convenient prototype for selecting the suitable functions — the atomic orbitals of the H atom, which are exact solutions of the Schrödinger equation for this *one nucleus–one electron system*. Organic experimentalists may realize that they recognize these solutions (functions) as the familiar non-hybridized $1s$, $2s$, $2p$, $3s$, $3p$, $3d$, $4s$, $4p$, $4d$, and $4f$ orbitals, and the energies of these atomic orbitals (the eigenvalues of the Schrödinger equation for the H atom) are the energies of the corresponding electronic states of the H atom.[b] These exponential functions [more accurately, the radial (non-angular) parts of the eigenfunctions Ψ_i for the H atom] are a reasonable starting point for the choice of the mathematical form of the basis set functions.

Why? One can define an atom-centered function, the radial distribution function (RDF) of the electron in space, as $\mathrm{RDF} = (\Psi_{el})^2 \times 4\pi r^2$. One can write for the i-th electronic state: $\mathrm{RDF}_i = (\Psi_{el\_i})^2 \times 4\pi r^2$. Clearly, each RDF_i function accurately indicates the probability of the electron location at a distance r in any direction from the nucleus in the i-th electronic state. Since functions Ψ_{el_i} for the hydrogen atom (the orbitals $1s$, $2s$, $2p$, $3s$, $3p$, $3d$, $4s$, $4p$, $4d$, and $4f$) are $-r$ exponential functions of the certain known form,[c] each RDF_i derived from the corresponding Ψ_{el_i} for this atom explicitly shows the *accurate* probability of the electron location at a distance r in any direction from the nucleus $^1\mathrm{H}$ in the i-th electronic state of the H atom. For instance, Fig. 90, showing the RDF for $2s$ and $2p$ orbitals of the H atom,[d] actually presents the 3D domain where the electron may be located with a non-negligible probability *exactly* known for each point, for the two first excited states of this "molecular" system. Hence, using such accurate electron "clouds" of the H atom as etalons, one

---

[a]In principle, many continuous algebraic functions could be adapted for the linear combination in Eq. 36.

[b]The registered frequencies of spectral transitions in atomic hydrogen confirm this exact QM model. Of course, the relative energies of electronic states (the relative energies of the $1s$, $2s$, $2p$, $3s$, $3p$, $3d$, $4s$, $4p$, $4d$, and $4f$ orbitals) for other atoms differ from those for the H atom.

[c]These exponents are given in many textbooks on theoretical chemistry.

[d]For this system, the $2s$ and $2p$ orbitals are exact solutions Ψ_2 and Ψ_3 (i.e., the fully specified exponents; not given here) of this equation.

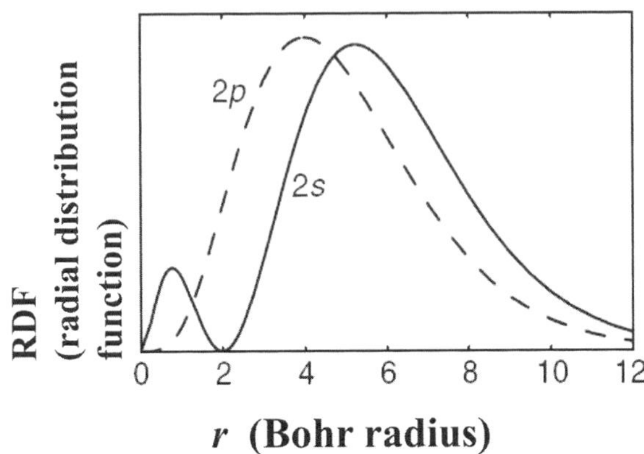

Figure 90. Example of RDF functions for the H atom: graphic representation of functions RDF$_{2s}$ and RDF$_{2p}$ whose parent functions are the 2s and 2p orbitals. The latter, together with other orbitals (*e.g.* 1s, 3s, 3p, and 3d), are obtained by exactly solving the Schrödinger equation for this simplest atomic system.

can estimate how successfully an algebraic (e.g., an exponential) function reproduces the electron distribution for an electronic state.

The exponential form of the orbitals for the H atom has prompted computational chemists to use two similar "H atom orbital-like" functions as basis functions μ_p (Eq. 36) in QM calculations. They are Slater-type orbitals (STOs) of the general form $exp(-\zeta r)$ and Gaussian-type orbitals (GTOs) of the general form $exp(-\alpha r^2)$ [more accurately, $exp(-\alpha r^2 f^2) \times x^l y^m z^n$; for exact expressions for these functions, see monographs on computational chemistry]; currently they are the most frequently used basis functions. The similarity here means that the corresponding RDF functions (i.e., the RDF functions derived from STOs and GTOs) represent the radial distribution of the electron in the H atom in nearly its true (exactly calculated) distribution. This representation of the radial electron distribution by STOs and GTOs is achieved by including empirically adjusted elements in these exponents where the general, oversimplified notations $exp(-\zeta r)$ and $exp(-\alpha r^2)$ mask the actual, complicated structure of the STO and GTO exponents.

It appears that *ab initio* methods are not absolutely non-empirical. The structure of their primary elements, i.e., basis set functions, does not follow from strict QM theory. Basis set functions as well as basis sets are constructed according to a single empirical criterion allowing the calculations to reproduce experimental data for different molecular quantities.

Nevertheless, in contrast to MM, the accuracy of the used physical theory (i.e., QM), but not this hidden empiricism, enables adequate modeling of molecular properties.

Let us briefly consider GTOs. They are atom-centered if the origin of the Cartesian coordinates is atom-centered. In Cartesian space, these functions are similar to atomic orbitals of the H atom (i.e., $1s, 2p_x, 2p_y, 2p_z$, etc.) because the sum $l + m + n$ of exponents at Cartesian coordinates x, y, and z (i.e., exponents at variables x, y, and z) in GTO functions may be viewed as an analog of the orbital quantum number ℓ for atoms ($\ell = 0, 1, 2, \ldots n - 1$, where n is the principal quantum number). For instance, the simplest GTOs ($l + m + n = 0$) have spherical symmetry and are therefore named s-type orbitals. Basis functions, which have axial symmetry in respect of a single Cartesian axis (one among the exponents l, m, and n is equal unity while the two others adopt zero values; thus, $l + m + n = 1$), are called, by an analogy with hydrogenic atomic orbitals, p_x, p_y, p_z.

The analogy between GTOs and hydrogenic orbitals is not absolute. There are six d-type GTOs (the H atom has five $3d$ orbitals) and 10 f-type GTOs *vs.* seven $4f$ orbitals for H. GTOs have a remarkable feature; they enable relatively easy computation of the integrals required to solve the HF equations. On the other hand, GTOs, being $-r^2$ exponential functions, have a faster radial fall from the nucleus than that of the $-r$ exponential functions of the H atom. Consequently, individual GTOs (called primitives) do not quite successfully reflect the radial distribution of electrons, and primitives alone are not used. In contrast, STOs, being $-r$ exponential functions, reflect the radial behavior of electrons more or less correctly. Why not employ STOs? The reason for neglecting these functions is technical: STOs are inconvenient for computation.

A satisfactory solution is to use GTOs to mimic STOs that are used in computations as basis functions defined by Eq. 36. This approach adopts the best features of STOs and GTOs — the resulting STOs have a proper radial distribution, while the GTOs provide computational efficiency. Therefore, most computational packages employ GTOs or, more accurately, both individual GTOs (primitives) and their linear combinations (called contracted GTOs or contracted basis functions) in forming one function in a basis set. For instance, basis set STO-3G is a combination of three GTOs for each STO. The acronym supplies information about the contraction scheme. The symbol G indicates the type of functions used (here: GTOs); the symbol

3 points out their number per STO. The more primitives are included in the newly formed STO (i.e., in the contracted GTO or, in other terms, in the contracted basis function), the more accurate this function is. In fact, the number of GTOs involved should not be too large because of the high computational cost of the basis set enlargement.

In constructing a basis set in this way, we use only one basis function (i.e., one STO) to define an orbital of a certain type (s, p, d, f, g, or h) in the basis set. For instance, each element from Li to Ne has five orbitals $1s$, $2s$, $2p_x$, $2p_y$, and $2p_z$. For constructing each of these five orbitals (the s orbital or any of the p-type orbitals), one certain, orbital-specific STO (i.e., one exponential function) is used. Such basis sets are called single-ζ (single-zeta, i.e., one-exponent) basic sets. In practical aspect, single-ζ basis sets are obsolete. They supply poor accuracy because they present distorted radial distributions for the outer (valence) as well as inner (core) electrons.

Further development of basis sets has led to the appearance of multiple-ζ basis sets. In this approach, an atomic orbital is modeled using *two or more different basis functions* that are constructed from different contracted or individual GTOs, instead of modeling the orbital using one basis function constructed from several primitives. Mathematically, each valence orbital in multiple-ζ basis sets is a linear combination of "STOs" (contracted GTOs) or "STOs" and individual GTOs;[a] the coefficients in these linear combinations vary as the SCF procedure (Section 5.2.2) proceeds. Terminologically, if each orbital in the basis set is formed by two basis functions (two STOs), it is a double-ζ (double-zeta, i.e., two-exponent) basis set; if three basis functions are included in each orbital, we have a triple-ζ (triple-zeta, i.e., three-exponent) basis set, and so on.

These basis sets are much more successful, and they are the most common basis sets used in contemporary applied QM calculations. In principle, by using more and more basis functions in orbital construction, one could reach the HF accuracy limit or the accuracy limit of the employed DFT method. Unfortunately, it is impossible even to double the number of basis functions that form each orbital in single-ζ basis sets since the computational cost becomes unproportionally high.

---

[a]Of course, genuine STOs may be used in these linear combinations. The computational cost for molecular modeling increases in this case.

An elegant solution to this problem was proposed by John Pople, who introduced in practical computational chemistry the basis sets known as 3-21G, 6-21G, 4-31G, 6-31G, and 6-311G. Their distinctive feature is that these basis sets are split-valence basis sets.[a] The term *split-valence* emphasizes the design idea of these basis sets; a distinct number of basis functions is used to compose core *vs.* valence orbitals of an atom. This approach is very fruitful. Valence electrons, in contrast to core electrons, participate in intra- and intermolecular interactions. Therefore, constructing outer atomic orbitals in greater detail (i.e., using many basis functions) is a prerequisite for increased accuracy in QM calculations. On the other hand, the simplified structure of inner orbitals (i.e., a minimal number of orbital-forming basis functions) leads to valuable economy in computational cost without appreciably decreasing the accuracy potential.

In Pople's basis set family, core orbitals are represented by one basis function because inner (core) orbitals are not influential. The numerical symbol before the hyphen in the acronyms of these basis sets (Table 4) indicates the number of primitives that construct this inner orbital (i.e., from three to six primitives for the corresponding basis sets).

Valence orbitals are crucial for molecular structure, and therefore, they are more finely composed using more basis functions. The main distinctive characteristic of a split-valence basis set is the number of basis functions (i.e., STOs) per atomic valence orbital. This number is not large; it runs over 1 (minimal basis sets), $2, 3, \ldots, 7$ (highly extended basis sets).[b] For instance, each atomic valence orbital of the double-ζ basis set 6-31G is composed of two basis functions, whereas that of the triple-ζ 6-311G contains three basis functions. Thus, split-valence basis sets represent each atom in the molecule as a set of atomic orbitals (one core and several valence orbitals) constructed from the basis functions in the above-explained manner. The number of basic functions per atom in different split-valence basis sets may vary from less than 12 to several tens. Clearly, when exploiting a very extended basis set, *ab initio* calculations applied to an organic molecule of medium size (20–30 atoms) deal with hundreds or

---

[a]Since then, the split-valence idea has been implemented in many basis sets, e.g., MIDI! (synonym: MIDIX), Dunning's basis sets (see below), TVZPP, and QZVP.

[b]Split-valence basis sets belong to the multi-ζ family of basis sets. Therefore, they are classified as single-ζ, double-ζ, triple-ζ, etc. basis sets.

Table 4. Most known basis sets employed in routine *ab initio* modeling of organic compounds containing elements from H to Kr (gray color indicates very rarely used, practically obsolete basis sets).

Basis set	Contracted GTOs for H*	Contracted GTOs for the second-row elements*
6-31G	$2s$	$3s2p$
6-31G(d)	$2s$	$3s2p1d$
6-311G(d)	$3s$	$4s3p$
6-31G(d,p)	$2s1p$	$3s2p1d$
6-31+G(d)[†]	$2s$	$1s2s_{diff}2p_{diff}1d$
6-31+G(d,p)[†]	$2s1p$	$1s2s_{diff}2p_{diff}1d$
6-31++G(d,p)[†]	$2s_{diff}1p$	$1s2s_{diff}2p_{diff}1d$
6-311++G(d,p)[†]	$3s_{diff}1p$	$1s3s_{diff}3p_{diff}1d$
6-311++G(2df,2pd)[†]	$3s_{diff}2p1d$	$1s3s_{diff}3p_{diff}2d1f$
cc-pVDZ	$2s1p$	$3s2p1d$
cc-pVTZ	$3s2p1d$	$4s3p2d1f$
cc-pVQZ	$4s3p2d1f$	$5s4p3d2f1g$
aug-cc-pVTZ[†]	$1s_{diff}2s1p_{diff}1p1d_{diff}$	$3s1s_{diff}1p_{diff}2p1d_{diff}1d1f_{diff}$

*Digits show the number of basic functions used in constructing orbitals of a certain type (s, p, d, or f).
[†]Subscript$_{diff}$ denotes the additional (diffuse) function to each corresponding orbital.

even thousands of basis functions. It is not surprising that such basis sets are currently applied to small molecular systems.

In acronyms of Pople's basis sets, the presence of two numerical symbols after the hyphen indicates a double-ζ basis set (e.g., 6-31G), while three such symbols show that the basis set is triple-ζ (e.g., 6-311G). Each numerical symbol after the hyphen denotes the number of primitives that form one basis function of the valence orbitals. To make this point clear, let's italicize this symbol in the acronym of the basis set 6-31G. Then, our dummy notation 6-*3*1G illustrates that the first basis function for the valence orbital of an atom is constituted by three primitives (three GTOs). Transparently, 6-3*1*G explains that the second basis function of this orbital is actually one primitive (one GTO).

The possibility to use different-sized orbitals allows the researcher to adequately model the valence shell of any atom; practically, we only need to select the proper basis set among those with differently represented atomic valence orbitals. Simple ideas in constructing these orbitals make this selection rational (see below). Moreover, basis functions model the radial electron distribution in atoms of different elements with distinct accuracy. The "per atom specification" in basis sets provides additional flexibility by giving us the possibility to apply different basis sets to atoms of different elements [e.g., the "light organic" (H through Kr) *vs.* post-Kr ones] in the molecule and, thus, enables to verisimilarly reproduce anisotropic electron distribution in molecular structures that contain both "light organic" and "organic post-Kr" elements (e.g., iodine).

Among Pople's basis sets, 6-31G and 6-311G augmented with polarization and diffusion functions (see below) until recently have been the main QM workhorses in organic practice that deals with elementary molecular modeling (e.g., calculations of molecular geometry, energy, conformational barriers, dipole moments, charge distribution, and NMR chemical shifts). Basis sets 6-31+G(d,p), 6-311+G(d,p), 6-31++G(d,p), as well as *aug*-cc-pVTZ (see the next paragraph for Dunning's basis sets) have probably been most frequently employed in routine (i.e., thermochemistry-and-electron-redistribution-not-including) modeling of organic molecules undertaken in the last decade.

Other multiple-ζ basis sets are Dunning's cc-pVxZ (where x = D, T, Q, etc.; V denotes their split-valence character), which are known as correlation-consistent polarized double, triple, quadruple, quintuple, sextuple, etc., respectively, basis sets (Table 4; the prefix 'cc' means correlation-consistent). These basis sets may become the most frequently used instruments from the basis sets' arsenal. The field of elementary QM molecular modeling (QM calculations targeted on the geometry and relative energy of isomeric structures) is currently divided approximately in half between Pople's and Dunning's basis sets, and one can expect that, with further progress in computer performance, the latter will push aside the former. The great advantage of the cc-pVxZ basis sets is that they provide the systematic approach to improving the accuracy of QM calculations. The quality of the cc-pVxZ-supported modeling consistently increases in parallel with increasing the ζ rank (i.e., *the higher the better* is a sufficiently reliable rule in this case). For molecular systems of minimal size, one can practically reach the accuracy limit of either a large Dunning's basis set by increasing the level of the calculation method (Section 5.2.3), or a post-HF calculation method by employing high-rank cc-pVxZ basis sets one after another in combination with the examined calculation method. Selection of the suitable pVxZ basis set is therefore more rational than such selection in the Pople's series since we are able to estimate (say, from the related benchmark studies) the accuracy order that will be provided by either Dunning's set for the chemical structure of interest.[a]

Dunning's basis sets possess a remarkable feature; they include polarization functions (this is indicated by the prefix "p" in their acronyms that means "polarization functions"). Their number increases with increasing "zeta" (formally, with ranking the symbol x higher; x = D, T, Q, 5, etc.; Table 5), and high "zeta" basis sets (with "post-Q" values of x) are very large basis sets (tens of basis functions for one atom of any second-row element).

Let us formulate a qualitative understanding of what polarization functions are. Constructed s and p orbitals, regardless of their number in the basis set, cannot model anisotropy of the distribution of electrons for second-row elements with sufficient accuracy. The inaccuracy is so critical that even in some trivial cases, such basis sets lead to a distorted molecular geometry, e.g., for ylides and pseudoylides (e.g., phosphates, sulfoxides, and sulfones). The next step in developing basis sets, which exploits the same strategy of increasing the number of basis functions, has significantly corrected this drawback of early non-empirical calculations: basis sets have been supplied with more exponential functions.

These additional functions (termed polarization functions) may be formally[b] considered as higher angular (e.g., $\ell + 1$) momentum functions. If mixed with the parent functions of lower angular momentum ℓ, the resulting hybrid function [e.g., called sp function when mixing the s ($\ell = 0$) and p ($\ell = 1$) functions] represents the anisotropic electron distribution in the atom much more accurately than the parent basis function alone. To a different extent, valence electron distribution is tuned by any polarized basis set. Concerning Dunning's or polarized Pople's basis sets, one can say that yet cc-pVDZ and 6-31G(d,p) do not accommodate an explicit roughness in the modeled geometries of neutral non-polar organic molecules.[c]

---

[a] Suitability of a basis set is crucial in energy calculations (Section 5.3.3).

[b] Basis functions of s, p, d, etc. types are not the s, p, d, etc. orbitals of the H atom.

[c] Assuming that a sufficiently accurate calculation method (e.g., a density functional from the HGGA group; Section 5.2.6) is used.

Simply put, polarization in basis sets consists of adding basis functions of the p type to the s functions of the valence shell, d functions to the p functions of this shell, and f functions to the d functions of the shell. For instance, in Pople's basis sets with polarization functions, second-row elements have a set of d functions, and first-row elements have a set of p functions. Both Dunning's cc-pVxZ and polarized Pople's basis sets contain more "orbitals" (additional sets of s, p, d, etc. functions) for each element than these elements have as occupied in different valence states.

Such basis sets, which include vacant "orbitals," are sometimes called extended basis sets. The other understanding of this term is that extended basis sets are those that use more than one basis function in constructing a valence orbital of the atom, i.e., mulitple-ζ basis sets. We will follow this ζ-associated terminology. Very extended basis sets (say, starting from cc-pVQZ) guarantee a high accuracy if complemented by a proper calculation method (e.g., a post-MP3 method; Section 5.2.3). However, computational cost rises very steeply with the increase of the size of the basis set. For instance, although the accuracy in energy computations is often appreciably improved when upgrading the calculation level from MP2/aug-cc-pVQZ to MP2/aug-cc-pV6Z, this improvement comes at the expense of a $\sim$30-fold time increase. Suppose that energy minimization (in other terms, geometry optimization; Section 5.2.2) at the MP2/aug-cc-pVQZ level takes 20–60 days for a 10–15-atom molecule when starting the optimization from a near-optimal molecular geometry.[a] For optimizing the geometry of such molecule at the MP2/aug-cc-pV6Z level, we would need years! Therefore, the use of highly extended basis sets is limited; for employing them in modeling even moderate-sized structures, exceptionally extensive computer resources are necessary to complete the computations in a reasonable time period. Fortunately, these extended sets with $x = 5$ and 6 (quintuple-zeta and sextuple-zeta, respectively) are unnecessary in elementary molecular modeling (i.e., in modeling the PES of an organic molecule in the ground state).

Thus, researchers must balance the size of the basis set and the reasonable computation cost. The need in this balance has led to the idea of economic basis sets, e.g., Truhlar's cc-pVxZ+ ($x = $ D, T, Q, etc.).[61a] In DFT calculations of energy, the accuracy of these new basis sets, which have been derived from the "noble" families of both Pople's and Dunning's basis sets, is comparable to that of significantly larger sets from the cc-pVxZ family.

In the acronyms of polarized Pople's basis sets, this refinement — addition of d functions to the p orbital set of basis functions of chemical $2p$ elements and addition of p functions to the s orbital set of basis functions of chemical $1s$ elements — is reflected by the addition of parenthesized symbols "d" and "p," respectively (Table 4). The inclusion of

---

[a]This is a verisimilar estimate of the time required by such computational run to be completed when reserving, say, eight parallel processors of a contemporary computer cluster.

more polarization functions, e.g., three sets of d functions, two sets of f functions, and one set of g functions for "post-first-row" element and two sets of p functions and one set of d functions for hydrogens, is denoted in this basis set series as (3d,2fg,2pd), e.g. as for the basis set 6-311++G (3d,2fg,2pd).

Nonetheless, even extended basis sets that include polarization functions have been shown to be incapable of even satisfactory modeling of certain molecular systems, e.g., electron-rich (e.g., carbanions), H-bonded (e.g., pairs of complementary nucleobases from nucleic acids), or highly polarized fragment-containing (e.g., fluorinated) ones. The improvement followed has been attained by introducing into basis sets new additional functions, the so-called diffuse functions. Again, we deal with a further expansion of the valence shell. Diffuse functions are specific exponential functions of slow decay that appear in addition to each outer orbital. For Pople's basis sets, this augmentation to both $2s$ and $2p$ orbitals constructed for second-row elements (so-called minimal augmentation) is indicated using one symbol +. The additional + symbol, if present, is related to a diffusion function that corrects the vacant $2s$ orbital of hydrogen (Table 4). Basis sets cc-pVxZ extended with diffuse functions are specified by adding the prefix *aug-* to their acronyms.

How successful are calculations using these improved basis sets? When combined with an electron correlation or proper DFT method, augmentation with diffuse functions enables valid estimation of the energy of structures with considerable localization of the negative charge (e.g., ylides and fluorinated systems), electron-rich molecules (anions, anion radicals), and weakly interacting systems (e.g., molecular complexes including solvates and structures with H-bonds). As mentioned, *ab initio* estimates of energy for such structures are absolutely irrelevant without incorporating diffuse functions into basis sets.

Another characteristic that is important for the applicability of a basis set is the range of chemical elements encompassed by the basis set of interest. With increasing quantum number n (in terms of the periodic table of chemical elements, the row number), the number of orbitals that describe real atoms jumps. Hence, the number of basis functions of constructed orbitals in an extended basis set for the next-row elements should jump even more. Starting with a certain n (i.e., some chemical element), the size of such a basis set becomes unmanageable for computational procedures. Moreover, relativistic effects in electron motion (increasing electron mass) should be taken into account when modeling molecular systems that include heavy atoms. These circumstances complicate approaches to constructing basis sets, and if we inspect existing basis sets, we would find that those that embrace elements up to Kr predominate. Fortunately, this is sufficient for almost all chemical classes of mainstream organic (i.e., not transition metal organic) chemistry. Only organic derivatives of Sb, Te, and I are not covered by these basis sets.[a]

Basis sets are under constant active development targeted on modeling structures that contain heavy atoms (e.g., atoms of a lanthanide element), decreasing computational cost, and/or improving the accuracy of non-empirical calculations. For instance, the basis sets cc-pVnZ-F12 (n = D, T, and Q) have been recently proposed[61b] for extending the use of new highly accurate QM methods, so-called explicit correlation methods, among which F12 methods are very promising.[53c] The CCSD(T)-F12 or MP2-F12 method in combination with a basis set from the cc-pVnZ-F12 series both improves the accuracy in energy estimation and lowers computational cost in respect of CCSD(T) or MP2 calculations that employ the Dunning's basis set of similar size.[61c,d] Sometimes, the accuracy for energy estimates by F12 appears similar in both the double-ζ and quadruple-ζ basis sets.[61d] This is an appreciable progress in practice of the accurate modeling of organic molecules by QM. Indeed, providing a high accuracy of *ab initio* modeling in energy calculations, the current reference method, CCSD(T) (Section 5.2.3), requires very extended basis sets (at least *aug*-cc-pVQZ). Owing to this computationally "overheavy" combination CCSD(T)/*aug*-cc-pVxZ (x = Q, 5, 6), these calculations are applied to structures that contain a few atoms.

Nonetheless, explicitly correlated methods themselves are computationally demanding. Although cc-pVnZ-F12 basis sets improve the situation in using high theory methods (here: F12) at the usual limitations of computer resources, also F12-treatable molecular systems have yet included structures of no more than 8–12 atoms.[61c] Whether, in the next decades, such basis sets (coupled with F12 methods) or the successful others will displace traditional Pople's and Dunning's basis sets and transform *accurate* modeling of organic molecules of "normal size" into routine calculations is difficult to predict. Besides, F12 methods and cc-pVnZ-F12 basis sets have yet to be implemented in either user-friendly QM computational package except Turbomole. The versatile, freely available package ORCA (developed at the Max Plank Institute; Section 6.3.2) also includes these methods; however, considerable computer skills are required for using it.

---

[a]Choosing a proper basis set for such compounds as well as organic complexes of heavy metals is not easy. There are no basis sets (even with relativistic effect corrections) that could be selected with confidence in an acceptable modeling accuracy.

At this point, after an overview of commonly used basis sets, we need no further discussion of other examples. We have understood what basis sets generally contain and which content is specified by a certain acronym. That is, we have learned to distinguish these basis sets according to their potential accuracy in *ab initio* calculations. What about selecting a concrete basis set for modeling certain chemical systems? Non-computational chemists face obvious difficulties in their judgment regarding the adequacy of the basis sets selected by other researchers or, more importantly, when selecting one by themselves.

A free web source of basis sets, EMSL,[a] contains nearly 475 published basis sets at the start of 2013; these basis sets may be downloaded and, when possessing considerable skills in programming, installed in some convenient calculation packages, e.g., Gaussian. Most organic experimentalists scarcely need to glance at even a part of them. One probably has only to give a formal simplistic insight into this pool.

The majority of basis sets are for general multipurpose use. Their design idea is to provide satisfactory or highly accurate molecular models of structures containing element(s) of the first four rows. For instance, the cc-pV(x+d)Z basis sets correct Dunning's cc-pVxZ ones whose d polarization functions are not sufficiently successful (an additional set of d functions is present in the cc-pV(x+d)Z series). Nevertheless, basis sets with a certain orientation are sufficiently represented. Orientations of these basis sets may be quite different. Some basis sets have been designed targeting the optimal balance between calculation accuracy and computational cost (e.g., the "minimally augmented" basis sets *maug*-cc-pVXZ). Some others are DFT-oriented, e.g., VTZP (Valence Triple Zeta + Polarization). Several basis sets, e.g., (*aug*-)cc-p(wC)VxZ-PP (x = D, T, Q, 4, 5), are intended to cover heavy element-containing compounds. Transparently, acronyms of basis sets do not specify their predestination.

There is no general specification concerning applicability and quality of basis sets. One-phrase characterizations of basis sets are given in Ref. 62a; some fragmentary information is available in Refs. 14, 33b, 62b,c, as well as at the Gaussian09 website.[b] Inexperienced QM users are in a puzzling situation and have to look for primary sources or reviews to be oriented in the basis set pool. This is definitely an undesirable alternative for them. In addition, compressed information about common basis sets would not enable synthetic specialists to rationally choose the proper case-specific basis set.

---

[a]https://bse.pnl.gov/bse/portal.
[b]http://www.gaussian.com/g_tech/g_ur/m_basis_sets.htm.

An ocean of discussions of the quality of different basis sets and their applicability to the modeling of various molecular properties would surely confuse their own selection. Two firm conclusions that can be deduced from this extensive literature are only that the basis set should be sufficiently extended, and concrete selection is finally a balanced compromise or even intuitive. These conclusions offer minimal assistance when beginning to use non-empirical computations.

Probably, until some experience in QM calculations is accumulated, it would be optimal to follow rigid prescriptions for the selection of the proper basis set to be used for relatively fast, but at least satisfactory accurate, non-empirical calculations. In this connection, it would be reasonable for beginners to be guided by a few concrete instructions. Assuming the recruitment of a relevant calculation method (say, either MP2 from the post-HF group, or B3LYP, XYG3, XYGJ-OS, MO6, M11, TPSSh from the DFT group; Section 5.2.6), basis sets can be regarded as follows:

(1) The basis set is selected from those mentioned in items *(3)–(8)* below. These basis sets are adequate for modeling organic structures that contain elements only from the first four rows.

(2) In selecting the basis set, one should obey two balance rules. *(i)* Well-extended basis sets should be used only in combination with moderate- to high-quality calculation methods. Coupling a non-extended or weakly extended basis set (e.g., 6-311G) with an accurate calculation method (e.g., a post-MP2 method or fifth-rung density functional) is meaningless. The opposite option (e.g., the use of the combination RHF/*aug*-cc-pVTZ) is undesirable but not forbidden. Items *(3)–(8)* illustrate this rule. *(ii)* If two basis sets are combined in a calculation procedure wherein the atoms of the molecule are "split up" between the sets, these basis sets should be extended similarly, e.g., one cannot combine minimally and highly extended basis sets. Item *(8)* illustrates this rule.

(3) When routinely modeling organic molecules (i.e., calculating geometry and relative energy of stereoisomers), Pople's basis sets 6-31+G(d,p), or 6-31++G(d,p), or 6-311+G(d,p) are the first choice if they are combined with the MP2 method. Of these basis sets, 6-31++G(d,p) is preferable since it often provides sufficient or higher calculation accuracy at a moderate computational cost. If computer resources allow,

use *aug*-cc-pVTZ in combination with MP2 or MP3 (for 15–20-atom molecules); however, a significantly higher accuracy relative to that provided by the indicated Pople's basis sets is not warranted.

(4) Both Pople's basis sets, e.g., 6-31+G(d,p) or 6-311+G(d,p), and Dunning's basis set *aug*-cc-pVDZ are suitable if HF calculations are used instead of electron correlation methods.

(5) In DFT calculations of geometry and relative energy of stereoisomers exploiting HGGA or HMGGA functionals (Section 5.2.6), any of the basis sets 6-31+G(d,p), 6-31++G(d,p), 6-311+G(d,p), *aug*-cc-pVDZ, *aug*-cc-pVTZ, or *aug*-cc-pV(T+d)Z are suitable. When using much more accurate DHGGA functionals (Section 5.2.6), Dunning's basis sets (not Pople's ones) are required; among them, *aug*–cc-pVTZ, or *aug*-cc-pV(T+d)Z are basis sets with the minimal zeta value which still enables support of DHGGA functionals. It is difficult to foresee which moderately extended basis set will yield the best accuracy in combination with a concrete density functional. From the practical point of view, it is worth trying first the indicated Pople's basis sets (when possible) since the computational cost of calculations with involving them is significantly lower than that of calculations with using the above Dunning's sets in combination and the same level of QM approximation.

(6) Recruitment of diffuse functions is necessary when optimizing geometry of open-shell (unpaired spin), anionic, highly polar, weakly interacting, lone electron pair(s)-bearing, large-radius atom (e.g., I)-containing or electron-dissipating systems.

(7) Basis sets *aug*-cc-pV(T+d)Z (for molecules containing atoms of the third and/or fourth rows) and *aug*-cc-pVTZ (for molecules containing atoms of elements from the first and second rows) are suitable for modeling the geometry of molecules with considerably proximal (with the separating distance of less than 0.4 nm) electron-rich molecular fragments, e.g., systems with π–π or n–π interactions.[a] For providing reliable energy estimates, *aug*-cc-pVQZ is highly preferable.

(8) For iodoorganic compounds, it is worth combining two basis sets cc-pVTZ (or, better, *aug*-cc-pVTZ) and MIDIX. The first of them is

---

[a]Do not forget to use electron correlation methods (Section 5.2.3) for such structures.

applied to all atoms except iodine atoms, and MIDIX is applied to iodine atoms. Energy estimates are not trustworthy, but the modeled geometry rather will be of a fairly well quality.

These recommendations are certainly not universal. Other basis sets may show better performance for either chemical structure. Nevertheless, by following this elementary guide, fledgling QM users will easily obtain results of acceptable accuracy for any organic system of regular size that is modeled for the gas or liquid phase conditions.

5.2.5 *Second division of routine QM calculations: DFT*

Because of the ubiquitous use of DFT methods, the term DFT has become recognizable to organic chemists, although not many experimentalists know exactly what it means. Let us now learn this in general. *Electron density* is a function $\rho(r)$ that depends on three spatial coordinates [depicted as a generalized coordinate (r)] of electrons and obeys the condition

$$\int \rho(r)dr = N \tag{37}$$

where N is the number of electrons in molecular system. Nuclei can be considered as point charges and, according to Eq. 37, $\rho(r)$ "prescribes" their location (i.e., the 3D geometry of the molecular backbone). For a given molecular structure (N electrons; specified nuclei), Eq. 37 holds for any spatial disposition of the point charges (the nuclei). If we know the value of the electron density $\rho(r)$ at each point r_i, we can easily reveal the relevant molecular geometry (the disposition of the nuclei that provides this spatial distribution of the density ρ of N electrons); values of the function $\rho(r)$ are equal to zero at those space points where the nuclei are located. Thus, it seems that it is worth utilizing this molecular quantity in determining molecular geometries.[a] The principal difficulty of this alternative[b] approach to molecular modeling is obvious. We have no idea (and even cannot have it) regarding the mathematical form of this function for either molecular system, and, therefore, are unable to calculate values ρ_i at points

---

[a] The electron density is a measurable quantity for molecules. The 3D distribution of electron density may be determined, e.g., by X-ray scattering.
[b] To mimicking the wave function (Section 5.2.2).

r_i, for any molecular geometry. In addition, it is still not clear how the electron density is related to the electron energy.

In view of the uncertainty in mathematically describing the function $\rho(r)$, let us shift focus to the relationship between this function (i.e., the electron density) and the energy E (or, more accurately, the electron energy E_{el}).[a] Can we relate a set of corresponding E values to different (say, arbitrary) functions $\rho_i(r)$, i.e., design a functional of density, $E_i[\rho_i(r)]$? As the Hohenberg–Kohn *existence* theorem proves, such a unique functional does exist. Moreover, it predetermines the Hamiltonian, and, thus, the energy E_0 of the ground state for non-generative ground state systems as well as the energies E_i of their excited states. This is the first milestone for DFT calculations; it turns out that electron density (a clear, observable molecular quantity) implicitly "contains" the wave function (a key, but in principle, obscure molecular characteristic). If computed, the electron density does not require an independent approximation of the wave function! To realize how admirable this feature of DFT is, one could recall how Christopher J. Cramer, in his famous monograph,[33b] begins the chapter "Density Functional Theory": "What a strange and complicated beast a wave function is." The existence theorem indicates that one can determine the electron energy E without designing a mysterious wave function.[b]

The Hohenberg–Kohn *variational* theorem states that the density functional satisfies the variational principle, so that

$$E_i[\rho_i(r)] \geq E_0[\rho_0(r)] \tag{38}$$

where E_0 is the energy of the ground state, $\rho_0(r)$ is the electron density in this state, and E_i is the value of the density functional for the corresponding arbitrary electron density $\rho_i(r)$, which is equivalent to the true density $\rho_0(r)$ if integrated over space. That is,

$$\int \rho_i(r)dr = \int \rho_0(r)dr = N \tag{39}$$

The second milestone of DFT (the variational theorem) provides the principal route to the electron energy. Equations 38 and 39 are exact (in contrast

---

[a] For briefness, E represents E_{el}, and Ψ represents Ψ_{el}.

[b] Notice that this laborious design is never perfect. Equation 30 (Section 5.2.2) is an example of approximation of Ψ.

to Eq. 30, the theoretical basis of HF calculations), and, hence, they also cover electron correlation. However, these equations alone are impractical; they do not guide us to the density functional E.

Thus, the next step is to design E. In their elegant suggestion, Kohn and Sham have proposed to consider a hypothetical system with *non-interacting* electrons for which the density $\rho(r)$ is equal to the electron density $\rho_0(r)$ of the same *real* system in its ground state. Then, the density functional for the *real* molecular system can be divided into separate components as

$$E_0[\rho_0(r)] = T_{\text{non-inter}}[\rho(r)] + V_{\text{nucl-el}}[\rho(r)] + V_{\text{el-el}}[\rho(r)] + E_{\text{xc}}[\rho(r)]$$

$$(40)$$

where $T_{\text{non-inter}}$, $V_{\text{nucl-el}}$, and $V_{\text{el-el}}$ refer to the kinetic energy of *non-interacting* electrons, the potential energy of attraction of nuclei and *non-interacting* electrons, and the classical potential energy of repulsion of electrons, respectively, while the term E_{xc} characterizes the non-classical interaction (exchange-correlation) energy of electrons. In other words, it contains the energy of all electronic interactions that are not described by classical physics.

Formally, Eq. 40 exactly represents the density functional for the *real* molecular system (i.e., the functional $E_0[\rho_0(r)]$). However, do we have exact "prescriptions" for obtaining each of the four summands in the r.h.s. of Eq. 40? The first three terms are exact expressions[a] since the energy is an additive quantity for physical particles considered as non–interacting objects. However, the correcting term E_{xc} is a weak link in DFT. We do not know how to design it in order to obtain the corresponding *exact* energy that makes the difference between the KS hypothetical system of non-interacting electrons and the real system. Solving this problem *approximately*, the density functional E_{xc} is mathematically constructed in different forms (Section 5.2.6) which usually incorporate empirical parameters. The difference between DFT calculation methods (specified in the chemical literature by certain acronyms; Table 5) is the difference in the structure of these functionals $E_{\text{xc}}[\rho(r)]$. In other words, approximations in DFT are post-KS approximations. They begin with the design of some "artificial" functional $E_{\text{xc}}[\rho(r)]$ that, in using it in the exact KS equation (i.e., Eq. 40) as a "true" functional $E_{\text{xc}}[\rho(r)]$, makes the resulting post-KS equation as accurate as possible.

To obtain the electron energy of a system with interacting electrons, how does DFT proceed with Eq. 40? For the KS abstract system that has

---

[a]The reader may find them in, e.g., Refs. 33b, 53a, and 58.

N electrons, one can relate the KS spin-orbitals χ_i^{KS} and electron density $\rho(r)$:

$$\rho(r) = \sum_i^N |\chi_i^{KS}|^2 \tag{41}$$

Similarly to HF MOs (Section 5.2.2), KS orbitals, in the form of the KS Slater determinant, determine the electron energy for the *real* system with interacting electrons. In this case, the Slater determinant exactly represents the wave function since KS orbitals are related to non-interacting electrons.[a]

When calculating the energy of this interacting system, why is it possible to use KS orbitals, which are related to a fictitious system with no electron interactions? In the Schrödinger equation for this abstract system (Eq. 42),

$$h^{KS}\chi_i^{KS} = E_i\chi_i^{KS} \tag{42}$$

one can define the one-electron operator (the KS operator) so that

$$h^{KS} = -\frac{1}{2}\nabla^2 + V_S \tag{43}$$

where the first term is the operator of the kinetic electron energy and the second term is an effective potential (operator) V_S that validates Eq. 41.

Let us cast a glance at the *real* system. One can show that if KS orbitals χ_i^{KS} are used to determine its energy in the Schrödinger equation, the corresponding expression takes a simple form,

$$\left(-\frac{1}{2}\nabla^2 + V_{eff}\right)\chi_i^{KS} = E_i\chi_i^{KS} \tag{44}$$

where operator V_{eff} is defined as

$$V_{eff} = \sum_k^N Z_k/|r_i - r_k| + \int \rho(r_j)/|r_i - r_j|dr_j + V_{xc} \tag{45}$$

where k runs over nuclei, i runs over electrons, and V_{xc} is a one-electron operator (functional derivative $V_{xc} = \delta E_{xc}/\delta\rho$) to which the exchange-correlation energy of electrons E_{xc} corresponds.

It follows from Eqs. 42–44 that $V_{eff} = V_S$, and we actually have a "bridge" between the hypothetical non-interacting and real interacting systems. Thus, one can apply KS orbitals to calculating the electron energy E of the real system if one can somehow determine V_{xc}.

---

[a]Note that these orbitals have no physical sense in describing real molecules. The square of such a function (a KS orbital) does not characterize the probability of the electron location in either point of space.

With a known V_S, Eq. 42 can be solved for an arbitrary molecular geometry of the non-interacting electron system, i.e., it is possible to obtain the KS orbitals and the energy using the KS Slater determinant. This opens the way to calculating the energy of the real system. Recall that, according to the choice of the hypothetical molecule with non-interacting electrons, the electron density is equal for the fictitious and real systems if they have the same 3D geometry. Using Eq. 41, we therefore obtain the electron density for the *real* system via the KS orbitals of the *reference* system.

Then, the energy of the real system may be calculated via the primary DFT equation, Eq. 40, if the mathematical form of functional $E_{xc}[\rho(r)]$, or, simply, a "dependence" of energy E_{xc} on density $\rho(r)$, is specified. As emphasized above, various mathematical forms of the density functional are actually different approximations in DFT; a separate functional $E_{xc}[\rho(r)]$ is precisely a certain DFT method.

KS orbitals are computed in an iterative SCF procedure, which is similar to that of the HF-SCF procedure (Section 5.2.2). As in HF calculations for MOs, basis functions μ_p as well as coefficients c_{ip} (Eq. 36) are employed to construct orbitals χ_i^{KS} in explicit algebraic form. In the DFT procedure for single-point energy calculations, coefficients c_{ip} are optimized in an analogous fashion to their optimization in the HF-SCF procedure. However, in contrast to this procedure for MOs, a guess electron density is used at the initial calculations to compute all the integrals (including V_{xc}-related integrals), which are required for the construction of the KS orbitals for the initial molecular geometry. Using these orbitals, the new density is calculated (Eq. 41), and then the new KS orbitals (i.e., recalculated coefficients c_{ip}) are obtained, and so on. Iteration cycles are continued until a convergence criterion is satisfied. The principal difference in energy calculations between the HF and DFT methods is that the HF-SCF procedure optimizes the wave function Ψ, approaching it iteratively and reaching a sufficiently accurate probe function Φ, while DFT iteratively optimizes the electron density, employing the physically meaningless probe function Φ^{KS} as an auxiliary technique.

Thus, from the QM viewpoint, DFT methods are various approximate methods for calculating the electron density;[a] from the viewpoint of organic experimentalists, these methods are tools of different accuracies for supplying the molecular geometry and energy. Their non-systematic acronyms (see Table 5 for the most recognizable DFT methods) should not disorient non-specialists; these methods actually differ only in the functionals E_{xc}. In such an acronym, one refers to a certain density functional, i.e., a certain

---

[a]As summarized in Ref. 53a: "HF and the various post-HF electron correlation methods provide an exact solution to an approximate theory, but DFT provides an exact theory with an approximate solution."

DFT method of calculations. Employing a certain basis set (Section 5.2.4) in combination with such a calculation method, we completely specify the calculation methodology used, or in accepted terms, the *level* of calculations. For instance, a typical expression, "Calculations have been performed at the B3LYP/6-31+G(d,p) level" means that the used calculation method is B3LYP (Section 5.2.6), which exploits basis set 6-31+G(d,p) (Section 5.2.4) to compute the energy.[a]

No universal theoretical approach to the design of density functionals exists. We briefly discuss these differently constructed methods in the next Section; at this point, we restrict the description to some general remarks.

(1) DFT is not a purely theoretical methodology. No explicit mathematical form of functional $E_x[\rho(r)]$ can be theoretically deduced from QM equations. Therefore, *DFT methods cannot be systematically improved for modeling any particular molecular property*, e.g., molecular geometry or NMR shieldings. The treatable (suitable for computations) forms of $E_x[\rho(r)]$, i.e., different DFT methods, are designed employing diverse reasonable, but pure theory-independent, approximations (Section 5.2.6). Moreover, a majority of designed density functionals contain empirical parameters. Because of a considerable freedom in the design of the density functionals as well as contribution of empiricism, the accuracy of a DFT method for different chemical structures may vary to a larger extent than the accuracy in similar HF calculations.

(2) For various reasons, some DFT methods are very popular, whereas others are practically abandoned. Selection of the most suitable DFT method is not problematic when modeling "normal-size" molecules whose atoms are not of heavy elements,[b] do not form electron-deficient multicenter bonds, and have no long-range electronic interactions.

---

[a] In relation to HF calculations, expressions such as "at the RHF/6-31+G(d,p) level" mean that the theoretical method is RHF approximation (Section 5.2.2), which employs the 6-31+G(d,p) basis set (Section 5.2.4) in energy calculations. Note that the term *at the level* used in conjunction with the acronym of a non-empirical calculation method has been accepted in computational chemistry for a long time.

[b] Heavy elements may be considered as the chemical elements that have occupied orbitals with the quantum number $l > 3$, i.e., the "post-Ca" elements. In computational chemistry, the term heavy element (or heavy atom) often indicates elements (atoms) other than H.

Table 5. The first four groups (rungs) of DFT functionals with their "design elements" (in parentheses*) and examples of these groups' members (these functionals are used with dissimilar frequency in computational studies of organic molecules).[a]

LDA/LSDA (ρ)	GGA $(\rho, \nabla\rho)$	HGGA $(\rho, \nabla\rho, HF)$	HMGGA $(\rho, \nabla\rho, HF, \tau)$
SVWN	BLYP	B1LYP	M05
SVWN5	BP86	B3LYP	M05-2X
	BPW91	B3PW91	M06
	PBE	B3P86	M06-2X
	PWPW91	PBE1PBE[‡]	M08-HX
	mPWPW91	mPW3PW91	M08-SO
	BVWN	B97-2	ωM05-D
	PKZB	O3LYP	MPW1B95
	BP86-D	X3LYP	BR89B94
	PBE-D	B3LYP-D	TPSSh
	ωB97X	B970-D	M11
	SOGGA	PBE1PBE-D[‡]	mPW1KCIS[§]
			PW6B95

*ρ, electron density; $\nabla\rho$, density gradient; HF, HF account of a portion of total electron density; τ, kinetic energy density.
[†] Synonyms: PBE0, PBEh.
[‡] Synonym: PBE0-D.
[§] Synonyms: mPW0, mPW1PW91, MPW25.

Numerous successful DFT models of different chemical structures permit to claim that, with a high probability, at least four methods of the third or fourth rung (e.g., B3LYP, B3PW91, PBE1PBE, mPW3PW91 and M11; Table 5, Section 5.2.6) will provide a satisfactory or accurate 3D geometry for the next such system. Most of the DFT methods of the fifth rung will model it with practically guaranteeing an accurate molecular geometry and at least satisfactory energy.[a]

---

[a] This does not mean that molecular systems with electron-deficient multicenter bonds (e.g., carbenium ions) cannot be successfully modeled using these DFT methods. DFT models of such molecules usually are sufficiently accurate when using a proper DFT method. The problem is that the selection of the successful DFT functional is not straightforward for these systems.

5.2.6 *Classification of DFT methods*

What exactly are DFT methods? Even organic and bioorganic researchers, who never deal with computational chemistry, often encounter the DFT terminology, because of the widespread use of these calculations. Terms such as *local density approximation* and *hybrid functional* intrigue advanced experimentalists. Certainly, a rigorous formalism of various DFT methods would be redundant information for them. To rationally use DFT calculations in an organic or bioorganic laboratory, one needs to know the general concepts latent in the frequently encountered acronyms for different DFT calculation methods. Let us consider these methods (i.e., different approximations of the $E_{xc}[\rho(r)]$ representation) in more detail.

One may classify DFT methods, which have practical significance in organic chemistry, according to the general design principles of these functionals. There are five groups (known as rungs in the DFT jargon; Table 5): (*1*) local density approximation (LDA/LSDA); (*2*) generalized gradient approximation (GGA); (*3*) hybrid GGA (HGGA; often called hybrid functional methods); (*4*) meta GGA (MGGA) and hybrid meta GGA (HMGGA); and (*5*) double hybrid GGA (DHGGA; usually called double hybrid functionals) methods.[a] The level of the complexity of functionals increases in the order LDA/LSDA < GGA < HGGA ≈ MGGA < HMGGA < DHGGA.[b] However, there is no exact correlation between the functional complexity of individual DFT methods from the neighboring groups and the resulting calculation accuracy.

It is difficult to reliably characterize the relative accuracy of DFT methods even with respect to the primary molecular quantities (molecular geometry and energy). It is not uncommon for a method from a lower-ranked group to "beat" a method from a higher-hierarchized group. For instance, SVWN5 from the lowest-accuracy group (LDA/LSDA) models

---

[a]The -D augment in the names of functionals indicates explicit treatment of dispersion (in simple terms, VDW interactions). These functionals calculate the dispersion contribution using *empirical* atom–atom dispersion corrections; thus, the total molecular energy is the sum of the DFT-supplied energy and the dispersion energy computed in the MM-like pairwise interaction manner. From a puristic point of view, such QM-based, supramolecular chemistry-oriented calculations cannot be characterized using either the *empirical* or *non-empirical* term.

[b]Development of density functionals is constantly progressing. One may expect the present as well as new fifth-rung density functionals to predominate among the DFT methods to be used in the near future.

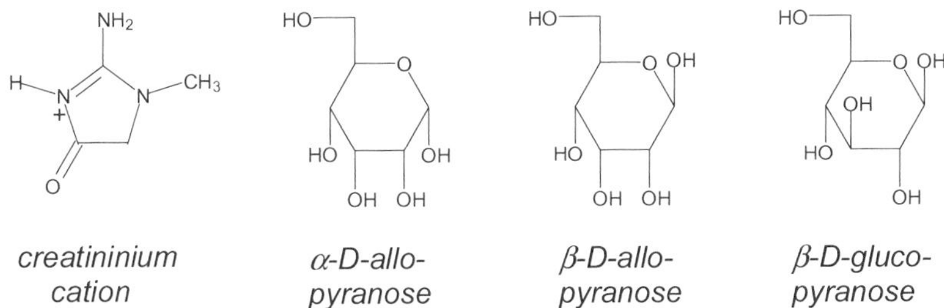

Figure 91. Examples of structures that have been used as patterns to examine the accuracy of some DFT methods.

the molecular geometry of protonated creatinine (Fig. 91) much better than does B3LYP, a method from the highly ranked accuracy group HGGA.[59] Another example: a GGA method, BLYP, can satisfactorily reproduce the geometry of many metal organic complexes whereas B3LYP fails. For such structures, BLYP is superior even to the HMGGA group.

Beginners, who are estimating the quality of different density functionals, should probably evaluate a DFT method according to its relationship with one of the five groups *(1)–(5)*. For calculating the molecular geometry and energy of organic molecules, a general accuracy order for these groups is roughly DHGGA > HMGGA > HGGA > MGGA > GGA > LDA/LSDA, implying calculations that use the same, sufficiently extended basis set (Section 5.2.4). As mentioned, one should regard this order with caution.

As noted, organic and bioorganic chemists do not need to go too deeply into the theoretical descriptions of these or other functionals. They need to at least recognize concrete DFT methods to be capable of estimating the quality of calculations performed by other researchers and, more importantly, to select the optimal method for their own DFT-based computations. Table 5 provides the acronyms of the best-known functionals and their group affiliations. Note that density functionals are usually referred in chemical publications in this form, with no additional explanations.[a]

---

[a]Functional acronyms do not follow either systematic nomenclature. Often, they are abbreviations derived from the last names of their developers, combined (sometimes) with the last two numbers of the publication year.

Nevertheless, what are the essential differences between these groups of DFT methods? In computational chemistry, the exchange-correlation functional $E_{xc}[\rho(r)]$ is usually divided into two components, the exchange functional $E_x[\rho(r)]$ and the correlation functional $E_c[\rho(r)]$,

$$E_{xc}[\rho(r)] = E_x[\rho(r)] + E_c[\rho(r)]$$

$$= \int \rho(r)\varepsilon_x[\rho(r)]dr + \int \rho(r)\varepsilon_c[\rho(r)]dr \qquad (46)$$

where each functional is a function of the electron density $\rho(r)$ and the so-called energy density $\varepsilon[\rho(r)]$ that, in turn, is divided into density-dependent components $\varepsilon_x[\rho(r)]$ and $\varepsilon_c[\rho(r)]$. The difference between the groups of functionals lies in how they treat $\varepsilon_x[\rho(r)]$ and $\varepsilon_c[\rho(r)]$.

LDA. In the simplest DFT approximation, LDA, ε_x and ε_c for coordinates r_i are assumed to depend only on the density $\rho(r_i)$ at this *local* position. This means that the electron density at other points of molecular space does not affect ε_x and ε_c. The "density by LDA" does not smoothly change in molecular space and may even have sharp drops which the real electron density does not have.

The local spin density approximation (LSDA) that also takes into account electron spins has a greater generality. This approach simply uses individual functionals for the α and β electron densities if the two are not equal (e.g., in open-shell systems).

Members of the LDA/LSDA group differ in the mathematical forms developed for one or both functionals ε_x and ε_c. Despite the obvious roughness of the LDA/LSDA model, it supplies fair and sometimes surprisingly excellent results.[59] The overall quality of LDA/LSDA is rather inadequate. Methods of this group overestimate bonding while significantly underestimating attractive London dispersion interactions of non-bonded molecular fragments or separate (chemically non-bonded) molecules.

GGA. The second group, GGA, contains a further assumption. That is, either the exchange functional ε_x or the correlation functional ε_c depends not only on the electron density, but also on local changes in $\rho(r)$, i.e., on the gradient of $\rho(r)$. This idea is implemented differently in different GGA functionals. In GGA models related to the exchange functional ε_x [e.g., B (the Becke's functional), PW, FT97, and mPW], a correcting term is added to the LDA term for ε_x. Other developed GGA functionals ε_x (e.g., B86, P, and PBE) use various function expansions for the gradient. Regarding the correlation functional ε_c, proposed GGA functionals (e.g., P86, PW91, and B95) use a simple approach of adding a gradient-related correcting term to the LDA term. LYP is a stand-alone correlation functional ε_c; it has been constructed without additional augmentation of the LDA term. It provides the correlation energy E_c using four empirical parameters.

Since the full electron density functional $E_{xc}[\rho(r)]$ contains both exchange and correlation functionals ε_x and ε_c (Eq. 46), GGA calculation methods are based on combinations of these individual functionals (one exchange and one correlation functional). For instance, BLYP is a combination of exchange functional B and correlation functional LYP, whereas combinations of B with the correlation functional P86 or W91 yield DFT functionals BP86 or BW91, respectively; the combination of mPW with PW91 affords the mPWPW91 functional, etc. (Table 5).

Although GGA methods are generally of better quality than LDA methods, they can still supply unsatisfactory estimates. One can say that an averaged *large* calculation error for LDA is replaced with a *moderate* one for the GGA group. The error trend is the same: overestimation of the interactions of bonded atoms and underestimation of the interactions of non-bonded atoms.

Hybrid functionals. Functionals that frequently provide acceptable accuracy for the main organic compounds,[a] are of the third group, HGGA. The idea in HGGA is to decrease the uncertainty in design of functionals by correcting functional $^{DFT}E_{xc}$ via an additional, *exactly calculated* term. This term, $^{HF}E_x$, is E_x of the fictitious (no electron correlation) system. The point is that, in HF calculations, KS orbitals form an *exact* wave function (the Slater determinant; Section 5.2.2) for this abstract system. In other words, the additional, HF-calculated term includes a portion of the *total* (not only local) HF-type exchange for the *real* system, and the energy for this portion can be calculated exactly. The next problem in constructing the functional is the proportion in which the functionals $^{GGA}E_{xc}$ and $^{HF}E_x$ should be mixed (or more strictly speaking, adiabatically connected) in order to yield their hybrid (i.e., a new functional) that provides the best

---

[a]Those do not contain "post-Ca" elements and have neither electron-deficient multicenter bonding nor long-range electron–electron interactions.

approximation of the real system. Since this problem cannot be solved theoretically, one resorts to an empirical approach. Numerical parameters are selected as coefficients of a linear combination of $^{GGA}E_{xc}$ and $^{HF}E_x$. Then, the general notation for the HGGA functional $^{Hyb}E_{xc}$ is

$$^{Hyb}E_{xc} = (1-a)\,^{DFT}E_{xc} + a\,^{HF}E_x \qquad (47)$$

where a is a parameter.

For instance, such an expression in the most widely used DFT method, B3LYP, is

$$^{B3LYP}E_{xc} = (1-a)\,^{LSDA}E_x + a\,^{HF}E_x + b\Delta^B E_x$$
$$+ (1-c)\,^{LSDA}E_c + c\,^{LYP}E_c \qquad (48)$$

where $a = 0.2$, $b = 0.72$, and $c = 0.81$ are parameters. The use of three parameters is reflected in the acronymic name of this functional (B*3*LYP).

Most of the routine DFT calculations applied to organic molecules have been yet performed using the HGGA arsenal. In this connection, one has to know that hybrid functional methods are not systematically free of inaccuracy. They poorly represent some non-covalent interactions, e.g., π–π (stacking) interactions, and "overdelocalize" electrons in conjugated systems. As a result, the modeled geometry is distorted. The major DFT flaw — inadequacy in estimating interactions of any non-bonded molecular fragments — is lessened in HGGA *vs.* GGA; however, this improvement is unsatisfactory. The accuracy of the HGGA methods diminishes in parallel with molecule enlargement, i.e., as the role of non-covalent interactions increases. Therefore, it is absolutely incorrect to use these functionals in modeling molecular systems of nano-size, e.g., in attempting to understand charge transfer in the nucleobase stack of double-strand DNA fragments. The general performance of HGGA methods *for the molecular geometry* of small and moderate size molecules may be characterized as good.[a]

B3LYP merits a separate note. As mentioned, this DFT method has been most often used in molecular geometry modeling and energy calculations for organic molecules consisting of elements of the first three rows. Although it is incorrect to accept any DFT method as

---

[a]This statement does not imply the above-mentioned systems of "complicated" chemical structure, e.g., "post-Ca" metal organic complexes, even if they are small-size molecules. Regarding such concrete systems, the accuracy of an HGGA method is *a priori* uncertain.

superior, the common belief of sporadic users of chemical computations is that B3LYP could afford the best accuracy for a given organic structure. This opinion has no solid basis. To prove it, comparative examination of a very broad set of different chemical structures by various DFT methods should indicate superiority of B3LYP for most of the structures tested. However, the situation is the opposite of what one may anticipate. For instance, the calculation accuracy degrades when B3LYP is applied to increasingly larger systems, and it is not uncommon for B3LYP to estimate the relative stability of isomers with a significant error. With screening HGGA methods, the evidence is growing that B3LYP is not the best choice for many compounds (see, e.g., Refs. 60a, b).

Hence, from the molecular geometry perspective, HGGA functionals are suitable for conformational analysis of "normal" organic systems. What about energy estimates by HGGA? It appears that sometimes they are "of the MP2 level" in modeling small molecular systems which do not contain weakly bonded atoms. One could conclude that this DFT approximation indeed takes into account the main portion of the electron correlation energy. However, if non-covalent interactions of weakly-bonded or distant non-bonded fragments are significant in the molecule or molecular complex, the relative stability predicted by an HGGA method may turn out to be computed with an intolerable error of more than $2\,\mathrm{kcal\,mol^{-1}}$. This systematic inaccuracy shows that the above conclusion regarding the ability of HGGA to "incorporate" the correlation energy significantly is too rigorous. For conformational research, this means that the use of MP2 for *energy calculations* is preferable to the use of hybrid functional methods if, of course, the available computational facilities are so extensive that the difference in computational cost may be neglected.[a]

HMGGA. The functional design level in HMGGA is higher than in HGGA; it extends the expansion of the functional $E_{xc}[\rho(r)]$ to include also the kinetic energy density $\tau(r)$. Rephrasing this in formal symbols, the third term $\tau(r)$ appears in the r.h.s. of Eq. 46. The practical result can be foreseen: energy predictions by HMGGA are typically somewhat more accurate. However, again, the higher complexity of the density functional does not guarantee more accurate results.

---

[a]Inexperienced researchers could simply check the time difference by mimimizing the energy for a probe structure of interest via an HGGA method and MP2, with the same basis set (Section 5.2.4). Such an examination is very fruitful. The researcher quickly learns the possibilities of the available computers for running QM calculations.

A comparative computational conformational analysis of selected monocarbohydrates[60a] probably supplies a representative example. One could think that HMGGA methods should show consistent superiority. In agreement with these expectations, PBE1PBE, B3PW91, and B3LYP from the HGGA group perform more poorly than M05-2X, as shown by calculations of relative energy for conformers of α- and β-D-allopyranose (Fig. 91), for vacuum conditions.[a] Surprisingly, for conformers of a very close analog, β-D-glucopyranose, the HMGGA method demonstrates approximately the same accuracy as B3LYP, whereas PBE, a method from the two-rung lower group (GGA) yields much better accuracy. In addition, B3PW91 and PBE1PBE (the HGGA group) are more accurate than M05-2X in estimating the relative stability of these conformers.

Further, the above-mentioned "chronic disease" of DFT — missing dispersion and other non-covalent interactions — does not disappear in HMGGA calculations. Despite some improvement in accounting for these interactions, the molecular size appreciably affects the reliability of modeling by HMGGA methods. One might admit that on the whole, the methods in this group do not persuasively advance the calculation accuracy relative to that of HGGA methods.

Thus, in contrast to the rather rigid hierarchy of post-HF methods of QM (Section 5.2.3), DFT practice does not supply researchers with information about the relative quality of individual density functionals. Indeed, dozens of benchmark studies have reflected on the practical aspects of different DFT methods, dealing mainly with geometry and energy calculations. However, this direct empirical examination of different density functionals has not revealed a method or a group of methods that could be reliably associated with high accuracy in modeling any chemical structures. How do non-computational chemists manage such a situation? An effortless way out does not exist; the forthright, affordable, and practical solution is to examine a widely used DFT method of the fourth rung, e.g., either M11 (implemented in Q-Chem), or M06-2X (implemented in Gaussian 09, Q-Chem and ADF), or PW6B95 (implemented in Jaguar), or TPSSh (implemented in Gaussian09, Turbomole and ADF). In this examination, the two first stereoisomeric structures *A* and *B* of interest are modeled using the selected DFT method and a sufficiently extended basis set, e.g., 6–31+G(d,p) or *aug*-cc-pVTZ (Section 5.2.4). The obtained geometries and the energy difference ΔE_{DFT} for *A* and *B* are compared with the corresponding molecular parameters for these 3D structures supplied by MP2 calculations that employ the same basis set. If

---

[a] B3LYP, a "favorite" among hybrid functionals till recent time, here yields worse accuracy than PBE1PBE and B3PW91.

the geometries provided by the DFT and MP2 calculations are essentially the same, and the energy differences ΔE_{DFT} and ΔE_{MP2} differ by less than 0.3 kcal mol^{-1}, the examined DFT method may be considered as valid for modeling the PES's of "normal"-sized organic molecules that have the same chemical functionality and the same relative positions of functional groups as the probe structures *A* or *B*.

DHGGA. Among the DFT methods that have become recognizable to non-specialists, the top group is DHGGA. The fifth-rung generation of DFT functionals has come into use relatively recently. These functionals (e.g., B2PLYP, B2PLYPD3, B2GP-PLYP, MC3BB, MC3MPW, DSD-BLYP, DSD-BLYP-D3, PBEO-DH, PTPSS-D3, PWPB95-D3, XYG3, XYGJ-OS, and TPSSO-DH) are thought to bring DFT calculations to "the heaven of chemical accuracy" (with errors in energy differences on the order of 1 kcal mol^{-1}).[60b–f] Indeed, they are intended to take into account non-covalent interactions with sufficient accuracy and, consequently, are absolutely preferable to other DFT methods for modeling large molecular systems. However, these promising functionals have not yet been sufficiently tested,[a] and it is difficult to "pick up" DHGGA functionals which can be reasonably assumed to outperform other methods from this group when applied to either concrete chemical structure. Because of the non-local character, DHGGA methods require the use of sufficiently extended basis sets, e.g., QZVP (a quadruple-ζ set; Section 5.2.4) or CQZV3P (a quadruple-ζ set with three sets of valence polarization functions augmented by core-polarization functions) and therefore are time-consuming in computations to approximately the same extent as MP2 calculations. Besides, only some of them have been implemented in the versions of commonly used QM calculation packages. Before 2013, Gaussian09 includes two double hybrid functionals, B2PLYP and PW2PLYP; Q-Chem offers B2PLYP, B97X-2, XYG3, and XYGJ-OS; Turbomole provides only B2PLYP, while Jaguar, ADF, Spartan10, and HyperChem8 all support neither DHGGA method. The assortment of the currently available double hybrid functionals is thus narrow for non-computational chemists.

---

[a]Nevertheless, the recent review that analyzes DHGGA functionals (Ref. 60f) throws some light on the relative quality of these contemporary DFT methods. Interestingly, B2PLYP, the first well-developed DHGGA method, appears remarkably successful: it does not considerably yield to later functionals of this rung in the modeling accuracy.

After considering the pool of density functionals, we can conclude which of them are suitable for routinely modeling geometries and relative energies of stereoisomeric structures of "normal"-sized organic molecules. They are functionals from the HGGA (e.g., B3LYP, B3PW91, and PBE1PBE), HMGGA (e.g., M11, M06-2X, and TPSSh), and DHGGA (e.g., B2PLYP, PW2PLYP, XYG3 and computationally less demanding XYGJ-OS) groups. HGGA and HMGGA functionals are still the practical choice for this elementary modeling, because of significantly lower computational cost and satisfactory or even higher accuracy obtained when exploiting these theoretical methods. Nevertheless, the third-rung functionals (i.e., those from the HGGA group) have begun to become obsolete, and one should minimize recruiting them. If computer resources enable to use DHGGA functionals, e.g. PW2PLYP, XYG3 and XYGJ-OS, one should follow this way when modeling at least 10–30 atom molecules: such functionals significantly increase the quality of DFT-derived molecular models relative to the lower rung functionals. If computer resources available do not allow to exploit such "heavy" DFT functionals, the remaining option is to resort to functionals from the HMGGA group.

5.3 Accuracy of QM Calculations

5.3.1 *Two routes to estimating the calculation accuracy*

The accuracy of molecular modeling may be understood as deviations of modeled molecular parameters (numerical values) from the corresponding parameters of real molecules; thus, the considered accuracy is a quantitative characteristic. In determining the modeling accuracy, we actually answer the quantitative question of how small these deviations are for the modeled structure. Qualitative terms such as *good, fair, excellent,* etc. are ubiquitous when characterizing this accuracy; these terms reflect some quantitative limits of this accuracy. Clearly, the accuracy limits cannot be strictly defined. The accuracy required in conformational analysis and the accuracy provided by routine QM calculations are discussed in Sections 5.3.2 and 5.3.3.

How can one then determine the accuracy of QM calculations performed for stereoisomers, (conformers implying rapidly interconverting stereoisomers; Section 2.6) of a given system? If experimental data are available for the relative stability of the conformers (ΔG^0 values) of this molecule or the related conformational barriers ($\Delta G^{\#}$ values), we already have the answer.[a] However, computational modeling of conformers and conformations is usually performed when experimental data for the system of interest are absent. In that case, the accuracy of the obtained results must be inspected. For this purpose, one needs to use either a reference system or a reference model that would demonstrate the accuracy of both the chosen model (the gas phase or solution) and the calculation method. The reference *system* is a structural analog (in fact, we need several analogs for more confidence), for which we have related experimental data, i.e., measured values of ΔG^0 and/or $\Delta G^{\#}$ for its conformational equilibrium. The reference *model* is a model of the studied system itself obtained by using calculations at a high level of theory (CCD, at least; Section 5.2.3). The applicability of the chosen model and calculation method are proven or disproven by comparing calculated ΔE_0 (or $\Delta E_{\min}$; see Section 5.3.3) values for the given system with either experimental data (ΔG^0 and/or $\Delta G^{\#}$ values) for reference compounds or the corresponding calculated ΔE_0 (or $\Delta E_{\min}$) values for the same stereoisomers of the reference model.

The reference model methodology is unpractical for organic experimentalists; they do not intend to spend too much time for theoretical modeling.[b] Calculations that provide the reference model are actually a separate, not quite trivial computational task. For instance, solvation should be taken into account wherein the selection of an adequate solvation model is a separate problem (see below). High-theory-level QM computations are so demanding of computer resources that such verifying calculations cannot be undertaken even in an unordinary well-equipped organic laboratory.

The reference system approach is preferable for synthetic chemists. Therein, calculations are also undertaken for the reference molecule using the same calculation method, basis set, and external parameters (medium, temperature, and pressure) that have been used in modeling the studied

---

[a]If molecular geometry alone is the focus, experimental data for the 3D molecular structure are sufficient.

[b]As a rule, non-computational organic chemists, who, nonetheless, exploit non-empirical calculations in their research, prefer to merely use "standard" DFT methods from the HGGA group (Section 5.2.6) without examining the quality of the resulting molecular models. One should note that this practice is careless.

system. Comparing the results of QM calculations with the relevant experimental data for the reference system (e.g., the experimental and calculated geometries or the measured and calculated barriers of the same conformational transformations), we determine the accuracy of the modeling of the reference molecule, and, thus, decently estimate the accuracy of the modeling of its close structural analog, our system of interest.

5.3.2 *Calculation accuracy: Molecular geometry*

As we know, conformational studies are explorations of the PES. In the language of molecular modeling, they deal with the location of minima and saddle points in the PES (Section 3.1). Sometimes, other molecular quantities of stereoisomeric structures, e.g., the oscillation frequencies, magnetic shielding, or dipole moments, are also calculated; they may be useful in conformational analysis (see, e.g., Chapter 6). Nevertheless, the primary target of non-empirical conformational studies is the PES, i.e., molecular geometries of stereoisomeric structures and corresponding molecular energies. Therefore, let us consider the accuracy of QM calculations related to these main attributes of the PES.

As one might expect, no modeling is perfectly precise. In particular, the 3D geometries of a real molecular system and its model are *always* non-identical. Surprisingly, in considerations of calculation accuracy for molecular geometry, we have to assess here *both* the model and the real system. It is trivial that different theoretical models of the same molecule are of different quality. However, the real molecular system, which at a passing glance should be an unaffected primary standard, appears to be not identically demonstrated by various *experimental* structural studies.

First, let us understand what molecular geometry is when modeled by locating the energy minimum in the PES with using the SCF procedure (Section 5.2.2). The MEP (not populated; Figs. 42 and 92) of a potential well in the PES corresponds to the geometry located upon energy minimization. Thus, this geometry is exactly what we call the calculated (modeled) geometry; in other terms, the equilibrium geometry. However, potential wells for real molecules are not symmetrical, and the r-coordinates r_{min} (the r-coordinate of this point) and r_{max} (the r-coordinate of the maximum of the probability for the geometry) do not coincide (Fig. 92). This means that the

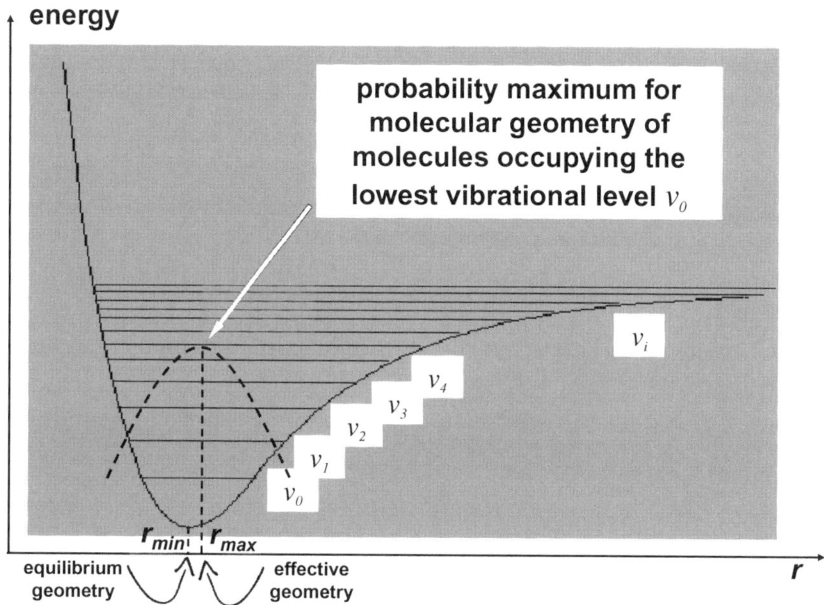

Figure 92. Potential energy well for a real molecule (r is the generalized geometry coordinate). Dotted curve schematically shows the probability of the geometry r_i for molecules at the lowest vibrational level v_0. Routinely modeled geometry corresponds to the r-coordinate r_{min}, while the experimentally detected geometry approximately corresponds to the r-coordinate r_{max} (effective geometry; the precise experimental geometry is an average of statistical weights of such r_{max} geometries "supplied" by all vibrational and rotational levels).

modeled geometry does not correspond to the potential well's uncentered, probability-spread, and vibration-level-averaged 3D geometry determined experimentally (effective molecular geometry). Strictly speaking, there is no exact molecular geometry at all, only some range of geometrical parameters predetermined by the width of the potential well.

Fortunately, potential wells are narrow at the depth of the zeroth vibrational level for most small molecules without weak bonds (e.g., long intramolecular H-bonds). In routine molecular modeling, the geometry of such a stereoisomer is *approximated* by the geometry corresponding to the MEP of the located energy minimum in the PES. Owing to this approximation, some principal inaccuracy is an attribute of the comparison of experimental and modeled molecular geometries.

Let us then address the second part in considering molecular geometry determined experimentally. To accept that the experimental geometry is not the absolute primary standard (see above) means to realize that we cannot speak about a unique experimental molecular geometry (even a virtual one averaged over vibrational and rotational levels) of a stereoisomer determined at concrete thermodynamic conditions. The first reason for this uncertainty is that different analytical physical methods furnish somewhat distinct molecular 3D structures. For instance, interatomic (i.e., internuclear) distances provided by X-ray diffraction are shorter than the distances measured by gas electron diffraction, microwave spectroscopy, solid-state NMR, or neutron scattering.[40a] This discrepancy ($\sim$0.01–0.03 Å) is too large to be neglected in examinations of the accuracy. There is no absolute experimental molecular geometry; there is an experimental method-variant molecular geometry.

In calibration studies, when referring to experimental data for molecular geometry to determine the accuracy of the obtained theoretical model, one should know which geometry is, in fact, extracted from experiments. For each pair of nuclei (in classical interpretation, an oscillator), X-ray diffraction and other methods (excluding electron diffraction) provide the distance between the time-averaged positions of each nucleus. Why does X-ray diffraction underestimate these distances? In X-ray diffraction analysis, nuclei are located as maxima of electron density. The averaged positions of nuclei and the maxima of electron density are not at precisely the same location. Therefore, distances determined by X-ray diffraction are not an entirely successful primary standard for QM calculations.

An example of the experimental method, which provides time-averaged interatomic distances without distorting these values, is solid-state NMR spectroscopy. By measuring the coupling constant of dipole–dipole interactions of a pair of *nuclei*, it measures the distance between their time-averaged positions with high accuracy.[63]

Electron diffraction is a separate case in measurements of molecular geometry. This method gives the average of internuclear distances due to molecular oscillations (averaging of time-independent geometries for molecules occupying different vibrational levels), and this value is not equal to the distance between the time-averaged positions of nuclei. One may conclude that the distance determined by this method is not an ultimate primary standard.

The second reason for the uncertainty in experimentally determining molecular geometry lies in the limited measuring possibilities of analytical physical methods. Any accurate spectral method has its own experimental error. This means that experimental methods supply somewhat blurred 3D geometries. For instance, the absolute error in measuring interatomic distances is of the order of 0.002 Å through 0.009 Å (or, as a relative error, on the order of 0.1% through 1%) for solid-state NMR.[63] The experimental errors of other methods are at least as large. In case of high-level theoretical QM calculations, we find that they may provide, e.g., a 0.001 Å difference relative to experimental values for different C–C bonds.[64] Since this value is less than the typical error of experimental methods, we do not know whether it represents the calculation accuracy. Perhaps the latter is even higher the experimental accuracy; in other words, the experimental accuracy, as a "crude" value relative to that of high-level theoretical calculations, restricts the estimation of the accuracy of these QM computations of molecular geometry. Paradoxically, accurate QM calculations may establish the 3D geometry at least as well as modern (expensive) spectral techniques do!

This statement is probably news to experimentalists. The common understanding is that calculated values, at best, could be close to experimental ones. In fact, the situation in structural chemistry is exactly the opposite. High-level theoretical modeling is a primary standard, while experiments may be close to it.

In our discussion of the principal accuracy of *ab initio* calculations in modeling 3D geometry, we actually have passed from considering experimentally determined geometries to considering the modeling quality. In conformational studies, high-level QM models (e.g., *ad hoc* multilevel models; see Ref. 33b) are not used because of their enormous computational cost. Therefore, non-computational chemists probably need information regarding the desired accuracy in routine, lower-level calculations of molecular geometry. We can speak in terms of numbers that are of a conventional nature. A *satisfactory* accuracy could be associated with values of ± 0.03 Å for bond lengths (relative error of 2–3%), $\pm 3°$ for bond angles (relative error of 2–3%), and $\pm 10°$ for dihedral angles (relative error of 15–25%).[65] The order of these numbers is remarkable. It originates from the energy, which is hidden in each of these structural fragments. The energies of covalent bonds are high (60–200 kcal mol^{-1}), and therefore even a small change in their length or directionality has a significant energy cost. The energies of non-covalent interactions are much lower, and some alterations in the mutual orientation of neighboring fragments (i.e., internal rotation for several degrees) result in insignificant changes in molecular energy as compared to the energy contribution due to deformation of the covalent framework. Since the energy is the parameter "monitored" in the SCF procedure of geometry optimization (Section 5.2.2), insufficiently accurate QM methods miscalculate the structural parameters of low energy contribution. Therefore, exactly the torsion angles are a weak point in non-quite-accurate geometries[a] modeled by QM calculations with using DFT methods of the third and fourth rung (Section 5.2.6).

If several torsion angles in a molecular model inappreciably deviate from their true values, the molecular shape is significantly distorted as a whole. Therefore, modeled 3D molecular structures of satisfactory quality[b] may be useful for qualitative (say, on-the-screen) estimations of steric hindrances, proximity or remoteness of functional groups, intramolecular cavity size, or orbital overlap, but they are not suitable for comparing molecular energy of isomers, biorecognition modeling, establishing QSAR, or calculations of chemical shieldings and vibrational frequencies.

An accuracy three times higher for these structural parameters (± 0.01 Å for bond lengths, $\pm 1°$ for bond angles, and $\pm 3°$ for dihedral angles) may be called *good* because a molecular geometry of this quality satisfies the modeling of reaction transition states, guest–host interactions for biomolecules, binding of small ligands to molecular surfaces, etc. Note that it is usually sufficient to use DFT methods from the HGGA group (Section 5.2.6) to reach such accuracy. However, good 3D geometries (modeled or even experimental) of organic molecules do not guarantee good values of the *calculated* energy of these 3D structures or their other molecular quantities such as chemical shieldings. Sometimes, a better molecular geometry is required for valid estimates of these properties.

Nevertheless, in the context of our main discussion — how to explore conformational equilibrium by theoretical computations — one should note that, in the above sense of the accuracy, *good* molecular geometries are sufficient for QM-based modeling of the PES. In this modeling, difficulties appear in energy estimation, and they are caused by factors that are not related to the quality of the modeled geometry (see the next Section).

---

[a]Geometries of satisfactory accuracy, according to the conventional view on the terminology for accuracy mentioned in this page.
[b]I.e., molecular geometries of satisfactory accuracy.

5.3.3 Calculation accuracy: Relative energy of stereoisomers

The absolute values for steric energy E_{ster} (Section 4.2.3) as well as adiabatic potential energy E_{el} (Section 3.1) supplied by MM and QM for a point in the SES and PES, respectively, are useless. Only the corresponding *differences* ΔE_{ster} and ΔE for two points on such surfaces are useable, since they precisely reflect the relative stability of the corresponding 3D molecular structures.[a] As explained in Section 4.2.6, an accuracy of minimum $0.5\,\mathrm{kcal\,mol^{-1}}$ for a relevant energy difference ΔE is desirable in conformational studies. From a rigorous viewpoint, this accuracy in the energy is insufficient, since it still represents a significant error of 15–20% in the relative quantity of conformers.

What convention could be used for the accuracy in calculating ΔE for the needs of conformational analysis? An accuracy of $0.2\,\mathrm{kcal\,mol^{-1}}$ is a reasonable value.[65] Let us consider the most uncertain case of calculated relative stability: the *calculated* energy appears equal for two conformers (ΔE is zero). We cannot determine which of them is favorable, but the degree of predominance at room temperature is less than 7% if the calculation accuracy is $0.2\,\mathrm{kcal\,mol^{-1}}$. In other words, the conformer ratio is estimated as $50\pm7\%$. The estimation error of conformer fractions decreases with an increase in calculated ΔE, owing to the exponential character of the Boltzmann distribution (Section 3.1, Eq. 6). For instance, if the calculated $\Delta E = -0.826\,\mathrm{kcal\,mol^{-1}}$, the content of the predominant conformer at room temperature is $80 \pm 3\%$ for the discussed accuracy of $0.2\,\mathrm{kcal\,mol^{-1}}$ in energy calculations.

With respect to MM, it would be naive to expect from current force fields a systematic accuracy for ΔE_{ster} of even $0.5\,\mathrm{kcal\,mol^{-1}}$. In contrast, QM calculation methods can in principle provide the required accuracy of $0.2\,\mathrm{kcal\,mol^{-1}}$ for ΔE. The problem is that it is extremely difficult to achieve this in short-term QM calculations, whose role in organic laboratories is only to support the main, i.e., experimental, research. One can inform

---

[a]The ratio of equilibrating non-interacting components p_1 and p_2 is determined by the difference in their free energies: $p_2/p_1 = \exp(-\Delta G^0/RT)$, where p_i is the molar content of the i-th component, ΔG^0 is the molar Gibbs energy, R is the universal gas constant, and T is the absolute temperature. Thus, we can freely operate further with ΔE or ΔG^0 values believing that the reader can easily associate these numbers with the relative quantities of equilibrating conformers.

the reader immediately that there is no prescription for such high accuracy for ΔE. Nevertheless, let us consider how one could increase the accuracy in routine energy calculations by QM.

Referring to Sections 5.2.2–5.2.6, the reader can note that tuning the accuracy is much more complicated in QM than it is in MM. Instead of the two rough accuracy "switches" in MM, the selection of a force field and the decision about vacuum *vs.* solvation model, several fine regulators of the accuracy are available in QM. On the other hand, options for rationally improving the calculation quality are more effective and profound in QM (which is a strict physical theory) than they are in MM (which is basically an empirical method based on calibration of somewhat "artificial" equations).

Interestingly, principal possibilities of MM modeling are not limited. Computational chemists may develop a new force field or extend parameterization of an existing fine force field, which could successfully model any particular class of molecular structures. However, in order to cover the unlimited diversity of organic structures, one should create an infinite number of force fields. Therefore, QM supplants MM in today's computational conformational analysis of organic compounds. This strict physical theory, minimally using empiricism (Sections 5.2.4–5.2.6), has introduced universal calculation methods for modeling various chemical structures. When applied to concrete molecular systems, QM modeling needs fine tuning to provide or approach the desired accuracy (Section 5.3.4).

5.3.4 *Improving calculation accuracy*

What are the main accuracy "regulators" in *ab initio* calculations? Let us arrange them in the order of the effectiveness of their tuning.[a]

(1) **Selection of a proper calculation method/basis set.** This is probably the main route for approaching the desired accuracy of $0.2\ \text{kcal mol}^{-1}$ for ΔE for a pair of stereoisomeric structures. This selection is generally the most complicated decision because it touches three accuracy "regulators." Calculation methods (electron correlation or DFT methods; Sections 5.2.3 and 5.2.4) are the first "regulator," and basis sets (Section 5.2.5) are the second. As we know, a calculation method and a basis set are combined in

---

[a]These items, excluding items (2) and (3), are also related to the accuracy in modeling molecular geometry.

forming an actual calculation tool. Varying such combinations is the third "regulator."

The scheme for using these "regulators" may be as follows. One should start from selecting the calculation method. To bypass the unavoidable difficulties in selection, beginners in routine QM modeling of the PES could take simple prescriptions (Sections 5.2.3 and 5.2.6) as a guide. If the selected calculation method is insufficiently fine, the other two regulators will not help attain good accuracy. After selecting the theoretical method, a basis set from the recommended ones (Section 5.2.4) may be chosen. The following calculations will show (Section 5.3.1) how successful the used calculation tool is in modeling the system of interest. If the resulting accuracy for ΔE is close to the desired threshold (suppose, we have defined this accuracy threshold to be 0.2–0.3 kcal mol^{-1}), it is worth examining a more extended basis set from the same series (e.g. Dunning's series; Section 5.2.4). If this accuracy is essentially low, one should start the accuracy tuning from the beginning by selecting a higher-theory calculation method.

(2) **Zero-point energy correction.** The minimum energy E_{min}, which is reached by SCF iterations, corresponds to the bottom of the potential energy well (r-coordinate r_{min} in Fig. 92; Section 5.3.2). As explained in Section 2.4, real molecules (i.e., a molecular ensemble) differently populate the vibrational levels in the energy well, while the well bottom is not occupied at all. This means that real molecules have more energy than the calculated E_{min} value indicates, i.e., some energy in addition to E_{min} is dissipated in irremovable molecular vibrations. For a harmonic potential energy well, this additional energy E_{ZPE} is calculated as

$$E_{ZPE} = \sum_i \frac{1}{2} h \nu_i \tag{49}$$

where index i runs over vibrational modes, h is the Planck constant, and ν_i are the calculated vibrational frequencies. This augmented E_{ZPE} is called the zero-point energy (ZPE) correction.

Thus, the ZPE-corrected energy E_{corr} of the modeled molecular ensemble is the sum $E_{corr} = E_{min} + E_{ZPE}$. Hence, when comparing the relative stability of two stereoisomeric structures, i.e., of two stable stereoisomers or a stable stereoisomer and the transition state of its 3D reorganization, one should consider the difference ΔE_{corr} and not the difference ΔE_{min}.[a]

However, one should calculate vibrational frequencies ν_i (arithmetically, they are numbers proportional to the energies of the corresponding vibrational levels) in order to derive the component E_{ZPE} of the energy E_{corr} using Eq. 49. *Ab initio* calculations of ν_i require additional computational time and more computer resources than required for energy minimization. In conformational studies, the question arises of whether it is worth calculating E_{ZPE} in order to increase the modeling accuracy or it is sufficient to obtain only E_{min} values in providing the same accuracy.[b] It seems at first sight that there is no need to calculate E_{corr}, but only E_{min}. Indeed, the difference ΔE_{corr} (and not the absolute values of E_{corr}) indicates the relative stability of stereoisomeric structures. The difference in vibrational energy seems to be insignificant for a pair of stereoisomers; in a classical analogy, the frameworks of the same chemical connectivity are of similar elasticity for small deformations of the molecular geometry. Therefore, one could think that the *difference* ΔE_{ZPE} is very small for stereoisomeric structures. The methodological conclusion could be to neglect the frequency calculations and use E_{min} to estimate the relative stability of conformers. However, classical analogies fail again. It turns out that the difference between the E_{ZPE}-corrected value ΔE_0 and the uncorrected value ΔE_{min} for a pair of stereoisomers is often on the order of several tenths of 1 kcal mol^{-1} for $\sim$20- to 40-atom organic molecules. Apparently, the E_{ZPE}-corrected values E_{corr} should be calculated when modeling stereoisomeric structures more or less accurately.

Organic experimentalists normally perform chemical reactions in solutions. Thus, modeling solvation should probably be an attribute of accurate *ab initio* calculations related to organic reactivity. The absolute solvation free energies even for $\sim$10- to 20-atom non-ionized organic structures are several kcal mol^{-1} or even more than 10 kcal mol^{-1} for organic solvents.

---

[a]Equation 49 is related to harmonic approximation of oscillations. Fortunately, low-lying vibrational levels at stationary points of the PES are approximated well by the harmonic law. Most of real molecules from the molecular ensemble populate these levels, according to Boltzmann statistics (Eq. 1, Section 2.4). Therefore, the simply defined E_{ZPE} (Eq. 49) is used in applied QM calculations.

[b]Research articles usually indicate whether the calculated molecular energies include the ZPE correction.

Calculating E_{ZPE} obviously would not help achieve the total goal accuracy of $0.2\,\text{kcal}\,\text{mol}^{-1}$ if solvation modeling does not have an accuracy of $0.1\,\text{kcal}\,\text{mol}^{-1}$. Such accuracy in modeling solvated molecules is a distinctly difficult task. Usually, QM solvation models have an accuracy of around $1\,\text{kcal}\,\text{mol}^{-1}$ (see below). Thus, one can come to an opposite conclusion: it is sufficient to use the uncorrected values ΔE_{min} when modeling molecules in solution. How can one decide in view of these contradictory conclusions? The final recommendation regarding E_{ZPE} is exacting; it is better to calculate it if the computer capacity and time frame allow. Although the desired accuracy of $0.2\,\text{kcal}\,\text{mol}^{-1}$ for ΔE_{corr} might be not attained, the accuracy will at least be higher than that in the absence of this correction.

(3) **Free energy.** QM modeling of an organic structure, if understood as only locating the relevant energy mininum, is targeted to one molecule. However, real systems are ensembles of molecules. An abstract ensemble of identical *modeled* molecules and an ensemble of *real* molecules are non-identical. Indeed, even molecules of the same 3D structure, being part of a real molecular ensemble, are not equal in their energy. They are distributed into different energy levels (electronic, vibrational, and rotational). Although higher-energy levels are minimally occupied at ambient temperature, some inaccuracy can occur in calculations if molecules at these states are ignored. Moreover, these energy levels are differently populated at different temperatures, and the temperature increase forces molecules to more and more occupy them. In the E_{corr} description, the temperature factor does not exist for the molecular systems. As we remember, the PES, and thus the E_{corr} values, are temperature-independent. This simple consideration leads us to the conclusion that the values of E_{corr} do not describe molecular systems (i.e., molecular ensembles) adequately; they do not include the entropy contribution.

An additional inaccuracy associated with the absence of the entropy contribution in values E_{corr} is introduced by intermolecular interactions in real systems. Molecules in solution form different short-lived associates, i.e., various interconverting solvates. In PES terms, many similar-energy, low-barrier-separated minima appear (as illustrated metaphorically in Fig. 93) because of such an aggregation, instead of one minimum for a stereoisomer in vacuum. Solute molecules are distributed into these multiple minima, and the entropy factor of this disorder becomes important. No single

Figure 93. Jagged mountain terrain: a metaphoric illustration of the huge complexity of PES's for multi-conformer organic systems in solution.

minimum-related molecular structure represents a molecular ensemble of a stereoisomer in the liquid phase. Hence, the difference ΔE_{corr} does not precisely characterize the relative stability of stereoisomers in solution. They are molecular ensembles!

As we know from thermodynamics, the main characteristic of such an ensemble is the Gibbs (free) energy G, which takes into account the entropy. In particular, the energy difference measured in equilibration experiments is exactly the Gibbs energy difference ΔG^0 (G^0 is the molar free energy). Hence, to estimate the relative stability of stereoisomers, the energy parameter that must be calculated is ΔG^0. This is not an easy task. To calculate G^0 for a molecular ensemble, one should consider the distribution of molecules into all states; more accurately, Gibbs energy G^0 for a molecular ensemble is derived by inserting the partition function Q into fundamental thermodynamic equations.[a]

---

[a] This function $Q = f(N, V, \mathrm{T}) = \sum \exp(E_i(N, V)/k_{\mathrm{B}}\mathrm{T})$, where E_i is the energy of the i-th state, k_{B} is the Boltzmann constant, N is the number of molecules in the ensemble,
In mathematical notation, this Gibbs energy G^0 is represented as follows: $G^0 = H^0 - \mathrm{T}S^0$, where $H^0 = U^0 + PV^0$ (H^0 is enthalpy, U^0 is molar internal energy, and P and V^0 are pressure and molar volume, respectively), $U^0 = k_{\mathrm{B}}\mathrm{T}^2(\frac{\partial \ln Q}{\partial T})$, and $S^0 = k_{\mathrm{B}} \ln Q + k_{\mathrm{B}}\mathrm{T}(\frac{\partial \ln Q}{\partial T})$ (S^0 is molar entropy).

V is the volume, and T is the absolute temperature, by considering the distribution of the molecules from the ensemble into energy levels of different natures (electronic, translational, rotational, and vibrational), describes the molecular ensemble as a whole (i.e., its distribution into N^n states, where n is the number of all energy levels for a molecule from the ensemble). Function Q is relatively easily deduced only for the case of non-interacting molecules, i.e., for the ideal gas approximation. In this case, one can introduce the molecular partition function q that is related to the ensemble function Q as follows: $Q = q(V, T)^N/N!$. Function q permits to consider molecules themselves (i.e., their energy levels and the related distributions) and not the ensemble as a whole. Then, the function q may be represented as the mathematical product $q_{el} \times q_{trans} \times q_{rot} \times q_{vib}$, where these q_i are partition functions for electronic, translational, rotational, and vibrational levels, respectively. These separate functions q_{el}, q_{trans}, q_{rot}, and q_{vib} may be derived as explicit algebraic functions whose independent variables are computable molecular quantities and parameters of external physical conditions, e.g., the principal moments of inertia and absolute temperature in expressing q_{rot} or vibrational frequencies and absolute temperature in expressing q_{vib} (see, e.g., Ref. 33b for details). Thus, these functions provide the explicit form of the partition function Q that in turn leads to the total molecular energy E_0, the QM model of G^0 in the harmonic approximation, for the molecular ensemble in the gas phase.

In practice, the total molecular energy E_0 that includes E_{corr} (as the main component) and these molecular statistics-related energy contributions (electronic, translational, rotational, and vibrational) is calculated, e.g., by the Gaussian program, as one computational task; the calculated value is the calculated free energy for the gas phase. This ease in computations should not mislead; non-empirical calculations of free energy difference ΔG^0 remain non-trivial due to inadequacy of the simplest model explained above. For instance, multiple interconverting complexes (in the first instance, different solvates) may be formed in the liquid phase, and the PES for a such system in solution may be likened to a mountain terrain (Fig. 93).

The abundance of the energy minima means that it is practically impossible to model a free energy surface of a multi-conformer system at any temperature of solution and directly determine partition function Q for the solvation-associated distribution of the solute molecules. Thus, in the worst-case scenario (formation of many solvates), the calculated differences ΔE_0 for stereoisomers do not accurately represent the related ΔG^0 values supplied by experiments. The "accuracy regulator" via free energy calculations is not fully functional for non-computational chemists undertaking conformational analysis by QM.

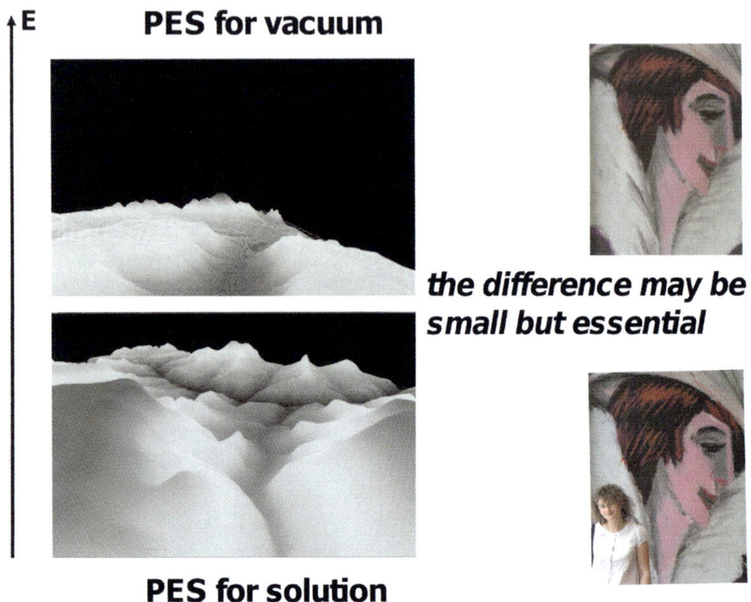

Figure 94. Two hypothetic PES's illustrate that the energy landscapes for a molecular system in vacuum and in solution are somewhat different.

(4) **Solvation.** Experiments in the gas or solid phase or in supercritical fluids are rare in most organic laboratories. Therefore, molecular modeling of organic compounds should, in principle, be performed for solvated molecules. Continuing the PES motif, we may state that a solvated molecule is characterized by its PES, which differs from the PES of this molecule in vacuum (Fig. 94). First, considering the potential energy, the new PES (for solution) falls lower than the parent PES for vacuum. Second, each point in the new PES (in structural terms, each 3D structure) is lower in energy than the same point in the parent PES, which is related to the same 3D geometry, by the value of the solvation energy for this concrete structure. The solvation shells of stereoisomeric structures are not identical. Therefore, their solvation energies differ. Any point of the new PES is distant from the corresponding point of the PES for vacuum by a different energy value than the distance of another point of the PES for solution from its corresponding point in the PES for the molecule in vacuum. This means that the PES's for a molecular system in vacuum and in solution differ in shape. Only in the

case of weakly interacting solutes and solvents, e.g., solutions of polar compounds in saturated hydrocarbons or solutions of non-halogenated organic species in polyfluorinated solvents,[a] are these PES's of similar form on average.

Also, the energy landscapes in vacuum and in solution may differ considerably. For instance, acyclic backbones with H-bond-forming or ionogenic functional groups adopt dissimilar geometries in vacuum and in polar solvents because the tendency to expose these groups to the outside is strong only in a charge-dissipating environment. This means that the PES's of such molecules in vacuum and in solution differ in more than the details. It may even appear that a minimum in the PES for vacuum corresponds to a saddle point in the PES for the polar solution. Hence, one cannot be sure of the 3D geometries of stereoisomers of such a system when they are modeled for vacuum conditions. The calculated ΔE_0 estimates for these 3D structures would definitely be unrealistic. Therefore, if non-empirical calculations are targeted to conformational analysis of, e.g., peptides, oligosaccharides, carbanions, or calixarenes, solvation models are undoubtedly necessary.

A general understanding of the PES's for vacuum and for solution is insufficient for practical selection of an adequate molecular model. Should it always be a solution model, or is a molecule-in-vacuum model a sufficient approximation for modeling the molecular geometry and even the molecular energy? It would be quite laborious in routine structural studies to model these two PES landscapes in order to answer this question. Therefore, in simple non-empirical studies (i.e., obtaining the geometry and energy) of organic molecules in solution, it is better to adhere to the maximalist strategy. *One has to prefer to use the solution model and not the vacuum model if computer resources and the time frame allow.*

Note that this strategy in selecting the model (vacuum or solution) does not delegitimate that of the minimalist, i.e., approximating molecular systems in solution via modeling of these systems in vacuum. For organic compounds of ordinary size, vacuum models sometimes turn out to have similar accuracy in estimating ΔE_0 for their stereoisomers as solution models. The applicability of calculations for vacuum conditions to conformational analysis of a compound of interest may be tested via comparison of, say, two calculated ΔE_0 values for vacuum with the two corresponding experimental ΔG^0 values for stereoisomers

---

[a] A significantly limited solubility indicates a weakness of solute–solvent interactions.

of rationally selected close analogs (e.g., with the same chemical functionality and similar backbone) in solution. A positive result almost surely indicates that the enthalpy of solvation of the stereoisomers of the studied compound is very similar. Also, their solvation includes a similar entropy change for each stereoisomer, moving from vacuum to solution.

As mentioned, involvement of solution models becomes necessary for modeling polyfunctional species, large molecules, new skeleton systems, as well as compounds in strongly coordinating solvents (e.g., pyridine, amides, alcohols, and water). On the other hand, one has to be aware that an improper solvation model may significantly distort solvent-sensitive molecular geometries when optimizing them even by an accurate QM calculation method.

In Section 4.2.6, MM solvation models are divided into two general groups, implicit and explicit models. Both groups are in the QM arsenal. Moreover, an additional group, hybrid explicit/implicit solvation models, is at researchers' disposal. The theory of QM solvation models is more complicated than it is in MM, and therefore concrete methods are not considered here (for further explanations on them, see Refs. 33b and 67a, b). Only the polarized continuum model (PCM), although very general, is mentioned because of its widespread use.

The PCM belongs to the self-consistent reaction field (SCRF) subgroup of the implicit model group. As we remember from Section 4.2.6, when modeling a solvent composed of neutral molecules, implicit models approximate it as a continuum dielectric with dielectric permittivity ε. The solute is represented by a charge non-uniformly dissipated in a 3D cavity. In reality, the solute reorients solvent molecules. Their reorientation indicates the appearance of an anisotropic electric field, which, in turn, changes the molecular properties of the solute (e.g., energy, geometry, charge distribution, and vibrational frequencies). In implicit models, this solute-induced electric field of the solvent is often termed the reaction field. Hence, as soon as the reaction field has been somehow determined, a route to calculating the molecular properties of the solute is open.

Methods of the SCRF subgroup determine this electric field in the form of the electrostatic potential $\phi(r)$. They are based on a numerical solution of the Poisson equation[a] and differ in two aspects. When modeling the cavity where the solute molecule is located, they approximate it as a sphere or a 3D domain formed by either spherical atoms of the solute skeleton (space-filling models) or an electron-density isosurface. When modeling the solute, they approximate it as either electric dipoles or multipoles or a continuous charge density distribution. The PCM constructs the cavity using a spherical collection of atoms

---

[a]The Poisson equation relates electrostatic potential $\phi(r)$ to charge density $\rho(r)$ (ϕ and ρ are dependent of the spatial coordinates r) at a given dielectric permittivity ε as $\nabla^2\phi(r) = 4\pi\rho(r)/\varepsilon$. In our case of a cavity filled by a dissipated charge of density $\rho(r)$ (i.e., the solute molecule), ε is the dielectric permittivity of the surrounding medium (approximated as a continuous dielectric).

of the solute molecule and models the solute at a high approximation level (a space-filling structure with a distinct charge density at each point).

For such a structure, the differential Poisson equation is solved numerically, yielding the electrostatic potential $\phi(r)$ at each cavity point; in other words, the reaction field becomes specified. Then, the Schrödinger equation may be modified by adding the reaction field term to the Hamiltonian. The solution of this Schrödinger equation for the solute, i.e., the molecular energy E_{min} [or E_0, if E_{ZPE} and free energy-related corrections are also calculated; see items *(2)* and *(3)* in this Section] is found in the usual SCF procedure (Section 5.2.2) of energy minimization in HF or DFT calculations.

Solvation modeling may supply its own inaccuracy in energy estimations. Implicit QM models usually do not consider dispersion interactions between the solute and the solvent. Moreover (see also Section 4.2), these models ignore *solute–solvent* chemical coordination, e.g., H-bonding. In such cases, it is reasonable to recruit the so-called mixed solvation models. They are very simple: the core structure is a solute molecule that bears several coordinated solvent molecules from the first solvation shell; actually, we deal with a molecular complex. Note that roughly designing such a complex is elementary using the GUIs in QM calculation packages. Several solvent molecules (designed on screen) are connected (on the screen) to selected fragments of the solute structure in an approximately reasonable orientation (e.g., provided by H-bonding, $\pi-\pi$ interactions, or steric hindrances) at approximately appropriate distances (from 2–4 Å). This molecular complex is treated by an implicit model, e.g., PCM, when performing energy minimization. In other words, we optimize the geometry of the structured core (the designed molecular complex that is a truncated explicit model of solvation of the structure of interest) placed into the unstructured continuous polarizable environment (an implicit model of solvation). Such mixed models of solvation are effective in calculating the molecular geometry of organic species in solution. However, these models are unreliable for estimating the relative stability even of stereoisomers; energy calculations obviously may be inaccurate for structures that include only fragments of the solvation shell.

QM modeling of solvation does not guarantee a reasonable accuracy of $0.2\,\text{kcal}\,\text{mol}^{-1}$ for ΔE_0 of stereoisomers. Let us glance at explicit QM models from this perspective. In practice, these models cannot be applied to organic molecules of medium or large size. By surrounding the solute molecule with one layer of solvent molecules, we construct only a static first solvation shell. Solvent molecules from this shell would not be modeled satisfactorily since they are exposed to vacuum in the model. In real solutions, the first shell itself is solvated by molecules of the second shell, and so on *ad infinitum*. Hence, an acceptable static explicit model of solvation should include several solvation shells, or, in other words, dozens or even hundreds of molecules of the solvent. QM calculations cannot be applied to such supramolecules because the performance and data storages of contemporary computers are insufficient for modeling large molecular systems. Thus, from the viewpoint of QM computations, too many atoms are in the supramolecular structure that would imitate a representative solvation shell. However, even if such calculations are possible, they would not be accurate. The solvation shell is a dynamic (i.e., changing in time) structure, and any static model does not adequately represent it. We come to the insoluble problem of an unspecified molecular ensemble (remarkably, a large one): we have no idea regarding the states and related partition function Q [item *(3)*] for this statistical ensemble.

Implicit models of solvation are much more widely used in *ab inito* calculations. Let us understand their accuracy in the context of computational conformational analysis. The most universal implicit QM model of solvation — the continuum SM8 model,[67a,b] which accurately calculates charge dissipation and also takes into account dispersion — achieves an accuracy of $\pm 0.6\,\text{kcal mol}^{-1}$ when estimating the solvation free energy for neutral molecules (for ionized species, the accuracy is lower); other implicit methods sometimes show an accuracy of roughly $\pm 0.5\,\text{kcal mol}^{-1}$.[67c] For contemporary QM calculations applied to ordinary-sized organic molecules, this accuracy in energy calculations (here: solvation energy) is high. It could be scarcely improved using higher-theory calculation methods or more-extended basis sets because solvation modeling becomes the accuracy-determining factor.

The accuracy of the calculated value of the free energy of solvation (see the previous paragraph) is a component of the accuracy of the total molecular energy E_0 calculated at the same theory level for a modeled *solvated* structure.[a] For two modeled *solvated* stereoisomeric structures,

---

[a]The values of total molecular energy E_0 (or E_{min}) in implicit models include a term for solvation energy as a summand.

the difference of the calculated solvation energies is a component of the difference ΔE_0 of their calculated molecular energies E_0. Thus, the solvation component-associated *accuracy* in the value ΔE_0 is the difference of the accuracies of the calculated solvation energies for the related individual structures.[a] Suppose for simplicity, that, in values of E_0 for two modeled solvated conformers, there is no additional inaccuracy except that of $\pm 0.5\,\mathrm{kcal\,mol^{-1}}$ originated from the solvation free energies calculated for room temperature. Does this fair accuracy for the free energy of solvation calculated for each conformer provide a fair accuracy for the *difference* ΔE_0 for these conformers in solution and, thus, for the related quantity, their relative content[b]? The answer is no — the corresponding estimates of the relative content of conformers are not within the accuracy limits of a few percent. Indeed, since we have no idea of the sign of this calculation error in E_0 for each conformer, we do not know whether the derived difference ΔE_0[c] includes a zero ($0.5\,\mathrm{kcal\,mol^{-1}} - 0.5\,\mathrm{kcal\,mol^{-1}} = 0$), or doubled ($0.5\,\mathrm{kcal\,mol^{-1}} \times 2 = 1.0\,\mathrm{kcal\,mol^{-1}}$), or an intermediate inaccuracy value. The inaccuracy of $1.0\,\mathrm{kcal\,mol^{-1}}$ for ΔE_0 is intolerable. The absolute error for the related quantity, the fraction of a conformer, may reach 34% for a pair of conformers equilibrating at room temperature (Section 4.2.6). Even in this idealized case (errors only from calculating solvation) and with fairly accurate quantitative estimates (E_0 values from calculations for vacuum) in hand, assuming satisfactory accuracy for the relative stability of stereoisomers (for their ΔE_0 value calculated for solution) would be speculative.

Similar to MM modeling of stereoisomers in solution (Section 4.2.6), one could assume that, because of the structural similarity of isomeric structures, solvation of both conformers is practically identical, and the difference in the absolute errors is near zero in the calculated ΔE_0 value (the modeled ΔG^0) for the conformers. However, the presumptive character of such a conclusion is not appropriate for computational chemistry. In quantitative modeling, numerical estimates (and not assumption–explained numerical estimates) are the sole basis for firm conclusions.

---

[a] Formally, the indicated calculation accuracy (calculation error) is an absolute error. The absolute error in an algebraic sum of two values determined with a known accuracy (i.e., with known absolute errors) is a sum of the *absolute* values of the absolute errors of the summands. Note that the term *absolute* has two meanings here (absolute errors in determination of values and mathematical absolute values).

[b] As noted in the beginning of Section 5.3.3, the ratio of two conformers is trivially derived from the corresponding free energy difference ΔG^0.

[c] It approximates the difference ΔG^0; see item *(3)* in this Section.

We have come to the final part of our brief discussion of QM solvation models which are relevant to uncomplicated computational conformational studies. Solvation models too often cannot provide the accuracy required in conformational analysis. In addition, when modeling stereoisomers for vacuum conditions, calculations are less time-consuming. Why is it better, nevertheless, to include solvation in modeling? First, proper modeling of solvation increases the QM calculation accuracy to some extent; when modeling polar molecules in polar solvents, the accuracy improvement is significant. Second, it is *a priori* unknown whether the molecule's PES's for vacuum and solution are essentially different (Fig. 94) or similar. If these PES's differ, one may miss some conformers that participate in the conformational equilibrium in solution (e.g., very minor but reactive conformers that cannot be detected by DNMR), when choosing the simpler way of modeling the PES of a molecule in vacuum. By using appropriate models of solvation (e.g., mixed explicit/implicit solvation models), one can locate them. Thus, the accuracy "regulator" by using solvation models is functional for synthetic chemists.

For beginners, PCM is the first choice when selecting solvation model. With further experience, they can select from the solvation models implemented in available calculation packages the model which provides the best results for their own compounds. In energy calculations in solution, it is worthwhile for organic experimentalists to follow a general methodological concept. It may be formulated as the primacy of structural models (structures and external conditions to be modeled) over calculation tools. *The quality of the structural model predetermines the highest accuracy that may be reached in modeling the molecular system by varying calculation methods of QM (this quality may be estimated in terms of the adequacy of molecular composition, molecular geometry, and external conditions of the structural model to represent the real molecular system after treating this concrete model by QM calculations).* According to this principle, if a structural model (here: the solvation model) is insufficient, the use of high-level calculations (e.g., advanced post-HF methods or very extended basis sets) would not provide correct estimates. Alternatively, if a structural model is successful (e.g., the structure of an explicitly modeled solvation shell is close to the real one, or the PES's for vacuum and solution are similar), unsatisfactory calculation results may be improved by using higher theory calculation tools of QM.

Foreseeing unavoidable situations of either limitations in computational resources, we can figure out that the principle of structural model primacy recommends the use of good structural models and even imperfect calculation tools rather than bad structural models and perfect calculation tools. Nevertheless, one should remember that good structural models are only a necessary prerequisite for successfully modeling organic molecules by QM; they do not guarantee the sufficiently high calculation accuracy until an appropriate QM method is applied to them.

In summary, routine QM modeling of the PES for an ordinary organic molecule may be performed as following:

- (*1*) Using MM combined with conformational search, locate low-energy stereoisomers (Section 4.3.2) and, using the NMA-based methodology (Section 4.4.3), locate the related transition states. For another methodology of locating transition states, see item (*4*) below.
- (*2*) Refine the geometry of these new molecular structures by using non-empirical calculations of relevant accuracy (Sections 5.2.3, 5.2.4, and 5.2.6) and adding a QM continuum solvation model (Section 5.3.4).[a] In other words, the MM-built conformational scheme (for such a scheme, see Section 2.8, Fig. 53) should be "QM-recomputed." One may remind the readers that MP2/6-31++G(d,p), MP2/*aug*-cc-pVTZ, B2PLYP/*aug*-cc-pVTZ, XYGJ-OS/*aug*-cc-pVTZ, or, in the worst case scenario of insufficiency of computer resources, M11/6-31++G(d,p), M06-2X/6-31++G(d,p), or M06-2X/6-311++G(2d,p), are their optimal calculation methods for modeling molecular geometries with good or at least satisfactory accuracy, and PCM is their fair choice of the continuum solvation model. In addition, the quality in modeling solvation may be somewhat improved by using the truncated solvation model in the "envelope" of the PCM or, better, SM8 solvation model (Sections 6.3.2 and 6.3.3).
- (*3*) By calculating vibrational frequencies for each of the resulting stereoisomeric structures, examine whether the located structures, which are supposed to be thermodynamically stable stereoisomers or transition states, indeed belong to PES energy minima or PES saddle points, respectively. All values for these frequencies should be positive for a

---

[a]Of course, calculation methods, basis sets, algorithms of energy minimization, frequency calculations, etc. are calculation options provided by currently main packages of QM computational chemistry (mentioned in Section 5.1).

stereoisomer at an energy minimum. In contrast, one frequency value should be negative if we deal with a first-order transition state. With calculated values of vibrational frequencies in hand, one can see straight away whether the executed energy minimization has supplied molecular structures that indeed belong to energy minima and whether the performed procedure of location of transition states has indeed located those of the first order. If the located structures pass the examination, we conclude that the PES has been successfully modeled. Using our calculated values of the interconversion barriers, we easily assign the located stable stereoisomeric structures to either conformers or stereoisomers rigid at temperature specified in modeling them (Section 2.6). If the signs of the frequency values do not correspond to the expected ones, try to locate the desired stationary points of the PES by optimizing the initial (pre-optimized) or resulting ("optimized") structures in the other way: involve an alternative method of geometry optimization from the assortment of the used QM calculation package (e.g., Newton-Raphson minimization in Cartesian coordinates instead of Berny optimization in internal coordinates).

- (*4*) There is no common successful strategy for locating transition states. Therefore, different approaches are used for this purpose. In practice, transition states are located by using either molecular energy- or geometry-manipulating algorithm implemented into versatile QM computational packages. For instance, assuming that the transition state geometrically is approximately in between the related stable molecular 3D structures, the latter are used in Gaussian for producing a molecular structure with parameters of the molecular geometry which are mean values of the corresponding geometrical parameters of the parent 3D structures (option specified QST2 in this computational package). This "near-transition-state" molecular structure is further subjected to a relevant procedure of geometry optimization [Newton-Raphson minimization, or, better, rational function optimization (RFO)[58b]]. While often failing to locate transition states of chemical reactions, this simple approach is, as a rule, effective in modeling conformational transition states. Other algorithms of locating transition states are much more complex.[58b] Nevertheless, non-computational chemists, not puzzling over mathematical principles of this modeling of the PES, may merely resort to the algorithm(s) implemented into the available QM computational package(s).

6

Calculated Chemical Shifts in Conformational Analysis

6.1 Experimental and Modeled Chemical Shifts

6.1.1 *Chemical shifts and chemical shieldings*

From the previous Chapter, we learn that it is not easy to obtain ΔE_0 values that reflect relative stability of stereoisomeric structures with 0.3–0.5 kcal mol^{-1} accuracy. This disappointing situation is particularly relevant to *ab initio* modeling of stereoisomers of a compound that have locally dissimilar solvation shells, e.g., conformers of peptides or enolates. Accurate theoretical estimation of relative stability of such "unfair" conformers in solution is beyond the competence of non-computational chemists.

Often, organic chemists deal with "fair" compounds with similarly solvated stereoisomers. Using accuracy "regulators" described in Section 5.3.4, it is possible to mimic experimental $\Delta G°$ values via calculated ΔE_0 values with 0.5 kcal mol^{-1} accuracy for stereoisomers of such compounds. However, it is unknown *a priori* whether a chemical system is "fair" or "unfair." Thus, even "fair" systems are not an easy object for conformational excurses.

Indeed, the first important methodological decision in routine QM modeling of organic molecules in solution is what structural model, for vacuum or solution (Section 5.3.4), should be chosen; in other words, one should decide whether solvation can be neglected without essentially decreasing the accuracy of energy estimates. Selection of the calculation method is the

455

next weighted decision. It is obvious that, when choosing, we do not know whether the decisions made do provide the desired accuracy of ΔE_0 for stereoisomers of a molecular system. An examination of the accuracy of obtained ΔE_0 values is necessary. Such examination involves additional, accuracy-verifying calculations, which should be performed for a close analog (reference system; Section 5.3.1) with *measured* $\Delta G°$ and/or $\Delta G^{\#}$ values. Unfortunately, the comparison of the ΔE_0 value calculated for the studied system of and the $\Delta G°$ value measured for the reference system is not always possible.

The first problem is that there are no reliable criteria for accepting a molecular system with an experimentally explored conformational equilibrium as a reference compound. Any discrepancy in chemical functionality, positions in the molecular backbone, or even spatial orientations of chemical moieties in molecules is associated with a different solvation, and the related energy cost is an unknown value. In addition, electronic effects may be different for apparently similar molecules. These "invisible" factors result in uncertainty regarding whether the conformational equilibrium for the reference system and that for the studied one are characterized by close $\Delta G°$ and $\Delta G^{\#}$ values, and, thus, whether the ΔE_0 value obtained for a pair of conformers of the reference system reflects the accuracy of the related ΔE_0 value for conformers of the studied system. Modeling of additional molecular reference systems is required to deliberately estimate the accuracy of the "parent" ΔE_0 value. Owing to these difficulties, examination of the accuracy of the energy calculations performed for the system of interest may transform into a separate study.

The second problem appears if neither compounds with known experimental $\Delta G°$ and/or $\Delta G^{\#}$ values for the conformational equilibrium may be accepted as a reference compound. Without undertaking experimental quantitative studies of this equilibrium for an appropriate structural analog and calculating the corresponding energy difference ΔE_0, the situation with the examination of the accuracy of the initial ΔE_0 value is deadlocked.

PES (or SES for MM) modeling is thus not a universal tool for conformational analysis. Ideally, theoretical calculations provide complete information regarding conformers and their transformations. In practice, PES modeling may lead to unreliable conclusions because routine (non-high-theory) theoretical calculations[a] cannot guarantee the required accuracy in estimating the relative stability of conformers. If routine theoretical estimates of molecular energy are not an absolute tool of conformational analysis, could another molecular quantity be considered? Such quantity should be sensitive to molecular 3D structure, easily measurable, computable by the developed QM methods, and familiar to organic chemists. Such a unique molecular quantity is, of course, NMR chemical shifts. Characterizing molecular ensembles, chemical shifts have the following desired properties: (*1*) they reflect tiny changes of molecular structure (chemical as well as stereochemical); (*2*) they are provided in seconds by contemporary NMR spectrometers;[b] (*3*) for all nuclei of any chemical structure, they may be supplied by QM computations more or less accurately; and (*4*) they are a daily analyzed material in organic/bioorganic laboratories.

NMR spin-spin coupling constants also fulfill requirements (*1*)–(*4*). Predominantly ^{1}H,^{1}H-, and sometimes ^{1}H,^{19}F-, ^{1}H,^{13}C-, as well as ^{1}H,^{31}P-couplings are the main basis of conclusions regarding molecular geometry that are made in *experimental* organic chemistry. *QM-calculated* NMR spin-spin coupling constants are less convenient for stereochemical assignments. As theory level dependent values, both calculated chemical shifts and NMR coupling constants are not perfectly accurate. However, the latter magnetic "fingerprints" of molecular geometry are changed in a relatively narrow numerical range, while ^{13}C or ^{19}F chemical shifts vary in wide limits of units used in practical NMR (hundreds of part per million, ppm). One may say that chemical shifts of such nuclei are numerically more sensitive to changes in molecular 3D geometry. Therefore, these numerical values are more convenient for assessments in computational conformational analysis than constants of internuclear spin-spin coupling interactions.

Calculations of molecular magnetic properties (chemical shieldings and spin-spin coupling constants) using QM methods are not news in computational chemistry. Several theoretical approaches to calculated chemical shieldings (see below for the relationship between chemical *shieldings* and chemical *shifts*) have been developed and their accuracy in combination with different *ab initio* calculation methods has been thoroughly inspected (see, e.g., Ref. 68a–d,69a,b). Theoretical basics of QM computations of

---

[a]See practical recommendations from Sections 5.2.3, 5.2.4, and 5.2.6 for such QM modeling.
[b]As known, these physical devices are a basic instrumental facility of every department of organic or bioorganic chemistry.

chemical shieldings are included in many modern monographs on computational chemistry (see, e.g., Refs. 33b, 62b, and 69a). Computational chemists have also found a very fruitful application of *ab initio* calculated chemical shifts (CCS). *CCS are a very effective tool for determining chemical and stereochemical structures of organic compounds.* The principle of this technique of structure determination is extremely simple. Chemical shifts (usually, for ^{13}C nuclei) are calculated for alternative structures of optimized geometry, e.g. for stereoisomers, with using the calculation method (Section 6.3.2) and the molecular model (Section 6.3.3) that can supply satisfactory accuracy of CCS for the molecular structure of interest. Then, these calculated spectra are compared with a routine experimental spectrum (i.e., a C,H-decoupled one in the case of ^{13}C NMR) of this compound.[a] The modeled spectrum (merely, a set of CCS), which corresponds to the experimental one, indicates the correct structure. Comparison of CCS and experimental chemical shifts is an independent and, in principle, errorless method of elucidation of organic compound structures. The computational CCS method is obviously complementary to ubiquitously used structure determination by experimental NMR.

There is, nevertheless, an essential difference between the theoretical and experimental methods. The CCS method of structure determination requires alternative structural hypotheses (alternative molecular structures) to be finished before analyzing modeled (calculated) NMR spectra. In principal, structural assignments via NMR experiments do not need preliminary structural assumptions, and alternative hypotheses appear and are examined when interpreting NMR spectra.[b] In other words, experimental NMR is an unbiased method for structural studies, in contrast to the applied "theoretical NMR" (i.e., theoretical calculations of chemical shieldings or constants of spin-spin interaction). However, chemists who are not very well versed in structure elucidation by NMR often find it difficult to make the correct

---

[a] In the case of a mixture of rigid stereoisomers, the experimental NMR spectrum is a superposition of their spectra. If stereoisomers are flexible under conditions of the NMR experiment, i.e., they are conformers, the spectrum is an averaged spectrum of the spectra of individual conformers (Section 3.3). Analysis of NMR spectra of conformationally flexible compounds by using CCS is considered in Section 6.2.2.

[b] In organic practice (for cardinally simplifying it), structural assignments based on experimental NMR spectra almost always are performed starting from the use of externally (NMR-independently) developed structural hypotheses (Section 6.1.3).

choice among alternative structural hypotheses for non-trivial compounds (e.g., new alkaloids), for which various NMR data are available or may be supplied by additional NMR experiments.[a] The CCS-based methodology is comprehensible to any organic chemist because it consists of selecting the correct structure from several alternatives by merely comparing sets of numbers (i.e., sets of chemical shifts). Not surprisingly, many structure determinations have been performed by computational chemists for organic compounds of different complexities by means of this simple methodology. Synthetic experimentalists have probably only recently realized that an additional robust method of structure determination has appeared, and following computational chemists, some of them have started to utilize or at least consider CCS (see, e.g., Refs. 71a–d).

These calculations, which are supplied by strict QM theory, should not be confused with other prediction methods that deduce chemical shifts for a system of interest by analyzing experimental chemical shifts of structurally similar compounds (reference structures). An example of such an approach is the methodology of additive chemical shifts, wherein the first appreciable success is the prediction of ^{13}C chemical shifts of methylsubstituted cyclohexanes.[70] Using empirical increments of chemical shifts for carbon atoms of differently positioned axial and equatorial methyl groups, good estimates of ^{13}C chemical shifts have become available for stereoisomers of these compounds. In the era of computerized databases, the ^{1}H and/or ^{13}C spectra for an organic compound specified as a Lewis structure are predicted by computationally analyzing a large database of experimental NMR spectra for chemically different compounds.[b] The obtained NMR spectra are called simulated spectra (one should not take them for *calculated* spectra considered in this Chapter). Sets of reference structures may be very large, and accurate (but not universal[c]) *structure–chemical shift* correlations indeed may be established because of the use of sophisticated algorithms implemented in the database-analyzing programs.

Despite successful simulations of NMR spectra for many compounds, there are always doubts regarding the verisimilitude of the simulated NMR spectrum when such *structure– chemical shift* correlations are applied to a new, out-of-set system. Inaccuracy of prediction may be so significant that the simulated spectrum and the experimental one seem relating to different compounds. The principal difficulties of this prediction method are clear. No compound set can represent the unlimited structural diversity of organic structures or, to an even greater extent, estimate the conformer ratio[d] at the conformational equilibrium

---

[a] For such problems in conformational analysis by NMR, see Section 6.1.3.
[b] Such databases are updated from time to time.
[c] See below in this digression.
[d] Theoretically expressed by Eq. 13 (Section 3.3).

for the molecular system in question. Chemical shifts, which characterize certain chemical compounds [i.e., the training (reference) set structures], are, in particular, associated with conformational averaging of molecular geometry [i.e., "averaged molecular geometry" of conformers; Eq. 9, Section 3.3] as well as solvation-dependent distribution of the electric charge in their molecules. However, "new" structural factors, e.g., additional and/or altered steric contacts, ground-state-associated (Mulliken) charge transfer, and long-range electron interactions, may be present in the out-of-set molecule and may disturb molecular geometries of conformers and/or intramolecular charge distribution "implied" by the training set. Obviously, simulated chemical shifts will be far from accurate in such cases. Even if "new" molecular skeletons are composed from structural fragments embarrassed by the training set, the predicted chemical shifts may still be unreliable. Organic molecules are not additive LEGO constructions of chemical fragments, due to cooperativity of intramolecular interactions. Addition of chemically trivial fragments to a molecule or their positional permutation in the molecular skeleton may twist the reference-dictated molecular geometry, invert the conformer ratio, or locally alter the solvation shell. It is clear that a relationship between chemical shifts of molecular structures from the reference set and chemical shifts of any examined molecule is the basic assumption in simulating NMR spectra (predicting chemical shifts) on the basis of empirical NMR data. However, the mentioned structural effects may make it irrelevant for either next molecular structure. For instance, one should not expect that simulated ^{13}C NMR spectra of "NMR-untouched" chemically isomeric oligosaccharides of different carbohydrate sequence would be as distinct as their experimental ^{13}C NMR spectra would be. Consequently, the chemical structure of even a linear oligosaccharide with the known sugar composition cannot be determined from comparing the simulated NMR spectra of alternative sequences for this set of the carbohydrate units and the experimental NMR spectrum of this compound. The methodology of empirical *structure–chemical shift* correlations is far from being universally applicable, and, thus, it is inadequate in structural studies of complex, or polymeric, or non-trivial, or structurally trivial, but insufficiently "NMR-covered" molecules.

As mentioned, such NMR spectra simulations are carried out nowadays by means of specialized computer programs aimed at analyzing databases for chemical shifts and spin-spin constants of different compounds. Organic experimentalists should have an insight into the applicability limits of these simulation engines (computer program + NMR database) when using them for structure elucidation. *Applied alone, computerized analysis of NMR databases is not valid for drawing ultimate conclusions regarding chemical or stereochemical structure.* Besides, designed to recognize molecules only accordingly to Lewis structures, some current commercially available engines for simulation of NMR spectra do not distinguish among diastereomers. In their projections, also diastereomeric conformers, which are characterized by very dissimilar ^{13}C NMR spectra, e.g., two conformers of alkaloid **44** (Section 6.3.1, Fig. 99), are identical.

In order to learn how unreliable these computer tools are, let us consider a simple case: simulation of chemical shifts for an unordinary structure. Arseno alkaloid **45** shows ^{13}C resonance signals at 17.03 and 23.05 ppm (Section 6.3.3; Fig. 101 and Table 8).[76] Having a small training set of reference structures, ChemDrawUltra14.0 (CambridgeSoft), a popular auxiliary non-computational chemical program, gives disappointing simulated values of 37 and 38, ppm for its ^{13}C chemical shifts. Using its own database, a specialized program, Mnova NMRPredict Desktop (Mestrelab Research[a]), incorrectly considers all three carbons of **45** as equivalent and returns an extremely inaccurate value of 5.1, ppm. If this program analyses the database of Modgraph,[b] which includes ^{13}C chemical shifts for nearly 200,000 structures, the predicted value is 95.6 ppm. The ACD/NMR predictor (ACD/Labs[c]), operating with the training set of a similar size, provides a better accuracy. However, incorrect values of 32.5 and 34.3 ppm are obtained. The simulation inaccuracy is higher than it may seem at first sight: the relative position of these resonance signals is reversed in the simulated ^{13}C NMR spectrum.

Thus, simulation of NMR spectra based on empirical *structure–chemical shift* correlations is not a resolving tool in structural research. In sharp contrast, non-empirical calculations of chemical shifts are a universal and self-sufficient methodology because it is based on an exact physical theory. CCS for a molecular structure are predetermined, in the first instance, by the structure itself, without any relation to other structures. Any molecular systems, even of a unique chemical or 3D structure, in principle are equal from the viewpoint of QM modeling of chemical shifts.

Nevertheless, despite various failures, NMR spectra-simulating (NMR database-analyzing) computer programs are indispensable in fast *preliminary* screening of structural hypotheses. Note that simulated proton-decoupled ^{13}C NMR spectra are generated by a relevant computer program supported by a solid NMR database in less than 1 sec. More important, such spectra *very accurately* reproduce experimental ones for compounds whose chemical structures are well recognizable for the database training set. No special skills are required for using such commercially available programs, and the determination of chemical structures of "normal" organic compounds may be significantly accelerated. Organizing work in a modern, well-supported synthetic research group, one cannot ignore these extremely useful (though expensive) auxiliary programs of quick probing of chemical structure. Otherwise, we come back to the notorious past with its effort- and time-consuming non-computerized analysis of each structural hypothesis when interpreting ^{1}H and ^{13}C NMR 1D spectra of reaction products or compounds isolated from the nature.

New, unusual, and diastereoisomeric molecules are an obvious target for structure determinations by CCS. Simple QM-based chemical computations (modeling of the ground state molecular geometry and computing of related molecular quantities) face a single serious problem; it is related to the practice of QM computations. Computer resources and/or the research timeframe are always limited and, thus, prevent these "all-embarrassing" theoretical calculations from being routinely carried out for large molecular

---

[a]Mestrelab Research: http://mestrelab.com
[b]Modgraph: http://www.modgraph.co.uk
[c]ACD/Labs: http://www.acdlabs.com/home

structures (including nanoobjects), many molecular structures of ordinary size (e.g., multi-conformer molecules), post-Kr element-containing molecular structures, and transition states of absolutely unknown geometry.

Nevertheless, CCS may be effectively obtained for most known and new organic compounds synthesized in organic laboratories. Importantly, one may rationally improve the numerical accuracy of QM calculations of chemical shifts. As discussed in Sections 5.3.4 and 6.3.2, accuracy in non-empirical molecular modeling is exclusively determined by both the quality of the structural model [e.g., a single molecule or molecular ensemble model, modeling for a solution or vacuum, a static or dynamic (time-dependent) model] and the quality/applicability of the theoretical method employed (e.g., HF approximation, QM approaches that take into account electron correlation and relativistic effects, various DFT methods). This indicates that unsatisfactory CCS may be replaced by newly obtained more accurate values; in principle, one may obtain sufficiently accurate CCS values for any chemical structure.[a]

At this point, the reader probably has realized that CCS may be very useful in his/her own experimental chemical research. Let us learn what, in essence, CCS are. To understand CCS better, one should refresh our understanding of some physical quantities used to describe the NMR phenomenon. In the presence of an unchanging external uniform magnetic field B_{ext}, the motion of the molecule's electrons induces an "additional" local magnetic field B_{ind}, which has distinct anisotropy for each nucleus in a molecule. The result is that each such nucleus is in an effective magnetic field B_{eff}, which is a superposition of the uniform B_{ext} and the nucleus-specific B_{ind}, i.e., $B_{eff} = B_{ext} - B_{ind}$. To characterize this response of nuclei surrounded by moving electrons to the external magnetic field, a physical quantity σ, called chemical shielding, is introduced. It is defined by a simple, well-known equation as follows:

$$B_{ind} = -\sigma B_{ext} \tag{50}$$

---

[a]This statement should be understood correctly. Calculation accuracy is indirectly "structure-dependent." For instance, molecular systems that are characterized by a very flat minimum in the PES (e.g., systems with elongated bonds) or are in an exited state, require more careful QM modeling than ordinary organic systems need. In addition, the impact of solvation effects may vary for different molecules, and the computational solvation model that is suitable for either molecular structure may appear disappointing if applied to another one. However, this only indicates that, in order to be modeled successfully, dissimilar molecular structures may require QM calculation tools of different "sharpness."

That is, the stronger the external magnetic field is, the stronger the magnetic response (i.e., the induced magnetic field B_{ind} of opposite direction) of the nucleus is, and chemical shielding σ is the dimensionless coefficient of proportionality.[a] This magnetic response is weak, but, nevertheless, measurable: σ values usually are of the $10^{-5} - 10^{-6}$ order of magnitude. In order to avoid operating with small numbers, σ values are multiplied by a factor of 10^{6}. Then σ values are supplied with physically dimensionless ppm units.

Clearly, the resulting magnetic field B_{eff} may be written as

$$B_{eff} = B_{ext} - \sigma B_{ext} = B_{ext}(1 - \sigma) \tag{51}$$

The chemical shielding σ is termed absolute chemical shielding. What does this term *absolute* mean here? It means that σ characterizes the "responding" nucleus itself, without referring to B_{ind} of any other nucleus (some "reference" nucleus). Chemical shielding σ (as indicated, a small numerical value) quantitatively describes the magnetic response of the nucleus *in the molecule* to the external magnetic field when comparing this nucleus with the same, but bare[b] nucleus in vacuum. Such a bare nucleus obviously does not have the magnetic response induced by electron motion, and it is the primary reference standard for chemical *shieldings* of nuclei of the same isotope which are "surrounded" by different electronic "clouds." In simple words, values of σ for a magnetic nucleus are different for different molecules or for different positions of the nucleus in a molecule, and the zero value of σ is related to the bare, chemically non-bonded nucleus.[b]

Equations 50 and 51 clarify the physical sense of σ. Both magnetic fields B_{ind} and B_{eff} are anisotropic (see below), and the resulting magnetic field B_{eff} differs from point to point in molecular space. In these equations, chemical shielding σ of a nucleus is a numerical parameter that adopts different values depending on whether the nucleus belongs to a neutral atom, or a related ion, or a molecule; σ is not a continuously changing independent variable. Thus, it reflects the strength of the local magnetic

---

[a]Linear Eq. 50 is valid for field B_{ext} that is not too strong. In this approximation, chemical shielding σ of a nucleus may be represented as the sum of two opposite sign increments, diamagenetic σ_D and paramagnetic σ_P. The former weakens the effective magnetic field B_{eff} while the latter strengthens it. Thus, σ_D and σ_P may be associated with shielding and deshielding, respectively. For more details, see standard texts on NMR basics or, e.g., Refs. 68a and 69b.

[b]Electron-deprived. Such "naked" nuclei may be generated in some physical experiments. Essentially distanced from any electrons (e.g., electrons of other atoms), such a positive particle is an isolate nucleus, i.e., is not a part of an atomic or molecular structure.

field B_{eff} (the magnetic field at a point of molecular space) associated with this nucleus in a concrete atomic or molecular structure.

Chemical shieldings may be experimentally determined for magnetic nuclei in simple compounds, though not trivially. For example, the 10^6 factor-scaled chemical shielding is 189.9 ppm for ^{19}F of $CFCl_3$ and 356.0 ppm for ^{31}P of H_3PO_4 (see also Table 7 in Section 6.1.2 for experimental σ_{ref} for ^{13}C TMS nuclei). These values are rarely known in synthetic laboratories although they are associated with the well-known reference compounds.

Thus, we have learned that, being in an external uniform magnetic field, magnetic nuclei in different molecules or in distinct chemical surrounding in a molecule actually are in the magnetic field of different strength or, in simple words, such nuclei are differently magnetically shielded from the external magnetic field. In σ terms, chemical shielding σ is dissimilar for (*1*) atoms of different chemical elements or different magnetically active isotopes of the same element in any molecular structure; (*2*) atoms of the same magnetic isotope in distinct molecular structures; and (*3*) either chemically, or positionally, or stereochemically distinct atoms of the same magnetically active isotope in a molecule. Such magnetic nuclei are easily distinguishable experimentally: as known, they have differently positioned resonance signals in NMR spectra of chemical compounds in solution or in the gas phase (in NMR terms, they are chemically non-equivalent). This spectral non-equivalency appears exactly because of the structure-specific differences in σ listed above. NMR spectrometers do not measure chemical shieldings of magnetic nuclei; they measure their resonance frequencies. These frequencies are suitable for faithfully determining the *difference* of chemical shieldings of such distinct nuclei (see Eqs. 52 and 53 below). Therefore, so-called secondary reference standards are of practical interest in experimental organic chemistry. They are compounds, e.g., ubiquitously used TMS, with nuclei whose absolute isotropic (see below) chemical shielding is accepted as the reference point for chemical shieldings of the same isotope in various molecular structures. Such a chemical shielding is referred to as σ_{ref}. For instance, values of chemical shieldings σ_{ref} of chemically equivalent nuclei ^{1}H and the ^{13}C nucleus of TMS correspond to the zero reference points of 0 ppm in the commonly known scales of chemical shifts for protons and ^{13}carbons, respectively.

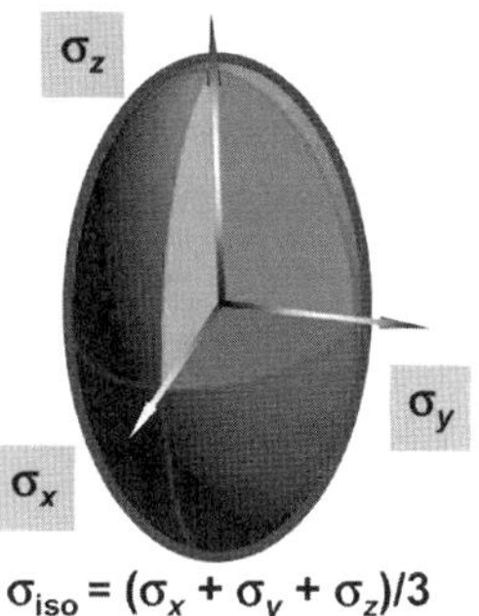

Figure 95. Representation of the chemical shielding tensor as a three-axial, nucleus-centered ellipsoid. Principal components σ_x, σ_y and σ_z of the tensor provide value σ_{iso} (isotropic chemical shielding). The sum $\sigma_x + \sigma_y + \sigma_z$ is called the trace of the tensor.

It is useful to know that chemical shielding σ, as a number, does not fully characterize the magnetic response of a magnetic nucleus. Chemical shielding is a tensor (Fig. 95) with components $\sigma_{i,j} = -\partial B_{\mathrm{ind}(i)}(r)/\partial B_{\mathrm{ext}(j)}(r)$ at any coordinate r. The tensor feature of chemical shielding indicates that, in principle, the *induced* magnetic field is anisotropic (numerically different in different directions).

In the gas phase or in solution, the anisotropy of the chemical shielding disappears because of the rapid tumbling of molecules in gases or non-viscous liquids. Then, the chemical shielding of the nucleus is characterized by an averaged scalar value σ_{iso}, called isotropic chemical shielding. This numerical quantity σ_{iso} is defined as one-third of the tensor trace [$\sigma_{iso} = 1/3 \times (\sigma_x + \sigma_y + \sigma_z)$; Fig. 95]. Owing to difficulties in experimental determination of σ or principal components σ_x, σ_y, and σ_z, absolute chemical shieldings themselves are not practical molecular characteristics for utilizing them in structural studies of organic compounds. The related, ubiquitously used quantity is the "organic-chemist-friendly" chemical *shift* δ. Through σ_{iso}, chemical shift δ (in ppm) of a nucleus for a molecule in *solution* or *gas* is expressed as

$$\delta = [(\sigma_{\mathrm{ref}} - \sigma_{iso})/(1 - \sigma_{\mathrm{ref}})] \times 10^6 \approx (\sigma_{\mathrm{ref}} - \sigma_{iso}) \times 10^6 \qquad (52)$$

where σ_{iso} is the isotropic chemical shielding (not multiplied by the factor 10^6) of the considered nucleus, and σ_{ref} is the isotropic chemical shielding (not multiplied by the factor 10^6 either) of the nucleus of the same chemical isotope in the reference compound. If chemical shieldings σ_{iso} and σ_{ref} are ppm values, i.e., they include the factor 10^6, Eq. 52 can be re-written as

$$\delta = [(\sigma_{ref} - \sigma_{iso})/(1 - \sigma_{ref}/10^6)] \approx (\sigma_{ref} - \sigma_{iso}) \tag{53}$$

Equivalent Eqs. 52 and 53 are important for understanding the importance of scalar quantity δ, which, as mentioned, is well-known to organic experimentalists as chemical shifts. Formally, chemical shift is a derivative of the primary physical quantity σ, i.e., chemical shielding. Why is this formal derivate so useful? Equations 52 and 53 establish that chemical shift is an approximate, but sufficiently accurate, estimate of the *difference* in chemical shieldings of these nuclei. The value δ reflects to what extent B_{ind} for a given nucleus, e.g., a certain proton in a molecule, is different in relation to B_{ind} for the corresponding nucleus in a reference compound, e.g., the protons in TMS. Whereas chemical shielding is a measure of the magnetic response of magnetic nuclei to the external magnetic field B_{ext} relative to the B_{ind}-non-generating nuclei (electronless ones), chemical shift is a measure of this response for any nucleus in any molecular structure relative to that for the nucleus of the same isotope in the reference molecular structure. Therefore, chemical shift is a very convenient molecular characteristic in experimental chemical research. In contrast to chemical shielding, this physical quantity is "object-oriented": organic chemists deal with only compounds and not with compounds and bare nuclei.

Moreover, as mentioned earlier, measurements of chemical shieldings are an exception rather than a rule; they are useful for only "extra small" molecules. Thus, direct measurements of the difference in chemical *shieldings* are practically always impossible. In contrast, isotropic chemical *shifts* for any soluble organic compound are readily and very accurately measured by radiospectrometry in the range of resonance frequencies of atomic nuclei, i.e., by NMR.

6.1.2 *Chemical shifts provided by non-empirical calculations*

Synthetic chemists, of course, are accustomed to using easily measurable chemical shifts δ; every NMR experiment supplies chemical shifts

$\delta_{exp\_1}, \delta_{exp\_2}, \ldots, \delta_{exp\_i}, \ldots, \delta_{exp\_N}$ for nuclei $1, 2, \ldots, i, \ldots, N$ of a molecule. From the viewpoint of unambitious organic experimentalists, chemical shielding σ is not relevant to their research, and it is unnecessary for them to appreciate this primary, but "impractical" physical quantity σ. However, σ is worth recognizing in advanced synthetic organic groups: this quantitative molecular characteristic is important in the context of CCS. Exact values of chemical *shieldings* σ_{x_i}, σ_{y_i}, σ_{z_i}, and thus σ_{iso_i} for each i-th nucleus of a molecular system, *not chemical shifts*, are modeled by QM calculations.

QM theory for calculating chemical shieldings is too mathematized to be "digested" by synthetic chemists (see Refs. 68a or 69b for a brief description of this theory). Therefore, QM methods of chemical shielding calculations are not assessed herein; they are only characterized in general. There are several QM approaches for calculating chemical shieldings. The most well-known methods for these calculations are gauge-invariant atomic orbitals (GIAO), individual gauge for localized orbitals (IGLO), continuous set of gauge transformations (CSGT), localized orbitals/localized origin (LORG), individual gauges for atoms-in-molecule (IGAIM), and gauge including projector augmented wave (GIPAW; it enables models of chemical shieldings for periodic supramolecular structures, e.g., crystals). As their names suggest, these approaches primarily differ by the treatment of the gauge origin.

Indeed, in QM calculations of the chemical shielding tensor by means of finite basis sets (Section 5.2.4),[a] the so-called calibration invariance problem appears. This methodological problem may be formulated as the dependence of the calculated values of chemical shieldings on the origin of the magnetic vector potential[b] (the gauge origin for the position vector r). The perturbation, which is caused by magnetic field B_{ext}, cannot be satisfactorily represented for the wave function with atomic orbitals, which have centers that do not lie in the gauge origin. In this case, the calculated chemical shieldings for atoms, which are located far from the gauge origin, become merely irrelevant. The larger the molecule, the worse the calculation results.

---

[a]Infinite basis sets are, of course, unreal.

[b]The magnetic field (mathematically, a vector at any point r) cannot be added to the Hamiltonian directly. It is inserted into the Hamiltonian as a certain vector potential.

The GIAO, IGLO, CSGT, LORG, IGAIM, and GIPAW methods for calculations of chemical shieldings solve the calibration invariance problem. For instance, GIAO eliminates the dependence of results on the gauge origin, incorporating this origin into basis set functions (as we remember, the latter are atom-centered). This enables the use of GIAO in combination with a wide range of QM calculation methods (HF as well as post-HF methods), including various DFT functionals. Because of this convenience, GIAO is probably the most widely used method in the modeling of chemical shieldings; the related method, which takes into account relativistic effects, is MB-GIAO (magnetically balanced gauge-including atomic orbitals). IGLO calibrates each localized MO by means of a specific exponential phase multiplayer and chooses the center of the MO to which the particular multiplayer is related, as the gauge origin. Therefore, IGLO, in contrast to GIAO, is not sufficiently universal because many wave functions cannot be mimicked using these spheroidal MOs.

Chemical shieldings may be non-empirically calculated for structures of any molecular geometry; obviously, it is reasonable to calculate them for structures of optimized geometry. Thus, before modeling chemical shieldings, one should optimize the geometry of the molecular structure studied; the use of the relevant solvation model is desirable. Calculations of chemical shieldings are then performed in the next, separate step.[a]

Technically, chemical shieldings are computed by combining a QM method of chemical shielding calculations (see above) with a QM method for mimicking the wave function (Sections 5.2.2, 5.2.3, and 5.2.5) coupled with a basis set (Section 5.2.4).[b] The related notation is, e.g., GIAO-MP2/aug-cc-pVDZ or GIAO/MP2/aug-cc-pVDZ. This example indicates that the GIAO method (calculation of chemical shieldings) is used with the MP2 method, which, in its own turn, constructs the probe function (a substitute of the wave function; Section 5.2.2) with borrowing the "building block functions" from the aug-cc-pVDZ basis set. There is no requirement to use the same couple *QM calculation method–basis set* in both energy minimization (i.e., in geometry optimization) and calculations of chemical

---

[a] QM calculation packages provide σ_{calc_i} values in one computational procedure for all nuclei of all chemical elements of the system.

[b] The GUIs of modern QM packages allow the user to combine these three components of chemical shielding calculations by merely specifying the names of the concrete theoretical methods and basis set chosen.

shieldings for the structure of optimized geometry. That is, if some certain couple *calculation method–basis set*, e.g., M06-2X/6-31+G(d,p), has been used in energy minimization for a molecule of interest, another couple, e.g., MP2/6-311++G(2d,p), may be employed in calculating chemical shieldings of the nuclei of this molecule if this change provides better accuracy of CCS elementarily derivable from the obtained values of calculated chemical shieldings (see Eq. 54 below).

In choosing the QM calculation method and the basis set for calculations of chemical shieldings of a molecular system, one should be guided by simple practical considerations. The simplest recommendation is to use the GIAO method in conjunction with the MP2/aug-cc-pVTZ, or MP2/6-31++G(d,p), or MP2/6-311++G(2d,p) approximations and involve a solvation model [e.g., PCM or S8; subsection (*4*) in Section 5.3.4]); these theoretical methods usually afford sufficiently accurate values of CCS (see Section 6.3.1) for non-aromatic organic systems.[a] If computer resources and/or the time frame do not allow application of these methods, one can resort to the combination of GIAO with a DFT method, which is at least of the third rung (e.g., B3LYP-D or PBE1PBE-D; Section 5.2.6); of course, involvement of a solvation model is highly desirable. In this case, selection of the suitable DFT method is often simplified by finding (in the literature) more or less accurate results for CCS modeled by either DFT method for structurally similar molecular systems. Concerning molecular geometry, any reasonable combination of a QM calculation method and a basis set (Section 5.2.4) may be used without forgetting to recruit a relevant solvation model; the only requirement is that the selected combination provides a high accuracy of the exploited resulting (optimized) geometry. Even MM may be employed if one can be sure in a high accuracy of the exploited force field in modeling the 3D geometry of the molecular structure of interest.

Desiring CCS values (in our consistent notation, values δ_{calc_1}, δ_{calc_2}, $\ldots$, δ_{calc_i}, $\ldots$, δ_{calc_N} for nuclei $1, 2, \ldots, i, \ldots, N$) and not those of chemical shieldings, we have to derive CCS from calculated isotropic chemical shieldings (in this notation, values σ_{calc_1}, σ_{calc_2}, $\ldots$, σ_{calc_i}, $\ldots$, σ_{calc_N}). How can one do this "conversion"? Equation 53 indicates the

---

[a]Aromatic rings have an essentially non-uniform solvation shell. Therefore, an explicit modeling of solvation of aromatic fragments should be involved for obtaining more accurate values of CCS for the nuclei of these molecular fragments.

obvious route: in expressing the CCS value δ_{calc_i} for the i-th nucleus in a molecule, one should replace (*1*) σ_{ref} in Eq. 53 with the *calculated* value σ_{calc_ref} for chemical shielding for an atom of the same isotope in a reference compound, and (2) σ_{iso} in this equation with the *calculated* value σ_{calc_i} of chemical shielding for this i-th nucleus from the considered molecule. That is, one can "rewrite" Eq. 53 by applying it to CCS and the calculated isotropic chemical shieldings as

$$\delta_{calc\_i} = [(\sigma_{calc\_ref} - \sigma_{calc\_i})/(1 - \sigma_{calc\_ref}/10^6)] \approx (\sigma_{calc\_ref} - \sigma_{calc\_i})$$

$$(54)$$

where δ_{calc_i} is the calculated chemical shift for the i-th nucleus, σ_{calc_i} is the calculated isotropic chemical shielding for this nucleus, σ_{calc_ref} is the calculated isotropic chemical shielding for a nucleus of the same isotope in the reference compound. Indices *iso* (isotropic) for δ_{calc_i} and σ_{calc_ref} are omitted for simplicity. Different non-empirical methods supply distinct values σ_{calc_ref} for nuclei of the same magnetic isotope of the same chemical element. Therefore, Eq. 54 implies that values of σ_{calc_i} and σ_{calc_ref} are obtained by using the same chemical shielding calculation method, same wave function-constructing method, same basis set, and same solvent model.

Thus, this equation shows how calculated values of isotropic chemical shieldings $\sigma_{calc\_1}, \sigma_{calc\_2}, \dots, \sigma_{calc\_i}, \dots, \sigma_{calc\_N}$ can be transformed to desired values $\delta_{calc\_1}, \delta_{calc\_2}, \dots, \delta_{calc\_i}, \dots, \delta_{calc\_N}$, which are exactly what we call CCS. It is unnecessary to use the exact expression in Eq. 54 for deriving δ_{calc} from σ_{calc}. Chemical shieldings are calculated by *ab initio* methods with inaccuracy that does not permit distinguishing between CCS values supplied by the approximate $(\sigma_{calc\_ref} - \sigma_{calc\_i})$ and the exact $[(\sigma_{calc\_ref} - \sigma_{calc\_i})/(1 - \sigma_{calc\_ref}/10^{-6})]$ expressions.

Equation 54 actually instructs us that, for obtaining CCS for a molecular structure, one should apply QM calculations of chemical shieldings also to a reference molecular system (e.g., TMS) in its optimized geometry. The obtained (calculated) values are chemical shieldings σ_{calc_ref} for the nuclei, whose chemical shifts are defined as zero and, in this capacity, used in the chemical shift scales for the related isotopes, e.g., as chemical shifts of ^1H and ^{13}C nuclei of TMS in the well-known scales of ^1H and ^{13}C chemical shifts, respectively. In fact, the need in these additional calculations that provide values σ_{calc_ref} is not absolute for the "most interesting" nuclei, e.g., for ^{13}C nuclei of TMS; such values of calculated chemical shieldings σ_{calc_ref} can be found in the literature for many different NMR reference standards. For instance, Table 7 shows values of chemical shieldings σ_{calc_ref} for ^{13}C atoms of TMS, which have been calculated by employing some commonly favored non-empirical methods and basis sets, and involving the PCM solvent model (Section 5.3.4) of the chloroform bulk in some cases.

At this point, we can summarize the principal technical route to CCS of a molecular 3D structure. Values $\sigma_{calc\_1}, \sigma_{calc\_2}, \ldots, \sigma_{calc\_i}, \ldots, \sigma_{calc\_N}$ for the N magnetic nuclei of the considered chemical element, e.g., ^{13}C nuclei, are calculated for the molecular structure of *optimized* geometry, within the limits of the relevant structural model (i.e., using a vacuum or solvent model). The same computations are performed for the reference standard of *optimized* geometry to obtain the value σ_{calc_ref} for nuclei of this chemical element[a] (we remember that both σ_{calc_i} and σ_{calc_ref} should be computed keeping calculation elements *the QM theoretical method, the basis set, and the structural model* the same). CCS $\delta_{calc\_1}, \delta_{calc\_2}, \ldots, \delta_{calc\_i}, \ldots, \delta_{calc\_N}$ for the molecule are easily derived by substracting each corresponding value σ_{calc_i} from σ_{calc_ref} (Eq. 54).

For instance, to obtain ^{13}C CCS of a rigid 16–carbon molecular structure in chloroform solution, its geometry is optimized with recruiting a solvent model (e.g., PCM, or, better, S8; Section 5.3.4). Next computations are calculations of the values $\sigma_{calc\_1}, \sigma_{calc\_2}, \ldots, \sigma_{calc\_i}, \ldots, \sigma_{calc\_16}$ related to the carbons 1, 2, …,i, …, 16. These 16 values are provided all together by applying a method for chemical shielding calculations, e.g., GIAO (in combination with MP2 or a DFT method and a relevant basis set), to the obtained optimized geometry of this molecule.[b] For increasing the accuracy of σ_{calc_i} values, a solvent model should be included in chemical shielding calculations. Also, four values $\sigma_{calc\_1}, \sigma_{calc\_2}, \sigma_{calc\_3}$, and σ_{calc_4} that are related to the four carbons of TMS are calculated in the same manner, i.e., using the optimized geometry of this molecule, calculation methods, and the structural model (e.g., the solvent model if used) that are the same as those applied to the examined 16–carbon molecule. For obtaining the σ_{calc_ref} value for ^{13}carbons from these values $\sigma_{calc\_1}, \sigma_{calc\_2}, \sigma_{calc\_3}$, and σ_{calc_4} calculated for

---

[a]If the desired (i.e., calculation methodology-relevant) value σ_{calc_ref} is unavailable. Notably, Table 7 provides a considerably large set of reference values σ_{calc_ref} for ^{13}C nuclei of TMS.
[b]Suitable computational packages are indicated in Section 6.2.1.

Table 7. Values of isotropic chemical shielding σ_{calc_ref} for carbon nuclei of TMS supplied by some commonly used non-empirical methods.[a]

Calculation methodology	σ_{calc_ref} (ppm)	Calculation methodology	σ_{calc_ref} (ppm)
GIAO-B3LYP/6-31G(d)[b]	189.7	GIAO-B3PW91/6-311+G(d,p)[b]	186.6
CSGT-B3LYP/6-31G(d)[b]	188.6	GIAO-PBE1PBE/MIDIX[b]	201.7
GIAO-B3LYP/6-31G(d,p)[b]	191.8	GIAO-PBE1PBE/SPV[b]	193.8
GIAO-B3LYP/6-311G(d,p)[b]	184.5	GIAO-MP2/6-31G(d,p)[b]	202.7
GIAO-B3LYP/6-31+G(d)[b]	190.9	GIAO-MP2/6-31+G(d,p)[b]	202.5
GIAO-B3LYP/6-31+G(d,p)[b]	192.6	GIAO-MP2/6-31++G(d,p)[b]	202.8
GIAO-B3LYP/6-31++G(d,p)[b]	192.7	GIAO-MP2/6-31+G(d,p)[c]	202.4
GIAO-B3LYP/MIDIX[b]	197.4	GIAO{PCM:CHCl$_3$}-MP2/6-31+G(d,p)[d]	202.9
GIAO-PBE1PBE/6-31+G(d,p)[b]	196.8	GIAO-MP2/aug-cc-pVTZ[c]	195.2
GIAO-PBE1PBE/6-311+G(d,p)[b]	188.3	GIAO{PCM:CHCl$_3$}-MP2/aug-cc-pVTZ[d]	195.6
GIAO{PCM:CHCl$_3$}-TPSSh/cc-pVTZ[d]	196.7	GIAO{PCM:CHCl$_3$}-TPSSh/6-31+G(d,p)[d]	195.5

[a]For the molecular geometry of TMS optimized using the same calculation method, the same basis set and the same structural model (vacuum conditions or the PCM solvent model).[9] Measured values for σ_{ref} of ^{13}C in TMS are 185.4[68a] and 188.1 ppm.[68e] [b]For vacuum conditions in both geometry optimization and chemical shielding calculations. [c]Using the PCM model (for CHCl$_3$) in geometry optimization. [d]Using the PCM model (for CHCl$_3$) in both geometry optimization and GIAO calculations.

TMS, we need to involve simple additional considerations. The four carbons of TMS are isochronous at room and even significantly lower temperatures; we remember that the C,H-decoupled ^{13}C NMR spectrum of TMS merely is one singlet, i.e., this reference compound shows one ^{13}C resonance signal (Me rotation is free in the TMS molecule in a very broad temperature range). All conformers of TMS (i.e., Si–C rotamers of this molecule) have the same molecular geometry. Hence, according to Eq. 9 (Section 3.3), the σ_{calc_ref} value is the arithmetic mean of σ_{calc_1}, σ_{calc_2}, σ_{calc_3}, and σ_{calc_4}, i.e., $\sigma_{calc\_ref} = 1/4 \times (\sigma_{calc\_1} + \sigma_{calc\_2} + \sigma_{calc\_3} + \sigma_{calc\_4})$. Each value δ_{calc_i} (among the 16 CCS) that corresponds to the i-th carbon among the 16 carbons of the molecule of interest is elementarily obtained by subtracting the related calculated value σ_{calc_i} from the calculated value σ_{calc_ref}, as Eq. 54 prescribes.

6.1.3 *Conformational analysis by NMR often requires theoretical modeling*

Specialists in organic synthesis usually do not realize that, in their practice, NMR is not an independent tool of conformational analysis. An exceptional role of this spectroscopy in conformational studies of non-polymeric molecules is actually based on its ability to distinguish (though not always routinely) between alternative structural hypotheses. The hypotheses themselves are outer to NMR. They are of NMR origin only if the researcher decides not to resort to other structural ideas. However, this rarely happens in the organic laboratory, and we, realizing or not realizing this, examine "extramurally" proposed molecular geometries when interpreting an NMR spectrum. They are supplied by theoretical models of different quality, from primitive Newman projections to *ab initio*-derived molecular geometries, or are prompted by results of other experimental structural studies. Let us understand this little-known view.

We start the discussion by recalling what experimental NMR spectra indicate about conformers which rapidly interconvert.[a] An NMR spectrum for a system at a rapid conformational equilibrium is not a superposition of spectra of individual conformers. Resonance signals of the same magnetic nucleus in the molecule, which have different resonance frequencies (in other terms, different δ values) for individual conformers, appear as one signal in such a spectrum. Moreover, resonance signals of usually detected isotopes (^1H, ^{13}C, ^{19}F, and ^{31}P) are sharp, since a fast kinetics of exchange of chemically non-equivalent positions for a nucleus does not appreciably contribute to the line width of its resonance signal (Section 3.3, Fig. 60). Such a spectrum image gives a full impression of detecting a molecular structure of individual 3D geometry. However, as repeatedly mentioned in this account, NMR spectra of such a conformationally mobile compound represent a virtual molecular geometry (spectral geometry; Section 3.3) that is a result of averaging structures of all conformers, proportionally

---

[a]As we know, at room temperature, a majority of organic compounds in the gas or liquid phase are at fast (in the NMR timescale) conformational equilibria.

to their fractions in the equilibrium. The "averaged" spectrum appears a well-resolved spectrum of a non-existing, virtual molecular 3D structure (structure of spectral geometry; Section 3.3); it does not represent either conformer.[a] Since such NMR spectra do not provide information regarding the number and 3D geometries of the real molecular objects (conformers), they themselves are not the source of either conformational hypothesis.

If there is no preliminary information regarding molecular flexibility and 3D structures of stereoisomers of the compound of interest, researchers are even more difficultly situated when trying to extract it from constant temperature NMR spectra with sharp lines. As emphasized in Section 3.3, such spectra do not show whether the molecular backbone is or is not flexible[b] at conditions of the NMR experiment. The situation that such NMR spectra with sharp resonance signals may be equally interpreted either as spectra of a rigid molecular 3D structure or as "averaged" spectra of interconverting conformers has negative practical consequences: in synthetic research groups, conformationally mobile organic molecular structures are often taken for rigid ones.

One can argue that, under conditions of conformational exchange that is fast in the NMR timescale, registered chemical shifts $\delta_1, \ldots, \delta_i, \ldots, \delta_N$ (lower indices indicate atom numbers) are ultimately predetermined by conformer structures as well as relative content of conformers, and, thus, these values, in an implicit form, carry the information about the conformational equilibrium. Nevertheless, one cannot "decode" this information (determine both the composition of the mixture of conformers and their 3D structures) by analyzing such experimental NMR spectra. Indeed, it is commonly known that the observed chemical shift δ_i of such a resonance signal of rapidly interconverting conformers $A, B, \ldots, K, \ldots, M$ is an average of their individual chemical shifts $\delta_A, \delta_B, \ldots, \delta_K \ldots, \delta_M$ that contribute proportionally to the statistical (Boltzmann) weights $p_A, \ldots, p_K \ldots, p_M$ of all M conformers. This is simple in mathematical notation (Eq. 55a):

$$\delta_i = p_A\delta_A + p_B\delta_B + \ldots p_K\delta_K \ldots + p_M\delta_M \tag{55a}$$

where $p_A + p_B + \ldots + p_M = 1$. Thus, Eq. 55a is a particular case of Eq. 9 (Section 3.3) applied to chemical shifts.

---

[a] See, e.g., the cyclohexane example (Section 3.3). Only if one conformer is highly predominant ($>98\%$), the considered spectrum may be related to this conformer (see Eq. 55a below), and the spectral geometry (introduced in Section 3.3) may be accepted as representing the molecular 3D structure of this conformer.
[b] In the timescale of NMR of the monitored isotope.

Equation 55a considers one experimental value δ_i (the measured chemical shift of the time-averaged NMR signal of the i-th nucleus). This value δ_i is a sole term of the linear Eq. 55a that is known (as a concrete number) from the NMR spectrum; all $2M$ variables in the r.h.s. of this equation are unknowns. For a molecular system with N chemically non-equivalent magnetic nuclei, no more than N equations of the Eq. 55a form with containing the numerically known l.h.s., i.e., numerical value δ_A, or δ_B, ... or δ_K ..., or δ_N, can be written. These N equations contain $M(N+1)$ unknowns (unknowns $p_A, p_B, \ldots p_K, \ldots, p_M$ are of the same values in each of these equations). To unambiguously determine the unknowns, $M(N+1)$ linear equations are required. However, as indicated, there are only $N+1$ such equations [N equations of form Eq. 55a and one trivial equation $p_A + p_B + \ldots p_K \ldots + p_M = 1$). It is clear that $N+1 = M(N+1)$ only if $M = 1$, i.e., if there is only one stereoisomer. Hence, if there are two or more conformers, no information about any of them can be extracted from δ_i values in NMR spectra recorded for conditions of fast conformational equilibrium.

After realizing that routine experimental NMR does not supply "conformationally telling" spectra of flexible molecular systems at conditions of fast conformational equilibrium, we may have another idea for "staying" with only NMR. We may think that one can determine the 3D structure and relative stability of conformers when having NMR spectra that indeed are a superposition (and not an average) of individual spectra of these stereoisomers that are in the same proportion as they are when rapidly interconverting. However, this NMR approach is absolutely not warranted, as we will see straight away below.

An obvious route to such NMR spectra of individual conformers is to abandon the deadlocked situation of a rapid conformational exchange. In principle, the way to detect these kinetically unstable stereoisomers by NMR is to "freeze" the conformational equilibrium by essentially decreasing the temperature of the NMR experiment. Uptake of internal energy from conformers decreases their conformational mobility (increases their lifetimes) so that they become rigid molecular structures in the NMR timescale. The NMR spectrum of rapidly interconverting conformers is split into superimposed individual spectra of these formerly undetectable species (Section 3.3, Fig. 60).[a]

---

[a]Excluding the case of DNMR-invisible equilibria. Low conformational barrier ($\Delta G^{\#} <\sim 4 - 5\,\mathrm{kcal\,mol^{-1}}$) as well as hidden partner systems (Section 3.3) fall under this category. Lifetimes of very papidly interconverting conformers are near the lower limit of the NMR time scale, and, therefore, these conformers are not detected by NMR as individual molecular structures. In the hidden partner

If the frozen stereoisomerization is a single one that may occur in the molecule, the detected components of the conformer mixture indeed are individual conformers (rigid structures under conditions of the NMR experiment, from the NMR timescale "viewpoint"). For instance, diatopomerization *apical phenylene-equatorial phenylene* is the sole 3D reorganization in selenurane **16** with homotopic bisphenylene fragments (Fig. 23). If this permutation of the Se ligands C_6H_4 is slow (e.g., at low temperatures), ^1H or ^{13}C NMR detects true individual conformers. This is easy to understand: the time-averaged resonance signals (registered at room temperature) of the phenylene groups are split into two sets upon lowering the temperature, signals of the apical phenylenes and those of the equatorial phenylenes. The resulting stereoisomers do not have structural fragments with additional rotational or inversional freedom. Hence, in contrast to room temperature NMR spectra of **16**, NMR spectra of this compound, which are registered at low temperatures, are "NMR fingerprints" of the true molecular geometry and not of a stereodynamics-averaged (virtual) one. Further lowering of the temperature would not split the resonance signals.

Two or more structural fragments often are conformationally mobile in organic molecules at room temperature. In many cases, 3D reorganizations of structurally dissimilar fragments of the molecular backbone [e.g., rotation around the amide bond C–N in a dialkylamido fragment and NIR (Fig. 21) in a trialkylamino one] occur with essentially different rates and, then, temperature decrease "freezes" these conformational transformations one after the other and not synchronously. Let us suppose that two different stereorearrangements, each with its own rate, occur in a molecule. Temperature-induced splitting of one or more resonance signals in the NMR spectra of this compound indicates deceleration of the slower stereoisomerization among the two inherent in the molecule. That is, doubling of some NMR signals in these low-temperature spectra tells us that two stereoisomeric molecular structures have been detected. However, these "NMR-caught" species may not be conformers. They may be new virtual molecular structures with a "frozen" fragment, and NMR spectra obtained at this temperature reflect the true geometry for this only fragment. In this case, if the temperature is decreased further, an additional signal splitting occurs when temperatures become sufficiently low to inhibit (in the NMR timescale; Section 3.3) the faster stereoisomerization. The resulting NMR spectra are a superposition of NMR spectra of real individual stereoisomeric structures since the latter, being deprived of a further stereodynamic freedom, are rigid *molecular* structures (in the NMR timescale) at very low temperatures. In simple words, NMR "exposes" conformers of the considered molecular system only when reaching the ultimate temperature-induced splitting of resonance lines.

---

case, slowing of conformational dynamics does not change NMR spectra to an appreciable extent. Besides, low-temperature NMR spectra of some compounds do not differ from their room temperature spectra, except for some effects (temperature-induced changes of chemical shifts, line broadening associated with viscosity increased with lowering the temperature). Those are molecular systems whose conformers are identical or enantiomeric structures without diastereotopic fragments that contain magnetic nuclei and exchange their stereochemical environments upon conformational transformation.

As known, the primary quantitative characteristics of a thermodynamic equilibrium are the number and relative content of molecular components involved in the equilibrium. Both these characteristics of the conformational equilibrium are available from low-temperature NMR spectra. Such spectra contain an entire set of resonance signals for each NMR-detectable "frozen" conformer, for the temperature of the NMR experiment.[a] The number of such sets is the number of conformers that have distinct molecular geometry. Determination of relative stability of NMR-visible conformers is not problematic either. The intensities of resonance signals of the "frozen" conformers reflect the proportion of these equilibrium participants at the temperature of the "freezing" NMR experiment. Integration of these intensities of signals of conformers I and J (as known, an elementary technical procedure in processing NMR spectra) supplies the conformer ratio p_i/p_j at this temperature. Thus, in contrast to NMR experiments at room or slightly lower temperatures, those conducted at low temperatures (say, 140–180 K) are capable of providing at least minimal information regarding conformational equilibria.

We remember that $p_i/p_j \equiv K_{eq}$, where K_{eq} is the equilibrium constant for the thermodynamic equilibrium of conformers I and J at absolute temperature T. Such measurements of intensities at several temperatures within a temperature interval of the 40–50 K range provide an empirical dependence of K_{eq} on temperature T, for these temperatures. This empirical dependence is further used as a basis for satisfactorily estimating K_{eq} (in other words, the conformer ratio p_i/p_j) for other temperatures that lie within this interval or are not essentially distanced from it (say, for less than 100 K). Using the experimental values of K_{eq} and the corresponding values T, the new dependence of K_{eq} on T is usually represented by a linear regression plot (for linear regression, see Section 6.2.2) in Cartesian coordinates 1/T (the x coordinate) and $\ln K_{eq}$ (the y coordinate). The reasoning for quantitatively approximating the relationship of K_{eq} and T as a linear dependence of $\ln K_{eq}$ on 1/T is that logarithmation of Eq. 12 (Section 3.3) gives linear equation $\ln K_{eq} = -\Delta G^0_{i-j}/RT$.

However, our idea under examination is that NMR may supply us with full information about conformers. This includes that NMR-detected conformers can be *identified*. In other words, our belief is that molecular 3D

---

[a]A nucleus may have the same or practically the same resonance frequency in different conformers of the molecule. Molecules of non-minimal size usually have many such nuclei. Therefore, only several resonance signals of the original NMR spectrum (say, a room temperature one) are usually observed split when lowering the temperature of the NMR experiment.

structures of individual conformers can be determined from NMR spectra recorded for conditions of slow conformational equilibrium. How informative, however, are such NMR spectra regarding molecular geometry of conformers?

"Frozen" conformers are rigid molecular structures from the viewpoint of the equilibrium-freezing NMR experiment. The above question is, therefore, equivalent to the question of whether a rough geometry of any rigid molecular backbone of closed electron shell or such a geometry of any rigid molecular fragment may be determined by only NMR, without involving any outer structural hypothesis (e.g., a conformon-recruiting hypothesis or that based on diminishing repulsive interactions of non-bonded atoms). A strict answer is no. For instance, [6]radialene carbocycles $[-C(= CR_2)]_6$, tetraaminosubstituted ethylenes $(R_2N)_2C = C(NR_2)_2$, and cyclic boroxines $[(AlkO)B-O-]_3$, are "inconvenient" structures for establishing the general motif of their molecular geometry [e.g., twisting around the C=C bond in ethylenes $(R_2N)_2C = C(NR_2)$ and the ring shape for the mentioned radialenes and boroxines] by only NMR. The reader may ask himself/herself whether he/she could determine the molecular geometry of the biscorannulene conformer shown in Fig. 96 (e.g., the mutual rotational orientation of the corannulene bowls) by using only NMR and forgetting the extramural idea of the bowl shape of the corannulene ring.

However, a flexible answer is yes. Determination of a rough molecular geometry solely by NMR is possible for rigid (at the temperature of the NMR experiment) molecules or rigid molecular fragments of most organic compounds. Unfortunately, determination of molecular shape by NMR is complicated even if applied to organic molecules which conformational equilibrium becomes slow at low temperatures. Experimental difficulties and some structure-related features of the studied system are factors that may keep synthetic specialists from solving such a conformational puzzle. The experimental difficulties to be encountered include overlapping of signals of individual conformers upon insufficiently high operating frequency of the NMR instrument, freezing of most organic solvents at low temperature NMR experiments, insufficient solubility of polyfunctionalized organic compounds in "last resource" solvents of low freezing point (freons), as well as decrease the line resolution at low temperatures owing to viscosity-caused line broadening. Molecular structural features, which complicate conformer identification, include low conformational barriers, non-minimal number M of NMR-detectable conformers ($M \geq 3$), and either the total or partial absence of vicinal aliphatic protons ("Karplus protons"), or magnetic nuclei with measurable NOE interactions (i.e., magnetic nuclei in essential spatial proximity), or non-quadrupolar magnetic nuclei (nuclei with spin I=1/2).

In view of these difficulties, we are forced to resort to extraneous conformational *hypotheses* or additional experimental data for molecular geometry (e.g., results of X-ray diffraction studies) when interpreting spectra of conformationally flexible compounds studied by NMR at conditions of slow conformational equilibrium. The textbooks-inspired approach to conformational analysis is to consider molecular 3D geometries of conformers or their molecular fragments as alternative geometries composed from relevant conformons (see Section 3.4 for critical remarks on this practice). Note that the role of NMR in this simple methodology is to indicate the correct molecular geometry among alternative ones for the frozen conformer (and not to "build" it from zero). While this approach is successful when applying it to conformationally trivial compounds, it is problematic in other cases. The below example of bicorannulenyl (**41**, Fig. 96), a compound with a convoluted conformational behavior, demonstrates that conformational studies by NMR may turn out non-telling if not supported by quantitative theoretical modeling.

Bicorannulenyl, a molecular system of both undetermined conformational space and conformational mobility, provides an example of how low-temperature ^{1}H and ^{13}C NMR spectra may only provide little information for identification of conformers as well as understanding of their interconversions even with involving reasonable conformational models. From accepting the convex molecular geometry of corannulene **4a** (Fig. 68) and the s-gauche molecular geometry of diaryls as conformons (see Section 3.4), one can assume regarding **41** that (*1*) the corannulene fragments have a convex (bowl) geometry; (*2*) these non-planar fragments undergo bowl-to-bowl inversion; and (*3*) there are conformers that arise from rotation around the central C-1−C-1' bond. In simple words, basic conformational models tell us that conformers of this compound result from occurring both two intramolecular motions, bowl inversion and C–C rotation. From considering the molecular symmetry in molecule of **41**, one can suggest that there are four possible combinations of relative orientations of the two bowl fragments in this bis-corannulene. This is all what one can say about conformers of this system without studying it. For instance, since, in contrast to simple diaryls, aromatic fragments of **41** are not planar, we have no idea about the number of thermodynamically stable C-1−C-1' rotamers and the rates of their interconversions. In a remarkable study of stereodynamics of this diaryl,[74a] NMR experiments have demonstrated that when the temperature is decreased from 320 K to 180 K, the set of time-averaged resonance signals is split into three sets. Three rates of conformational transformations, which cause these changes in NMR spectra of **41** upon decreasing the experimental temperature, have been measured.

What stereochemistry-related information can be gained from these temperature-variable NMR experiments? They show that this molecular system is indeed flexible at room temperature.[a] The number of the signal sets registered at 180 K indicates that NMR "observes"

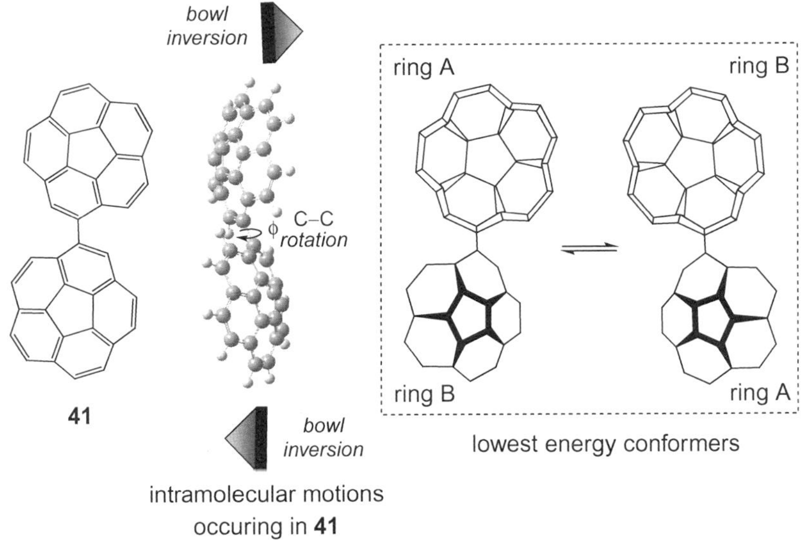

Figure 96. Two isolate intramolecular motions ($C_{arom} - C_{arom}$ rotation and bowl-to-bowl inversion) in bicorannulenyl **41**. The modeled 3D structure[74a] (shown in the ball-and-stick graphics) is related to one of two enantiomeric stereoisomers of minimal energy; they have no symmetry elements, but possess helix chirality of the molecule as a whole. Their interconversion (see the structures in dotted rectangle) is NMR-undetectable in achiral solvents. Theoretical calculations[74a] demonstrate that it occurs as a sequence of three rotations around the central C–C bond; the first rotation is separated from the second by inversion of one of the bowls **A** and **B**, and the second rotation is separated from the next one by inversion of the other bowl.

---

[a] For general understanding of the temperature impact on NMR spectra of interconverting stereoisomers, see Fig. 60, Section 3.3.

at least three diastereomeric structures that, in the NMR timescale (Section 3.3), are rigid at this temperature and interconvert at higher temperatures. The three measured rates are of interconversions of these three structures; the molar ratio of these diastereomers at 180 K is supplied by integrating the signal intensities at this temperature.

What do these experiments *not* show in respect of the issue interesting for us? They do not reveal both the conformer number and conformer structures of **41** (i.e., the relative direction of convexity and exact mutual rotational orientation of the corannulene polycycles in the detected diastereoisomeric conformers). They do not identify diastereoisomeric structures whose interconversion rates have been measured; consequently, one cannot assign a certain intermolecular motion (i.e., either the bowl inversion, or rotation around the C-1−C-1' bond, or a concerted motion) to a certain measured rate.

Indeed, the temperature-induced signal splitting does not indicate whether any of the resulting three sets of partially overlapping resonance signals (*1*) belongs to two enantiomeric 3D structures (e.g., enantiomeric conformers shown in Fig. 96) that, in the NMR timescale, do not interconvert at 180 K;[a] or (*2*) belongs to a pair of enantiomeric conformers that are conformationally mobile at this temperature; or (*3*) arises from averaging the signals of the still rapidly interconverting diastereomeric conformers with some fragments (i.e., the corannulene bowls *or* dihedral C−C-1–C-1'−C units) that are rigid at 180 K in the NMR timescale (then, these spectra indicate a virtual, time-averaged 3D structure). In other words, the obtained NMR spectra do not show how many stable rotamers of **41** result from rotating one bowl ring relatively to the other one, and whether a 360° turn is possible without inverting one corannulene bowl or both of them. It is meaningless to try to extract structural information regarding conformers from a set of NMR signals if there is no evidence that this set belongs either to individual ("frozen") diastereomeric or enantiomeric (interconverting or "frozen") conformers. Thus, the conducted NMR experiments[74a] alone do not transform the uncertainty regarding conformers of 1,1'-bicorannulenyl and their interconversions into a quantitative detailed description anticipated from spectroscopic measurements. The general conformational models (the conformons mentioned for **41**), of course, do not assist both in identifying the detected diastereomers and deciding whether they are true conformers (fully "frozen" molecules) or are locally "frozen" structures with the still flexible rest of the molecule. These primitive models only permit to start the interpretation of the NMR spectra from using some "of-the-shelf" molecular geometries (i.e., the bowl and s-gauche conformons) for molecular fragments of **41** and not from trying to establish molecular geometries of these fragments by NMR.

The studied molecular system is associated with a broad set of factors that complicate interpretation of experimental NMR data, such as a sufficiently large number of conformers; the presence of two intramolecular motions that may couple in interconverting some conformers; many pathways of conformational transformations; low conformational barriers for interconversion of some conformers. One may assume that more sophisticated NMR

---

[a]As known, NMR spectra of enantiomers in an achiral medium are identical.

experiments (e.g., examination of stereodynamics of bicorannulenyl **41** in a chiral solvent including experiments at essentially lower temperatures; calibrated measurements of NOE for "frozen" conformers) could provide a worthwhile understanding of stereoreorganizations occurring in this molecule. However, it is obvious that such or other quite non-routine NMR experiments and, more importantly, the experimental strategy in such NMR studies are out of professional competence of specialists in organic synthesis.

One should stress that there are no preliminary data (e.g., established molecular geometries for close analogs, or estimates/measurements of kinetic barriers of rotation around the central C_{arom}–C_{arom} bond for such systems with non-planar aromatic fragments) that could assist in interpreting the registered temperature-variable NMR spectra of the exotic diaryl compound **41**. As mentioned, conformationally well-studied diphenyls are not close analogs of this two-bowl-shaped molecule. We are able to indicate the above mentioned complicating factors in NMR probing of **41** only because this study has recruited theoretical modeling (a fifth-rung DFT method, Section 5.2.6) to determine the structures of individual conformers of this system, their relative stabilities, and kinetic barriers of their interconversions. Exactly theoretical modeling of this conformationally non-trivial system permitted to relatively easily build a quantitative conformational scheme (for such schemes, see Sections 2.7 and 3.5.2) that includes six pairs of enantiomeric conformers and sixteen transition states of their interconversions.[a] Experimental measurements (the stereoisomer structures obtained X-ray analysis and kinetic barriers provided by NMR)[74a] ultimately serve only as a confirmation of this detailed, computationally derived conformational description of diaryl **41**. Note that, by combining different experimental measurements and extensive theoretical modeling, this now classical investigation[74a] of the molecular system with absolutely unexplored conformational space and flexibility establishes the quality standards in contemporary conformational studies.

Thus, we have learned that (*1*) without involving extraneous conformational hypotheses, a NMR spectrum recorded for conditions of a fast conformational equilibrium does not provide any information about the conformers, and (*2*) when involving such hypotheses (e.g., conformon-based), one may indicate the correct one by using NMR, if the conformational behavior of the compound is not too complicated, and its low-temperature NMR experiments succeed to provide well-resolved spectra of individual conformers. However, the real NMR practice in organic laboratories is almost always to conduct NMR experiments at ambient conditions. Concerning conformational analysis, the practice of NMR-non-professionals includes exploiting conformon-based hypotheses for the

---

[a] Torsion angle Φ (Fig. 96) is changed by 11–15° for the modeled bowl inversions.[74a] This indicates that there is no isolate bowl inversion in **41**; there is a concerted motion *bowl inversion*–C-1–C-1'*rotation*.

number, 3D structures, and relative stability of interconverting conformers; as a rule, theoretical calculations are not involved in inferring molecular structure from such spectra. Is this way valid? The answer is as follows: if recorded for conditions of fast conformational equilibrium, NMR spectra do not permit to select the veracious conformational hypothesis. The below example of structurally simple compounds **42a** and **42b** (Fig. 97) illustrates this.

The room temperature ^{1}H NMR spectrum is a single unsplit ^{1}H resonance signal for cyclooctane (**42a**); cyclooctatetraene (**42b**) shows similar NMR "fingerprints." For each

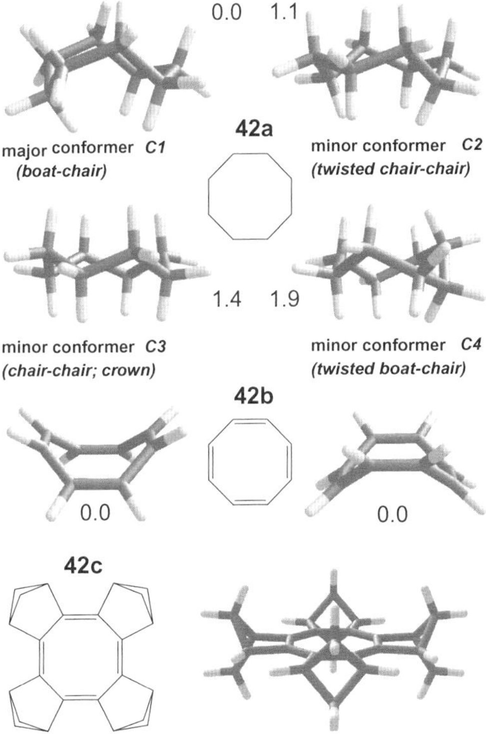

Figure 97. Stereoisomers of carbocycles **42a-c**; for **42a** and **42b** at room temperature, the stereoisomers fall under category *conformers* (Section 2.5). Structures optimized[9] by MM3(96) and MMFF94 are shown for **42a** and **42b,c**, respectively. Numbers indicate relative steric energy ΔE_{ster} (kcal mol^{-1}) for conformers of each compound.

compound, two conformational hypotheses may explain the appearance of a single ^{1}H singlet. The simplest one is that the molecule is rigid at conditions of the NMR experiment, and all of its protons are undistinguishable due to the molecular symmetry both for **42a** and **42b**. The second conformational hypothesis is that there is a fast interconversion of conformers, and all the protons are isochronous.[a]

Both hypotheses are equally valid to an untrained eye. Suppose that we have no information regarding the molecular geometry of cycles **42a** and **42b**. Then, we could think that both **42a** and **42b** are planar. A planar ring geometry of each cycle implies chemical equivalence of its protons and, thus, satisfies the rigidity hypothesis for molecules of these monocycles. Within the flexibility hypothesis, it is not difficult to suggest non-planar structures, where the time-averaged (virtual) structure is a planar ring. In this circumstance, the ring protons are isochronous. We quickly meet difficulties in analyzing molecular flexibility in **42a** and **42b** by solely using their room temperature NMR spectra. These spectra, showing the chemical equivalence of magnetic nuclei, do not indicate whether this equivalence results from symmetrical disposition of these nuclei in a static 3D structure or from their isochronism. We certainly need extraneous (NMR data-independent) information regarding molecular geometries of cycles **42a** and **42b** to select between these hypotheses.

As indicated, structural hypotheses commonly used in unsophisticated modeling of 3D molecular backbones are conformons (Section 3.4). When considering molecular geometry in synthetic laboratories, room temperature NMR spectra usually serve as tests for conformon-suggested geometrical alternatives for molecules of interest. From the viewpoint of many synthetic chemists, such a NMR spectrum should correspond to one of such "hand-drawn" molecular 3D structures of the examined compound; in other words, it should indicate the conformation favored by the molecule. Let us turn to cycles **42a,b** with this simple auxiliary tool.

The extraneous structural hypothesis for cyclic hydrocarbon **42a** is supplied by the common conformational rule for alkanes — the preference for the staggered geometry *vs.* the eclipsed one (see the preferred conformons for butane systems, Fig. 61). All ring substituents are eclipsed in the planar ring geometry for **42a**, and thermodynamic stability of this molecular geometry may be thought hardly probable. On this ground, the rigidity hypothesis can be tentatively abandoned for this cycle.[b]

---

[a]If interconversion of stereoisomers leads each of two 3D non-equivalent atoms or groups to occupy a spatial position that is identical to the former spatial position of its counterpart, the atoms become equivalent in time when rapidly (from the viewpoint of the observer) interchanging their 3D environments. For interconverting conformers that have shorter lifetimes than the NMR time resolution (Section 3.3, Table 1), such atoms or groups are called isochronous. Clearly, isochronous nuclei have the same resonance frequency. Because each proton of cycle **42a** in turn occupies all non-equivalent positions in conformers *C1–C4* (Fig. 97) upon their interconversion, all protons of **42a** are isochronous under assuming that this interconversion is rapid for the ^{1}H NMR spectroscopy.

[b]As more than once noticed in this account, primitive molecular models (including conformon-based ones) should be verified by involving true research tools. NMR experiments, which have been conducted for nematic solvents,[73a] variable-temperature NMR studies,[73b] as well as theoretical modeling[74b] have

Both suggestions — a rigid planar (symmetry group D_{4h}) and a flexible, invertible tub-shaped ring (Fig. 95; symmetry group D_{2d}) — explain the chemical equivalence of all protons in the ^{1}H NMR spectrum of cycle **42b**. All the protons are identically oriented in the rigid D_{4h} structure. Flipping of the flexible D_{2d} ring leads to that the time-averaged (i.e., NMR-"detected") geometry which also is of the D_{4h} symmetry. We have to admit that room temperature NMR spectra of **42b** provides no clue as to whether this molecular skeleton is flexible or rigid at this temperature. However, the next extraneous conformational hypothesis appears fruitful.[a] With borrowing the only possible conformon from the set of 1,3-butadiene conformons, *s*-gauche (Fig. 61), we can claim that the flexibility suggestion for this molecule is correct.[a]

Suppose that the above conclusions regarding molecular flexibility of these compounds which are based on both external (conformon) hypotheses and room temperature NMR spectra satisfy us. We do not know yet whether the employed conformons do reflect molecular geometries of conformers of **42a** and **42b**. The methodological assumption, which we examine, is that room temperature ^{1}H (or ^{13}C) NMR spectra of these compounds should deliver the answer. As one can see, our assumption does not pass the test "proposed" by these spectra of **42a** or **42b**. The sole ^{1}H singlet does not provide any clue no the 3D structures of conformers of these compounds. That is, it does not carry any information about the number of low-energy (i.e., NMR-detectable) conformers, their relative stabilities, or their structural features. Should we not rationalize that the practice ubiquitously enrooted in organic laboratories (room temperature NMR spectra + conformon-based hypotheses) of identifying conformations of organic molecules in solution is inadequate?[b]

Let us strengthen our starting position in conformational analysis by NMR. Suppose that we have some extraneous hypotheses that are supplied by theoretical calculations regarding molecular geometries of conformers of these compounds. Could we examine them by means of these NMR spectra? For instance, let us assume that only low-energy conformers *C1–C4* (Fig. 97) are involved in the conformational equilibrium for cycle **42a**. The ^{1}H spectrum

---

demonstrated that compound **42a** is conformationally flexible at room temperature. Conformer geometries, e.g., those of conformers *C1–C4* (Fig. 97), have been determined by modeling these molecules by *ab initio* methods.[74b] Also, by applying *ab initio* calculations to a small molecule **42b**,[74c] the tub geometry suggestion and the flexibility hypothesis for this chemical structure have been proven to be correct.

[a]Both **42a** and **42b** may be considered as parent molecular systems in the series of cyclooctanes and cyclooctatetraenes, respectively. Then, the tub geometry is a conformon for cyclooctatetraenes, while boat-chair, twisted chair-chair, chair-chair, and twisted boat-chair (Fig. 97) are conformons for cyclooctanes. Similarly to using other conformons (Fig. 61), one should involve them into predicting conformations with caution. For instance, *ab initio* calculations have revealed that the eight-membered ring of cyclooctatetraene **42c** is planar (Fig. 97),[74d] exposing the unreliability of the "conformon tool" (see Section 3.4 for other examples). To a greater extent, 3D structures of stereoisomers of [*n*]annulene with $n > 8$ are not deducible from any conformons.

[b]Excluding anancomeric equilibria (in NMR jargon, hidden partner exchange; see below). Concerning molecular geometry, NMR spectra of such systems may be considered as the spectra of one (the favored) conformer.

of **42a** — one singlet — does not disprove this hypothesis. Let us assume that there are four geometrically identical tub-shaped conformers for cycle **42b** (two of them are shown in Fig. 97) that rapidly interconvert in the NMR time scale. The related spectrum does not contradict the hypothesis of a rapid interconversion of these conformers. However, if the hypothesis of the molecular tub geometry is applied to **42a**, the considered ^{1}H spectrum of this saturated cycle is also in accordance with this now incongruous assumption!

It is worth indicating that kinetic methods of quantum coherence NMR, e.g., EXSY (*E*xchange *S*pectroscop*Y*), are useless if applied to unfrozen conformational equilibria. We may intuitively feel that one cannot register a 2D NMR spectrum (including that reflecting spin-spin cross-relaxations) of a conformer which, due to its instability in the NMR timescale, does not show an individual 1D NMR spectrum. For instance, in EXSY, one cannot saturate the resonance of a magnetic nucleus in molecular structure A and, thus, transfer the resonance saturation to molecular counterpart B if the lifetime of A is shorter than the Δv_{A-B}-dictated lifetime τ from Eq. 8 for the nucleus of interest (Section 3.3; Δv_{A-B} is the difference between resonance frequencies of the nucleus in structures A and B). Since Δv_{A-B} is equal to zero for any geometrically identical conformers of **42b** (including the tub-shaped conformers; Fig. 97), EXSY is absolutely inapplicable in exploring conformational freedom of this molecule. Values of Δv_{A-B} for a proton or carbon in diastereomeric conformers *C1–C4* of **42a**, in principle, differ from zero. However, they are obviously so small that EXSY is incapable of "resolving" the conformers of this compound either.

"Time-averaged" NMR spectra supply information about a 3D molecular structure only for conformational equilibria with one *highly predominated* conformer (anancomeric or near-anancomeric equilibria), e.g., that with equatorial orientation of the *tert*-Bu substituent in cycle **18b** (Fig. 31). An NMR spectrum for such a mixture of equilibrating conformers is identical to that of the static major conformer. Therefore, temporal signal averaging does not cause confusion when interpreting measured constants of nuclear spin-spin interactions, NOE enhancements, chemical shifts, etc. The *conformer* appears in the NMR spectra as a single rigid stereoisomer.[a]

In testing our conformational hypotheses by means of routine NMR spectra, the last resource is to quantitatively analyze them. The first quantitative NMR data that we have are chemical shifts. The relationship between the time-averaged chemical (i.e., registered at room temperature) shift and chemical shifts of the individual interconverting molecular structures is

---

[a]In discussing molecular geometry of either compound in publications on synthetic organic chemistry, a single geometry is usually in the focus; brief NMR-related considerations supply evidence for it. They tranquilly presuppose that precisely the NMR-analyzable case of conformational equilibria, i.e., an anancomeric equilibrium, occurs. This assumption may not always be correct. Theoretical modeling, e.g., non-empirical calculations of the ^{13}C chemical shifts (Section 6.2.1) of the major, computationally located stereoisomer, is an easy method to verify the hypothesis.

given by Eq. 55a (this Section); the latter is the single mean in utilizing values of experimental chemical shifts for our purpose. Regrettably, this quantitative relationship is useless for testing molecular geometry hypotheses for conformers, as explained in this Section when introducing Eq. 55a. Any chemical shift for individual conformers is unknown to the interpreter, and, hence, Eq. 55a does not help us in "extracting" these stereoisomeric 3D structures from NMR spectra registered for conditions of a fast conformational equilibrium.

The parent Eq. 9 (Section 3.3) may be applied to other NMR observables, e.g., vicinal constants $^3J_{H,H}$ of proton spin-spin interactions. To the interpreter of "averaged" NMR spectra, this technique diminishes the uncertainty regarding conformers if considered as molecular frameworks assembled from alternative conformons. The rational here is that such spin-spin interaction constants for M conformons (i.e., certain 3D geometries of molecular fragments) may be associated with certain characteristic values (i.e., numerical parameters $^3J_{par-1}, ^3J_{par-2}\ldots,^3J_{par-i}\ldots,^3J_{par-M}$). Then, equations of type

$$^3J_i = p_1^3 J_{par-1} + p_2^3 J_{par-2} \ldots + p_i^3 J_{par-i} \ldots + p_M^3 J_{par-M} \qquad (55b)$$

which are a particular case of Eq. 9 (p_i panelizes the relative fraction of the *i-th* conformer, and M is the number of used conformons), become more certain for numerically solving them in respect of p_i. Our conformon-based structural hypotheses may be examined by simulating the experimental (i.e., averaged) vicinal constants $^3J_{H,H}$ using Eq. 55b. Then, the correct geometrical hypothesis (a set of "correct" conformons) resembles the measured values of $^3J_{H,H}$.

However, it is often impossible to succeed in such identification of conformers since a molecular structure of interest may have both too many statistically weighty conformers and too few vicinally positioned protons. For simplicity, let us consider saturated six-membered cycles. Thermodynamic preference of the chair geometry (it is a conformon; Fig. 61) of the six-membered ring has been proven for many basic mono- and polycyclic systems of different chemical structures. However, different saturated six-membered ring compounds have favored conformers that adopt the *twist* geometry (another conformon; Fig. 61), e.g., 1,4-cyclohexanedione, cyclic dimers of aromatic amino acids (2,6-diketopiperazine derivatives), cyclohexanes (and their heteroanalogs) with two bulky *trans* substituents, and many 1,2,3-dioxaphosphorinanes (cyclic trialkyl phosphites) as well as 1,2,3-dioxaphosphorinane-2-oxides (cyclic trialkyl phosphates). This indicates that twist conformers cannot be immediately excluded from consideration of a new molecular system that incorporates a six-membered saturated ring if the spectra do not indicate a high predominance of a chair-shaped conformer.[a] For example, in the case of 4-monosubstituted 1,2,3-dioxaphosphorinanes, two chair- and three twist-shaped conformers should be considered (i.e., $M=5$ for the corresponding Eq. 55b). As one can readily see, it is impossible to determine relative fractions of these conformers from time-averaged coupling constants ^3J_i. There are two non-equivalent pairs of vicinal ring protons in these molecular systems; thus, there are only two numerically unequal constants $^3J_1$ and $^3J_2$ measured by NMR. Each experimental ^3J_i satisfies the equation $^3J_i = p_{chair1}{}^3J_{chair1} + p_{chair2}{}^3J_{chair2} + p_{twist1}{}^3J_{twist1} + p_{twist2}{}^3J_{twist2} + p_{twist3}{}^3J_{twist3}$. Hence, there are five unknowns p_{chair1}, p_{chair2}, p_{twist1}, p_{twist2}, and p_{twist3}; to find them, we need five linear equations, where all other parameters are known. When adding the obvious equation $\Sigma p_i = 1$, we need four equations of type Eq. 55b with four distinct known (here: experimental) values of ^3J_i to determine these five unknowns p_{chair1}, p_{chair2}, p_{twist1}, p_{twist2}, and p_{twist3}. With only two known (i.e., experimental) unequal values $^3J_1$ and $^3J_2$, one can set up only two linear equations of type Eq. 55b, where p_{chair1}, p_{chair2}, p_{twist1}, p_{twist2}, and p_{twist3} are unknowns.

Furthermore, conformers of organic compounds may have significantly distorted canonic, atypical (belonging to conformons of another chemical structure), or even new (never described) molecular geometries. It is clear that the parametric Eq. 55b would mislead: characteristic values of ^3J_i associated with the related conformons obviously are irrelevant for using them in equations of type Eq. 55b if applying this equation to analysis of molecular systems that are not conformon-shaped. Hence, the analysis of

---

[a]This means that there are no large (11–14 Hz) coupling constants ^3J for vicinal protons in ^{1}H NMR spectra. As is known, such values are typical for antiperiplanar 1,2-positioned protons (Fig. 11). Detection of such a ^{1}H–^{1}H spin-spin constant, ^3J, for a conformationally mobile fragment CH–CH indicates significant predominance of one conformer with antiperiplanar geometry for these vicinal protons, i.e., a chair-shaped conformer. In other terms, there is an anancomeric equilibrium.

nuclear spin-spin interaction constants is useless for conformon-irrelevant molecular systems that are conformationally mobile in the NMR timescale.

When studying conformational behavior of a new (i.e., non-trivial) molecular system by NMR, we have no idea whether or not it is conformon-shaped. This situation poses a dilemma to a forethoughtful researcher: is it or is it not worth trying to judge the geometry of a molecular system of interest without varying the temperature of NMR experiments? At this point, recall that all these difficulties associated with the interpretation of NMR spectra can be avoided by theoretically modeling the conformational equilibrium (i.e., conformers and conformational transition states) for this molecule.

Are there additional NMR-supplied observables that may reflect conformer geometries? At first glance, they appear to be NOE enhancements. As known, they quantitatively depend on nucleus–nucleus distances,[a] and changes in NOE enhancements associated with changing these distances are measurable. However, the molecular structure, which is "extracted" by means of quantitative correlations of *NOE enhancements* and *interatomic distances*,[b] has a time-averaged (virtual) 3D geometry; it does not contain analyzable information regarding individual conformers.[74e]

We ultimately realize that one cannot study rapidly interconverting conformers by experimenting with only NMR of non-viscous solutions of organic compounds at ambient temperature. The statement "An exceptional role of NMR in conformational studies of organic molecules is actually based on its ability to distinguish (though not always routinely) between the alternative structural hypotheses" from the beginning of this Section should be supplemented with the condition that, to be successive, conformational analysis by NMR requires the situation of either anancomeric or frozen conformational equilibrium. In principle, *no hypothesis regarding 3D structures of conformers and their relative content in the liquid or gas phase, which explains an NMR spectrum of rapidly interconverting*

---

[a]Organic researchers may recognize this relationship in its qualitative form well known to them: the less distance the two nuclei are, the larger the related NOE enhancements.

[b]Quantitative correlations *NMR observable–geometry parameter* use intricate algorithms and therefore require considerable specialization. This computerized tool of analysis of NMR spectra [computer programs for biomolecular NMR, e.g., QUEEN (http://www.cmbi.ru.nl/software/queen/), Felix Model (http://www.felixnmr.com), and AQUA (www.bmrb.wisc.edu/~jurgen/aqua/)], is practically unknown in synthetic organic groups. Note that, in treating the data of different NOE measurements, these programs analyze, but do not create geometrical hypotheses for biomolecules.

(non-frozen) conformers in a non-anancomeric equilibrium (equilibrium with more than ~5–10% of minor conformers), is trustworthy if the spectrum is a single argument in supporting the hypothesis.[a]

This verdict regarding conformational analysis and routine NMR spectra exposes a serious problem in drawing stereochemical conclusions by NMR non-professionals. A frequent misbelief in organic laboratories is that relative configuration of two chiral or prochiral elementary molecular fragments in the molecule and determination of the molecular conformation can be determined by NMR independently of each other.[b] An example of this misunderstanding is a widely exploited idea that observation of an appreciable NOE interaction of two protons, which are 1,n-positioned (n $\geq$ 3) substituents in a ring of a stereoisomer of a cyclic compound, indicates that we deal with the *cis* isomer. An equivalent misinterpretation would be that such observation reveals predominance of a certain conformer of the corresponding *trans* isomer. The truth (well-known to NMR specialists[c]) is that registration of such a NOE interaction is not indicative. It may be equally interpreted as a predominance of either a conformer with the *cis* orientation of these ring protons, or a conformer of the *trans* isomer. Let us discuss this in detail.

A plain, intuitive view on distances between *cis* and *trans* ring substituents is that 1,n-positioned *cis* substituents, e.g., protons, of any conformer of a non-planar cycle are less distanced than the equally positioned *trans* substituents in *any* conformer. This view is far from reality. In fact, the shape (i.e., the conformation) of the ring predetermines whether the equally positioned substituents of a certain relative stereochemistry (*cis* or *trans*) are proximal or distal. This is not surprising. Geometrically, interatomic distances between non-bonded atoms are determined

---

[a]As we remember, the PES (Sections 3.1, 3.2 and 4.2.3) of a molecular structure is exactly that theoretical model which we need for accurately understanding a flexible shape (i.e., the conformation) of the molecule. NMR experiments conducted at conditions of fast conformational exchange do not supply information about any point of the PES. In sharp contrast, quantitative molecular modeling (see Chapters 4 and 5) targets this "telling" surface. Should organic experimentalists not prefer this successive research tool?

[b]For molecular systems that are flexible in the NMR time scale under experimental conditions (e.g., for most open chain systems as well as saturated or partially saturated cycles).

[c]Emphasizing the methodological importance of integrality of configurational and conformational determinations by NMR, the authors of a remarkable monograph on NOE measurements for organic compounds[72] have titled the corresponding section therein "Why Structural and Conformational Problems Are the Same."

by both relative stereochemical configurations of elementary molecular units "embedded" into the chemical backbone (what we call relative stereochemistry) and mutual rotational orientation of all elementary units (what we call rotational isomerism). If we consider a pair of atoms L_i and L_j, which belong to different elementary molecular units A_1L_N and A_2L_N sufficiently distanced in a non-short open chain of bonded atoms, we can find out that rotational flexibility of the chain may bring these atoms so close that their VDW spheres contact regardless of stereochemical configuration of the related units A_1L_N and A_2L_N. The longer the chain, the more obvious this geometrical possibility. Indeed, by differently rotating all units $A_1L_N, \ldots A_iL_N, \ldots, A_ML_N$ relative one to another, the chain may be twisted to a large extent and adopt almost any smooth desired form. Now, let us include units A_1L_N and A_2L_N into a *ring* (A_1 and A_2 are ring atoms separated by two or more bonds). A closed structure of the chain puts more restrictions on the shape of the ring: tolerable values of endocyclic torsion angles lie in narrow intervals. Nevertheless, the remaining rotational freedom for ring bonds may bring substituents L_i and L_j of non-small rings fairly close, irrespective of the relative stereochemistry of units A_1L_N and A_2L_N.

A schematic example (Fig. 98) shows that *trans* substituents L_i and L_j as well as L_j and L_l, which are at a large distance in a flat ring, come in

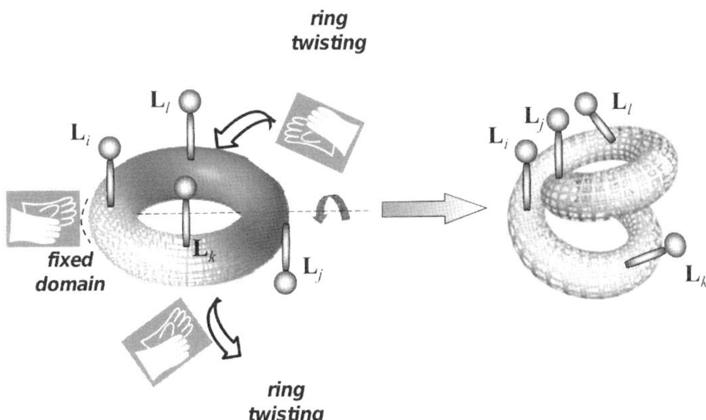

Figure 98. Transformation of an abstract flat ring with substituents L_i, L_j, L_j, and L_l, to a twisted ring with sterically contacting *trans* substituents. This illustration demonstrates that *trans* substituents in cyclic compounds may be spatially proximal if the ring is twisted.

steric contact with one another after drastically twisting the ring. At the same moment, the ring twisting increases the distance between *trans* substituents L_j and L_k. *cis* substituents L_i and L_l become closer to each other, while *cis* substituents L_i and L_k as well as L_k and L_l, diverge in space. Although this abstract example represents the extreme case of ring twisting, which is irrelevant to real non-polymeric molecules, it illustrates the tendency of certainly positioned *trans* substituents to be spatially proximal in twisted rings. It also indicates the tendency of remoteness associated with *cis* substituents, which have certain relative positions in the ring. Thus, we see that the plain view — proximity for *cis* substituents and remoteness for *trans* substituents — is true only when considering planar or flattened rings. Distances between *cis* substituents and distances between *trans* ones are determined both by the shape of the ring and relative positions of the substituents there. Should one notice again that *ring shape* is a synonym of *ring conformation*?

The distances between remote substituents in *cis/trans* stereoisomers of many cyclic compounds do correspond to the primitive view criticized above. However, the opposite situation is not an abstract possibility. For instance, the *trans* bridging in 13-aza-*trans*-bicyclo[9.3.1]pentadecane (**43**, Fig. 99) twists the dodecane ring. Its endocyclic torsion angles are changed so that the 15-positioned protons are proximal to some *trans* protons, in the preferred conformation of this ring. They are appreciably closer to these *trans* protons than to the geminal counterparts of the latter (i.e., to the related *cis* protons).

Smaller cycles, due to geometry constraints, may be bent to a lesser extent. However, this geometrical effect of decreased "*trans* distances" and increased "*cis* distances" may manifest itself even in five- and six-membered rings. For example, in the classical twist conformon of cyclohexane **18a** (Section 3.4), the 1,3-positioned, pseudoaxially oriented protons, (i.e., *trans* substituents; Fig. 99) are somewhat less distanced than *cis* 1,3-protons. Interatomic distances between the 2'- and 4'-protons of the deoxyribose ring of thymidine (**43b**), which adopts the so-called C-2'-*endo*-3'-*exo* conformation,[a] have the same geometrical relations: the *trans* protons are in closer proximity to each other than the *cis* protons.

---

[a] In stereochemical jargon of nucleoside chemistry, the South conformation. It corresponds to the twist geometry with pseudorotation angle $\phi = 270°$ (Fig. 63, Section 3.4). Note that there is no general systematic nomenclature for indicating molecular geometry of conformers; therefore, various "nicknames" are used in describing conformers of different classes of organic compounds (see Fig. 61 for the commonly known names). For instance, the shown rotational orientation of the pyrimidine nucleobase (Fig. 99) is called *syn* (the opposite orientation is *anti*) by specialists in nucleoside, nucleotide, and nucleic acid chemistry.

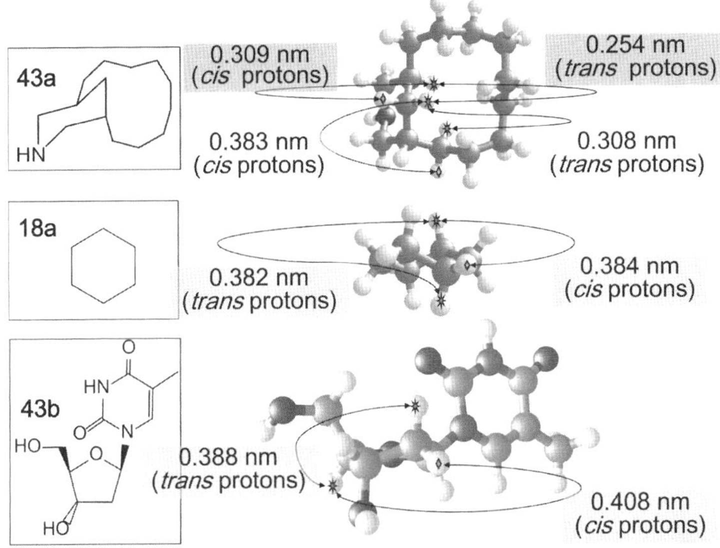

Figure 99. Some conformers of *trans*-bridged bicycle **43a**, nucleoside **43b** and monocycle **18a** as examples of cycles with *trans* protons, which are less distanced than the corresponding *cis* protons. The MM3(96)-optimized 3D structures are shown for the lowest energy conformer of **43a** and twist-shaped conformer of **18a**, while the shown molecular geometry for a low energy conformer of **43b** is supplied by DFT (B3LYP/6-31G+d,p; Section 5.2.6).[9] H· · ·H distances given in the boxes are related either to a selected pair of *trans* protons (arrow-connected, star-labeled) or to the corresponding pair of *cis* protons (arrow-connected, star-rhomb-labeled).

These examples demonstrate that *ring conformation* governs distances between ring substituents regardless of governing them by the relative configuration of the substituents-bearing ring atoms. Therefore, assignments of NOE enhancements according to the "simple" view on interatomic distances may result in incorrect conclusions, i.e., *trans* substituents of a cyclic system may be incorrectly recognized as *cis* if they are proximal, and *vice versa*. Stereochemical dilemma *certain relative stereochemistry or certain conformational preference* cannot be solved by resorting to only constant temperature NMR spectra of cyclic systems that have twisted ring conformations and rapidly change these conformations in a non-anancomeric equilibrium at the temperature of the NMR experiment.

Most stereochemical assignments, which are provided even by routine, synthesis-accompanying NMR experiments, are, nevertheless, correct. Intentionally or not, they are based on verisimilar conformational assumptions, and therefore, the relative stereochemistry deduced from such NMR spectra usually does not require to be re-assigned. Nevertheless, because, in organic laboratories, inadequate conformational hypotheses regarding either molecular structures may seem solely possible, incorrect stereochemical interpretations of NMR spectra appear from time to time in the related periodicals. To manage the most frequently occurring problem with stereochemical determinations — establishment of relative stereo-chemistry in a new organic compound — by means of NMR and within the limits of NMR non-professionals' competence, speculations in relation to conformers should be replaced with a trustworthy conformational hypoth-esis. The source of such hypotheses is, of course, quantitative molecular modeling (Sections 4 and 5) of conformational equilibria.

Nevertheless, by selectively exciting any magnetic nucleus in the molecule, NMR does provide a unique, though "encrypted," information about the molecular structure. Can we not leave this fine experimental tool of structural analysis of molecules and, at the same time, avoid difficulties in interpreting NMR spectra as well as minimize the use of time-consuming DNMR, NOESY, or ROESY experiments, when undertaking conforma-tional analysis? The next Section deals with an emerging methodology of conformational analysis, which is based on non-empirical modeling (the-oretical calculations) of *chemical shifts* and provides self-verification of calculation results by involving these instantly obtainable experimental NMR data.

6.2 Conformational Analysis using CCS

6.2.1 *Case in which experimental chemical shifts of individual conformers are known*

At this point, we return to CCS (Section 6.1.2). As indicated in Sections 3.3 and 6.1.3, conformational transformations slow down as temperature decreases (Eq. 14, Section 3.5.3). At certain low temperatures, the slow-est conformational transformation becomes so sluggish that NMR, with its "unsuccessfully" large time constant τ_{method} (Section 3.3), has suf-ficient time to register NMR signals of individual conformers or NMR signals of the same molecular fragment in distinct conformations. From the NMR perspective, those are rigid ("frozen") molecular backbones or their "frozen" fragments, respectively.[a] Let us assume that it is not easy to extract detailed information regarding 3D structures of such individual conformers by analyzing such a time-resolved NMR spectrum of an organic compound. However, we are situated well: the undertaken low-temperature NMR experiment delivers a few sets of incredible numerical data — experimental chemical shifts of each detected conformer. Non-empirical calculations for modeled conformers offer equivalent sets, sets of CCS, where *each CCS set is explicitly related to a concrete 3D structure*. With these additional data, a direct method for determining 3D structure of conformers of the studied compound (in other words, for identifying its NMR-registered conformers) is obvious. Experimental chemical shifts of an individual conformer should be approximately equal to related CCS from the CCS set of only one certain conformer of optimized molecular geometry; CCS sets for other modeled conformers should not show such a correspondence.[b] The set of experimental chemical shifts of another "frozen" conformer should numerically correspond to the CCS set for another certain modeled conformer. This should hold true for each NMR-registered conformer from the "frozen" conformer mixture. In this manner, each conformer detected by low-temperature NMR is identified through showing numerical equivalence of the set of its experimental chemical shifts and the set of CCS of a certain modeled conformer (i.e., a concrete conformer of known 3D geometry). Of course, this identification implies that spectra of individual conformers are appreciably different in values of chemical shifts; otherwise, one cannot distinguish these conformers by means of any NMR spectra at all.

This CCS-based conformational analysis is methodologically simple. For performing it, operations (*1*)–(*4*) should be sequentially executed as follows:

(*1*) Record an NMR spectrum, e.g., a C,H-decoupled ^{13}C spectrum, at temperatures of a slow conformational exchange (in the NMR timescale). In simple words, take NMR spectra while stepwise decreasing the

---

[a]As we know, only individual conformers, which relative steady concentrations are not below the threshold of the sensitivity of the used NMR spectrometer ($\sim$ 2–3% for serial NMR instruments), are detected.

[b]The temperature impact on chemical shifts of organic compounds in solution is almost always neglected in CCS-based conformational analysis.

temperature of the NMR experiment and register the final (desired) spectrum at temperatures after observing that the further temperature decrease practically does not change resonance frequencies of split signals. As we know, such a NMR spectrum is a superposition of the spectra of individual conformers, for which interconversion is "frozen," and is not an average (Eq. 55a and 55b; Section 6.1.3) of their individual spectra.

(2) By means of MM, locate low-energy stereoisomers (see for conformational search in Section 4.3.2) and reoptimize their geometry using *ab initio* calculations of relevant accuracy (Sections 5.2.3, 5.2.4, and 5.2.6), with adding a solvation model (Section 5.3.4).

Among located stereoisomers, those with ΔE_0 [Section 5.3.4, subsection (*3*)] lying in the range 4–5 kcal mol^{-1} over the global minimum (i.e., are less stable by this value than the located stereoisomer of minimal energy) can be considered as low-energy stereoisomers (here: low-energy conformers). This energy range, at first glance, may seem too wide. As indicated, serial NMR instruments do not register resonance signals of molecular species in low concentrations, e.g., stereoisomers with Gibbs energy that is higher by ~ 2 kcal mol^{-1} than that of the most stable stereoisomer. However, this 4–5 kcal mol^{-1} range is not wide when considering non-empirically modeled molecules in solution. Because of possible inaccuracy in calculations of ΔE_0, the cutoff value of ΔE_0 should be at least two times more than 2 kcal mol^{-1} for selecting low-energy structures that could be NMR detectable.

(*3*) Obtain CCS for each localized stereoisomer by means of *ab initio* calculations (Section 6.1.2).

(*4*) Identify conformers by comparing CCS and experimental chemical shifts for all obtained CCS sets (Section 6.2.2).

In other words, before executing the final step (*4*), we should have M experimental sets of type $\delta_{\exp\_1}, \ldots, \delta_{\exp\_i}, \ldots, \delta_{\exp\_N}$, where N is the number of resolved resonance signals (signals with the peak not overlapped by other signals) for a detected conformer, and M is the number of detected conformers [operation (*1*)]. We should locate conformers of low energy (say, Q ones) and succeed in modeling their accurate molecular geometries in solution [operation (*2*)]. Then, we should obtain CCS sets of type $\delta_{\text{calc}\_1}, \ldots, \delta_{\text{calc}\_i}, \ldots, \delta_{\text{calc}\_N}$ for all Q located conformers [operation (*3*)].

Each CCS set belongs to a certain stereoisomer among Q modeled stereoisomeric structures (typically, $Q > M$). In the final step (*4*), a set

M_i of experimental chemical shifts for an arbitrarily selected "frozen" conformer (e.g., a major one) is compared to each CCS set from Q sets of CCS (Section 6.2.2). Correspondence of this experimental set $\delta_{exp\_1}, \ldots, \delta_{exp\_i}, \ldots, \delta_{exp\_N}$ to only one CCS set $\delta_{calc\_1}, \ldots, \delta_{calc\_i}, \ldots, \delta_{calc\_N}$ indicates successful identification of this NMR-detected conformer. The same correspondence for another pair of the experimental and calculated sets indicates that the next "frozen" conformer has been identified, and so on.[a] Thus, with experimental chemical shifts for *individual* conformers, the general computational scenario of conformational analysis (direct analysis by CCS) is as follows: *conformational search → geometry optimization for located stereoisomeric structures → calculations of chemical shifts → comparison of CCS and experimental chemical shifts*. Current conformational studies by means of CCS follow the outline of this scenario.[26,43,75a−j]

Several commercial multipurpose packages of computational chemistry may provide CCS. These tools of molecular modeling, possessing distinct functionality, are also non-equivalent with respect to modeling of chemical shifts: their CCS-related options vary from package to package. For instance, the option of modeling chemical shifts within a solvation model is not an attribute of most programs; sets of methods for calculating chemical shifts (e.g., GIAO, IGLO, and CSGT) are different in different packages. Gaussian (e.g., versions 03 and 09), Turbomole, Q-Chem (starting from version 3.0) and Jaguar (late versions) computational suites are probably the first choice for modeling chemical shifts because of their broad assortment of computational methods and proven calculation accuracy. These GUI-equipped workhorses of non-empirical molecular modeling provide procedurally simple "one roof" work: both geometry optimization and calculations of chemical shifts. Turbomole and Gaussian are more versatile in CCS-related options than Jaguar. Turbomole, Q-Chem and Jaguar's calculations are appreciably faster than those by Gaussian. A brief, but very useful, characterization of the essential features of these and other calculation packages that deal with non-empirical modeling of NMR spectral parameters is given in Ref. 76. Usually, the best sources of general information about such programs are commercial distributors' websites. This is

---

[a]Very often, a two-position conformational exchange, i.e., interconversion of only two conformers, is the subject of temperature-variable NMR experiments.

also the case for a freely distributed program, CFOUR, developed by academic experts (J. F. Stanton, University of Texas at Austin, and J. Gauss, Universität Mainz; http://www.cfour.de/).

6.2.2 *Comparison of experimental chemical shifts and CCS*

The reader who has perused Chapters 4 and 5 does not need further explanation of computational steps (*2*) and (*3*). However, step (*4*) merits separate consideration. As pointed out in the previous Section, in performing each comparison, a CCS set $\delta_{calc\_1}, \ldots, \delta_{calc\_i}, \ldots, \delta_{calc\_N}$ is weighted against a set of experimental $\delta_{exp\_1}, \ldots, \delta_{exp\_i}, \ldots, \delta_{exp\_N}$ values. That is, an arranged numerical set is compared with another arranged numerical set. This provides an admirable advantage to conformational analysis by CCS. A closeness of two values, e.g., a QM-calculated value ΔE^0 or MM-calculated ΔE_{ster} to experimental value $\Delta G°$, may be accidental. Comparison of several arranged values reveals contrastive discrepancy. From the basics of the probability theory, we know that the probability of simultaneous occurrence of n independent non-deterministic events is the product of n probabilities for separate occurrence of each individual event. If accidental, tight numerical closeness of each number from a N-membered set of specified numbers to the corresponding (equally specified) number from another arranged set of the same cardinality has the probability to occur which barely differs from zero, since this event is simultaneous occurrence of N independent non-deterministic events. Hence, if obtained, such a mutual one-to-one correspondence of two sets of numbers $\delta_{exp\_1}, \ldots, \delta_{exp\_i}, \ldots, \delta_{exp\_N}$ and $\delta_{calc\_1}, \ldots, \delta_{calc\_i}, \ldots, \delta_{calc\_N}$ is not accidental, i.e., the two sets *should* mutually correspond due to some reason. The reason is absolutely clear: the sets describe the same conformer. Thus, a positive result of a comparison of the experimental chemical shifts of an individual conformer and CCS for a conformer of certain 3D geometry (tight numerical closeness of δ_{exp_i} and δ_{calc_i} for each i-th pair) reliably identifies the NMR-detected conformer.

Probably, the best way to compare Q alternative *sets* of CCS with a set of experimental chemical shifts δ_{exp_N} and $\delta_{calc\_1}, \ldots, \delta_{calc\_i}, \ldots, \delta_{calc\_N}$ is to use a simple statistical tool, bivariate regression analysis (usually called regression analysis). In mathematics, bivariate regression is a function that approximately relates two random variables (here: experimental values $\delta_{\exp\_i}$ and calculated values δ_{calc_i}); the keyword *random* indicates that these variables have a stochastic (i.e., non-deterministic) relationship. Regression analysis is a technique that "measures" the association between a random independent variable X (in terms of mathematical statistics, an explanatory variable; in our case, a set of experimental chemical shifts) and a random dependent variable Y (a response variable; in our case, a set of CCS). Often, it is assumed that the relationship between X and Y is linear, referred to as linear regression analysis. In order to correlate these two data sets, the *linear* regression approximates the unknown stochastic relationship between X and Y as *linear* function $\tilde{Y}(X)$, which is defined as

$$\tilde{Y} = aX + b \tag{56}$$

Equation 56 is derived to find parameters a and b, which provide the smallest possible sum $\sum(\tilde{Y} - Y)^2$ (in statistics, the acronym for this sum is *SSE*, i.e., the sum of the squared errors). For x, y coordinates, this equation specifies a straight regression line (also called the best fit or trend line, Fig. 100). This line is a convenient graphical illustration of linear regression. In non-statistical terms, regression line is a straight line in the x, y plane that provides the best closeness to it for a finite set of any concrete points scattered in the plane. It is not difficult to figure out what *the best closeness* is for one point; however, what does this qualitative meaning say for a set of points? Let us consider *the best closeness (the best fit)* in more detail.

Any Y value from the Y set may be represented as $Y = aX + b + e_{err}$ and $e_{err} = Y - \tilde{Y}$, where e_{err} (called fitting error or residual, Fig. 100) is the deviation of a Y value relative to the perfect linear correlation $Y = aX + b$, which occurs when $\tilde{Y} = Y$ (i.e., $e_{err} = 0$). According to the linear regression definition, the sum of the squares of N residuals (N values of e_{err}), i.e., *SSE*, is minimal for the regression line compared to any other straight line. Residuals e_{err} are different for different values of X. That is, linear regression maximally decreases residuals *on an average* by using the least square criterion applied to all N values $e_{err-1}, e_{err-2}, \ldots, e_{err-N}$. Therefore, for linearly associated random variables X and Y, predicted values $\tilde{Y}_i$ can be considered as "improved" values of variable Y, where each Y_i value is "corrected" by all other values $Y_1, \ldots, Y_i, \ldots, Y_{N-1}$.

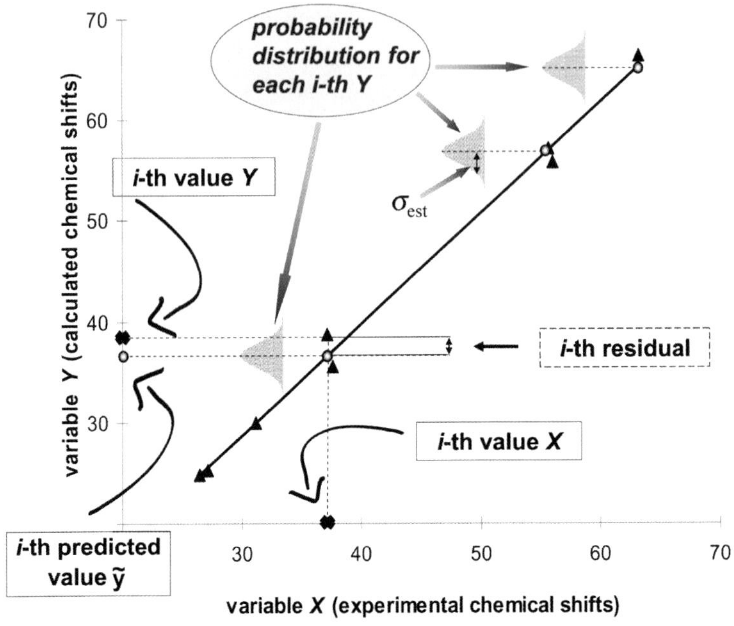

Figure 100. Graphical representation of linear association of random variables X and Y (linear regression). Scattered points (shown as triangles) correspond to pairs of stochastically related values X_i, Y_i. The regression line lies between scattered points so that the distances along the Y axis (residuals e_{err}) between these points and the line are minimal on the average. Graphically, the standard error σ_{est} is the distance between the symmetry axis and the inflection point in the probability distribution (here: Gaussian distribution) plot.

A random character of variable Y means that, for a value adopted by variable X, Y may adopt a concrete value with some probability.[a] That is, values, which variable Y may adopt at any point in the x axis, may be distributed along the y axis according to some statistical law. In linear regression, distribution of probability for Y obeys the Gaussian (normal) distribution law (schematically shown in Fig. 100 as a bell-shaped curve); it is clear that the probability distribution for random variable Y is the same for each X value, and exact $\tilde{Y}$ values (values of Y that correspond

---

[a] Interestingly, in our case of analyzing chemical shifts, X is not an "ordinary" random variable that adopts values unknown *a priori*; see below.

to the regression line) correspond to the maximal probability. These pieces of basic knowledge in statistics are important to us. Normal distribution contains numerical parameter σ (standard deviation) that, in each concrete case, shows how narrow/broad the distribution is. Thus, we are able to estimate how closely variables X and Y are associated in either case. That is, by calculating this parameter (of course, by means of a computer program for statistical analysis[a]), we may select the "best" CCS set when comparing experimental chemical shifts and CCS (see below).

Why should CCS and experimental chemical shifts be considered as random values that are stochastically related? One can claim that there is no deterministic relationship between them. With a set of δ_{exp_i}, it is impossible to predict exact values δ_{calc_i} and *vice versa*. On the other hand, we do not assume modeled chemical shifts adopt any arbitrary values. We expect that each value δ_{calc_i} for the "correct" molecular structure is approximately close to the corresponding given value δ_{exp_i} because our *ab initio* calculations *model* this structure. There are some factors (e.g., the factor of general structural similarity and some local dissimilarities of the true and modeled molecular geometries) which "force" δ_{calc_i} values to deviate from fixed δ_{exp_i} values, but do not allow large deviations. No exact (deterministic) law can predict the magnitude of these deviations. This indicates that δ_{calc_i} and δ_{exp_i} are stochastically related.

Furthermore, it can be asked why we consider chemical shifts given by experiments (i.e., values that are assumed to be known *a priori*) as a random variable. At first sight, it seems that a set of experimental values $\delta_{exp\_1}, \ldots, \delta_{exp\_i}, \ldots, \delta_{exp\_N}$ should not be considered as a random variable. Nevertheless, we are able to use regression analysis when comparing sets $\delta_{exp\_1}, \ldots, \delta_{exp\_i}, \ldots, \delta_{exp\_N}$ and $\delta_{calc\_1}, \ldots, \delta_{calc\_i}, \ldots, \delta_{calc\_N}$, since there is no deterministic relationship between values δ_{exp_i} and δ_{calc_i} for each i-th pair. In dealing with δ_{exp_i} and δ_{calc_i} values, regression analysis does not consider the general case of both "normal" random variables X and Y. It deals with a particular case, where the response variable (δ_{calc_i} values for a molecular structure) adopts random values while the random explanatory variable X(δ_{exp_i} values for this structure) has already adopted certain values, i.e., X values are exactly known.

How do we identify a NMR-detected conformer by applying regression analysis to Q sets of obtained CCS, where each set is related to a certain computationally located (modeled) stereoisomer? In other words, what do

---

[a]Computer programs of different complexity, which are oriented on mathematical statistics or include it [e.g., XLSTAT (a multifunctional statistical software supplement to MS Excel), SIMSTAT, Statistica Base, MathLab, Mathematica], quickly construct this equation for given sets of numbers X and Y, i.e., deliver the related values a and b (Eq. 56). In addition, such programs supply estimates σ_{est}, r, and r^2 of the quality for the obtained linear regression (see below).

we intend to reveal by conducting linear regression analysis of Q sets of type $\delta_{calc\_1}, \ldots, \delta_{calc\_i}, \ldots, \delta_{calc\_N}$ obtained for alternative stereoisomeric structures *vs.* a set of $\delta_{exp\_1}, \ldots, \delta_{exp\_i}, \ldots, \delta_{exp\_N}$ related in the experimental NMR spectrum to a "frozen" conformer? In the hypothetical case of an ideal molecular model and absolutely accurate *ab initio* calculations, the corresponding values for the experimental set and the set calculated for this NMR-detected stereoisomeric structure should be equal, i.e., $X = Y$. Then, the regression of Y (i.e., the δ_{calc_i} set) from X (i.e., the δ_{exp_i} set) is perfectly linear, with coefficients $a = 1$ and $b = 0$ for the regression line. Therefore, it is reasonable to assume that among all *ab initio* derived Q sets of CCS (i.e., Q sets of type $\delta_{calc\_1}, \ldots, \delta_{calc\_i}, \ldots, \delta_{calc\_N}$), the set, which is related to the "proper" conformer, demonstrates a near-linear correlation with the set of experimental chemical shifts $\delta_{exp\_1}, \ldots, \delta_{exp\_i}, \ldots, \delta_{exp\_N}$. CCS sets of "improper" modeled stereoisomers have no correlation with experimental NMR data, and the linear regression model should show a lower linearity of statistical correlation. Thus, the "proper" CCS set should be identified by the best linear regression (see below for parameters of the linear regression quality).[a] By performing *linear* regression analysis for the δ_{exp_i} set *vs.* alternative δ_{calc_i} sets, we actually answer the question "How good is the *linear* regression model for each set of CCS?"

In applying this statistical approach in step *(4)*, we do not analyze the quality of linear approximation of regression in CCS analysis (e.g., we do not try to examine another, non-linear, regression for obtaining better results), since, by deciding to use linear regression, we actually presuppose that the statistical relationship between the experimental chemical shifts and CCS of the corresponding ("proper") stereoisomer is linear. Our goal is to estimate the *relative* quality of correlations of Q sets of CCS and a set of $\delta_{exp\_1}, \ldots, \delta_{exp\_i}, \ldots, \delta_{exp\_N}$ values *within the limits of the linear regression model*. Thus, by comparing CCS and experimental chemical shifts via linear regression analysis, we intend to reveal the CCS set with the best linear regression among linear regressions obtained for all alternative sets of CCS. The "best" CCS set indicates the stereoisomeric

---

[a]The best linearity is associated with the closest (on avarage) similarity of the set of CCS of the "proper" stereoisomeric structure and the set of experimental values. For each CCS set, this closeness is "measured" by the parameters of the corresponding linear regression (see below).

structure, which corresponds to the individual conformer with resonance signals δ_{exp_1}, δ_{exp_i}, ..., δ_{exp_N} in the NMR spectrum of the "frozen" mixture of conformers.

Nevertheless, it is important to know that, in strict sense, there is a nonlinear stochastic relationship between an entire set of N values of CCS and the corresponding set of N values of δ_{exp_i} even for the "correct" molecular structure. Different inaccuracies such as in geometry, charge distribution, and solvation, are associated with chemically distinct fragments of *modeled* molecular systems. For instance, some molecular fragments (e.g., aromatic, H-donating, charged) of organic compounds interact with molecules of many solvents (e.g., aromatic, chelating, highly polar non-ionic, and ionic solvents) more tightly than others, i.e., they have thermodynamically and kinetically more stable association with solvent molecules than other molecular fragments. Implicit models of solvation do not reproduce these effects. Translating to the language of basic statistics, we say that there are some factors that affect only some CCS from the CCS set of the entire molecular structure, and one cannot consider all values for random variable Y (all values δ_{calc_1}, ..., δ_{calc_i}, ..., δ_{calc_N} from a CCS set) as possessing the same stochastic relationship (here: linear) with variable X. Linear regression is not a perfect statistical model if applied to CCS of chemically heterogeneous molecules.

If this approximation loses its validity for either studied chemical system (statistical association between these sets of δ_{exp_1}, ..., δ_{exp_i}, ..., δ_{exp_N} and δ_{calc_1}, ..., δ_{calc_i}, ..., δ_{calc_N} is far from linear), organic researchers face a difficult problem — finding a non-linear regression model that is applicable to a case-specific stochastic relationship. Clearly, for solving this problem, considerable skills in mathematical statistics are required. A much simpler and, nevertheless, effective solution is "to split" the structure (formally, a set of enumerated atoms) into atom subsets in such a way (e.g., a subset of aliphatic carbons and a subset of aromatic carbons) that the mutually corresponding values δ_{calc_i} and δ_{exp_i} for each atom subset could be successfully correlated linearly.

How is regression analysis performed in practice? There are many statistical and mathematical computer programs that, among other options, provide linear regression analysis; some of them are indicated in the first footnote in this Section. In fact, "minimally equipped" software packages that may use methods of mathematical statistics in treating numerical data are sufficient for performing this simple statistical analysis. For instance,

MS Excel, version 2003, is such a tool. However, later versions of this ubiquitously used office program should be augmented with a separate, complementary computational module for statistics, e.g., an advanced, versatile XLSTAT (Addinsoft, http://www.xlstat.com/en/) or a simple XLstatistics (Deakin University, http://www.deakin.edu.au/~rodneyc/XLStatistics/).

The first technical procedure in step (*4*) is to apply linear regression to each set $(\delta_{calc\_1}, \ldots, \delta_{calc\_i}, \ldots, \delta_{calc\_N})_A, \ldots, (\delta_{calc\_1}, \ldots, \delta_{calc\_i}, \ldots, \delta_{calc\_N})_K, \ldots, (\delta_{calc\_1}, \ldots, \delta_{calc\_i}, \ldots, \delta_{calc\_N})_Q$ and the same experimental set $\delta_{exp\_1}, \ldots, \delta_{exp\_i}, \ldots, \delta_{exp\_N}$, i.e., obtain a set of Q different regression lines (their parameters a and b are different). Regression analysis of each CCS set $\delta_{calc\_1}, \ldots, \delta_{calc\_i}, \ldots, \delta_{calc\_N}$ *vs.* the experimental δ_{exp_i} set provides a set of chemical shifts $\delta_{pred\_1}, \ldots, \delta_{pred\_i} \ldots, \delta_{pred\_N}$ (predicted or scaled chemical shifts). In graphical representation (Fig. 100), values $\delta_{pred\_1}, \ldots, \delta_{pred\_i} \ldots, \delta_{pred\_N}$ are associated with the regression line, pairs of values $\delta_{exp\_i}, \delta_{calc\_i}$ are coordinates of initial points that predetermine the regression line, and each δ_{pred_i} value corresponds to a certain δ_{exp_i} value. As a set, the δ_{pred_i} values are *on an average* close to the set of the δ_{exp_i} values. One may understand scaled chemical shifts $\delta_{pred\_1}, \ldots, \delta_{pred\_i} \ldots, \delta_{pred\_N}$ for the parent set of CCS values $\delta_{calc\_1}, \ldots, \delta_{calc\_i}, \ldots, \delta_{calc\_N}$ as statistically improved, "more accurate" values of these CCS.

There is some inconsistency regarding the use of linear regression in treating CCS. For instance, δ_{exp_i} values are sometimes used as the explanatory variable X (this is correct) and sometimes as the response variable Y (this is not quite correct). It is difficult to accept experimental chemical shifts as a variable that stochastically depends on the corresponding CCS values obtained for the related stereoisomeric structure (then CCS are the variable X, which has adopted certain values). Nevertheless, classification of δ_{calc_i} as the variable X and δ_{exp_i} as the variable Y offers an attractive advantage. In this case, predicted chemical shifts δ_{pred_i} are systematic error invariant (*systematic error*: the same magnitude, method-specific deviation present in numerical results obtained by either theoretical or experimental method).[a] Due to this advantage, this "antisense" permutation of variables

---

[a] Indeed, linear regression ignores any number C if added to variable X (here: to each value δ_{calc_i}), i.e., it provides the same predicted values $\delta_{pred\_1}, \ldots, \delta_{pred\_i}, \ldots, \delta_{pred\_N}$ (graphically, the same regression line) for both sets $\delta_{calc\_1}, \ldots, \delta_{calc\_i}, \ldots, \delta_{calc\_N}$ and $\delta_{calc\_1} + C, \ldots, \delta_{calc\_i} + C, \ldots, \delta_{calc\_N} + C$ irrespective of the magnitude of C. Obviously, values $\delta_{calc\_1}, \ldots, \delta_{calc\_i}, \ldots, \delta_{calc\_N}$ calculated for each located steroisomeric structure may be expressed as $\delta_{corr\_1} + S, \ldots, \delta_{corr\_i} + S, \ldots, \delta_{corr\_N} + S$, where S is the systematic error, and δ_{corr_i} is the unknown value of δ_{calc_i} that does not include the systematic error. Thus, treating δ_{calc_i} and δ_{exp_i} in the manner indicated above, systematic error S does not affect values of predicted chemical shifts $\delta_{pred\_1}, \ldots, \delta_{pred\_i} \ldots, \delta_{pred\_N}$.

X and Y is practical in identifying conformers by CCS; one should only remember that the parameters of the quality of such "reversed" linear regression, which values we use for identifying the NMR-"displayed" conformer (see below), have no real sense.

The next procedure reveals which CCS set among Q obtained ones delivers the maximal averaged closeness of its values δ_{pred_i} to experimental chemical shifts $\delta_{exp\_1}, \ldots, \delta_{exp\_i}, \ldots, \delta_{exp\_N}$ of a "frozen" conformer. In other words, one should select the "best" regression line among Q ones. We need to estimate the extent of regression linearity (the quality of linear regression) for each CCS set. A convenient estimate of the quality of linear regression is the *standard error of the estimate* $\sigma_{est} = [SSE/(N-2)]^{1/2}$ (also, though not quite systematically, called standard error of regression; N is the number of observations; here, the number of CCS $\delta_{calc\_1}, \ldots, \delta_{calc\_i}, \ldots, \delta_{calc\_N}$ from one set subjected to linear regression analysis).[a] In treating CCS by regression analysis (Fig. 100), N residuals are N differences $\delta_{calc\_i} - \delta_{pred\_i}$ (further depicted as $\Delta\delta_{calc\_i-pred\_i}$), and thus, the corresponding SSE is $\sum(\Delta\delta_{calc\_i-pred\_i})^2$. Then, re-writing the expression for σ_{est}, we have $\sigma_{est} = [\sum(\Delta\delta_{calc\_i-pred\_i})^2/(N-2)]^{1/2}$.

Estimate σ_{est} characterizes the spread of the probability distribution for CCS (Fig. 100); values of σ_{est} are equal for each value δ_{calc_i} from a set of CCS obtained for a stereoisomer. Low values of σ_{est} indicate a narrow distribution of probability for a CCS value δ_{calc_i} (a narrow "bell" of the Gaussian distribution), and high σ_{est} values are associated with a broad probability distribution for this value (a broad "bell" of the Gaussian distribution). Thus, the lowest value of σ_{est} among Q such values obtained for Q sets of CCS characterizes the "best" regression line among Q such lines obtained in the regression analysis of these Q sets of CCS *vs.* a set of experimental chemical shifts (from the NMR spectrum of the "frozen" conformer mixture) belonging to one selected (e.g., major) conformer. That is, this lowest value of σ_{est}, being associated with a concrete molecular geometry (a concrete modeled conformer), indicates the molecular 3D structure of this (e.g., major) experimentally detected conformer, i.e., identifies it. Note that σ_{est} is an expedite estimate because it displays the averaged closeness of Y (δ_{calc_i} values) and $\tilde{Y}$ (δ_{pred_i} values) in familiar ppm units.[b] Importantly,

---

[a] For the general case of n random variables, standard error of regression is expressed as $\sigma_{est} = [SSE/(N-n)]^{1/2}$.

[b] Small differences in these values for alternative CCS sets (say, less than 0.3 ppm for σ_{est}) cannot be accepted as indicative for selecting the "correct" structure. The magnitude of non-linearity effects (not

standard error of regression σ_{est} is not a quantitative estimate of the accuracy of CCS values. That is *MAD* (or the related *RMS*; see below), and one should not present the obtained value of σ_{est} as the value of the accuracy of CCS for either molecular structure. The value of σ_{est} *indirectly* indicates the accuracy of CCS (but never displays its value) only if there is a linear stochastic relationship between experimental chemical shifts $\delta_{exp\_1}, \ldots, \delta_{exp\_i}, \ldots, \delta_{exp\_N}$ and CCS $\delta_{calc\_1}, \ldots, \delta_{calc\_i}, \ldots, \delta_{calc\_N}$ for the related molecular structure.

Estimates of the linear regression quality are frequently supplied with non-systematic names. It is, therefore, useful to systemize here our knowledge of this terminology.

The simplest estimate of the linear regression quality is *mean absolute error (MAE)*. While σ_{est} is a root mean square (*RMS*)-based estimate of residuals $\Delta\delta_{calc\_i-pred\_i}$, *MAE* is the arithmetic mean of their absolute values, i.e., $1/N \times \sum |\Delta\delta_{calc\_i-pred\_i}|$, where N is the number of CCS $\delta_{calc\_1}, \ldots, \delta_{calc\_i}, \ldots, \delta_{calc\_N}$ in the set. *MAE* shows the average distance (in ppm) for a set of CCS values from the regression line.

A ubiquitously used estimate is *RMS error (RMSE)* or *RMS deviation (RMSD)*; in mathematical language, *standard deviation* (σ_{strd}). This common estimate is defined as $\sigma_{strd} = [\sum(y\_i - A_{mean})^2/N]^{1/2}$, where A_{mean} is some numerical mean. This σ_{strd} technique may be applied to statistically estimating three different quantities present in linear regression; then, one can speak about three different estimates of the σ_{strd} type. Let us glance at them very briefly. The first σ_{strd} is the above explained σ_{est}. That is, residuals (numerical differences) $\Delta\delta_{calc\_i-pred\_i}$ are in the focus.[a] The second considered σ_{strd} is associated with another sum, $\sum(\delta_{pred\_i} - \delta_{calc\_mean})^2$ [sometimes abbreviated as *SSR* (*Sum of Square Regression*)], where the arithmetic mean δ_{calc_mean} mentioned in the previous paragraph is the "reference" value. Rewriting for *SSR*, we have this next *RMSE*, $\sigma_{strd} = [\sum(\delta_{pred\_i} - \delta_{calc\_mean})^2/N]^{1/2}$. It indicates how tightly (or roomy) scaled chemical shifts $\delta_{pred\_1}, \ldots, \delta_{pred\_i}, \ldots, \delta_{pred\_N}$ group around the mean value for CCS. The third standard deviation considered in regression analysis is associated with the sum *SST* (*Total Sum of Squares*). The sum is used as a measure of total variation of random variable Y: $SST \equiv SSE + SSR$. By considering *SST* as a sum that figures in the general expression for σ_{strd} (see above), we have another σ_{strd}. Let us denote it as $RMSD_{CCS}$; then, $RMSD_{CCS} = [\sum(\delta_{calc\_i} - \delta_{calc\_mean})/N]^{1/2}$. That is, we "redirect" the standard deviation technique to estimate to which extent CCS values δ_{calc_i} disperse from the value of their arithmetic mean.

---

perfectly linear stochastic relationship between δ_{calc_i} and δ_{exp_i}) for the "correct" set may be larger than these small differences.

[a] As indicated *supra*, $SSE = \sum(\Delta\delta_{calc\_i-pred\_i})^2$, according to the general expression $SSE = \sum(\tilde{Y} - Y)^2$. This *SSE* shows why both names *standard error of the estimate* and *standard error of regression* used for σ_{est} are not quite successful. *Standard deviation* (σ_{strd}) implies the presense of a numerical mean A_{mean} in summed differences $\sum(y_i - A_{mean})$.

The latter standard deviation does not deal with predicted chemical shifts $\delta_{pred\_1}, \ldots, \delta_{pred\_i}, \ldots, \delta_{pred\_N}$, and, thus, ignores linear regression analysis at all. All other estimates (including *MAE*) reflect in ppm units how accurately the "improved" values of CCS — predicted chemical shifts, i.e., $\tilde{Y}$ values — represent experimental chemical shifts, assuming a linear correlation between X and Y. Which of them should one use for selecting the "best fit" set of CCS? Obviously, the required estimate is standard error of regression σ_{est}. It measures the spread of the probability distribution for each i-th value δ_{calc_i} (Fig. 100) that belong to the examined CCS set. This distribution with the δ_{pred_i}-centered mean is equal for all values $\delta_{calc\_1}, \ldots, \delta_{calc\_i}, \ldots, \delta_{calc\_N}$ of the same set. The related value of σ_{est} appears a characteristic of the CCS set as a whole and, therefore, it is a measure (as we know, the lower, the better) of the set quality in reproducing the set of experimental chemical shifts of a "frozen" conformer. Thus, exactly standard error of the estimate σ_{est} judges alternative CCS sets in their competition for being related to this conformer.

Other often employed parameters are dimensionless coefficients r^2 (*determination coefficient*, $0 \le r^2 \le 1; r^2 = SSR/SST$) and r (*Pearson coefficient*, $-1 \le r \le 1$). In bivariate linear regression, these coefficients are trivially related, i.e., $r = (r^2)^{1/2}$. The more convenient estimate r^2 shows the degree of linear association of variables X and Y. This degree can be represented as a percentage if r^2 is multiplied by 100. For instance, $r^2 = 0.49$ (the corresponding r value is 0.7) means that 49% of the variation in Y is because of linearity in the stochastic relationship between X and Y. Coefficient r is similar to r^2 in a qualitative sense: the higher its absolute value, the stronger the linear association.

Nonetheless, coefficients r or r^2 may mislead in identifying the "correct" conformer by analyzing CCS. Values of CCS for alternative sets are often very close, e.g., 0.997685 and 0.997543 for r. A "better" CCS set can hardly be reliably selected using these values: values of r may be non-descriptive owing to some factors [e.g., an undue impact on value r of large positive or negative values (outliers) adopted by variable Y]. Also, it is unknown whether the linear regression model is *equally* suitable for analyzing CCS of any stereoisomers of a chemical compound. Therefore, the obtained small differences in r are a dubious criterion for discriminating either set of CCS.

Frequently, estimates, which consider experimental chemical shifts δ_{exp_i} and CCS δ_{calc_i} as *independent* variables X and Y, are used. From this perspective, δ_{exp_i} and δ_{calc_i} are independent arranged sets of N members each, and an averaged numerical closeness between the related members (i.e., numbers) from these sets is the characteristic of the set similarity. This simplified consideration is insufficient in conformational studies via CCS: it ignores an additional source of accuracy, information "hidden" in inaccuracies. Deprived of statistical pre-treatment, values of such estimates incorporate absolutely cryptic random errors, which spring from *ab initio* calculations of chemical shieldings σ_{calc_i} for both the molecular structure in question and the reference molecule, as well as from the non-ideality of the molecular model itself. Regression analysis to some extent neutralizes these errors by scaling CCS (i.e., supplying "corrected" values $\delta_{pred\_1}, \ldots, \delta_{pred\_i}, \ldots, \delta_{pred\_N}$).

An example of such an estimate is *mean absolute difference* (*MAD*, also called *mean absolute deviation*). In our comparison of calculated and experimental chemical shifts δ_{calc_i}

and $\delta_{\exp\_i}$, *MAD* is the arithmetic mean of absolute differences $|\delta_{\exp\_i} - \delta_{\text{calc}\_i}|$, i.e., *MAD* $= 1/N \times \sum |\delta_{\exp\_i} - \delta_{\text{calc}\_i}|$.[a] *MAD* is actually what we mean when saying *averaged accuracy* of CCS, or, in short, accuracy of CCS.

Despite its obvious clarity, this non-statistical estimation may implicitly lead to incorrect conclusions. For instance, one or two outliers, i.e., largely deviated values of differences $|\delta_{\exp\_i} - \delta_{\text{calc}\_i}|$, significantly increase the *MAD* when estimating the similarity of two 10–15-membered sets $\delta_{\exp\_1}, \ldots, \delta_{\exp\_i}, \ldots, \delta_{\exp\_N}$ and $\delta_{\text{calc}\_1}, \ldots, \delta_{\text{calc}\_i}, \ldots, \delta_{\text{calc}\_N}$ for ^{13}C resonance signals by means of this technique. The obtained large *MAD* does not certainly mean that the two compared sets are "dissimilar." If, by discarding the outliers, *MAD* becomes small, the 3D molecular backbone, to which the CCS set belongs, is rather related to the sought-for conformer. From the statistical viewpoint, for a set with sufficiently large N, very few deviations are rather accidental. In fact, they appear because of "unpredictable" factors, which are ignored in the used molecular model, e.g., specific local solvation (e.g., a strong H-bonding for an OH group) or molecular fragments that require a higher level of *ab initio* modeling (e.g., a heavy atom-bearing one). The negative impact of these factors on the value of *MAD* may be dismissed by using a better molecular model or increasing the level of theory in non-empirical calculations, respectively. If our guess is correct, such a *MAD* disturbance will disappear for the re-modeled stereoisomer molecule.

A related, regression-ignoring estimate is the *RMS* of N differences $\delta_{\exp\_i} - \delta_{\text{calc}\_i}$, which is the quadratic mean of CCS inaccuracies $\delta_{\exp\_i} - \delta_{\text{calc}\_i}$ for individual atoms in the molecule. This *RMS* is calculated as $[\sum(\delta_{\exp\_i} - \delta_{\text{calc}\_i})^2/N]^{1/2}$. This quadratic mean should not be confused with another quadratic mean, standard deviation of linear regression σ_{est} explained above. Numerically, *RMS* $\geq$ *MAD*. Probably, *RMS* is used only because many researchers prefer to operate with differences $\delta_{\exp\_i} - \delta_{\text{calc}\_i}$ and not with absolute differences $|\delta_{\exp\_i} - \delta_{\text{calc}\_i}|$, which are used in deriving *MAD*. If we apply the standard deviation technique (see above for definition of σ_{strd}) to the considered absolute differences $|\delta_{\exp\_i} - \delta_{\text{calc}\_i}|$, we obtain estimates of the next *RMSD*, which may be denoted as $RMSD_{error}$; $RMSD_{error} = [\sum[|\delta_{\exp\_i} - \delta_{\text{calc}\_i}| - MAD]^2/N]^{1/2}$. Values of $RMSD_{error}$ are usually small, and they mislead some chemical researchers to claim that obtained CCS are high accuracy values. One should learn that either value of standard deviation $RMSD_{error}$ does not reflect the accuracy of related CCS. It shows how dense individual deviations $|\delta_{\exp\_i} - \delta_{\text{calc}\_i}|$ lie around the *MAD* value.

In addition to standard error of the estimate σ_{est} used for discovering the CCS set related to the "proper" stereoisomer, one can use the "statistically improved" *MAD*; it may be designated MAD_{pred}. This estimate is defined as arithmetic mean $MAD_{\text{pred}} = \sum |(\delta_{\text{pred}\_i} - \delta_{\exp\_i})|/N$. Assuming the linear regression model most suitable for CCS of exactly the "proper" stereoisomer, the related predicted chemical shifts $\delta_{\text{pred}\_1}, \ldots, \delta_{\text{pred}\_i}, \ldots, \delta_{\text{pred}\_N}$ may be considered as statistically (i.e., on average) refined CCS values. Under these circumstances, absolute differences of type $|(\delta_{\text{pred}\_i} - \delta_{\exp\_i})|$ are

---

[a] Sometimes, *MAD* is incorrectly called *MAE*, mean absolute error.

indicative (Section 6.3.1). The corresponding MAD_{pred} obviously should have a minimal value for the set of predicted chemical shifts of the "true" stereoisomeric structure relative to such MAD_{pred} values for the other, "improper" modeled stereoisomers.

Estimates of linearity in linear regression analysis, i.e., σ_{est}, r and r^2, have a remarkable beneficial feature. The linearity in correlating two numerical sets X and Y does not change, if the same arbitrary number is added to each value from the X or Y set. Consequently, σ_{est}, r and r^2 keep their values. For instance, let us consider two apparently different sets of CCS. In the first set, each value δ_{calc_i} is near the corresponding experimental value δ_{exp_i}. In the second set, all CCS are larger, say, by 1000 ppm each, than the corresponding chemical shifts δ_{calc_i} from the first set. If we compare each set of CCS with the same set of experimental chemical shifts with using linear regression analysis, the estimates of this quality (σ_{est}, r, and r^2 values) remain unchanged. We have only shifted the regression line along the Y-axis by value 1000 ppm relative to the regression line for the first δ_{calc_i} set.

This feature of linear regression estimates well benefits to our "conformer-targeting" analysis of chemical shift sets. *Ab initio* calculated chemical shieldings σ_{calc_i} (the primary magnetic molecular quantity supplied by non-empirical calculations, Section 6.1.1) incorporate *systematic* error, which results from both *ab initio* calculations of chemical shieldings and imperfectness of the modeled molecule itself (e.g., somewhat shortened atom-atom distances for chemically bonded atoms in a molecule modeled with using a DFT method). Of course, we have no idea on the magnitude of this error for a concrete chemical structure. However, as explained in the previous paragraph, even if the systematic error is large, it does not affect σ_{est} and other estimates of the relative quality of linear regression for alternative sets of CCS of this molecular system. That is, with linear regression analysis of CCS, structure identification by means of statistical estimates σ_{est}, r and r^2 is not tainted by the systematic error of any magnitude.

Consequently, the linear regression-mediated comparison of experimental chemical shifts with calculated chemical *shieldings* and that for the same experimental chemical shifts with calculated chemical *shifts* (i.e., CCS) are equivalent. Values of chemical shieldings σ_{calc_i} and CCS δ_{calc_i} differ only by constant value σ_{calc_ref} (Eq. 54, Section 6.1.2). Values of statistical estimates σ_{est}, r, and r^2 of the quality of linear regression are therefore the same, when correlating sets of values σ_{calc_i} *vs.* δ_{exp_i} and δ_{calc_i} *vs.* δ_{exp_i}. Nevertheless, the use of CCS in computational conformational analysis is probably preferable despite the additional arithmetic manipulation required to derive them (Eq. 54). It is convenient to operate with familiar chemical shifts rather than chemical shieldings, which are not "telling numbers" for synthetic chemists.

The value of the i-th individual chemical shielding σ_{calc_i} also contains its own non-systematic (random) error. The presence of this numerical error of different magnitude in each i-th value of chemical shielding σ_{calc_i} is actually the reason for the appearance of deviations of different magnitude of calculated values $\sigma_{calc\_1}, \ldots, \sigma_{calc\_i}, \ldots, \sigma_{calc\_N}$ from the corresponding genuine, but unknown, values $\sigma_1, \ldots, \sigma_i, \ldots, \sigma_N$. Non-systematic errors appear owing to local inaccuracies of modeled geometry, distinct accuracy of the same basic set functions in modeling light *vs.* heavy atoms, impossibility of precisely reproducing the non-uniform solvation shell, etc. This non-equivalence in modeling individual atoms in a molecule indicates that association of experimental and calculated values even in the best-case scenario may be only near linear. "Improving" CCS values to their scaled values $\delta_{pred\_1}, \ldots, \delta_{pred\_i}, \ldots, \delta_{pred\_N}$ by linear regression, we, to some extent, on average diminish the inaccuracy that is due to atom-specific random errors.

Linear regression does not maximally neutralize random errors in values of CCS $\delta_{calc\_1}, \ldots, \delta_{calc\_i}, \ldots, \delta_{calc\_N}$ of the "proper" conformer relative to CCS of other conformers. As mentioned, regression analysis may only, to some extent, diminish inaccuracy effects by supplying such new, "better" values (i.e., predicted chemical shifts $\delta_{pred\_1}, \ldots, \delta_{pred\_i}, \ldots, \delta_{pred\_N}$), according to the least square criterion for residuals. This simple mathematical manipulation with residuals is not an algorithm for annulling or minimizing random errors; such an algorithm probably does not exist. That is, statistical estimates σ_{est}, and *MAE* in no way display either systematic or random errors "hidden" in values δ_{calc_i} and δ_{pred_i}. For instance, the same level *ab initio* calculations produce the same systematic and random errors in modeling chemical shifts both of the "proper" and "improper" stereoisomers modeled for vacuum conditions. However, the mentioned statistical estimates would be certainly dissimilar for the sets of CCS of these two stereoisomers when subjecting these sets to regression analysis *vs.* the same set of experimental chemical shifts.

Surprisingly, CCS values $\delta_{calc\_1}, \ldots, \delta_{calc\_i}, \ldots, \delta_{calc\_N}$ as well as their scaled values $\delta_{pred\_1}, \ldots, \delta_{pred\_i}, \ldots, \delta_{pred\_N}$ may be free from the systematic error of non-empirical modeling. Simple methodological requirements should be fulfilled to provide cancellation of this error. The first is that calculated chemical shieldings σ_{calc_i} and σ_{calc_ref} should be supplied by the same theoretical approach to the calibration invariance problem (Section 6.1.2), e.g., by GIAO, as well as by the same method of *ab initio* calculations, e.g., by MP2/6-31+G(d,p), for both the examined molecular system and the reference molecule, e.g., TMS. The next requirement is that molecular modeling of both molecules should be of the same quality. That is, the same theoretical method, the same basis set, numerically the same convergence criteria, as well as the same structural model (e.g., the same approximation for solvation) should be used in optimizing geometries of both molecular systems. The calculation method and basis set used for calculating chemical shieldings for a molecular structure may differ from those used in optimizing the molecular geometry for it, without deteriorating cancelation of the systematic error for CCS values. The use of a solvation model when optimizing the geometry together with the use of a vacuum condition approximation or a lower quality solvation model when calculating chemical shieldings is possible, but rather undesirable.

These requirements adjust the method-and-model-specific systematic error to be the same in values of calculated chemical shieldings $\sigma_{calc\_1}, \ldots, \sigma_{calc\_i}, \ldots, \sigma_{calc\_N}$ of the molecule of interest and the value of σ_{calc_ref} of the reference molecule. The point in canceling the systematic error in CCS values is that CCS are relative values (Eq. 54; Section 6.1.2). The constant (systematic) error is annulled by deriving CCS via arithmetic subtraction with keeping both the above requirements. Thus, the resulting CCS (let us call them consistent CCS, or cCCS) do not include the systematic error of the calculation method and molecule approximation, while values of chemical shieldings $\sigma_{calc\_1}, \sigma_{calc\_2}, \ldots, \sigma_{calc\_i}, \ldots, \sigma_{calc\_N}$ and σ_{calc_ref} contain it.

These methodological precautions are important for obtaining accurate values of CCS since solvent-induced shifts of NMR signals may be significant for many organic compounds, even not functionalized. For instance, the chemical shift of the ^{13}C atom in TMS varies in the range of 4 ppm, if measured in different solvents.[74] It would be inconsiderate to derive ^{13}C chemical shifts (Eq. 54; Section 6.1.2) for an organic compound dissolved in some solvent, e.g., $CHCl_3$, by using that value of the ^{13}C chemical shielding of TMS, which is calculated with ignoring the solvent (in this example, $CHCl_3$) used in measurements of ^{13}C resonance frequencies for the studied compound. In contrast, by both modeling these molecules "enveloped" by a suitable model of the solvent and calculating ^{13}C chemical shieldings for the resulted molecular structures in such an "envelope," we free Eq. 54 from operating with numerical data that may be inconsistent.

Regrettably, direct conformational analysis by using CCS described in this Section is far from being a universal methodology. As we know, fast conformational interconversions at room temperature are inherent in most organic compounds. Conformational transformations for some molecular fragments [e.g., internal rotation in sterically unhindered C–C fragments, concerted rotation around ring bonds in five-membered saturated cycles (pseudorotation; Fig. 63)] essentially slow down only at very low temperatures. NMR experiments for solutions of compounds with such molecular fragments become very problematic. In the first instance, it is difficult to select the solvent for NMR experiments that should deal with solutions of polar compounds at very low temperatures (see subsection *Solvation* in Section 6.3.2). In addition, temperature-variable NMR experiments are rarely conducted in organic laboratories; as a rule, an assistance of a specialist in NMR is required.[a] Could conformational analysis be performed by applying almost always available CCS to familiar, routine room temperature NMR spectra that "expose" virtual, time-averaged molecular structures? A principal route to this dream of conformational analysis is described in the next Section.

6.2.3 *Case in which experimental chemical shifts of individual conformers are unknown*

Formally, isotropic chemical shifts are numbers. Let us try to utilize the potential of exact relations hidden in their numerical nature. With increasing the quality of non-empirical modeling (Section 6.3.2), cCCS values $\delta_{calc\_1}, \ldots, \delta_{calc\_i}, \ldots, \delta_{calc\_N}$ for a stereoisomer tend to approach the corresponding experimental values $\delta_{exp\_1}, \delta_{exp\_i}, \ldots, \delta_{exp\_N}$. One can discern that this tendency, to some extent, equalizes the calculated and experimental values: highly accurate cCCS $\delta_{calc\_1}, \ldots, \delta_{calc\_i}, \ldots, \delta_{calc\_N}$ may replace the related values $\delta_{exp\_1}, \delta_{exp\_i}, \ldots, \delta_{exp\_N}$ in any equation that includes these experimental values. This idea leads us to the commonly known Eq. 55a (Section 6.1.3) which relates the experimental

---

[a]For an experienced NMR researcher, such ^{13}C NMR experiments usually take 7–8 hours for a series of 2–4 compounds.

in-time-averaged chemical shift and chemical shifts of "NMR-invisible," rapidly interconverting conformers. As we remember, this equation is useless when trying to analytically extract values of the conformer fractions p_A, p_B, .., p_K, . . . , p_M from NMR spectra registered at temperatures of fast conformational exchange. With additional related numerical data, i.e., cCCS, can we use this linear polynomial equation for analyzing at least simple conformational equilibria, e.g., interconversions of two or three conformers?

In this algebraic problem, our starting data are a routine 1D NMR spectrum (e.g., a ^{13}C NMR spectrum) recorded at conditions of fast conformational equilibrium and Q computationally located stereoisomeric structures with N corresponding values of cCCS each. By using Eq. 55a and cCCS, we intend to identify M conformers, which essentially contribute to the spectrum, and determine yet unknown fractions p_A, p_B, .., p_K, . . . , p_M of these conformers.

The next point in our consideration is as follows. Suppose that considered cCCS are absolutely accurate values i.e., values of ^{13}C cCCS obtained for a conformer are one-to-one correspondingly *equal* to measured (or measurable) values of its ^{13}C chemical shifts. Then, Eq. 55a may be rewritten for an i-th experimental chemical shift (the l.h.s.) of any conformer among located (modeled) conformers $A, \ldots, K, \ldots, Q$ by replacing chemical shifts $(\delta_{exp\_i})_1, \ldots, (\delta_{exp\_i})_K, \ldots, (\delta_{exp\_i})_M$ in the r.h.s. with the corresponding calculated values $(\delta_{calc\_i})_K, \ldots, (\delta_{calc\_i})_K, \ldots, (\delta_{calc\_i})_M$, and, in the rewritten form, we have N equations:

$$\delta_{exp\_1} = p_A(\delta_{calc\_1})_A + p_B(\delta_{calc\_1})_B \cdots + p_Q(\delta_{calc\_1})_Q \tag{57_1}$$

$$\cdots$$

$$\delta_{exp\_i} = p_A(\delta_{calc\_i})_A + p_B(\delta_{calc\_i})_B \cdots + p_Q(\delta_{calc\_i})_Q \tag{57_i}$$

$$\cdots$$

$$\delta_{exp\_N} = p_A(\delta_{calc\_N})_A + p_B(\delta_{calc\_N})_B \cdots + p_Q(\delta_{calc\_N})_Q \tag{57_N}$$

where N indicates the number of resolved ^{13}C resonance signals and Q is the number of located conformers (in other words, the number of the related sets of ^{13}C cCCS). cCCS of high energy conformers may not be included in equations of type Eq. 57_i. These equations consider all *modeled* conformers; obviously, those of these modeled stereoisomeric structures, which correspond to high energy conformers, have the related coefficients $p_i = 0$ there. Under such high energy *modeled* stereoisomeric structures,[a] we understand those, which are energetically higher (by our calculations) more than by 4–5 kcal mol^{-1} from the energy calculated for the located minimal energy conformer. By this auxiliary, cCCS-independent criterion, we reject some located conformers from further consideration and consider F remaining modeled conformers (F sets of cCCS; $F < Q$) instead of Q ones. This means that F terms, instead of Q ones, figure in the r.h.s. of Eqs. 57_i.

In contrast to equations of type Eq. 55a, simplified Eqs. 57_i have less unknowns: there are only F unknowns $p_A, .., p_K, \ldots, p_F$, since theoretical calculations supply values of δ_{calc_i}. With these numerical values in hand, these consistent linear equations are soluble relative to F unknowns $p_A, ., p_K, \ldots, p_P$, if the number of the equations is not less than the number of the unknowns, i.e., $N \geq F$. However, the situation with these equations and unknowns is quite opposite. Number F of located low energy conformers (in the sense explained above) of organic molecule of non-minimal size, as a rule, exceeds 8–10. For Eqs. 57_i considering ^{13}C chemical shifts, number N usually may be 1 or 2; sometimes, it may be 3 or, more rarely, 4 (higher N values are a very rare case to apply). The reason is that almost each magnetic nucleus shows almost equal resonance frequencies when appearing in different conformations of an organic molecular backbone (see below for meaning *characteristic nuclei*). Equations of type 57_i, each with equal coefficients $(\delta_{\text{calc}\_i})_A = (\delta_{\text{calc}\_i})_B = \ldots = (\delta_{\text{calc}\_i})_F$, degenerate.[b]

Hence, Eqs. 57_i seem useless. Let us exclude from consideration all *modeled* structures from F ones that are irrelevant for describing the conformational equilibrium reflected by NMR spectra. Then, forgetting for a moment that M modeled conformers of low energy that do contribute to

---

[a]We reasonably assume that, when locating Q stereoisomers of low and high energies for an organic molecule, we well reproduce molecular geometries of its real conformers, but estimate their relative thermodynamic stability with a lower accuracy. Therefore, our methodological prerequisite is that relative *calculated* energy of modeled stereoisomeric structures is not a criterion for arranging, according to the relative stability, conformers that are characterized by the relative energy values lying in this 4–5 kcal mol^{-1} interval.

[b]Approximately, into the trivial equation $\sum p_K = 1$.

the equilibrium have not been identified yet, we may write for them:

$$\delta_{\text{exp\_1}} = p_A(\delta_{\text{calc\_1}})_A + p_B(\delta_{\text{calc\_1}})_B \cdots + p_M(\delta_{\text{calc\_1}})_M \qquad (58_1)$$

$$\cdots$$

$$\delta_{\text{exp\_}i} = p_A(\delta_{\text{calc\_}i})_A + p_B(\delta_{\text{calc\_}i})_B \cdots + p_M(\delta_{\text{calc\_}i})_M \qquad (58_i)$$

$$\cdots$$

$$\delta_{\text{exp\_}N} = p_A(\delta_{\text{calc\_}N})_A + p_B(\delta_{\text{calc\_}N})_B \cdots + p_M(\delta_{\text{calc\_}N})_M \qquad (58_N)$$

Equations 58_i seem more successful. They do not contain too many unknowns: those only are M unknowns $p_A, .., p_K, \ldots, p_M$ ($M < F$ since F modeled structures also include those which reproduce conformers possessing insufficiently low relative energy to manifest themselves in NMR spectra). Nevertheless, the problem of too many unknowns has been replaced with another one. We have F sets of ^{13}C cCCS and do not know which among these F sets correspond to M conformers $A, B, \ldots, M$ that contribute to the NMR spectrum, the source of the considered experimental values of chemical shifts $\delta_{\text{exp\_1}}, \delta_{\text{exp\_}i}, \ldots, \delta_{\text{exp\_}N}$. That is, we still cannot write Eqs. 58_i with coefficients $\delta_{\text{calc\_}i}$ as concrete numbers and, thus, cannot solve these simple equations numerically.

In addition, we meet familiar restrictions in constructing these consistent equations. The same carbon atom in stereoisomers of an organic compound may have practically the same ^{13}C resonance frequencies; this situation takes place for most of carbons in organic molecules. As explained, CCS values for these carbons are irrelevant for solving equations of type Eqs. 58_i. Some carbons in an organic molecule, e.g. the i-th and the j-th carbons, may have essentially different chemical shifts $(\delta_{\text{exp\_}i})_A$ and $(\delta_{\text{exp\_}i})_B$ (the i-th carbon), and $(\delta_{\text{exp\_}j})_A$ and $(\delta_{\text{exp\_}j})_B$ (the j-th carbon) at least for two lowest energy conformers A and B. Only CCS of conformers of such nuclei (here: carbons) $(\delta_{\text{calc\_}i})_A$ and $(\delta_{\text{calc\_}i})_B$ (the i-th carbon), and $(\delta_{\text{calc\_}j})_A$ and $(\delta_{\text{calc\_}j})_B$ (the j-th carbon) provide to the related Eqs. 58_i and 57_j the applicability to solving them in respect to unknowns p_A and p_B (fractions of conformers A and B; see below). Often, an organic molecule has one or a few such carbons. Let us call them *characteristic* carbons. Because number of characteristic nuclei (Z) is much less than number N of all magnetic nuclei of the molecule ($Z < N$), the condition for consistent Eqs. 58_i to be solved becomes harder to satisfy. *The number of characteristic nuclei (characteristic carbons and characteristic nuclei of other magnetic isotopes) in a molecule should be not less than the number of its spectrally observable (NMR-detectable) conformers, i.e., Z > M (instead of N > M), to enable identifying conformers of this molecular structure by means of their CCS.*

Most frequently, there are not more than two, three or four characteristic carbons in an organic molecule (i.e., $Z = 2$, or 3, or 4). This means that only 2–4–position conformational equilibria may be analyzed by means of the considered approach since one can write only two, or three, or four consistent Eqs. 58_i for ^{13}C carbon resonances which may be solved with respect to conformer populations p_A and p_B; p_A, p_B and p_C; or p_A, p_B, p_C and p_D, respectively. This scenario, nevertheless, is not bad: there are from two to four conformers in typical NMR-analyzable conformational equilibria of non-polymeric as well as non-macrocyclic organic molecules.

Thus, what is the solution for finding M numerical unknowns $p_A, .., p_K ..., p_M$ by means of Eqs. 58_i? Let us assume that we deal with the most widespread case of conformational equilibria — two interconverting conformers A and B (i.e., $M = 2$) — and suppose for simplicity that there are three characteristic carbons ($Z = 3$) in the examined molecule. Then, the equations are as follows:

$$\delta_{\text{exp\_1}} = p_A(\delta_{\text{calc\_1}})_A + p_B(\delta_{\text{calc\_1}})_B \tag{58_{1-3}}$$

$$\delta_{\text{exp\_2}} = p_A(\delta_{\text{calc\_2}})_A + p_B(\delta_{\text{calc\_2}})_B \tag{58_{2-3}}$$

$$\delta_{\text{exp\_3}} = p_A(\delta_{\text{calc\_3}})_A + p_B(\delta_{\text{calc\_3}})_B \tag{58_{3-3}}$$

Recall that we have F sets of ^{13}C cCCS and do not know which two from them. ($F > M$; here, $M = 2$) figure in Eq. 58_{1-3}, Eq. 58_{2-3}, and Eq. 58_{3-3}. Involving all these $A, \ldots, E, G, \ldots, F$ sets, one can form (*1*) $F!/2(F-2)!$ pairs of cCCS values $(\delta_{\text{calc\_1}})_E$, $(\delta_{\text{calc\_1}})_G$, where index *1* depicts characteristic carbon *1*, and a pair of indexes E, G indicate any pair of non-identical cCCS sets from all such possible pairs from the pool of cCCS sets $A, \ldots, E, G, \ldots, F$ (pairs formally related to Eqs. 58_{1-3}), and (*2*) $F!/2(F-2)!$ pairs of cCCS values $(\delta_{\text{calc\_2}})_E$, $(\delta_{\text{calc\_2}})_G$, where index *2* is related to characteristic carbon 2, and indexes E, G are as they are in item (*1*) (pairs formally related to Eq. 58_{2-3}). Let us substitute each numerical pair $(\delta_{\text{calc}\_1})_E$, $(\delta_{\text{calc}\_1})_G$ into Eq. 58_{1-3} and each numerical pair $(\delta_{\text{calc}\_2})_E$, $(\delta_{\text{calc}\_2})_G$ into the corresponding Eq. 58_{2-3}. That is, by substituting these values in related Eq. 58_{1-3} and Eq. 58_{2-3}, we transform these formal mathematical expressions into linear algebraic equations with numerical coefficients $(\delta_{\text{calc}\_i})_E$ and $(\delta_{\text{calc}\_i})_G$. Each two obtained consistent equations may be numerically solved with respect to two unknowns p_A and p_B; recall that the l.h.s. of each such an equation is also a known value (experimental chemical shift δ_{exp_i}).

Thus, there are $F!/2(F-2)!$ equations of type Eq. 58_{1-3} with certain coefficients in the r.h.s and the same number of equations of type Eq. 58_{2-3}; the unknowns in both equations are p_A and p_B. By numerically solving each two equally E, G-indexed consistent equations (one of type Eq. 58_{1-3} and the other of type Eq. 58_{2-3}) from this equation pool, we obtain $F!/2(F-2)!$ numerical solutions, i.e., $F!/2(F-2)!$ different pairs of values of p_A and p_B. Of course, only one of these formal solutions (one pair of obtained values p_A and p_B and two cCCS sets) is a true solution. It includes both a pair of "correct" values (let us call them p_{A_corr} and p_{B_corr}) that are related to real conformers A and B (registered in the considered NMR spectrum as a virtual molecular structure with time-averaged values δ_{exp_1}, δ_{exp_i}, ..., δ_{exp_N} of experimental chemical shifts), and two sets of cCCS associated with values p_{A_corr} and p_{B_corr}. Other solutions are, in fact, fictive: they cannot be assigned to the conformational equilibrium of *the same* real conformers A and B. Let us rename any pair of such values p_A and p_B to p_{A_incorr} and p_{B_incorr}. How can one distinguish true values p_{A_corr} and p_{B_corr} and fictive values p_{A_incorr} and p_{B_incorr}; i.e., identify the "correct" solution among $F!/2(F-2)!$ solutions obtained?

For this purpose, the third equation, Eq. 58_{3-3} (we may call it the adjusting equation), is involved into consideration. Equations 58_{1-3}, 58_{2-3}, and 58_{3-3} all imply conformers, which have a non-negligible impact on values of experimental chemical shifts δ_{exp_1}, δ_{exp_i}, ..., δ_{exp_N}. That is, they all together are not valid for "improper" conformers. Therefore, the "correct" values p_{A_corr} and p_{B_corr}, when being associated with cCCS values from two "correct" sets of cCCS, lead Eq. 58_{3-3} to numerically hold, while fictive solutions p_{A_incorr} and p_{B_incorr} that have no relationship to conformers A and B disturb this numerical equality.[a]

This reasoning prompts us to apply the following methodology. One should substitute into this formal equation Eq. 58_{3-3} all $F!/2(F-2)!$ pairs of obtained numerical solutions p_A and p_B (see the previous paragraph). The result is $F!/2(F-2)!$ expressions of form $\delta_{exp\_i} = p_A(\delta_{calc\_3})_E + p_B(\delta_{calc\_3})_G$, where all symbols p_A, p_B, $(\delta_{calc\_3})_E$ and $(\delta_{calc\_3})_G$ depict exact numbers. Numerical equality in these arithmetic expressions should be kept for only one pair of values p_A and p_B, while this numerical expression does not appear to hold true (the l.h.s. $\neq$ the r.h.s.) for other pairs p_A and p_B. Hence, the first pair is the "correct" values p_{A_corr} and p_{B_corr}, and all others are fictive solutions p_{A_incorr} and p_{B_incorr}. Performed selection of correct solution p_{A_corr} and p_{B_corr} also means identification of two "proper" cCCS sets: we remember that each pair of numerical solutions p_A and p_B is associated with two own concrete cCCS sets E and G. Thus, the two identified "proper" cCCS sets show what the NMR-spectrum contributing conformers A and B are, and the related values p_{A_corr} and p_{B_corr} indicate fractions of these conformers in the "NMR-photographed" fast equilibrium.

There is a test for correctness in selecting this concrete pair of values $p_{A\_corr} + p_{B\_corr}$ from the pool of all pairs p_A and p_B. It is provided by the more than once mentioned condition $\sum p_K = 1$ that may considered as an additional equation consistent with Eqs. 58_i. By using this obvious equality, one should find for our selection that both numerical expressions $p_{A\_corr} + p_{B\_corr} = 1$, and p_{B_incorr}, $p_{A\_incorr} + p_{B\_incorr} \neq 1$ hold.

The above consideration in general explains the methodology of identification of two rapidly interconverting conformers by using a routine experimental ^{13}C spectrum and a set of cCCS for individual alternative stereoisomeric structures located by means of sufficiently accurate non-empirical calculations. There are some additional methodological remarks related to analyzing most often occurring conformational equilibria, i.e. those with $M = 2$ and 3.

---

[a] Recall that we assume that cCCS $(\delta_{calc\_1})_K \cdots, (\delta_{calc\_i})_K, \ldots, (\delta_{calc\_N})_K$, are perfectly accurate values for each K-th conformer. Section 6.3 explains how actual imperfect CCS values may be used.

Note that, before starting to construct and numerically solve the above considered equations, the number of characteristic carbons, i.e., Z, should be estimated to reveal whether the necessary condition $Z > M$ is satisfied for the molecular system with the *assumed* number M of NMR-weighty conformers. In other words, one should understand whether we can construct the sufficient number of non-degenerated linear Equations of type 58_i to solve them with obtaining values of $p_A, ..., p_K, ..., p_M$. A trivial comparison of cCCS of each carbon for all modeled low-energy stereoisomers of the molecule provides the number Z of its characteristic carbons (for such a carbon, $\Delta\delta_{calc\_i} > 1$ ppm for any two modeled stereoisomeric structure of low energy).

If $M = Z = 2$, trivial condition $p_A + p_B = 1$ may serve as the adjusting equation itself. However, it is undesirable to use this "trick." When using Eqs. 58_i, values of $p_1..., p_K..., p_M$ are actually derived with a low accuracy because of imperfectness of cCCS values supplied by non-empirical calculations (Section 6.3). Numerical differences between proper and improper solutions p_{A_corr}; p_{B_corr} and p_{A_incorr}; p_{B_incorr} may be less than the accuracy interval for cCCS, and the adjusting equation loses its resolving power.

For $Z = 4$ and $M = 3$, equations of type Eq. 58_i have three terms in the r.h.s. We should similarly obtain and test all $F!/6(F-3)!$ triads of numerical solutions p_A, p_B, and p_C. Obviously, an appropriate computer program could quickly perform the required arithmetic calculations.

Equations of type Eq. (58_i) may be written for experimental chemical shifts and cCCS of other nuclei, which have spin 1/2 and a broad interval of resonance frequencies when included into different chemical structures (e.g., ^{19}F, ^{31}P, ^{29}Si, and ^{77}Se). For molecular organic systems that contain both carbons and such nuclei, number Z of characteristic nuclei may appear larger than 3 or 4. This CCS-based conformational analysis may then be freed from numerically adjusting solutions $p_A, p_B, ..., p_M$ (see above), and "correct" solutions $p_{A\_corr}, p_{B\_corr}, ..., p_{M\_corr}$ may be directly found by solving seemingly useless consistent Eqs. 57_i, if the condition $Z \geq F$ is fulfilled.

The idea of being able to identify conformers and determine their fractions from routine NMR spectra is extremely attractive: the attentive reader probably has conjectured that CCS of non-large molecular systems are not a difficult goal to reach. Unfortunately, one cannot smoothly realize this idea. There are three major difficulties with this unexamined methodology: (*1*) a low value of Z for many organic compounds; (*2*) the absence of preliminary information regarding the number of interconverting "not hidden partners" (i.e., number M in Eqs. 58_{i-3} is *a priori* unknown); and (*3*) insufficient accuracy of cCCS values for many compounds to successfully perform conformational analysis of them. Let us review these obstacles briefly.

(*1*) Satisfying the condition $Z > M$ is necessary for the use of this methodology of conformational analysis by cCCS. Unfortunately, too often, stereoisomers have only one characteristic carbon, or (rarely) even none.

(*2*) We do not know *a priori* how many conformers of an unexplored compound essentially contribute to its NMR spectra (Section 6.1.3). Only in some cases, e.g., for N-inversion in 2-azaadamantane (a structure with one degree of conformational freedom), one can claim that only two conformers are in the equilibrium. When studying conformational equilibria, organic researchers usually consider the compound in question as an analog of a parent system of known conformational behavior (Section 3.4). Furthermore, for the parent compound, the actual multi-position conformational equilibrium is usually reduced to a two-position conformational exchange. For instance, saturated six-membered rings are most often discussed in terms of interconversion of two chairs with equatorial and axial substituents, while twist forms (Fig. 53, Section 2.8) are neglected as conformers of much higher energy. However, as stressed in previous Chapters, structural analogies in conformational analysis may be misleading.

To determine the number and molecular structure of conformers of a small organic molecule by using routine NMR spectra and cCCS, the trial and error approach may be used. The starting approximation hypothesizes a two-position equilibrium ($M = 2$) for such an unexplored molecule. If examination of this hypothesis is successful, i.e., one pair of found numerical solutions p_A and p_B satisfies the adjusting equation while other pairs lead the equality the l.h.s = the r.h.s. to fail, this approximation is adequate to describe the equilibrium reflected in "averaged" NMR spectra. If none satisfies the adjusting equation, one should examine the hypothesis of a three-position conformational equilibrium. However, if no set of solutions p_A, p_B, and p_C, is successful, it is unclear whether there are more conformers in the solution studied by NMR, or the accuracy of the obtained cCCS values is too low.

Some researchers analyze "averaged" NMR spectra by means of equations of type Eq. 55a with CCS of modeled conformers there, and the Boltzmann equation (Section 2.4, Eq. 1). There, conformer fractions p_A, ..., p_K, ..., p_M are supplied by the Boltzmann equation. For delivering these values, this equation requires the relative energy ΔE_{K-L} for the related conformers K and L (more precisely, the free energy difference) to be known. At this point, this methodology actually loses its validity. Values of ΔE_{K-L} in these studies are determined by either MM or QM calculations. As we know (Section 6.1), such calculated energy values may include inaccuracy that devaluates conformational analysis when using them. Thus, the verisimility of obtained values p_A, ..., p_K, ..., p_M is illusive; the use of them in conjunction with CCS is rather meaningless.

(*3*) Up to this point, we considered Eqs. 58_i (e.g., including Eq. 58_{i-3}) as exact equations, similar to Eq. 55a. Because cCCS are calculated with some inaccuracy, these equations cannot be numerically exact. For simplicity, let us return to Eqs. 58_{1-3}, 58_{2-3}, and 58_{3-3} discussed for the two-position conformational equilibrium. As explained, the r.h.s of the adjusting equation Eq. 58_{3-3} adopts different values, when substituting $F!/2(F-2)!$ sets of numerical solutions p_A, p_B, ..., p_M of consistent Eqs. 58_{1-3} and 58_{2-3} into this third equation in turn; numerical equality the l.h.s. = the r.h.s of this equation indicates the "correct" values of p_A and p_B (i.e., values p_{A_corr} and p_{B_corr}) as well as "correct" conformers themselves. However, exact numerical equality of any measured value δ_{exp_i} and the corresponding modeled chemical shift δ_{calc_i} scarcely can take place, owing to the considerable inaccuracy of cCCS values (Section 6.3.1). Therefore, one should "equip" the l.h.s (i.e., number δ_{exp_3}) of the adjusting equation with certain numerical limits $\pm P$, which indicate whether the calculated value of the r.h.s is "equal" or "unequal" to the l.h.s. "Falling" of the calculated value of the r.h.s. of the adjusting equation within limits $\delta_{exp\_3} \pm P$ indicates that a certain numerical solutions p_A and p_B has passed the test. That is, this pair of numbers is what we depict p_{A_corr} and p_{B_corr}, and the related modeled stereoisomers are NMR-weighty conformers A and B. Despite this positive result (identification of "proper" conformers), some inaccuracy in their fractions p_{A_corr} and p_{B_corr} is still present because cCCS, which we substituted in Eq. 58_{3-3}, are not accurate values.

How can these important numerical limits $\pm P$ be rationally chosen? Obviously, we cannot say *a priori* whether limits, e.g., ± 0.5 or ± 1 ppm, are sufficient for correctly identifying conformers as well as their fractions in the equilibrium. Accuracy of cCCS values is a chief prerequisite, and this issue is discussed in the next Section. Till we understand which cCCS accuracy justifies the use of adjusting equations, e.g., Eq. 58_{3-3}, we will be unable to "see" conformers through the prism of CCS.

6.3 Accuracy in CCS Calculations

6.3.1 *Requirements for CCS accuracy in conformational analysis*

In Section 6.2.3, we have formally singled out some nuclei from the molecular framework and called them characteristic nuclei. According to this qualitative understanding, each characteristic nucleus has *essentially* distinct chemical shifts in experimental spectra of individual diastereoisomeric structures. That is, if the molecular system has even a very few characteristic nuclei, i.e., NMR spectra of alternative stereoisomeric structures A and H have different resonance frequencies for even a few their nuclei, NMR can distinguish between A and H. However, qualitative criteria are useless in our quantitative analysis. Such an experimentally evident difference is insufficient to start structural identification with CCS. It is clear that if cCCS of A and H are very close, non-empirical calculations may not distinguish between A and H by cCCS of their characteristic nuclei owing to some inaccuracy of any cCCS values. If inaccuracy is large, we cannot even indicate which cCCS value is related to the higher/lower field signal. It is clear that the "resolving power" of CCS-based methodologies in structural studies is predetermined by both the absolute difference $\Delta\delta_{chrct}$ between experimental chemical shifts of a nucleus for stereoisomeric structures A and H ($\Delta\delta_{chrct} = |\delta_A - \delta_H|$; a structure-specific, unmanageable factor) and the accuracy of CCS (a manageable factor). Thus, the absolute difference $\Delta\delta_{chrct}$ should be numerically defined before considering this nucleus as a characteristic one.

We have two quantitative factors to address the question of what should be the minimal accuracy of two CCS values that permits recognizing two corresponding resonance signals with chemical shifts δ_{exp_i} and δ_{exp_j} as separate signals (to "resolve" them by only using CCS). For answering this question, we accept for simplicity that averaged inaccuracy of a CCS set, *MAD* (Section 6.2.2), represents individual inaccuracies for these two CCS values. Then the answer is as follows. The absolute difference between these experimental chemical shifts $|\Delta\delta_{exp}|$ ($|\Delta\delta_{exp}| = \delta_{exp\_j} - \delta_{exp\_i}$) should be larger than twice the accuracy of CCS values, i.e., $|\Delta\delta_{exp}| > 2 \times MAD$ (the condition for reproducing two separate experimental signals by CCS). For chemical shifts of characteristic nuclei, this condition is written as $\Delta\delta_{chrct} > 2 \times MAD$. Let us call this condition the first requirement for CCS accuracy in conformational analysis, or, in short, the first requirement for CCS accuracy.

In CCS projection, two experimental values δ_{exp_i} and δ_{exp_j} may be considered as $\delta_{exp\_i} \pm MAD$ and $\delta_{exp\_j} \pm MAD$, where *MAD* is the CCS inaccuracy equal for these two experimental NMR signals. Then, the absolute difference $|\Delta\delta_{exp}|$ of chemical shifts δ_{exp_i} and δ_{exp_j} may be written as $|\Delta\delta_{exp}| \pm 2 \times MAD$. *MAD* is doubled because the absolute error of the algebraic sum of inaccurate values, which are characterized by their own absolute errors, is the sum of these errors. Therefore, the "resolution condition" for CCS is $|\Delta\delta_{exp}| > 2 \times MAD$, as indicated above.

Now, when we can see the "signal resolution condition" from the general perspective of CCS, it is worth glancing at the magnitude of $\Delta\delta_{chrct}$ for carbon atoms in organic compounds. For characteristic carbons of organic compounds in solution, absolute $\Delta\delta_{chrct}$ values most often are less than 3 ppm, and typical $\Delta\delta_{chrct}$ values lie between 1 and 2 ppm. Local 3D surroundings of stereochemically mobile fragments should be very dissimilar in stereoisomeric structures to provide characteristic carbon values $\Delta\delta_{chrct} > 5$ ppm, e.g., in alkaloid **44**[75j,77a] (Fig. 101) or amine **15**[26] (Fig. 22). Therefore, such large values for difference $\Delta\delta_{chrct}$ are relatively rarely discovered.

What can we infer about the manageable factor (the CCS accuracy) in this "signal resolution condition"; that is, what is the magnitude of *MAD* values? Normally, these values are large in cost-effective CCS calculations (i.e., calculations that do not use solvation modeling or MP2 or higher theory calculation methods), e.g., from 2 to 8 ppm.[68e,69b,77b–d] For instance, *MAD*

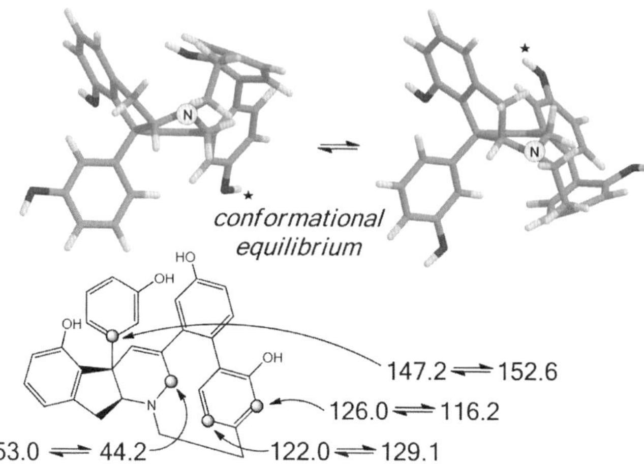

Figure 101. *Ab initio* derived geometries[75j] of two NMR-detected conformers of alkaloid **44**.[76] Star (*) is a formal label for the same OH group in these structures. Numbers show experimental chemical shifts (ppm) for characteristic ^{13}C atoms (indicated as gray spheres) for both conformers.

is 3–4 ppm for carbons of various methylated 1,3-dioxanes,[75f] 2.0 and 2.2 ppm for carbons of *cis*-1,4-di-*tert*-butylcyclohexane of twist and chair geometries, respectively,[43] and 5.3–5.4 ppm for carbons of two predominated conformers of dibenzocyclooctene.[75b]

The next step in our discussion of CCS, NMR and conformational analysis puts us in the difficult position of realizing that we try to exploit an unviable idea. Comparison of the typical values of $\Delta\delta_{chrct}$ and *MAD* shows that conformational analysis by CCS of carbons is impossible for a vast majority of organic compounds. According to the condition for reproducing two separate experimental signals by CCS, *ab initio* modeled chemical shifts are rather unable to "resolve" resonance signals of stereoisomers: the typical empirical $\Delta\delta_{chrct}$ values appear to be disappointingly small to be reliably reproduced by CCS.

Fortunately, the situation is not deadlocked. A simple solution is provided by regression analysis of CCS (Section 6.2.2). As explained, the "proper" conformer is identified exclusively according to the *relative* quality of linear regression for alternative CCS sets. A fruitful assumption in this selection is that CCS of the "proper" conformer should be linearly correlated with the set of experimental chemical shifts while CCS of other, i.e., "improper," 3D structures should lack such a correlation. Statistical estimates σ_{est} and r^2 reveal the linearity of the stochastic relationship between δ_{exp_i} and δ_{calc_i} (Section 6.2.2); however, they do not display random inaccuracies in δ_{calc_i} values, which result from *ab initio* calculations as well as from the molecular model used. In this connection, other regression-provided values, i.e., predicted chemical shifts δ_{pred_i} (values $\tilde{Y}$), are remarkably important. Statistical data treatment, assuming the presence of information in inaccuracies, suggests these new, improved values of CCS, predicted chemical shifts δ_{exp_i}. Generated by statistical treatment of CCS inaccuracies, δ_{pred_i} values should be on an average closer to their δ_{exp_i} values than CCS. Indeed, this CCS improvement does not appear to be illusory. For instance, absolute differences $|\Delta\delta_{pred\_i-exp\_i}|$ between δ_{pred_i} and δ_{exp_i} values ($|\delta_{pred\_i} - \delta_{exp\_i}| = |\Delta\delta_{pred\_i-exp\_i}|$) for characteristic ^{13}C nuclei of organic molecules of "normal size" are $\sim 0.5\text{–}1$ ppm in routine *ab initio* calculations.[26,75b] From the viewpoint of accuracy, these differences are much better than the above discussed differences $|\Delta\delta_{calc\_i-exp\_i}|$. Therefore, it is rational to recruit "improved" values of chemical shifts, i.e., δ_{pred_i}, instead of CCS to distinguish close resonance signals (singlets) of a characteristic nucleus in a *modeled* NMR spectrum. We may use predicted chemical shifts $(\delta_{pred\_i})_A$ and $(\delta_{pred\_i})_H$ of the i-th characteristic nucleus for modeled stereoisomers A and H, respectively, to assign their CCS values $(\delta_{calc\_i})_A$ and $(\delta_{calc\_i})_H$ to experimental values $(\delta_{exp\_i})_A$ and $(\delta_{exp\_i})_H$, respectively. By using δ_{pred_i} in the expression for the first requirement for CCS accuracy, we can rewrite the condition for the characteristic nucleus as

$$\Delta\delta_{chrct\_i} > 2 \times |\Delta\delta_{pred\_i-exp\_i}| \tag{59}$$

where absolute difference $|\Delta\delta_{pred\_i-exp\_i}|$ for the i-th characteristic atom is defined as the difference $|\delta_{pred\_i} - \delta_{exp\_i}|$ (Fig. 102).

Equations 59 and 60 (see below) are methodologically important. As we learned, loose–accuracy values of CCS more often are not suitable for needs of structural analysis of organic molecules. These simple expressions, when substituted with concrete numbers into, indicate whether one can perform a valid conformational analysis for a concrete chemical compound by using, directly or indirectly, *scaled* chemical shifts. That is, when

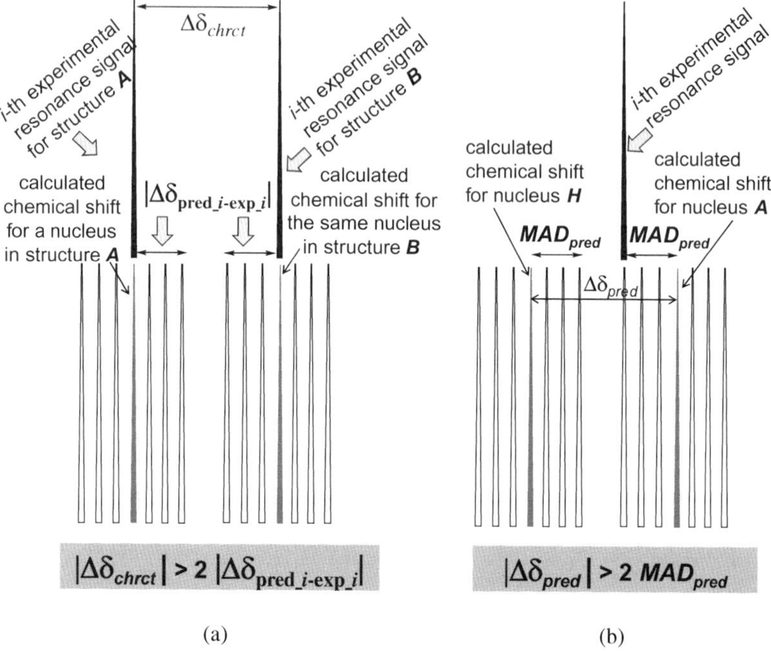

Figure 102. NMR singlets with numerically close experimental [case (*a*)] and calculated [case (*b*)] chemical shifts. For case (*b*), each group of vertical lines indicates some continuous interval in the scale of chemical shifts; the gray vertical line points out the center of the interval. Such an interval may be thought as a modeled NMR singlet with some uncertainty in location in this scale, due to unavoidable inaccuracy of non-empirical calculations. The first and the second requirements for CCS accuracy *with using predicted (scaled) chemical shifts*, which provide a correct assignment of such blurred CCS to the corresponding experimental chemical shift(s), are shown written in gray rectangles. If the shown resonance signals [the experimental ones in case (*a*) and the calculated ones in case (*b*)] shift toward each other, one can discern that the related experimental and calculated signals lose their previously unambiguous correspondence at the distance indicated by these requirements.

performing conformational analysis by, e.g., ^{13}C NMR together with CCS for these nuclei (Section 6.2.1), one should not operate with values of cCCS $\delta_{calc\_1}, \ldots, \delta_{calc\_i}, \ldots, \delta_{calc\_N}$ themselves, if these values are not of very high accuracy of 0.1–0.2 ppm. One should only use the corresponding scaled chemical shifts $\delta_{pred\_1}, \ldots, \delta_{pred\_i}, \ldots, \delta_{pred\_N}$.

What do predicted (scaled) chemical shifts additionally contribute to conformational analysis by CCS? It seems, at first sight, that, after examining CCS and experimental chemical shifts of the molecule of interest by Eqs. 59 and 60, values of $\delta_{\text{pred\_1}}, \ldots, \delta_{\text{pred\_}i}, \ldots, \delta_{\text{pred\_}N}$ are useless. This is not true: they prompt us to use MAD_{pred}, in addition to standard error of regression σ_{est} (Section 6.2.2), in identifying the "proper" modeled stereoisomer. Values $|\Delta\delta_{\text{pred\_}i-\text{exp\_}i}|$ are accuracies of concrete *predicted* (scaled) chemical shifts; a simple estimate for the averaged accuracy of predicted chemical shifts is MAD_{pred}. The lower the MAD_{pred} value is, the more accurate predicted chemical shift values are on average. Thus, in selecting the "proper" set of cCCS from the pool of alternative ones by means of linear regression analysis, the quantitative criterion in the selection may be both MAD_{pred} and σ_{est}: both estimates are "predicted-chemical-shifts-based." Similarly to the minimal value for σ_{est}, the minimal value for MAD_{pred} among those related to alternative regression lines identifies the sought-for, "NMR-photographed" conformer. A lower MAD_{pred} value shows a better correspondence to the linear regression approximation, and, as we remember, this statistical model is supposed to fit that cCCS set which belongs to the experimentally detected conformer.

Equation 59 retains the practical significance of CCS in conformational analysis. Compounds other than those with large $\Delta\delta_{chrct}$ values for their stereoisomers (e.g., alkaloid **44**, Fig. 101) can be within cCCS scope. Calculated differences $|\Delta\delta_{\text{pred\_}i-\text{exp\_}i}|$ usually lie in the 0.5–1 ppm range for*"proper"* stereoisomeric structures. When recalling that experimental differences $\Delta\delta_{chrct}$ for ^{13}C nuclei in different organic molecules often are of a 1–2 ppm magnitude, we discover that Eq. 59 holds true for such typical values of $\Delta\delta_{chrct}$. With some relief, we infer that, if supported by regression analysis, even cost-effective non-empirical calculations of CCS permit to "resolve" resonance signals of characteristic carbons for stereoisomers of most organic compounds and, thus, to successfully perform conformational analysis of them (Section 6.2.1).

Equation 59 expresses a necessary, but insufficient, condition for the accuracy of CCS required for conformational analysis. The complementary condition is to distinguish "proper" (NMR-detectable) stereoisomer A and any "improper" (NMR-undetectable) stereoisomer H in comparing their CCS and experimental $\delta_{\text{exp\_}i}$ values. Modeled stereoisomeric structures A

and H may have numerically close CCS for each carbon atom. Suppose that the absolute difference of CCS values $|(\delta_{calc\_m})_A - (\delta_{calc\_m})_H| = |(\Delta\delta_{calc\_m})|_{(A-H)max}$ for the carbon m in stereoisomers A and H (for simplicity, shortened notation $|\Delta\delta_{calc}|$ is further used in indicating this maximal difference) is larger than the similarly constructed difference $|(\delta_{calc\_i})_A - (\delta_{calc\_i})_H| = |(\Delta\delta_{calc\_i})|_{A-H}$ for any other i-th carbon atom of this molecule. What should the minimal value of this difference $|\Delta\delta_{calc}|$ be that only one of the located stereoisomeric structures can be reliably assigned to the experimentally detected conformer?

By theoretically modeling chemical shifts, we always insert numerically distinct inaccuracies $|\delta_{exp\_i} - \delta_{calc\_i}|_A$ and $|\delta_{exp\_i} - \delta_{calc\_i}|_H$ into the value of each i-th calculated chemical shift for both stereoisomers A and H. One can reasonably make an approximation that such inaccuracies are equal for the values of CCS for any i-th nucleus, which characterize its magnetic resonance in different stable conformations of the molecule, i.e., $|\delta_{exp\_i} - \delta_{calc\_i}|_A = |\delta_{exp\_i} - \delta_{calc\_i}|_H$.[a] This approximation also includes the nucleus m, which is characterized by the above indicated maximal absolute difference $|(\Delta\delta_{calc\_m})|_{(A-H)max}$. Then, the condition for CCS accuracy, which permits to select the CCS set of the "proper" stereoisomer A from a pair of CCS sets of stereoisomers A and H, is $|\Delta\delta_{calc}| > 2 \times |\delta_{exp\_m} - \delta_{calc\_m}|$ (the second requirement for CCS accuracy), where δ_{exp_m} is the experimental chemical shift of the nucleus m for either conformer A or H, δ_{calc_m} is the calculated chemical shift for this nucleus at either modeled conformer A or H, while $|\Delta\delta_{calc}|$ is the absolute difference of CCS values $(\delta_{calc\_m})_A$ and $(\delta_{calc\_m})_H$ for the magnetic nucleus m in stereoisomers A and H, respectively; as defined, the value of the difference associated with this nucleus m is maximal among such differences for all atoms of the same isotope in the molecule.

As in the case of the first requirement for CCS accuracy, not many *modeled* stereoisomers appear to be so dissimilar as to satisfy the second requirement for accuracy of CCS supplied by routine non-empirical calculations (Section 6.1.2). Therefore, instead of CCS themselves, we can again use

---

[a]We imply that the same level of theoretical molecular modeling and the same method of calculations of chemical shifts are used for both stereoisomers.

"better" values, predicted (scaled) chemical shifts δ_{pred_i}. Then, we can replace the difference $|\Delta\delta_{\text{calc}}|$ with the absolute difference for predicted chemical shifts, $|\Delta\delta_{\text{pred}}| = |(\delta_{\text{pred}\_m})_A - (\delta_{\text{pred}\_m})_H|$, where index m indicates the same nucleus m, which is characterized by the maximal magnitude for such differences $= |(\Delta\delta_{\text{pred}\_i})|_{A-H} = |(\delta_{\text{pred}\_i})_A - (\delta_{\text{pred}\_i})_H|$ for any i-th nucleus of the same magnetic isotope in modeled stereoisomers A and H. Absolute difference $|\Delta\delta_{\text{pred}\_i - \exp\_i}|$ from Eq. 59 is the accuracy of *ab initio* modeling for the i-th *predicted* chemical shift δ_{pred_i}. Assuming that the related averaged accuracy MAD_{pred} (Section 6.2.2) successfully represents such i-th accuracies for values of $\delta_{\text{pred}\_1}, \ldots, \delta_{\text{pred}\_i}, \ldots, \delta_{\text{pred}\_N}$ for all nuclei of the same magnetic isotope in the molecule, we can rewrite the second requirement for CCS accuracy as

$$|\Delta\delta_{\text{pred}}| > 2 \times MAD_{\text{pred}} \tag{60}$$

where $|\Delta\delta_{\text{pred}}|$ is the absolute difference between the predicted chemical shift and the experimental chemical shift for the nucleus m (Fig. 102). Values of MAD_{pred} usually are small. Equation 60 informs us that, with using linear regression analysis, it is practically always possible to successfully compare CCS sets of all computationally localized stereoisomers of an organic compound with sets of experimental chemical shifts of its conformers (Section 6.2.1). Of course, the best estimates in such a comparison of alternative sets of CCS are "predicted-chemical-shifts-based" ones, e.g., MAD_{pred} or σ_{est} (Section 6.2.2).

This condition is complementary to the first requirement for CCS accuracy. Both requirements are related to CCS accuracy, which imparts validity to the comparison of a pair of theoretically modeled NMR spectra with experimental ones. The first requirement for CCS accuracy considers two experimental and two calculated chemical shifts. For a given experimental value $\Delta\delta_{chrct}$, the first requirement indicates the minimal CCS accuracy required to "resolve" two molecular structures in modeled NMR spectra. In other words, this requirement establishes whether *experimental* NMR spectra of individual conformers are sufficiently dissimilar in chemical shift values to be analyzed by using CCS of non-superior accuracy. The second requirement for CCS accuracy considers two calculated and one experimental chemical shifts. It establishes the minimal CCS accuracy (in terms of MAD_{pred}, Section 6.2.2), which is sufficient to identify an NMR-detected conformer from a pair of modeled stereoisomers by assessing their calculated NMR spectra. In other words, this requirement indicates how different *modeled* NMR spectra with non-excellent accuracy of CCS values should be that only one among them can be reliably associated with the experimental spectrum of the related "frozen" conformer.

Of course, serious methodological difficulties arise from imperfect accuracy of non-empirical molecular modeling, and regression analysis is not a universal rescue tool. For instance, any statistical regression is inapplicable to conformer-oriented analysis of routine NMR spectra by means of CCS (Section 6.2.3). Thus, this "any-small-molecule-resolving" tool of conformational analysis cannot be undertaken at all with using CCS values of non-outstanding accuracy. As mentioned there, Eqs. 58_{i-i} operate with unavoidably inaccurate values of cCCS, and therefore their numerically calculated l.h.s and r.h.s. are *approximately* equal for the "proper" conformer. Thus, to identify this conformer, we should know to which extent cCCS values may be inaccurate that one can still accept the numerically non-holding adjusting equation of type Equation (58_{i-i}) with "correct" numerical solutions $p_{A\_corr}, \ldots, p_{K\_corr}, \ldots, p_{M\_corr}$ as that indicating correctness of these solutions exclusively (Section 6.2.3). Because the number of summands in the r.h.s. of the adjusting equation is $M = 1$, or 2, or 3, and each summand may be approximately characterized by inaccuracy MAD (Section 6.2.2), it is clear that the limits $\pm P$ (acceptance limits) for the numerical difference between the l.h.s and r.h.s. of this equation should be $\pm[M \times (MAD)]$ to satisfy the required approximate equality. Recalling that MAD values for CCS are usually large (see above), we realize that these limits are so broad that, probably, all improper solutions also satisfy the adjusting equation. Thus, values of, e.g., ^{13}C or ^{19}F cCCS, should have excellent accuracy of ~ 0.1 ppm to make Eqs. 58_{i-i} usable.

6.3.2 *Targeting CCS accuracy in routine calculations*

As we have learned, a considerably high calculation accuracy is absolutely necessary in conformational analysis by CCS. The following discussion deals with how we may reach the required accuracy without recourse to "heavy" calculation methods and complicated molecular models, when undertaking CCS-based analysis of "frozen" conformational equilibria (Section 6.2.1).

It is clear that, in the first instance, we should operate with cCCS values because they do not include the systematic error of calculations

(Section 6.2.2). However, cCCS values contain an irremovable random error, i.e., the magnitude of error is different for each δ_{calc_i} value, and these magnitude variations are unpredictable. How can we minimize the negative impact of this error? Excluding statistical treatment, no numerical manipulations with cCCS can decrease original inaccuracy "embedded" in these values. Predicted chemical shifts $\delta_{\text{pred}\_1}, \ldots, \delta_{\text{pred}\_i}, \ldots, \delta_{\text{pred}\_N}$ derived by regression analysis of cCCS (Section 6.2.2) are "better" values: they do not include the systematic error and contain somewhat decreased random errors. Nevertheless, no further treatment of scaled chemical shifts can diminish inaccuracies inherent to them. To lend a higher reliability to structure identification via CCS, it is necessary to improve the quality of non-empirical modeling itself. Let us assess possible fruitful strategies.[a]

Statistical analysis is a well-defined methodology. It uses exact laws of probability distribution and, by means of these laws, predicts the most probable values for a random variable. For real or computational experiments, these predicted (high probability) values are, of course, associated with higher accuracy. Non-statistical methodologies (e.g., the so-called multistandard approach[77a] that, fortunately, has had no time to spread) can scarcely deliver a true improvement. They merely modify obtained inaccurate values without a clear idea of how *to target the factors beyond inaccuracies (inaccuracy-causing factors)*.

Non-statistical numerical treatment of CCS may be fruitful only in one case: it should selectively "engage" CCS inaccuracies. However, these inaccuracies are random. As mentioned, in principle, it is impossible to identify the inaccuracy increments in obtained cCCS because the magnitude of this error in the modeling is *randomly* different for values δ_{calc_i}. Then, how can we deal with this "estimation-escaping" error? This is possible if we make reasonable *assumptions* regarding the error instead of irrelevant non-statistical attempts to estimate it.

As an example, we may consider an improved approach to structure identification via CCS: the use of CCS differences. It is based on a general idea that local solvation is structurally and dynamically similar for Q stereoisomers $A, B, C, \ldots, Q$ with the same solvent-accessible functional groups. This idea permits us to assume that the Q calculated values $(\delta_{\text{calc}\_i})_A, (\delta_{\text{calc}\_i})_B, \ldots, (\delta_{\text{calc}\_i})_Q$ for the i-th atom in these stereoisomers include the same random error, which arises from impossibility of perfectly modeling solvation shells and their intermolecular dynamics. According to this verisimilar assumption, *differences* such as $(\delta_{\text{calc}\_i})_A - (\delta_{\text{calc}\_i})_B$, $(\delta_{\text{calc}\_i})_A - (\delta_{\text{calc}\_i})_C$, and $(\delta_{\text{calc}\_i})_B - (\delta_{\text{calc}\_i})_C$ between cCCS of each atom do not incorporate this random error. These differences,

---

[a]In addition, it is important to remember that QM calculations without including relativistic corrections deliver the required accuracy of calculated chemical shieldings σ_{calc} to chemical elements of only the first four rows of the periodic table.

which may be considered as free from the error induced by solvation modeling, are excellent starting data for conformational analysis by cCCS. Let us identify two conformers "frozen" A and B of a molecule by means of regression analysis of routine accuracy cCCS (Section 6.2.2) with replacing comparison of cCCS values themselves with comparison of their differences. For example, for carbon atoms $1,2,\ldots,N$ of a pair of modeled stereoisomers I and J, we derive $N(N-1)/2$ differences of cCCS values $(\delta_{calc\_1})_I - (\delta_{calc\_1})_J, (\delta_{calc\_2})_I - (\delta_{calc\_2})_J, \ldots, (\delta_{calc\_N})_I - (\delta_{calc\_N})_J$. After performing this arithmetic operation with cCCS of stereoisomer I and other stereoisomer, we obtain $Q-1$ sets of the cCCS differences of the above form. They are values for random variable Y. In the same manner, we derive $N(N-1)/2$ differences $(\delta_{exp\_i})_B - (\delta_{exp\_i})_A$ for N experimental chemical shifts of two "frozen" stereoisomers A and B, intending to use these values as variable X. If we undertake linear regression analysis (Section 6.2.2) for all $Q-1$ sets of cCCS differences *vs.* the experimental shift differences in turn, one set of cCCS differences associated with a pair of modeled stereoisomeric structures will obviously be the best from the viewpoint of linear regression. What is the difference between this selection technique and that of correlating experimental chemical shifts and cCCS? By rationally choosing variables X and Y here, we strive that a worst quality of linear regression (non-linearity) is associated only with "worst" 3D structures and not with factors that affect only one random variable (here, insufficient modeling of solvation). Therefore, molecular 3D structures from the "best" pair are *more* reliably assigned to NMR-detected conformers A and B. A modification of this method of chemical shift differences is described in Section 6.3.3.

An additional example of purposeful treatment of CCS inaccuracies is separation of CCS for different types of atoms of a molecular system. For instance, non-functionalized aliphatic, aromatic, carbonyl, electron-poor, and electron-donating molecular fragments are dissimilarly solvated. CCS values for carbons of these fragments are of different accuracy. CCS values for, e.g., carbonyl or aromatic carbons, i.e., atoms from molecular fragments with strong anisotropic solute-solvent interactions, are, as a rule, very inaccurate, while CCS of aliphatic carbons, i.e., atoms of weakly solvated fragments, are of much higher accuracy. Hence, one cannot expect that CCS and experimental chemical shifts for molecular systems bearing both aliphatic and aromatic fragments have a linear stochastic relationship. Such a use of the linear regression model (a single statistical tool within the competence of organic experimentalists) becomes problematic for analyzing CCS.

On the other hand, we can assume that the impact of solvation on chemical shifts is practically identical for molecular fragments of the same structural type, e.g., unhindered non-functionalized methylene units or amide fragments in α-helical protein chains. The contribution of solvation-associated inaccuracy into the calculation error seems, therefore, a constant value in modeling chemical shifts of the same type atoms. This intuitive consideration suggests the methodology of CCS separation. The same isotope atoms of a molecular system, e.g. ^{13}C atoms, should be formally grouped depending on which type of structural fragments they belong to, and the linear regression procedure should be executed for the CCS of each group of atoms. By comparing of the extent of linearity in regression for alternative sets of CCS for chemically similar carbons, we will obtain telling statistical estimates σ_{est}

or MAD_{pred} (Section 6.2.2), while when comparing "mixed" CCS sets with carbons of distinct chemical functionality such estimates will be rather loose. Indeed, formal grouping of atoms of a molecular system accordingly to their chemical similarity is effective because this "trick" separates certain factors which unequally affect random variable Y. Consequently, the values of predicted chemical shifts obtained in this manner will be significantly "better" than such values delivered by regression analysis of the CCS of the molecule as a whole.

As mentioned in Section 6.2.2, the random error in calculated chemical shieldings (and, thus, in CCS) has two sources: inaccuracy of theoretical calculations that provide these values and inadequacy of the constructed molecular model. The route to minimizing the first inaccuracy seems obvious. It seems that, as far as possible, one should use accurate methods of non-empirical calculations. Although there is no "best" *ab initio* method, general hierarchy of theoretical approaches and basis sets with regard to accuracy is explicit (Sections 5.2.3, 5.2.4 and 5.2.6). Indeed, also in calculating chemical shieldings (and consequently, chemical shifts), post-HF methods are significantly more accurate than the HF approximation, higher rung density functionals, as a rule, provide better accuracy than lower rung ones, and highly extended basis sets may deliver results that may be missed when using less extended basis sets. Therefore, it would be preferable in routine calculations to use, e.g., the MP2/6-31++G(d,p), MP2/*aug*-cc-pVDZ, or B2PLYP/*aug*-cc-pVTZ level of calculations with respect to the B3LYP/6-31++(d,p) level or, certainly, with respect to the RHF/6-31++G(d,p) level, in both optimizing molecular geometry and modeling chemical shieldings σ_{calc}. In addition, for calculating chemical shieldings, it would be reasonable to choose, e.g., GIAO and not CSGT (for the main QM methods of chemical shielding calculations, see Section 6.1.2).

Many studies compare the accuracy of CCS in different non-empirical calculations for small inorganic and organic molecules (see, e.g., Refs. [68e], [69a,b], [75a], and [77c,d]). Of course, they have not revealed "the most accurate method" although the accuracy of obtained CCS values is certainly improved, though not always easily, with applying more advanced calculation methods. One can presume that MP2 and fifth-rung DFT methods, on average, provide the best accuracy for CCS, if only considering currently routine methods of non-empirical calculations and moderately extended basis sets, e.g., *aug*-cc-pVDZ or *aug*-cc-pVTZ.

The maximalist approach to selecting the calculation method is, nevertheless, unnecessary. Reflecting this approach, high theory *ab initio* methods, e.g., coupled cluster theory methods (Section 5.2.3), are used to provide a desired accuracy of 0.05–0.1 ppm for ^{1}H, a few tenths of 1 ppm for ^{13}C, and 1 ppm for ^{19}F chemical shifts. However, such "heavy" computations (see, e.g., Ref. 77e), even if practically possible for non-minimal-sized molecules, would be unjustified waste of effort and time in modeling of chemical shifts oriented to routine structure determination. The idea of the primacy of structural models over calculation tools in applied computational chemistry (Section 5.2.7) — to use a high quality structural model and a moderate quality calculation method — is also practical regarding CCS. No calculation method can supply CCS values that accurately reproduce experimental chemical shifts for a compound in highly polar solution, if its structural model is irrelevant, e.g., it represents a single molecule in vacuum. In contrast, one can reproduce these experimental values at least satisfactory, when using an adequate molecular model for solution and routine non-empirical calculations, e.g., MP2. Efforts to reach higher CCS accuracy should be therefore focused on selection of a successful structural model of the molecular system, "neutralizing" the second source of CCS inaccuracy as much as possible. At the same time, a sufficiently accurate method of *ab initio* calculations should be used in order to exploit the accuracy potential of the structural model as much as possible.

As mentioned, GIAO is usually the preferred methodology in modeling of chemical shifts. Which approximations of *ab initio* theory can provide an acceptable CCS accuracy and, at the same time, not put too hard requirements for computer resources? It is worth repeating that MP2 is the highest priority method for routine for *routine* modeling of CCS. In combination with medium-sized basis sets (Section 5.2.5), e.g., Pople basis set 6-31+G(d,p), 6-311++G(2p,d), or 6-31++G(d,p) (for 30–60 atom molecular systems), or Dunning's basis set with either *aug*-cc-pVDZ or *aug*-cc-pVTZ (for 10–30 atom molecular systems), MP2 or a fifth-rung DFT method (Section 5.2.6), e.g., B2PLYP, or XYG3 (in combination with the *aug*-cc-pVTZ basis set), should be selected. Then the CCS accuracy of 2–3 ppm (i.e., *MAD*; Section 6.2.2) is almost warranted for ^{13}C nuclei of organic molecules that do not include heavy ("post-Kr") atoms and are not solvated strongly. If, unfortunately, the use of MP2 or a DHGGA method is not possible, the fourth-rung DFT methods, e.g., M06-2X, M11, or TPSSh, used in combination with any one of the indicated basis sets are the next best option. In order to obtain considerably accurate CCS values with using such DFT methods, one should examine 3–5 combinations of the calculation method and the basis set, e.g., combinations M06-2X/6-311++G(2d,p)//GIAO, M11/6-311++G(2d,p)//GIAO, M11/*aug*-cc-pVDZ//GIAO, TPSSh/6-31++G(d,p)//GIAO, and M06-2X/*aug*-cc-pVDZ//GIAO by applying calculations either to the system of interest, or, for simplicity, to its structurally close and absolutely rigid molecular analog for which the related experimental NMR spectrum is available.

Let us now note some methodological points regarding the quality of models of organic molecules, which should be addressed when using "light" calculations[a] of chemical shifts with medium accuracy.

Molecular geometry. Chemical shieldings and, consequently, chemical shifts are very sensitive to the 3D geometry of the molecular backbone including its local distortions.[68a] The following example gives an idea about this sensitivity. A 0.016 Å elongation of the N–C bond shifts by 12 ppm the value of the chemical shielding of the α-carbon atom of alanine.[78] That is, inaccuracy of approximately 1% in the lengths of covalent bonds may result in inaccuracy of several ppm for CCS of carbons of the modeled hydrocarbon backbone. This inaccuracy in calculated bond lengths is quite ordinary, e.g., it appears when using the popular third-rung DFT hybrid functionals (Section 5.2.4). It is clear in this light that the quality of the *modeled* molecular geometry should be relatively high for structures subjected to chemical shift calculations. In the past decade, the compromise in routine, time-saving theoretical modeling of organic molecules between using very accurate, but "heavy" calculation methods and less accurate, "light" ones has been exactly such (third-rung) DFT methods. Since then the computer technology has progressed so far ahead that one should leave this becoming obsolete compromise and optimize molecular geometry of 20–50 atom structures with only applying MP2 or MP3 [in combination with either 6-31++G(d,p), 6-311++G(2d,p), or *aug*-cc-pVDZ, or *aug*-cc-pVTZ basis set] or DHGGA methods of DFT (e.g., B2PLYP, or XYG3; both in combination with the *aug*-cc-pVTZ basis set). Among these methods, calculations at the MP2/6-31++G(d,p) level have lesser demand to hardware and are less time-consuming; nevertheless, this level of approximating QM theory provides sufficiently accurate molecular geometry.

As pointed out in Section 5.3.2, the SCF procedure of energy minimization yields the geometry that corresponds to the non-populated, lowest

---

[a]I.e., not including rovibrational corrections as well as temperature, isotope, and relativistic effects into the molecular model and not using high theory *ab initio* methods and essentially extended basis sets in calculations.

energy point of the located PES minimum (Fig. 42, Section 2.3); this geometry (equilibrium geometry) is not that of a real molecule (Fig. 92, Section 5.3.2). Molecules retain some kinetic energy even at 0 K, when populating only the zeroth vibrational level. In classical representation, the molecular backbone still oscillates at this temperature, and its geometry may be characterized as some effective geometry. This means that, in principle, CCS for molecular structures that are modeled by simple *ab initio* calculations cannot be absolutely accurate (recall susceptibility of resonance frequencies to tiny perturbations of molecular geometry). In order to account for this effect on CCS, so-called rovibrational corrections are added to calculations of chemical shifts (for the related theory, see Ref. 68c). Their values are not small: typically, they are of a 2–4 and 5–20 ppm of magnitude for absolute chemical shieldings of ^{13}C and ^{19}F nuclei, respectively.

Calculations of these corrections encumber modeling of chemical shifts due to hardening requirements for computer resources. Besides, not too many computational chemistry packages possess this functionality. For instance, CFOUR provides it, while, even in the eve of 2015, Gaussian, Q-Chem and Turbomole do not have this option, and rovibrational corrections are yet not widely included into routinely modeling chemical shifts. When not calculating ro-vibrational corrections, a satisfactory accuracy of CCS is still achievable because CCS are relative values derived from calculated absolute chemical shieldings (Eq. 54). However, a higher accuracy (e.g., 0.5–1.0 ppm for CCS of ^{13}carbons in a molecule) requires consideration of the effective molecular geometry (and not the equilibrium geometry) and, thus, to include rovibrational corrections into the molecular model.

MM calculations using MM3 or CHARMM force fields (Section 4.2.5) seem very fascinating because of their quick delivery of optimized molecular geometry and energy (as steric energy E_{ster}). Indeed, they may be sometimes successful[41] because they tend to supply the oscillations-averaged (i.e., effective) molecular geometry, in contrast to that modeled by simplest non-empirical calculations (the equilibrium geometry, as mentioned; Fig 92). However, because of non-universality of parameterization in any force field (Section 4.2.5), researchers should rather evade to optimize geometry of the studied molecular structure by means of MM if comparison for CCS accuracy for its MM and QM-derived molecular geometries is not planned.

Solvation. Practically all NMR experiments are performed by synthetic chemists for compounds in solution. The usual solvent is $CDCl_3$, $CD_3C(O)CD_3$, CD_3OD, C_2D_5OD, $CD_3S(O)CD_3$, or D_2O; sometimes, it is 1,4-dioxane$_{d8}$, C_6D_6, C_5D_5N, CD_3COOD, or $(CD_3)_2NC(O)D$. Freons $CDClF_2$ and CDF_3, as well as deuterated dichloromethane $C_2D_2Cl_2$, are used in low-temperature NMR experiments, while C_6D_5Br, or $C_6D_5CD_3$ are non-expensive NMR solvents for increased temperatures. Except freons, all these solvents are unsuccessful, from the viewpoint of theoretical modeling of chemical shifts.

Three molecular features of solvents for NMR explain why they may be ranked so low. Excluding deuterated cycloalkanes (see below), they are polar, most are capable of forming strong H bonds with either or both electron-donating and electron-accepting systems, and some are aromatic.

First, let us consider H-bonding relevant to NMR solvents. Haloforms are effective H-donors. For instance, they bind amines and alcohols through H-bonds of $\sim 4\,\text{kcal mol}^{-1}$ energy. Even dichloromethane forms H-bonds of considerable strength (energy of this bonding with aliphatic amines is $\sim 2\,\text{kcal mol}^{-1}$). $CD_3C(O)CD_3$, 1,4-dioxane$_{d8}$, $CD_3S(O)CD_3$, C_5D_5N, and $(CD_3)_2NC(O)D$ are H-acceptors; the latter three are strong ones. In addition, $(CD_3)_2NC(O)D$ is a weak H-donor. Water, methanol, ethanol, and acetic acid possess dual activity in the formation of H-bonded structures. Strong bridges $O \cdots H-O \cdots H$ assemble these molecules in oligomeric 3D networks. The solvation shell of the concurring H-bond-forming molecule (solute) is the case-specific local distortion of such a regular network of H-bonds. One cannot guess the molecular structure and stability of this locally distorted solvent domain that includes an organic molecule; however, we certainly know that formation of H-bonding between the solute and solvent molecules means formation of a chemically new, though rapidly changing, molecular structure. It is obvious that the solvated molecule, i.e., many similar rapidly interconverting intermolecular associates, is characterized by its own resonance frequencies.

Second, almost all NMR solvents (including $C_2D_2Cl_2$) are polar molecules. To an appreciable extent, they change intramolecular charge distribution in solute molecules and transform them into molecular systems with somewhat disturbed magnetic shielding/deshielding of their nuclei. 1,4-Dioxane$_{d8}$, $CD_3C(O)CD_3$, $CD_3S(O)CD_3$ CD_3CN, CD_3NO_2 as well as aromatic NMR solvents, e.g., C_6D_6, and C_5D_5N, introduce an additional effect. Each molecule of such a solvent significantly disturbs the isotropy of the external magnetic field in some contiguous space. Surrounding the solute molecule, the multi–molecule solvation shell positions its core (i.e., the solute molecule) in a local anisotropic magnetic field. This local magnetic anisotropy significantly and differently alters chemical shifts of magnetic nuclei of the solute.[a]

With several exceptions, early and followed structure determinations of organic compounds by means of CCS have been carried out with considering molecular structures modeled for vacuum conditions. Inaccuracies of obtained CCS values cast doubt on the reliability of some of these structure assignments. Minimizing the possibility of misinterpretation of CCS to occur in molecular structure elucidation, incorporation of solvation in molecular models is nowadays an explicit trend in such studies.

Under these circumstances, negligence of solvation in molecular models may result in serious inaccuracies when theoretically calculating chemical shifts of organic compounds. However, in contrast to modeling of molecules at vacuum conditions, non-empirical modeling of solvated molecules is not easy (Section 5.3.4). "NMR solvent-associated" inaccuracy is an overt methodological problem in structural analysis by CCS. How can this problem be overcome? There are two obvious practical solutions. A simple one is to use in NMR experiments "coupled" with calculations of chemical shifts a perdeuterated solvent that minimally interferes with dissolved compounds. In other words, the solvent should be nonpolar, non-dissociating, and non-coordinating. In addition, it should not have a strong molecular diamagnetic anisotropy, be chemically inert, have a low freezing point to enable NMR experiments also at low temperatures, and be commercially available.

Deuterated methylcyclohexane C_7D_{14} well suits this role [F.p. $= -126°C$ (147 K), b.p. $= 101°C$ (374 K)]. Molecules of polar compounds in C_7D_{14} solution (weakly interacting solute–solvent molecules) verisimilarly imitate *non-interacting* solute–solvent molecules, i.e., molecules in vacuum. Then, molecular modeling of stereoisomers of a compound at vacuum conditions affords a fair ground for comparing the obtained CCS

---

[a]In terminology of experimental NMR, such a chemical shift alteration for aromatic solvents has a certain name: this particular effect is called aromatic solvent-induced shifts (ASIS).

with experimental chemical shifts registered (as resonance frequencies) for C_7D_{14} solutions. The comparative linear regression analysis of alternative CCS sets (Section 6.2.2) becomes reliable: an important random factor (strong anisotropic solvation) that affects only one variable (experimental chemical shifts) and, thus, disturbs the stochastic relationship between the two variables (CCS and experimental chemical shifts) disappears.

However, the deuteromethylcyclohexane technique to fitting calculated and experimental chemical shifts is not suitable for many organic compounds. Solvents that weakly interact with a chemical compound weakly dissolve it. For instance, morphine, carbohydrates, or purine nucleobases are polyfunctional polar compounds and do not have an appreciable solubility in freons, tetraalkylsilanes, or alkanes, including methylcyclohexane. If intending to examine molecular flexibility of a new or little-known polar compound (e.g., internal rotation in iminobismuthane ylide $Ar_3^+Bi–N^-Ac$) by NMR, one cannot expect that such solvents will dissolve it to the concentration which reaches the sensitivity threshold of NMR instruments. Moreover, if a polar compound is somewhat soluble, e.g., in methylcyclohexane at room temperature, its solubility there rather significantly decreases with dropping the temperature below 200 K. NMR experiments for "freezing" conformer interconversion obviously become impossible for such solutions.

Deuterodichloromethane $C_2D_2Cl_2$ delivers some compromise. This non-dissociating solvent with the freezing point of 97°C (177 K) has a low polarity; nevertheless, it is sufficient to provide fair solubility to polar non-ionized organic compounds even at low temperatures. However, there is a non-negligible cost of this compromise. It is clear that the values of chemical shielding of a magnetic nucleus in an organic molecule are somewhat different for vacuum conditions and the dichloromethane liquid phase. One cannot guess *a priori* whether this difference does not disturb the accuracy of chemical shifts calculated for the modeled molecule-in-vacuum to that extent that these calculated values may be unmistakably related to the corresponding experimental chemical shifts of this chemical compound in dichloromethane, i.e., in the solvent which certainly cannot be likened to vacuum and, at the same time, is not strongly associated with the solute.

The second solution of the "solvent-associated CCS inaccuracy" problem is computational: it lies in improving the theoretical model. In computational chemistry, the general features of any molecular model (one molecule or molecular ensemble, vacuum or a condensed phase, the ambient temperature, etc.) are specified before submitting this abstract conglomerate of the 3D molecular backbone and conditions describing it to theoretical calculations. That is, it is technically not difficult to include solvation into the theoretical molecular model. Using a GUI-based program of computational chemistry,[a] synthetic chemists may "augment" the on-screen 3D molecular backbone with both a selected "light" QM approximation (Sections 5.2.3, 5.2.4 and 5.2.6) and a solvation model (Section 5.3.4, subsection 4).

In organic laboratories, this solution to increase the CCS accuracy requires to answer to the question "Which concrete models of solvation can non-theoreticians employ in order to obtain more accurate CCS values without extensive examination of them as well as enormous computational cost in using the selected model?" The answer is, of course, ambiguous. An unpretentious method, PCM (Section 5.3.4), is more or less sufficient for modeling polar molecular structures, which are exclusively involved in VDW interactions with solvent molecules. That is, such solutes and solvents do not interact strongly, e.g. via formation of H-bonds or another coordination between electron–poor molecule (solute) and electron-donating molecules (solvent). Solvent-induced shifts of resonance signals are not large for such compounds in appropriate aprotic, "no magnetic anisotropy" solvents, c.g., $CDCl_3$, $C_2D_2Cl_2$, anhydrous 1,4-dioxane$_{d8}$, or THF_{d8}. For instance, they do not exceed 0.5 ppm for most aliphatic carbons of an organic molecule. One can see that PCM does effectively consider non-specific solvation effects: it provides a $\sim 0.3–0.5$ ppm improvement in the CCS accuracy for ^{13}C nuclei. This refinement of cCCS values is beneficial even if the accuracy of the "parent" CCS for vacuum only is fair, say, from 2 ppm to 3 ppm for carbon cCCS: the improvement of the accuracy values is from 10% to 25%. Consequently, reliability of estimates of linear regression of cCCS *vs.* experimental chemical shifts and the accuracy of resulting predicted chemical shifts δ_{pred_i} are improved.

The PCM model is readily available. This solvation model is a "one click" option in some basic research-oriented QM calculation packages, e.g., Gaussian, Jaguar, and Q-Chem.

---

[a]The most often used QM programs are mentioned in Section 5.1.

Another dielectric continuum solvation model, COSMO, is an attribute of computational packages Q-Chem, ADF and Turbomole. Also it is capable of improving the accuracy of, e.g., carbon CCS, to the same extent. Probably, Cramer–Truhlar model SM8 (implemented into late versions of Q-Chem and Jaguar) provides the best CCS accuracy, among the currently used continuum solvation models. The regrettable circumstance is that, practically, only PCM can be used by non-computational chemists when modeling chemical shifts. This solvation model as implemented into Gaussian09 may be applied both to geometry optimization and chemical shift calculations. The other commercial packages of computational chemistry, if including either solvation model, support combined option *geometry optimization (energy minimization)–solvation modeling*, but do not calculate chemical shifts in couple with the implemented method of modeling solvation. Some other QM packages (e.g., ORCA, DIRAC)[a] that are Linux-resident software free for academic researchers are capable of calculating chemical shifts in the "envelope" of a solvation model. However, one should possess some skills in software in order to install and use them.

In many solvents, functionalized organic molecules are involved in specific intermolecular interactions. Concretely, a weak directional chemical bonding occurs between solute and solvent molecules, e.g., H-bonding or coordination of electron-donating heteroatomic fragments with electron-deficient fragments. Formation of even a weak chemical bond may largely alter chemical shifts of the atoms that are near neighbors of the tightly contacting functional group of the solute and solvent molecules. For instance, the change of ^{14}N resonances of nitrogen-containing functional groups may be tens of ppm if these groups participate in H-bonding. A 5–10 ppm alteration for the chemical shift of a carbon, which bears a heteroatom (e.g., the carbon in the C=O group of a ketone), is not rare if the heteroatom is H-bonded in solution.

Constructing the out-the-molecule space as an abstract continuous medium, implicit models of solvation are not suitable for treating molecular systems with a highly structured solvation shell. In this case, such methods of solvation modeling improve the quality of CCS only insignificantly. We certainly need to model some relevant *molecular structure*, i.e., to explicitly model the solvent bulk or solvation shell. Unfortunately, there is no common *simple* approach in molecular modeling that is capable of adequately reproducing a structurally anisotropic solvation shell. With which solvation model can organic experimentalists proceed? Being focused on

---

[a]ORCA: http://www.freechemical.info/freeSoftware/ORCA.html;
DIRAC: http://www.diracprogram.org/doku.php.

experimental work, they can use only *rough* explicit models of solvation; for instance, modeled molecules with the solvation shell that is both cardinally truncated and "frozen." That is, only a few solvent molecules augment the initial (the solute) molecular backbone, and the resulting "supramolecule" is considered as a static structure, i.e., intermolecular dynamics (positional reorganization of solvent molecules and their exchange with outer solvent molecules) is neglected.

There is no standard route for designing verisimilar truncated solvation shells. However, beginners can be guided by a simple approach: solvation of only functional groups capable of weakly bonding solvent molecules is explicitly modeled. According to this mechanistic approach, one should "secure" solvent molecules to those functional groups of the modeled solute molecule that form weak chemical bonds with this solvent in real solutions. For instance, with using a 3D molecular graphics program (a component of every professional or educational QM package), chloroform molecules should be attached (through their H atoms) on-the-screen to those heteroatoms of the modeled solute molecule, which have lone electron pair(s), while water and methanol should be molecular "partners" of both such heteroatoms and H-donating groups of the solute. Because there are only a few functional groups in an ordinary-sized organic molecule, the number of solvent molecules that are "added" to such a solute molecule is small. Nevertheless, the methodology of truncated solvation shell does not represent explicitly structured domains of the extramolecular space as a non-structured anisotropic continuum, in contrast to implicit solvation models.

The molecular structure in this simple methodology is merely an enlarged molecule (supramolecule). It may be subjected to any routine non-empirical calculations, including geometry optimization and calculations of chemical shifts.[a] By applying these computations to the designed supramolecule, we obtain the *molecular structure* of a solvate that includes domains of the solvation shell which significantly contribute to the energy of solvation (the molecular structure of a solvate with rationally truncated solvation shell). Indeed, by "hanging" solvent-binding functional groups with

---

[a]Such "supramolecules" are not oversized and, therefore, are treatable by common computational facilities (HPC clusters) of chemical departments or such a cluster shared between a few research groups.

solvent molecules, we model the chemical connectivity in energetically essential solute–solvent contacts. By performing energy minimization for this roughly designed supramolecule, we reliably model one of its similar molecular geometries (the latter differ one from the other by various mutual orientations of the core molecule and solvent molecules). No model may be better for modeling molecules by QM than molecular structure *nuclei +electrons* (i.e., a large molecular backbone and not a smaller molecular backbone surrounded by some no nuclei, no electrons rest). It is not surprising therefore that disappointing CCS values calculated within the limits of a continuum solvation model for atoms of coordinating functional groups of a molecule are appreciably improved both for these group atoms and their near neighbors if the relevant supramolecule is the source of CCS.

6.3.3 *In search of the CCS accuracy: an indicative example*

Let us "feel" the original CCS accuracy (not improved by statistical treatment) of the often used non-empirical methods by discussing an easily understandable example of ^{13}C chemical shifts for a rigid structure **45** ("one conformer" compound; Fig. 103). This "adamantane" skeleton has

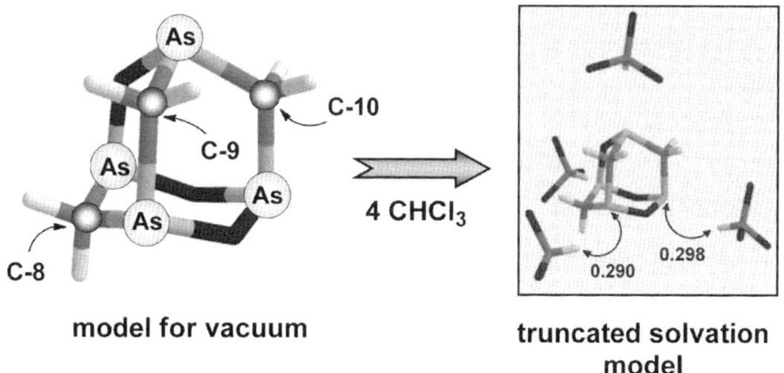

Figure 103. Molecular structure **45** (left) established for the As$_4$C$_3$H$_6$O$_3$ alkaloid.[79a,b] *Ab initio*-optimized geometries[9] derived for vacuum (as **45**; left) and for a molecule with a truncated solvation shell of CHCl$_3$ (as **45** • 4 CHCl$_3$; right) are shown. The gray background depicts a continuous medium (CHCl$_3$ in the PCM representation), and the arrows indicate the lengths (nm) of intermolecular H-bonds As$\cdots$HCCl$_3$ in the modeled solvate.

been proposed on the basis of indirect evidences for an alkaloid of molecular formula $As_4C_3H_6O_3$.[79a] Because resonance signals, NOE enchancements and spin-spin interactions all are few in number, it is risky to accept that this backbone is a single structure that may satisfy the registered NMR spectra[79a] for a compound of this molecular composition. It is not surprising that the CCS-based methodology of structure determination is among quantitative tools that have been recruited to prove molecular structure **45**.[9,79b]

This structure has appeared a hard nut for simulating their chemical shifts. Two ^{13}C NMR signals registered for this compound (Table 8) indicate that, among three, there are two chemically equivalent carbon atoms. One may think that experimental chemical shifts for this compound can be easily reproduced by theoretically modeling them for the proposed[79a] molecular structure **45** (Fig. 103). Indeed, there are only two types of chemically non-equivalent carbons in this molecule (C-8 and C-10 *vs.* C-9); they all have the same chemical functionality (i.e., they should have a similar local solvation), and the chemically non-equivalent carbons show well resolved ^{13}C NMR signals ($\Delta\delta = 6.3$ ppm). The molecule is rigid, and thus, only one stereoisomer (i.e., a single one) may be modeled.

However, with a "standard" DFT method, B3LYP, and a moderate basis set, cc-pVTZ, the first non-empirical calculations for this backbone[79b] have led to CCS values of intolerable inaccuracy (Table 8, first and second entries). There is no possibility of resorting to a trustworthy statistical treatment of the obtained CCS because of the minimally-sized sample (there are two ^{13}C NMR signals). Which modeling approaches can non-computational chemists follow in hunting for a reliable structure elucidation for alkaloid **45** by means of CCS?

This question is addressed in Table 8 that presents cCCS for the carbons of chemical structure **45** obtained with using some routine QM calculation methods; those are considered in this and previous Chapters as more or less suitable for applying by organic experimentalists. Some of these methods have been used both without and with involving simple solvation models. The calculation methodologies are approximately arranged in this Table according to increasing their complexity.

Let us summarize the modeling results. Changes of cCCS for **45** appeared with varying calculation methods and solvation models permit to conclude: (*1*) using of MP2 in calculations of chemical shifts is crucial: MP2-supplied

Table 8. Values of cCCS (ppm) for ^{13}C nuclei of compound **45** provided by different non-empirical calculations.

Geometry optimization	Chemical shift calculations (GIAO)	δ_{calc}, C-9	δ_{calc}, C-8, C-10	Δ_{calc}
B3LYP/6-311G(2d,2p)	B3LYP/cc-pVTZ	27.3[a]	30.3[a]	3.0
B3LYP/6-311G(2d,2p)	BP86/TZ2P[g]	27.0[a]	31.1[a]	3.1
B3LYP/6-31+ G(d,p)	B3LYP/6-31+G(d,p)	26.1	29.0	2.9
MP2/6-31+G(d,p)	B3LYP/6-31+G(d,p)	25.4	28.2	2.8
B3LYP/6-31+G(d,p)	MP2/6-31+G(d,p)	21.5	26.3	4.8
MP2/6-31+G(d,p)	MP2/6-31+G(d,p)	20.5	25.2	4.7
{PCM:CHCl$_3$}-MP2/6-31+G(d,p)	MP2/6-31+G(d,p)	20.0[b]	24.8[b]	4.8
{PCM:CHCl$_3$}-MP2/6-31+G(d,p)	{PCM:CHCl$_3$}-MP2/6-31+G(d,p)	20.1[c]	25.2[c]	5.1
{PCM:CHCl$_3$}-MP2/6-31+G(d,p)	B3LYP/6-31+G(d,p)	25.4[d]	28.2[d]	2.8
{PCM:CHCl$_3$}-MP2/6-31+G(d,p)	{PCM:CHCl$_3$}-TPSSh/6-31+G(d,p)	22.0[e]	25.3[e]	3.3
{PCM:CHCl$_3$}-MP2/6-31+G(d,p)	{PCM:CHCl$_3$}-TPSSh/cc-pVTZ	31.0[f]	34.4[f]	3.4
{PCM:CHCl$_3$}-MP2/6-31+G(d,p)	{PCM:CHCl$_3$}-MP2/6-31+G(d,p)	20.0[c,d]	25.2[c,d]	5.2
Experimental data (for CHCl$_3$ solution)[h]		*17.03*	*23.05*	*6.02*

[a] From Ref. 79b. Data from other entries (obtained using Gaussian09) are related to Ref. 9. [b] Derived using value 202.4 ppm of σ_{calc_ref} for TMS (Table 7, Section 6.1.2). [c] Derived using value 202.9 ppm of σ_{calc_ref} for TMS (Table 7, Section 6.1.2). [d] For the truncated solvation model used (**45**·4 CHCl$_3$). [e] Derived using value 195.5 ppm of σ_{calc_ref} for TMS (Table 7, Section 6.1.2).). [f] Derived using value 196.7 ppm of σ_{calc_ref} for TMS (Table 7, Section 6.1.2). [g] With adding relativistic corrections (not specified herein). [h] From Ref. 79a.

cCCS are appreciably more accurate than cCCS derived by means of the B3LYP or BP86 functional of DFT; the chemical shift difference $\Delta\delta$ may be reproduced by MP2 with a fair accuracy even for the molecular model for vacuum; (2) both these calculation methods afford the molecular geometry that is sufficient for modeling chemical shifts of fair accuracy; and (3) the accuracy of the obtained cCCS values is poor if no solvation model is employed.

The new obtained values of cCCS eventually confirm the structure **45** proposed[79a] for this alkaloid. These results probably are didactic for modeling CCS of organic compounds. *By qualitatively analyzing a concrete molecular structure, one may often foresee whether either QM method rather will or will not show a considerable accuracy in modeling it.* Indeed, from assessing observations (*1*)–(*3*) together with the structural features of this molecule, we can learn when a "standard" calculation tool can provide at least a fair accuracy for CCS. The assessment and the lessons are as follows:

Appropriate calculation method. The backbone of **45** is a medium-sized cage structure enriched by atoms with one or two lone electron pairs. There are four such atoms of relatively large covalent radius, i.e., four 1,3-positioned As atoms, as well as three atoms of smaller covalent radius, i.e., three O atoms. By a very qualitative representation, a cloud of their lone electron pairs covers and slightly penetrates almost the entire outer surface of the molecular cage of **45**. One can reasonably assume that electron correlation effects should be very strong in this molecule. The size of the heterocyclic skeleton in **45** is not as large as these effects could not affect chemical shielding of any of its atoms.[a] In solution, H-bonds $As\cdots HCCl_3$ bind solvent molecules (they are H-donors) to the solute (that, as mentioned, has an electron-rich molecular surface). Thus, supramolecule **45** $\cdot$ 4 $CHCl_3$ (Fig. 103) is a minimal molecular model with truncated solvation shell in explicitly representing the solvate of **45**.

From this perspective, we can expect that DFT calculations from any rung of the DFT ladder excluding double-hybrid (fifth-rung) functionals[b] (Section 5.2.6) would not demonstrate an acceptable accuracy of chemical shieldings for the carbons of **45** for both the molecular model for vacuum and solution. As we learned from Section 5.2.6, such methods do not "feel"

---

[a]However, it is questionable whether correlation of lone pair electrons of the "adamantane" skeleton-forming atoms significantly affects chemical shieldings of the electron-poor skeleton substituents, i.e., the H atoms.

[b]Those are designed to target long-range electron correlation (Section 5.2.6). The failure of density functionals B3LYP, BP86, and TPSSh in modeling chemical shifts for the "DFT-unfriendly" molecule **45** suggests that currently predominated (i.e., the third and forth rung) DFT methods could be not suitable for modeling organic systems that bear electron-rich fragments at distances 0.3–0.6 nm.

long-range electron correlation. In contrast, a better accuracy is anticipated for CCS of this system if *ab initio* methods that reckon with electron correlation (Section 5.2.3) are used. As Table 8 shows, these considerations have turned out to be true. The MP2 approximation that takes account of electron correlation to a significant extent is impressively better in modeling ^{13}C chemical shifts for **45** than any other examined DFT method. If the molecular structure of this compound is even primitively considered (see *supra*), attempts to use a second- or third-rung DFT method for modeling its ^{13}C chemical shifts would be rather unnecessary. Thus, *when choosing the calculation method for chemical shift modeling, it is worth paying attention to chemical structure as well as 3D geometry of the compound of interest. If computer resources available in your organic or bioorganic laboratory are sufficiently extensive, MP2 is your optimal choice for modeling chemical shifts of acceptable accuracy.*

Appropriate basis set. Dunning's basis sets (Section 5.2.4) "win" if their zeta is high ($x \geq Q, 5, etc.$); the use of a Dunning's basis set with a lower zeta entails some risk of obtaining inaccurate results for molecular systems containing electrons not involved in the σ bonding. Chemical structure **45** suggests the possibility of such an ill scenario, and the "CCS-shaming" inaccuracy (entries 1, 2, and 11 in Table 8) confirms our worst expectations. When undertaking their own computations, organic experimentalists may hold that *low extended Pople's basis sets [e.g. 6-31++G(d,p)], on average, are more relevant in routine non-empirical modeling of chemical shifts of organic molecules than non-augmented low-zeta Dunning's basis sets [e.g., cc-pVDZ].*[a]

Model for vacuum or model for solution? Solvent–solute interactions for **45** are not strong in chloroform; however, they are not negligible. Indeed, according to our *ab initio* model, H-bonding takes place; however, H-bonds As$\cdots$HCCl$_3$ are long (Fig. 103). We can presume therefore that involvement of a continuum solvation model in theoretically approaching this molecular system will increase the CCS accuracy to some considerable, but not crucial, extent, and the further accuracy improvement will not be significant if H-bonding is explicitly modeled. Exactly such results

---

[a]However, starting from *aug*-cc-pVTZ, Dunning's basis sets are preferable in modeling both molecular geometry and CCS, in respect to low extended Pople's basis sets.

have been obtained when the proper calculation method has been used (see MP2-related entries in Table 8). Confirmation of this prognosis advances an obvious idea that *the PCM model of solvation is sufficient in modeling chemical shifts for a chloroform solution of a compound*[a] *which molecular structure does not include electron-donating fragments*, e.g., heteroatoms with lone electron pairs, aryl rings, ylide moieties. For instance, trialkylboranes in $CDCl_3$ may be addressed as "PCM-compatible" structures when modeling their chemical shifts, while triarylboranes, or phosphine oxides, or trialkylamines would require explicit modeling of the solvent if their modeled NMR spectra are intended to be quantitatively compared with relevant spectra recorded for $CDCl_3$ solutions.

Consequently, the truncated solvation model is useful for non-empirically obtaining verisimilar values of chemical shifts for compounds that expose electron-donating moieties to protic solvent molecules. H-bonds in such solute–solvent complexes are short. This simple move toward CCS accuracy would require to model, e.g., a monosaccharide in water, as a supramolecule (a core molecule with hydroxyl groups connected to water molecules *via* H-bonds) in the "PCM water" bulk.

Chemical shift differences. The calculation results shown in Table 8 for structure **45** demonstrate that differences $\Delta\delta_{exp\_i,j}$ ($\Delta\delta_{exp\_i,j} = \delta_{exp\_j} - \delta_{exp\_i}$) for experimental chemical shifts of the *i*-th and *j*-th atoms *of the same chemical functionality* in a molecule are reproduced by the related differences $\Delta\delta_{calc\_i,j}$ ($\Delta\delta_{calc\_i,j} = \delta_{calc\_j} - \delta_{calc\_i}$; for simplicity, denoted in Table 8 as Δ_{calc}) essentially more accurately than the values of individual chemical shifts δ_{exp_i} and δ_{exp_j} are reproduced by the values of individual cCCS δ_{calc_i} and δ_{calc_j}, respectively. The reason is transparent: the $\Delta\delta_{calc\_i,j}$ values, being numerical differences of the values with a similar calculation error, are more accurate than the parent values δ_{exp_i} and δ_{exp_j}. The considered three-carbon molecule **45** is characterized by only one difference $\Delta\delta_{expi,j}$ for ^{13}C nuclei. With increasing the quality of the structural model, a satisfactory accuracy for the corresponding *calculated* value $\Delta\delta_{calc\_i,j}$ (Table 8, last entry for calculated values) is reached. This

---

[a]Exceptional ubiquity of $CDCl_3$ as an NMR solvent often requires the modeling of chemical shifts for chloroform solutions.

accuracy shown even by this single difference $\Delta\delta_{\text{calc}\_i,j}$ of cCCS values supplies the considered structural assignment with additional reliability.

One may notice that the accuracy of the obtained difference $\Delta\delta_{\text{calc}\_i,j}$ for structure **45** is two–three times higher than the accuracy of parent cCCS values δ_{calc_j} and δ_{calc_i}. This example indicates that, using cCCS differences $\Delta\delta_{\text{calc}\_i,j}$ instead of cCCS themselves, one may impart additional possibilities to structure determinations by CCS.

Let us see what this statement means regarding conformational analysis. cCCS of characteristic carbons are most interesting for us, when comparing cCCS sets of alternative stereoisomeric structures with the set of experimental chemical shifts in CCS-based conformational analysis (Section 6.2.1). As explained in Section 6.2.3, the number of characteristic nuclei in the molecule (Z) is, as a rule, very small, and this circumstance essentially restricts a very inviting conformational analysis by "resolving" trivial NMR spectra with Eqs. 57_i. The indicated increased accuracy of values of differences $\Delta\delta_{\text{calc}\_i,j}$ permits us to consider more magnetic nuclei in the molecule as characteristic ones. Indeed, the practical criterion for assigning, e.g., the i-th carbon nucleus of the molecule to characteristic, is $\Delta\delta_{\text{exp}\_i} > 1$ ppm (Section 6.2.3). If we consider differences $\Delta\delta_{\text{calc}\_i,j}$ for carbons of modeled stereoisomer A and such differences for carbons of alternative modeled stereoisomer B, and find that the value of $\Delta\delta_{\text{calc}\_i,j}$ for i-th and j-th carbons in stereoisomer A differs from the corresponding value of $\Delta\delta_{\text{calc}\_i,j}$ for stereoisomer B by more than 0.5 ppm, we, assuming the 0.5 ppm accuracy of values of such calculated differences, can modify Eqs. 57_i and 57_j for these carbons and write a new consistent Eq. 61

$$\Delta\delta_{\text{exp}\_i,j} = p_A(\Delta\delta_{\text{calc}\_i,j})_A + p_B(\Delta\delta_{\text{calc}\_i,j})_B \ldots + p_F(\Delta\delta_{\text{calc}\_i,j})_F$$

$$(61)$$

where F is the number of low-energy (NMR-spectra-contributing) conformers, indexes i and j depict nuclei with the values of $\Delta\delta_{\text{calc}\_i,j}$ that differ by more than 0.5 ppm at least for one pair of modeled stereoisomers, $\Delta\delta_{\text{exp}\_i,j}$ is the difference of chemical shifts for these nuclei in the related experimental NMR spectrum registered for conditions of fast conformational equilibrium, and $p_A, p_B, \ldots, p_F$ are fractions of the conformers.

By this technique, we actually weaken the criterion for assigning carbons to characteristic ones, or, equivalently, with employing cCCS differences

$\Delta\delta_{\text{calc}\_i,j}$, we increase number Z for the same molecule. Equations 57_i become possible to be solved if, by replacing them with new consistent equations of type Eq. 61, one may fulfill the condition $Z \geq F$ (Section 6.2.3). As mentioned in that Section, the number of NMR-weighty conformers is usually small for non-polymeric and non-macrocyclic organic compounds. Since the number of new consistent equations of type Eq. 61 obviously is larger than the number of Eqs. 57_i for the same molecule, condition $Z \geq F$ will be rather satisfied. Differences $\Delta\delta_{\text{calc}\_i,j}$ do appear useful in CCS-based conformational analysis.

In summary, we can learn from this Chapter that non-empirical calculations of NMR chemical shifts are an effective, broadly applicable methodology for exploring conformational space of organic molecules of ordinary size. Components of this computational methodology are conformational search (Section 4.3.2), geometry optimization with using routine QM methods (Sections 5.2.3, 5.2.4 and 5.2.6), QM modeling of chemical shifts (Section 6.1.2), and identification of NMR-detectable conformers by comparing experimental chemical shifts and cCCS obtained for alternative stereoisomeric structures (Sections 6.2.2 and 6.2.3). If derived and treated thoughtfully, cCCS are capable of providing the desired reliability to computational conformational analysis.

Appendix: Glossary

(The glossary does not include abbreviations of names of chemical compounds, spectral methods, theoretical methods of computational chemistry, basis sets, and related computer programs)

AO	atomic orbital
cCCS	consistent calculated chemical shifts
CCS	calculated chemical shifts
DFT	density functional theory
ESD	Euler steepest descent
GUI	graphical user interface
HBinMEP	highest barrier in the minimal energy path
HPC	high performance computer
ISR	isolate rotation
MAE	mean absolute error
MAD	mean absolute difference
MEP	minimal energy path
MD	molecular dynamics
MM	molecular mechanics
MO	molecular orbital
NIR	nitrogen inversion–rotation
NMA	normal mode analysis
PCM	polarized continuum model
PES	potential energy surface
QM	quantum mechanics

RI   ring inversion
RICIR   ring inversion–carbon inversion–rotation
RINI   ring inversion–nitrogen inversion
RINIR   ring inversion–nitrogen inversion–rotation
RMS   root mean of squares
RMSD   *RMS* deviation
RMSE   *RMS* error
SES   steric energy surface
VDW   van der Waals
VSEPR   valence shell electron pair repulsion

Bibliography

1. Gillespie, R. J. *Molecular Geometry*; 1972, Van Nostrand Reihold, Londres.
2. Gillespie, R. J., Robinson, E. A. *Chem. Soc. Rev.*, **2005**, *34*, 396–407.
3. a) Hall, C. D. In *Organophosphorous Chemistry*, 2002, pp. 74–90. b) Akiba, K. (Ed.) *Chemistry of Hypervalent Compounds*, John Wiley-VCH, New York, 1998. c) Mikhailov, B. M. *Pure Appl. Chem.*, **1980**, *52*, 691–704. d) Bersuker, I. B. *Chem. Rev.*, **2013**, *113*, 1351–1390.
4. a) Belostotskii, A. M., Gottlieb, H. E., Hassner, A. *J. Am. Chem. Soc.*, **1996**, *118*, 7783–7789. b) Arnason, I., Gudnason, P. I., Björnsson, R., Oberhammer, H. *J. Phys. Chem. A*, **2011**, *115*, 10000–10008.
5. a) Bock, H., Goebel, I., Havlas, Z., Liedle, S, Oberhammer, H. *Angew. Chem., Int. Ed. Engl.*, **1991**, *30*, 187. b) Fritz, G., Matern, E., Krautscheid, H., Ahlrichs, R., Olkowska, J. W., Pikies, J. *Z. Anorg. Allg. Chem.*, **1999**, *625*, 1604–1607.
6. Bürger, H., Niepel, H, Pawelke, G., Oberhammer, H. *J. Mol. Struct.*, **1979**, *54*, 159–174.
7. Gaensslen, M., Gross, U., Oberhammer, H., Rüdiger, S. *Angw. Chem., Int. Ed. Engl.*, **1992**, *31*, 1467–1468.
8. Klaptke, T. M., Schulz, A. *J. Fluor. Chem.*, **1997**, *82*, 181–183.
9. Belostotskii, A. M. (unpublished data).
10. Calhorda, M. J., Krapp, A., Frenking, G. *J. Phys. Chem. A*, **2007**, *111*, 2859–2869.
11. Jemmis, E. D., Gopakumar, G. In *The Chemistry of Organolithium Compounds*; Rappoport, Z., Marek, I., Eds.; John Wiley&Sons, New York, **2004**, 1–45.
12. Mulvey, R. E. *Chem. Soc. Rev.*, **1998**, *27*, 339–346.
13. Dmitriev, I. S. *Electron through the Eyes of The Chemist*; 1986, Chimiya, Leningrad.
14. Rauk, A. *Orbital Interaction Theory of Organic Chemistry*; 2001, John Wiley & Sons, New-York.

15. Mullin, W. J., Blaylock, J. *Amer. J. Phys.*, **2003**, *71*, 123–1231.

16. Gillespie, R. *J. Struct. Chem.*, **1998**, *9*, 73–76.

17. Popilier, P. L. A., *Coor. Chem. Rev.*, **2000**, *197*, 169–189.

18. Rolke, J. Zheng, Y., Brion, C. E., Shi, Z., Wolfe, S., Davidson, E. R. *Chem. Phys.*, **1999**, *244*, 1–24.

19. a) Orwille-Thomas, W. J. *Internal Rotation in Molecules*; **1974**, John Wiley & Sons, London. b) Bachrach, S. M. *J. Phys. Chem. A.*, **2011**, *115*, 2396–2401. c) Michl, J., West, R. *Acc. Chem. Res.*, **2000**, *33*, 821–823. d) Maekawa, H., Ge, N.-H. *J. Phys. Chem. B.*, **2012**, *116*, 11292–11301. e) Varghese, O. P., Barman, J., Pathmasiri, W., Plashkevych, O., Honcharenko, D., Chattopadhyaya, J. *J. Amer. Chem. Soc.*, **2006**, *128*, 15173–15187. f) Neumann, F., Teramae, H., Downing, J. W., Michl, J. *J. Am. Chem. Soc.*, **1998**, *120*, 573–582. g) Bertolucci, M., Harris, D. *Symmetry and Spectroscopy*, **1978**, Oxford University Press.

20. a) Ōsawa, E., Shirahama, H., Matsumoto, T. *J. Am. Chem. Soc.*, **1979**, *101*, 4824–4832. b) Jalluri, R. K., Yuh, Y. H., Taylor, E. W. In *The Anomeric Effect and Associated Stereoelectronic Effects*; Tratcher, G. R. J., Ed.; American Chemical Society, Washington, **1993**, 277. c) Saya, S., Wang, F., Falzon, C. T., Brunger, M. J. *J. Chem. Phys.*, **2005**, *123*, 124315–124327. d) Pemberton, R. P., MsShane, C. M., Castro, C., Kamey, W. L. *J. Amer. Chem. Soc.*, **2006**, *128*, 16692–16700.

21. a) Brocas, J. *The Permutational Approach to Dynamic Stereochemistry*; **1983**, McGraw-Hill, Cambridge. b) Couzijn, E. P. A., Slootweg, J. C., Andreas W. Ehlers, A. W., Lammertsma, K. *J. Amer. Chem.. Soc.*, **2010**, *132*, 18127–18140.

22. a) Casanova, D., Cirera, J., Llunell, M., Alemany, P., Avnir, D., Alvarez, S. *J. Amer. Chem. Soc.*, **2004**, *126*, 1755–1763. b) Alvarez, S., Alemany, P., Casanova, D., Cirera, J., Llunell, M., Avnir, D. *Coordination Chem. Rev.*, **2005**, *249*, 1693–1708.

23. a) Belostotskii, A. M., Hassner, A. *J. Phys. Org. Chem.*, **1999**, *12*, 659–663. b) Lambert, J. B. *Top. Stereochem.*, **1971**, *6*, 19–105. c) Bach, R. D., Raban, M. In *Organonitrogen Stereodynamics*; Lambert, J. B., Takeuchi, Y., Eds.; VCH Publishers, New-York, **1992**, 63–103. d) Dixon, D., Arduengo, A. J. *J. Phys. Chem.*, **1987**, *91*, 3195–3200. e) Creve, S., Nguyen, M. T. *J. Phys. Chem. A*, **1998**, *102*, 6549–6557.

24. López, C. S., Faza, O. F., Gregersen, B. A., Lopez, X., de Lera, A. R., York D. M. *Chem Phys Chem.*, **2004**, *5*, 1045–1049.

25. a) Buchweller, C. H. In *Acyclic Organonitrogen Stereodynamics*; Lambert, J. B., Takeuchi, Y., Eds.; VCH Publishers, New-York, **1992**, 1–55. b) Belostotskii, A. M., Aped, P., Hassner, A. *J. Mol. Struct. (THEOCHEM)*, **1997**, *398–399*, 427–434.

26. Sebag, A. B.; Forsyth, D. A.; Plante, M. A. *J. Org. Chem.*, **2001**, *66*, 7967–7973.

27. Belostotskii, A. M., Gottlieb, H. E., Shokhen, M. *J. Org. Chem.*, **2002**, *67*, 9257–9266.
28. Sato, S., Takahashi, O., Furukawa, N. *Coor. Chem. Rev.*, **1998**, *176*, 483–514.
29. a) Adzima, L. J., Martin, J. C. *J. Org. Chem.*, **1977**, *42*, 406–416. b) Gottlieb, H. E., Hoz, S., Elyashiv, I., Albeck, M. *Inorg. Chem.*, **1994**, *33*, 808–811. c) Mauksch, M., Schleyer, P. von R. *Inorg. Chem.*, **2001**, *40*, 1756–1769.
30. a) Gordon, M. S., Schmidt, M. W. *J. Amer. Chem. Soc.*, **1993**, *115*, 7486. b) Yoshizawa, K. *J. Organomet. Chem.*, **2001**, *635*, 100–109.
31. a) IUPAC Recommendations: Preferred IUPAC Names, Chapter 9, 2004. b) Testa, B., Vistoli, G., Pedretti, A. *Helvetica Chem. Acta*, **2013**, *96*, 564–623.
32. a) Sánchez-Gonzáles, A., Martinez-Garcia, H., Melchor, S., Dobado, J. A. *J. Phys. Chem. A*, **2004**, *108*, 9188–9195. b) Brand, H., Liebman, J. F., Schulz, A., Mayer, P., Villinger, A. *Eur. J. Inorg. Chem.*, **2006**, 4294–4308. c) Gakh, A. A., Bryan, J. C., Burnett, M. N., Bonnesen P. V. *J. Mol. Struct.*, **2000**, *520*, 221–228.
33. a) Eliel, E. L., Wilen, S. H., Mander, L. N. *Stereochemistry of Organic Compounds*; **1994**, John Wiley & Sons Inc., New-York. b) Cramer, C. J. *Essentials of Computational Chemistry*; **2004**, John Wiley & Sons, Chichester. c) Seeman, J. I. *Chem. Rev.*, **1983**, *83*, 83–134.
34. a) Gawley, R. E., Aubé, J. *Principles of Asymmetric Synthesis*; **1996**, Pergamon, New-York. b) Nógrádi, M. *Stereoselective Synthesis: A Practical Approach*; **1995**, VCH, Weinheim; New-York. c) P. S. Kalsi *Stereochemistry. Conformation and Mechanism*; **1990**, Wiley Eastern, New-Delhi. d) Eliel, E. L., Allinger, N. L., Angual, S. J., Morrison, G. A. *Conformational Analysis*; **1965**, Wiley, New-York. e) Robinson, M. *Organic Stereochemistry*; Oxford University Press, **2000**, Oxford. f) Nasipuri, D. *Stereochemistry of Organic Compounds: Principles and Applications*; New Age International, **1994**, New Delhi. g) Wolf, C. *Dynamic Stereochemistry of Chiral Compounds*; **2008**, RSC Publishing, London. f) Edenborough, M. *Organic Reaction Mechanisms*; **1998**, CRC Press, London. g) Carey, F. A., Sundberg, R. J. *Advanced Organic Chemistry*; **2007**, Springer, New-York.
35. a) Drozd, V. N., Zefirov, N. S., Sokolov, V. I., Stankevich, I. V. *Zh. Org. Khim.*, **1979**, *15*, 1785–1790. b) Sokolov, V. I. In *Introduction to Theoretical Stereochemistry*, Gordon & Breach, London. c) Sokolov, V. I. *Russ. J. Org. Chem.*, **2002**, *38*, 1553–1563.
36. a) Raban, M., Kost, D. In *Acyclic Organonitrogen Stereodynamics*; Lambert, J. B., Takeuchi, Y., Eds.; VCH Publishers, New-York, **1992**, 57–88. b) Jennings, W. B, Wilson, V. E. In *Acyclic Organonitrogen Stereodynamics*; Lambert, J. B., Takeuchi, Y., Eds.; VCH Publishers, New-York, **1992**, 177–243. c) Blanco, F., Alkorta, I., Elguero, J. *Croat. Chem. Acta*, **2009**, *82*, 173–183. d) Cahoon, J. F., Sawyer, K. R., Schlegel, J. P., Harris, C. B. *Science*, **2008**, *319*, 1820–1823. e) Zewail, A. H., Polanyi, J. C. *Acc. Chem. Res.*, **1995**, *28*, 119–132.

f) Penner, G. H., Chang, Y.-C. P., Nechala, P., Froese, R. *J. Org. Chem.* **1999,** *64,* 447-452. g) Belostotskii, A. M., Markevich, E. *J. Org. Chem.*, **2003**, *67,* 3055–3063. h) Huang, H., Stewart, T., Gutmann, M., Ohara, T., Niimura, N., Li,Y.-X., Wen, J.-F., Bau, R., Wong, H. N. C. *J. Org. Chem.*, **2009**, *74*, 359–369.

37. Belostotskii, A. M., Gottlieb, H. E., Aped, P., Hassner, A. *Chem. Eur. J.*, **1999**, *5*, 449–455.

38. a) Cremer, D., Szabo, K. J. In *Conformational Behavior of Six-Membered Rings*; Juaristi, E., Ed.; VCH Publishers, New-York, **1995**, 59–136. b) Mekenyan, O. G., Veith, G. D. In *From Chemical Topology to Three-Dimensional Geometry*; Balaban, A. T., Ed. Kluver Academic Publishers, New-York, **2002**, 43–72. c) Freeman, F., Hwang, J. H., Junge, E. H., Parmar, P. D., Renz, Z., Trinh, J. *Int. J. Quant. Chem.*, **2008**, *108*, 339–350. d) Rabideau, P. W. In *Conformational Analysis of Cyclohexenes, Cyclohexadienes, and Related Hydroaromatic Compounds*; Rabideau, P. W., Ed., VCH Publishers, New-York, **1989**, 91–126. e) Lipkowitz, K. B. *ibid*, 212–265. f) Cremer, D., Pople, J. A. *J. Am. Chem. Soc.*, **1975**, *97*, 1354–1358. g) Zubkov S. V., Chertkov, V. A. *Int. J. Mol. Sci.*, **2003**, *4*, 107–118. h) Belostotskii, A. M., Aped, P., Hassner, A. *J. Mol. Struct. (THEOCHEM)*, **1998**, *429*, 265–273. i) Makarewicz, J., Ha, T.-K. *J. Mol. Struct.*, **2001**, *599*, 271–278. j) De Leeuw, F., Altona, C. *J. Comput. Chem.*, **1983**, *4*, 428–437. k) Hendrickx, P., Martins, J. C., *Chem. Central J.*, **2008**, *2*, 20.

39. a) Testa, B. In *New Comprehensive Biochemistry*; Neuberger, A., van Deenen, L. L. M., Eds.; Elsevier Biomedical Press, Amsterdam, **1982**, v. 3, 9–10. b) Testa, B. *Principles of Organic Stereochemistry*; **1979**, Marcel Dekker, New-York.

40. a) Burkert, U., Allinger, N. L. *Molecular Mechanics*; **1982**, ACS, Washington, p. 3. b) Csizmadia, I. G. *NATO ASI Series, Series C,* **1988**, *267*, 1–32.

41. a) Martin, R. H., Marchant, M.-J. *Tetrahedron*, **1974**, *30*, 347–349. b) Wolf, C. *Dynamic Stereochemistry of Chiral Compounds*; **2008**, RCS, Cambridge, p. 54. c) Forsyth, D. A.; Zhang, W.; Hanley, J. A. *J. Org. Chem.*, **1996**, *61*, 1284–1289.

42. Sartillo-Piscil, F., Cruz, S., Sánches, M., Höpfl, H., de Parrodi, C. A., Quintero, L. *Tetrahedron*, **2003**, *59*, 4077–4083.

43. Gill, G., Pawar, D. M., Noe, E. A. *J. Org. Chem.*, **2005**, *70*, 10726–10731.

44. Nelsen, S. F., Ippoliti, J. T., Frigo, T. B., Petillo, P.A. *J. Am. Chem. Soc.* **1989**, *111*, 1776–1781.

45. a) Schneider, H.-J.; Sturm, L. *Angew. Chem.*, **1976**, *88*, 574–575. b) Belostotskii, A. M., Goren, Z., Gottlieb, H. E. *J. Nat. Prod.*, **2004**, *67*, 1842–1849.

46. a) Belostotskii, A. M., Hassner, A. *J. Comput. Chem.*, **1998**, *19*, 1786–1794. b) Mislow, K. *Chemtracts — Org Chem.*, **1989**, *2*, 151–174. c) Metzger, J. V., Chanon, M. C., Roussel, C. M. *Rev. Hetrocycl. Chem.*, **1997**, *16*, 1–45. d) Dobrowolski, M. A., Ciesielski, A., Cyrański, M. K. *Phys. Chem. Chem.*

Phys., **2011**, *13*, 20557–20563. e) Squillacote, M. E., DeFellipis, J., Shu, Q. *J. Amer. Chem. Soc.*, **2005**, *127*, 15983–15988.

47. a) Sygula, A., Rabideau, P. W. In *Carbon-Rich Compounds*; Halley, M. M., Tykwinsky, R. R., Eds.; Wiley-VCH, New-York, **2006**, 529–566. b) Zywietz, T., Jiao, H., Schleyer, P. R., Meijere, A. *J. Org. Chem.*, **1998**, *63*, 3417–3422. c) Yanney, M., Fronczek, F. R., Henry, W. P., Beard, D. J., Sygula, A. *Eur. J. Org. Chem.*, **2011**, 6636–6639.

48. Shepard, M. J., Paddon-Row, M. N. *Aust. J. Chem.*, **1993**, *46*, 547.

49. a) Aped, P., Senderowitz, H. In *The Chemistry of Amino, Nitroso, Nitro and Related Groups*; Patai, S., Ed.; John Wiley & Sons, **1996**, 1–84. b) Rappê, A. K., Casewit, C. J. *Molecular Mechanics Across Chemistry*; **1997**, University Science Books, Sausalito, CA. c) Brooks, B. R., Brooks, C. L., MacKerell, A. D., Nilsson, L., Petrella, R. J., Roux, B., Won, Y., Archontis, G., Bartels, C., Boresch, S., Calflisch, A., Caves, L., Cui, Q., Dinner, A. R., Feig, M., Fischer, S., Gao, J., Hodoscek, M., Im, W., Kuczera, K., Lazaridis, T., Ma, J., V. Ovchinnikov, V., Paci, E., Pastor, R. W., Post, C. B., Pu, J. Z., Schaffer, M., Tidor, B., Venable, R. M., Woodcock, H. L., H. L., Wu, X., Yang, W., York, D. M., Karplus, M. *J. Comput. Chem.*, **2009**, *30*, 1545–1614. d) Ponder, J. W., Case, D. A., *Adv. Prot. Chem.*, **2003**, *66*, 27–85. e) Vanommeslaeghe, K., Hatcher, E., Acharya, C., Kundu, S., Zhong, S., Shim, J., Darian, E., Guvench, O., Lopes, P., Vorobyov, I., MacKerell A. D. *J. Comput. Chem.*, **2010**, *31*, 671–690. f) Druzdzel, M. J. *Artificial Intelligence Simulation Behaviour Quarterly*, **1996**, *94*, 43–54.

50. a) Mackerell., A. D. *J. Comput. Chem.*, **2004**, *25*, 1584–1604. b) Liljefors, T., Gundertofte, K., Pettersonn, I. In *Computational Medicinal Chemistry for Drug Discovery*; Bultinck. P., Ed.; CRC Press, London, **2004**. c) Gerbst, A. G., Grachev, A. A., Shashkov, A. S., Nifantiev, N. E. *Russ. J. Bioorg. Chem.*, **2007**, *33*, 24–37. d) Hemmingsen, L., Madsen, D. E., Esbensen, A. L., Engelsen, S. B. *Carbohydr. Res.*, **2004**, *339*, 937–948. e) Stortz C. A., Johnson G. P., French A. D., Csonka, G. I. *Carbohydr. Res.*, **2009**, *344*, 2217–2228. f) Foley, B. L., Tessier, M. B., Woods, R. J. *WILEY Interdiscipl. Rev. Comput. Mol. Sci.*, **2012**, *2*, 652–697. g) Guvench, O., Mallajosyula, S. S., Raman, E. P., Hatcher, E., Vanommeslaeghe, K., Foster, T. J., Jamison, F. W., MacKerell, A. D. *J. Chem. Theory Comput.*, **2011**, *7*, 3162–3180.

51. Gillis, E. P., Burke, M. D. *J. Amer. Chem. Soc.*, **2007**, *129*, 6716–6717.

52. a) Wiberg, K. B., Bailey, W. F. *J. Mol. Struct.*, **2000**, *556*, 239–244. b) Golębiewski, W. M. *Magn. Resonance Chem.*, **1986**, *24*, 105–112.

53. a) Bachrach, S. M. *Computational Organic Chemistry*; **2007**, Wiley-Interscience, New-York. b) Rzepa, H. S. *Phys. Chem. Chem. Phys.*, **2009**, *11*, 1340–1345. c) Kong, L., Bischoff, F. A., F. Valeev, E. F. *Chem. Rev.*, **2012**, *112*, 75–107.

54. Zhang, W., Hou, T., Qiao, X., Xu, X. *J. Phys. Chem.*, **2003**, *107*, 9071–9078.

55. Roat-Malone, R. M. *Bioinorganic Chemistry*; **2007**, Wiley-Interscience, New-York.

56. Belostotskii, A. M., Gottlieb, H. E., Aped, P. *Chem. Eur. J.*, **2002**, *8*, 3016–3026.

57. Mock, W. L. *J. Amer. Chem. Soc.*, **1970**, *92*, 7610–7612.

58. a) Hinchliffe, A. *Molecular Modelling for Beginners*; **2008**, John Wiley & Sons, Chichester. b) Schlegel, H. B. *WIREs Comput. Mol. Sci.*, **2011** *1*, 790–809.

59. Settergren, N. M., Bülmann, P., Amin, E. A. *THEOCHEM.*, **2008**, *861*, 68-73.

60. a) Csonka, G. I., French, A. D., Johnson G. P., Stortz, C. A. *J. Chem. Theory Comput.*, **2009**, *5*, 679–692. b) Göricke, L., Grimme, S. *Phys. Chem. Chem. Phys.*, **2011**, *13*, 6670–6688. c) Göricke, L., Grimme, S. *J. Chem. Theory Comput.*, **2011**, *7*, 291–309. d) Perdew, J. P., Ruzsinszky, A., Constantin, L. A., Sun, J., Csonka, G. I. A. *J. Chem. Theory Comput.*, **2009**, *5*, 902–908. e) Burns, L. A., Vázquez-Mayagoitia, Á, Sumpter, B. G., Sherrill, C.D., *J. Chem. Phys.* **2011**, *134*, 84–107. f) Goerigk, L., Grimme, S. *WIREs Comput. Mol. Sci.*, **2014**, *4*, 576–600.

61. a) Papajak, E., Leverentz, H. R., Cheng, J., Truhlar, D. G. *J. Chem. Theory Comput.*, **2009**, *5*, 1197–1202. b) Adler, T. B., Werner, H.-J. *J. Chem. Phys.*, **2008**, *128*, 84–102. c) Copeland, K. L., Tschumper, G. S. *J. Chem. Theory Comput.*, **2012**, *8*, 4279–4284. d) Sherrill, C. D. *J. Chem. Phys.*, **2010**, *132*, Article ID 110902.

62. a) Young, D. *Computational Chemistry*; **2001**, John Wiley & Sons, New-York. b) Lewars. E. *Computational Chemistry*; **2003**, Kluwer Academic Publishers, Norwell. c) Feller, D., Davidson, E. R. *Rev. Comput. Chem.*, **1990**, 1–43. d) Müller, T. In *Computational Nanoscience: Do It Yourself!*; Grotendorst, J., S. Blügel, S., Marx., D., Eds.; NIC Series, **2006**, *31*, 19–43.

63. Lee, J.-C., Khitrin, A. K. *Concepts Magn. Resonanace*, Part A, **2008**, *32A*, 56–67.

64. Margulès, L., Demaison, J., Boggs E. B. *Struct. Chem.*, **2000**, 145–154.

65. Oberhammer, H. *J. Comput. Chem.*, **1998**, *19*, 123–128.

66. a) Tomasi, J., Mennucci, B., Cammi, R. *Chem. Rev.*, **2005**, *105*, 2999–3094. b) Ohrn, A., Karlstorm, G. In *Solvation Effects on Molecules and Biomolecules*; Canuto, S., Ed.; Springer Verlag, Berlin, **2008**.

67. a) Cramer, C. J., Truhlar, D. G. *Acc. Chem. Res.*, **2008**, *41*, 760–768. b) Cramer, C. J., Truhlar, D. G. *Acc. Chem. Res.*, **2009**, *42*, 493–497. c) Klamt, A., Mennucci, B., Tomasi, J., Barone, V., Curutchet, C., Orozco, M., Luque, F. G. *Acc. Chem. Res.*, **2009**, *42*, 489–492.

68. a) Facelli, J. C. *Concepts Magn. Resonance*, **2004**, *20A*, 42–69. b) Casabianca, L. B., de Dios, A. C. *J. Chem. Phys.*, **2008**, *128*, 52201–52210. c) Jain, R., Bally, T., Rablen, P. R. *J. Org. Chem.*, **2009**, *74*, 4017–4023. d) Alkorta, I., Elguero, *J. Struct. Chem.*, **2003**, *14*, 377–389. e) Cheeseman, J. R.;

Trucks, G. W.; Keith, T. A.; Frisch, M. J. *J. Chem. Phys.*, **1996**, *104*, 5497–5509. f) Gauss, J., Stanton, J. F. *Adv. Chem. Phys.*, **2002**, *123*, 355–422.

69. a) Gauss, J., Stanton, J. F. In *Calculation of NMR and EPR parameters*; Kaupp, M., Bühl, M., Malkin, V. G., Eds.; Wiley-VCH, Weinheim, **2004**. b) Mulder, F. F. A., Filatov, M. *Chem. Soc. Rev.*, **2010**, *39*, 578–590. c) Ruden, T. A., Ruud, K. In *Calculation of NMR and EPR parameters*; Kaupp, M., Bühl, M., Malkin, V. G., Eds.; Wiley-VCH, Weinheim, **2004**.

70. Sergeev, N. M., Subbotin, O. A.., *Russ. Chem. Rev.*, **1978**, *47*, 265–280.

71. a) Heydenreich., M, Koch, A., Sarodnick, J., Kleinpeter, E. *Tetrahedron*, **2005**, *61*, 2373–2385. b) Zhanoušek, Z., Holub, J., Hnuk, D., Londesborough, M. G. S., Shoemaker, R. K., *Polyhedron*, **2003**, *22*, 3541–3545. c) Balandina, A., Saifina, D., Mamedov, V., Latypov, S. *J. Mol. Struct.*, **2006**, *791*, 77–81. d) Burns, N. Z., Krylova, I. N., Hannoush, R. N., Baran, P. S. *J. Amer. Chem. Soc.*, **2009**, *131*, 9172–9173.

72. Neuhaus, D. Williamson, M. P. *The Nuclear Overhauser Effect In Structural and Conformational Analysis*, **1989**, VCH, New-York.

73. a) Meiboom, S., Hewitt, A. C., Luz, Z. *J. Chem. Phys.*, **1977**, *66*, 4041–4051. b) Anet, F. A. L., Basus, V. J. *J. Amer. Chem. Soc.*, **1973**, *95*, 4424–4426.

74. a) Eisenberg, D., Filatov, A. S., Jackson, E. A., Rabinovitz, M., Petrukhina, M. A., Scott, L. T., Shenhar, R. *J. Org. Chem.* **2008**, *73*, 6073–6078. b) Wiberg K. *J. Org. Chem.*, **2003**, *68*, 9322–9329. c) Bachrach, S. M. *J. Org. Chem.*, **2009**, *74*, 3609-3611. d) Nishinaga, T., Uto, T., Inoue, R., Matsuura, A., Treitel, N., Rabinovitz, M., Komatsu, K. *Chem. Eur. J.*, **2008**, *14*, 2067–2074. e) Bürgi, R., Pitera, J, van Gunsteren, W. F. *J. Biomolec. NMR*, **2001**, *19*, 305–320. f) Hoffman, R. E. *J. Magn. Resonance*, **2003**, *163*, 325–331.

75. a) Forsyth, D. A.; Sebag, A. B. *J. Amer. Chem. Soc.*, **1997**, *119*, 9483–9494. b) Jimeno, M. L.; Alkorta, I.; Elguero, J.; Anderson, J. E.; Claramunt, R. M.; Lavandera, J. L. *New J. Chem.*, **1998**, 1079–1083. c) Costa, V. E. U.; Grunewald-Nichele, A.; Carneiro, J. W. M. *J. Mol. Struct.*, **2004**, *702*, 71–76. d) Wang, B.; Miskolizie, M.; Kotovych, G.; Pulay, P. *J. Biomol. Struct. Dyn.*, **2002**, *20*, 71–80. e) Přecechtělová, J.; Munzarová, M. L.; Novák, P.; Sklenář, V. *J. Phys. Chem. B*, **2007**, *111*, 2658–2667. f) Migda, W.; Rys, B. *Magn. Resonance Chem.*, **2004**, 459–466. g) Belostotskii, A. M., Goren, Z., Gottlieb, H. E. *J. Nat. Prod.*, **2004**, *67*, 1842–1849. h) Vrček, V.; Kronja, O.; Siehl, H.-U. *J. Chem. Soc. Perkin Trans. 2*, **1999**, 1317–1321. j) Belostotskii, A. M. *J. Org. Chem.*, **2008**, *73*, 5723–5731.

76. L. Garrido, E. Zubía, M. J. Ortega, J. Salvá, *J. Org. Chem.*, **2003**, *68*, 293–299.

77. a) Sarotti, A. M., Pellegrinet, S. *J. Org. Chem.*, **2009**, *74*, 7254–7260. b) Smith, S. G., Goodman, J. M. *J. Org. Chem.*, **2009**, *74*, 4597–4607. c) Costa, F. L. P., deAlbuquerque, A. C. F., Santos, F. M., deAmorim, M. B. *J. Phys. Org. Chem.*, **2010**, *23*, 972–977. d) Auer, A. *Chem. Phys. Lett.*, **2009**, *467*, 230–232.

78. de Dios, A. C., Pearson, J. G., Oldfield, E. *J. Amer. Chem. Soc.*, **1993**, *115*, 9768–9773.

79. a) Mancini, I., Guella, G., Frostin, M., Hnawia, E., Laurent, D., Debitus, S., Pietro, F. *Chem. Eur. J.*, **2006**, *12*, 8989–8994. b) Tähtinen, P., Saielli, G., Guella, G., Mancini, I., Bagno, A. *Chem. Eur. J.*, **2008**, *14*, 10445–10452.

Index